Up and Running
with
Autodesk Advance Steel 2016
Volume: 1

Deepak Maini
Product Manager - BIM/MFG Solutions
Cadgroup Australia

Copy Editor
Pragya Katariya, Owner and Lead Editor, Ocean Blue Communications
www.oceanbluecommunications.com

Technical Editor
Kurt Jones, Product Manager, Cadgroup Australia

Cover Designer
Dushyant Chauhan, DDM Designs Australia

Cover Illustration of the Building Model
Dushyant Chauhan, DDM Designs Australia.

ISBN-13: 978-1530286089
ISBN-10: 1530286085

Dedication

*To all engineers and designers who create innovative
products and make this world a better place to live*

*To my mum and dad
who have always supported me unconditionally in my endeavors*

*To my wife Drishti and my son Vansh
whose motivation, support, and inspiration made this textbook possible*

Foreword

Autodesk Advance Steel is a purpose-built software for steel detailing industry. With country-based standards and tools for placing steel sections and generating documentation, this program is becoming a preferred product for the steel detailing industry around the world.

The two volumes of this textbook cover Autodesk Advance Steel in detail and empower users to reap full benefits of the software. The users can create a complex steel model by placing beams, columns, connections, portal and gable frames, purlins, trusses, stairs, railings, bracing, and so on and then generate 2D documentation from the 3D model. With every chapter containing "real-world" tutorial for the Structural and BIM industry, the two volumes of this book provide the users ample amount of practice time and hands-on experience.

Ilko Dimitrov
IDC-1 / M. Eng.
ACBS
http://www.acbs-usa.com

Acknowledgments

I would like to thank the following people for helping me throughout the process of writing the two volumes of this textbook:

Kurt Jones
Product Manager, Cadgroup Australia Pty Ltd

Adam Jackman
Supervisor Design Services, SOTO Consulting Engineers

Ilko Dimitrov
Managing Director, ACBS Structural Steel Detailing

Muni Vimawala, P.E., S.E.
PSM Engineers

Philippe Bonneau
Technical Marketing Manager, Autodesk

Stephan Gumpert
Sr. Technical Sales Specialist – Advance Steel, Autodesk Australia / Pacific Region

Stephanie Hoerndler
Technical Specialist Structural Fabrication, Autodesk Inc.

About the Author

Deepak Maini (Sydney, Australia) is a qualified Mechanical Engineer with more than 18 years of experience in working with various CAD software. He has been teaching various CAD software for more than 17 years and has authored the "***Up and Running with Autodesk Navisworks***" series of books. He is currently working as the Product Manager BIM/MFG Solutions with Cadgroup Australia, the first Platinum Autodesk Partner in the Australia/New Zealand region.

He is also a Guest Lecturer at the University of Technology Sydney (UTS) and the University of New South Wales (UNSW), two of the leading universities in Australia. In addition, he is one of the lead presenters at various events showcasing the latest Autodesk technology all around Australia, he is regularly invited to present at various User Group events around the country.

Deepak is a regular speaker at Autodesk University in Las Vegas and was awarded the "Top Rated" speaker status at Autodesk University 2014 and 2015.

Deepak's Contact Details

Email: *deepak@deepakmaini.com*
Website: *http://www.deepakmaini.com*

Accessing Tutorial Files

The author has provided all the files required to complete the tutorials in this textbook. To download these files:

1. Visit http://www.deepakmaini.com/AS/AS.htm

2. Click on cover page of the book whose tutorial files you want to download.

3. On the top right of the page, click on **ACCESSING TOC/TUTORIAL FILES**.

4. Click on the **Tutorial Files** link.

Free Teaching Resources for Faculty

The author has provided the following free teaching resources for the faculty:

1. PowerPoint Slides of all chapters in the textbook.
2. Teacher's Guide with answers to the end of chapter **Class Test Questions**.
3. Help in designing the course curriculum.

To access these resources, please contact the author at **deepak@deepakmaini.com**.

Dimension Units and Dialog Box Captures

*Autodesk Advance Steel can be installed based on a number of different country standards. For example, in the United States, you will install it based on **English US** standards, in United Kingdom, you will install it based on the **English UK** standards, in Australia, you will install it based on **English Australia** standards, and so on.*

*Every tutorial in Volume 1 and Volume 2 of this textbook provides the dimensions in the Imperial units based on the **English US** standards and Metric units based on **English Australia** standards.*

*In all these installations, some terminology used in the dialog boxes are different. For example, while editing grids in the **English US** installation, the dialog box that is displayed is called **Advance Steel - Axis, parallel** and the second tab on the left is called **Groups** tab, as shown in the figure below.*

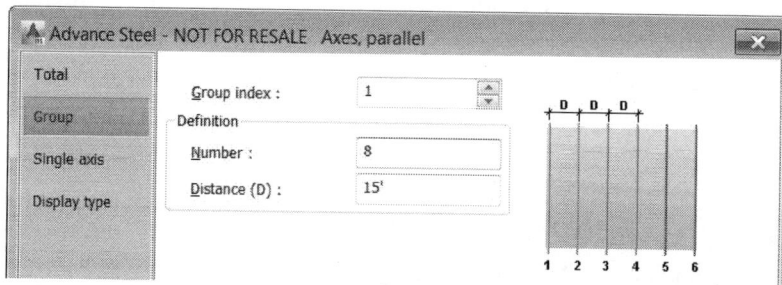

*However, while editing the same grids in the **English Australia** installation, the dialog box that is displayed is called **Advance Steel - Grid lines, parallel** and the second tab on the left is called **Sequence** tab, as shown in the figure below.*

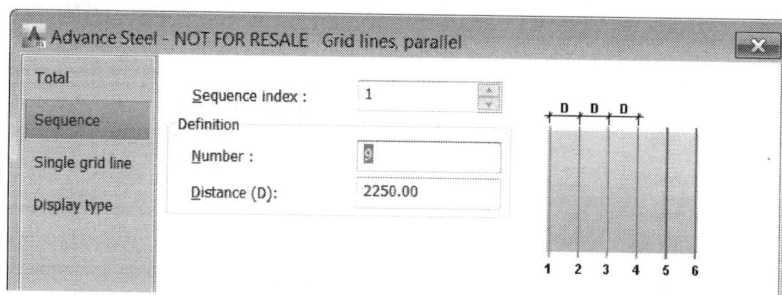

*In this book, most of the screen captures are taken from the **English US** installation. However, the tutorial steps will include the terminology in both **English US** and **English Australia** installation.*

Preface

Welcome to Volume 1 of Up and Running with Autodesk Advance Steel 2016.

*This textbook consists of ten chapters for the Structural and Building Information Modeling (BIM) industry, covering in detail the process of creating the 3D structural model. In all the chapters, I have used both Imperial and Metric units in these chapters. The Imperial units are based on Advance Steel **English US** installation and the Metric units are based on Advance Steel **English Australia** installation.*

The process of detailing and documenting these 3D models is discussed in Volume 2 of this textbook.

The chapters in this textbook start with the detailed description of the Autodesk Advance Steel tools and concepts. These are then followed by the detailed Structure and BIM tutorials. Every section of the tutorials starts with a brief description of what you will be doing in that section. This will help you to understand why and not just how you have to do certain things.

Real-world Structural and BIM models have been carefully selected to discuss the tools and concepts in the tutorials of every chapter. You will be able to find various similarities between the models used in this textbook and your current projects. This will allow you to apply the concepts learned in this textbook to your day-to-day work.

*I have also added the **"What I Do"** sections in most chapters. In these sections, I have discussed the approach I take while working with Autodesk Advance Steel. You will also find a number of **"Notes"** and **"Tips"** that discuss additional utilities of various concepts.*

I would like to take this opportunity to state that my intention of writing this textbook is not to teach you steel detailing. My intention is to teach you Autodesk Advance Steel. Therefore, you might find some dimensions that I have used in the book are different from what you generally prefer. In that case, you are more than welcome to use your preferred dimensions instead of the ones given in the book.

I hope you find learning the software using Volume 1 and Volume 2 of this textbook an enriching experience and are able to apply the concepts in real life situations.

If you have any feedback about this textbook, please feel free to write to me at the following email address:

deepak@deepakmaini.com

TABLE OF CONTENTS

Chapter 2 - Inserting and Editing Structural Sections

Chapter 3 - Advanced Structural Elements - I

Chapter 4 - Inserting the Plates at Beam and Column - Beam Joints

Chapter 5 - Inserting the Beam End to End, Platform Beam, and Purlin Joints

Chapter 6 - Advanced Structural Elements - II

Chapter 7 - Inserting the Bracing, Tube, and Stair Joints

Chapter 8 - Inserting Plates and Gratings and Controlling Object Visibility

Chapter 9 - Extended Modeling and Productivity Tools

Chapter 10 - Adding Custom Connections

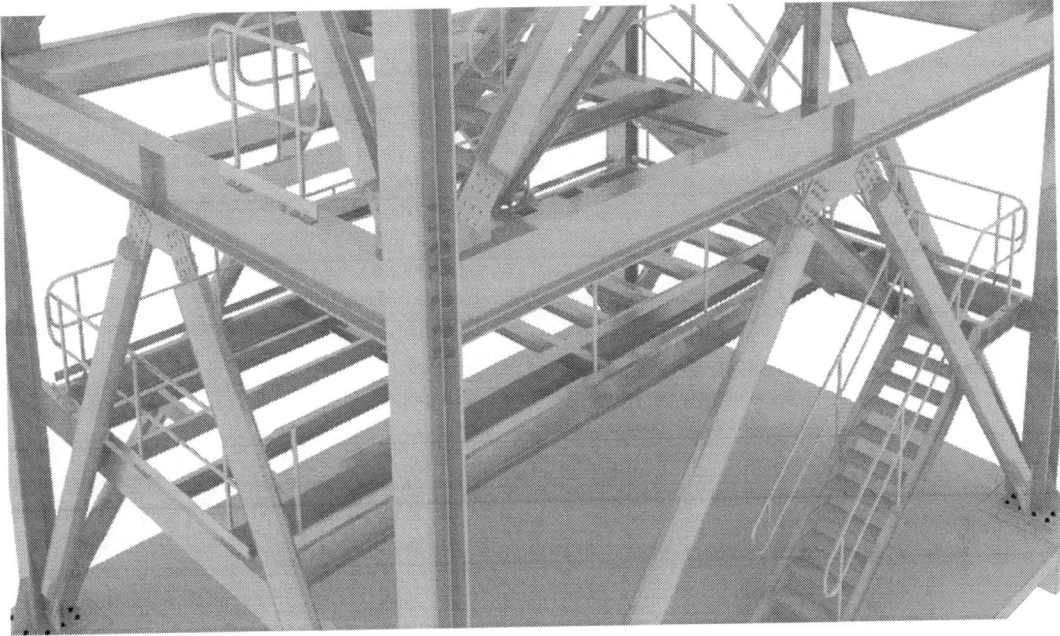

Chapter 1 – Introduction to Autodesk Advance Steel

The objectives of this chapter are to:

√   *Introduce you to Advance Steel*
√   *Explain the Advance Steel workflow*
√   *Explain various Advance Steel model items types*
√   *Familiarize you with the process of creating a new Advance Steel project*
√   *Explain the folder structure of the new project*
√   *Familiarize you with Advance Steel interface*
√   *Explain how to insert various grid types in Advance Steel*
√   *Explain how to insert various concrete elements in Advance Steel model*

AUTODESK ADVANCE STEEL

Autodesk Advance Steel is purpose-built software for structural engineers and steel detailers. Running on the AutoCAD platform, this software caters to the needs of the BIM and Plant/Mining industries by providing specialized tools to automate the process of creating complex structural models and connections. It also increases the user productivity by automatically generating detailed fabrication and shop drawings, reports, bill of materials (BOM), and NC/DSTV/DXF files for steel cutting.

While working with Building Information Modeling (BIM) data, Autodesk Advance Steel has bidirectional workflow with Autodesk Revit. Using the Autodesk Advance Steel add-in for Autodesk Revit, you can directly import Autodesk Revit models into Autodesk Advance Steel for creating connections and generating detailed documentation. You can also import Autodesk Advance Steel models into Autodesk Revit for design verification.

While installing Autodesk Advance Steel, you can select your country settings. As a result, the structure members, fasteners, and plates for that country standard are used by default while creating structural model. This allows you to start using the program immediately after installing without spending much time in configuring it to suit your country standards.

Figure 1 shows a structural steel model created in Advance Steel for the mining industry.

Figure 1   Structural steel model created for the mining industry

Figure 2 shows a structural steel model created in Advance Steel for the BIM industry.

Figure 2 Structural steel model created for the BIM industry

AUTODESK ADVANCE STEEL WORKFLOW

Before you start creating models in Autodesk Advance Steel, it is important for you to understand the workflow to be used. The following flowchart explains the general workflow in Autodesk Advance Steel:

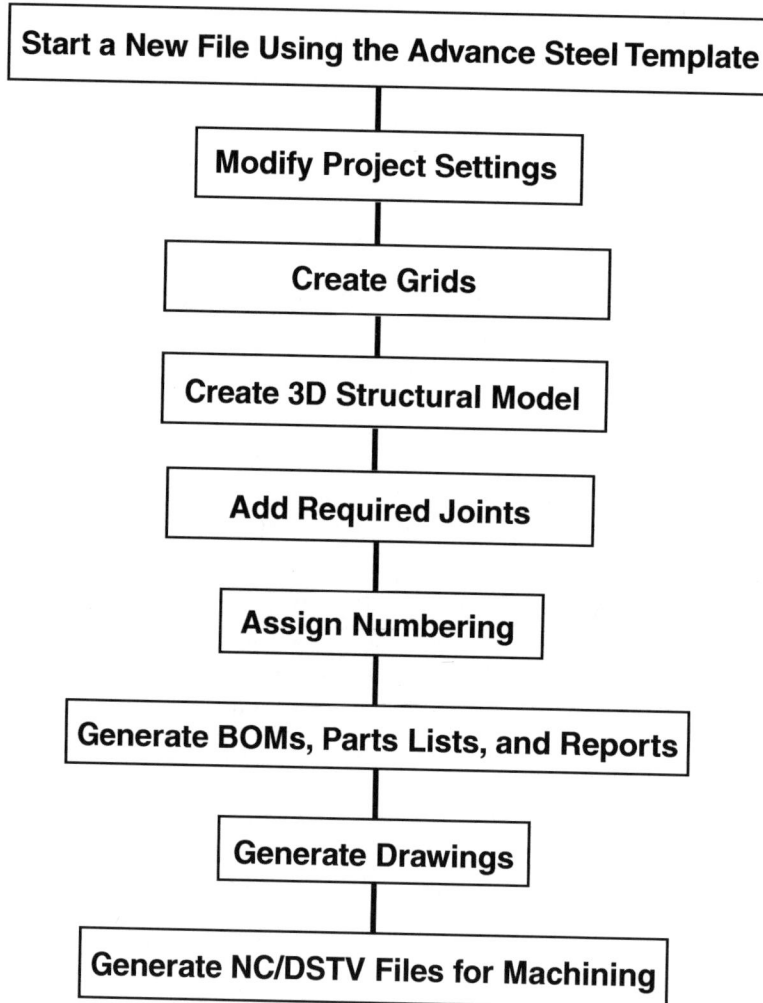

```
┌─────────────────────────────────────────────────┐
│  Start a New File Using the Advance Steel Template │
└─────────────────────────────────────────────────┘
              │
      ┌──────────────────────────┐
      │   Modify Project Settings  │
      └──────────────────────────┘
              │
       ┌────────────────────┐
       │     Create Grids     │
       └────────────────────┘
              │
     ┌──────────────────────────┐
     │  Create 3D Structural Model │
     └──────────────────────────┘
              │
      ┌──────────────────────┐
      │   Add Required Joints  │
      └──────────────────────┘
              │
       ┌────────────────────┐
       │  Assign Numbering    │
       └────────────────────┘
              │
   ┌──────────────────────────────────────┐
   │  Generate BOMs, Parts Lists, and Reports │
   └──────────────────────────────────────┘
              │
      ┌──────────────────────┐
      │   Generate Drawings    │
      └──────────────────────┘
              │
   ┌──────────────────────────────────────┐
   │  Generate NC/DSTV Files for Machining   │
   └──────────────────────────────────────┘
```

These workflow items are explained in brief below.

Start a New File Using the Advance Steel Template

With the installation of Autodesk Advance Steel, a default Imperial or Metric template is automatically copied on your computer. To start any new structural project, you will use this template. You can also create your own template, if required.

Modify Project Settings

Before you start working on any structural project, you will have to specify the project settings. These settings include project information, country profile, various units such as weight, length, area, and so on. You will learn how to do this later in this chapter.

Create 3D Structural Model

In this step, you will create the required structural model that could include the concrete slabs, concrete footings, structural beams, structural columns, stairs, handrailings, gratings, plates, and so on. Autodesk Advance Steel provides a number of specialized tools to add these structural members. You will learn about these tools in later chapters.

Add Required Joints

Once you have added the structural members, you need to connect them to each other. This is done using the **Connection Vault**, which provides a number of pre-configured joint types. You can also create your own custom joints and save them, if required. Adding joints and creating custom joints is discussed in later chapters.

Assign Numbering

After completing the structural model, you will have to assign numbering to various components of the model. Every single part, assembly, and preliminary part needs to be numbered before you can generate their drawings.

Generate BOMs, Parts Lists, and Reports

On completion of the model, you will be able to automatically generate Bill of Materials (BOMs), Parts Lists, and Reports. This is done using the **BOM Templates palette**, which is discussed in later chapters.

Generate Drawings

Autodesk Advance Steel allows you to automatically generate various types of drawings such as General Arrangement (GA) drawings, assembly drawings, anchor plans, single part drawings, and so on. All this is done using the **Drawing Styles Palette** or **Drawing Process Palette**, which are discussed in later chapters.

Generate NC/DSTV/DWF Files for Machining

You can automatically generate the NC/DSTV/DWF files of the structural members using the tools available in Autodesk Advance Steel. These files can be directly used in the CNC machines to cut structural members to exact shape and sizes.

AUTODESK ADVANCE STEEL INTERFACE

Figure 3 shows the default Autodesk Advance Steel window. Various components of this window are discussed next.

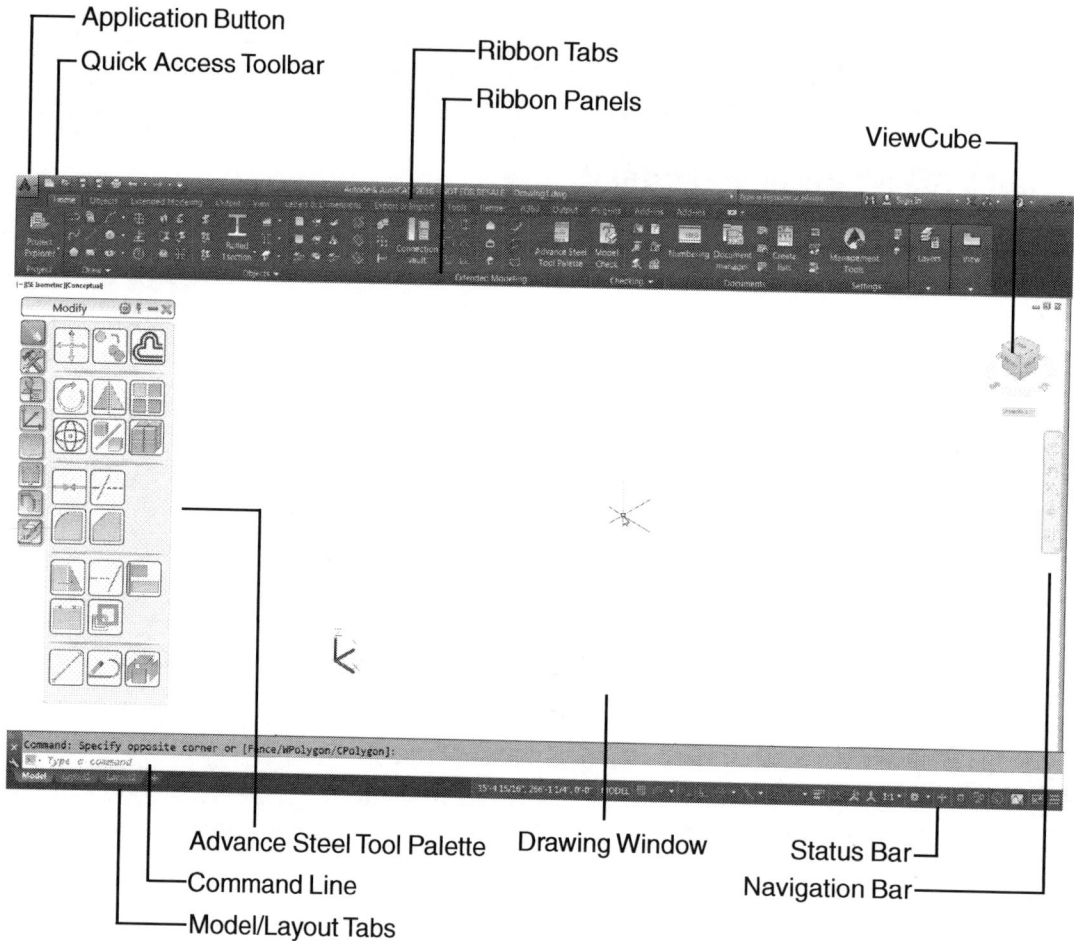

Figure 3 Autodesk Advance Steel interface

Application Button

This button is used to display the application menu, which is divided into two areas, as shown in Figure 4. The area on the left shows standard tools such as **Save**, **Save As** etc. The area on the right shows the recent files that you opened in Autodesk Advance Steel. The **Options** button at the bottom can be used to change the standard AutoCAD settings.

Quick Access Toolbar

This toolbar has the buttons that you often use, such as **New**, **Open**, **Save**, **Undo**, **Redo** etc. The down arrow (called flyout) on the right of this toolbar shows more tool buttons that you can add to this toolbar.

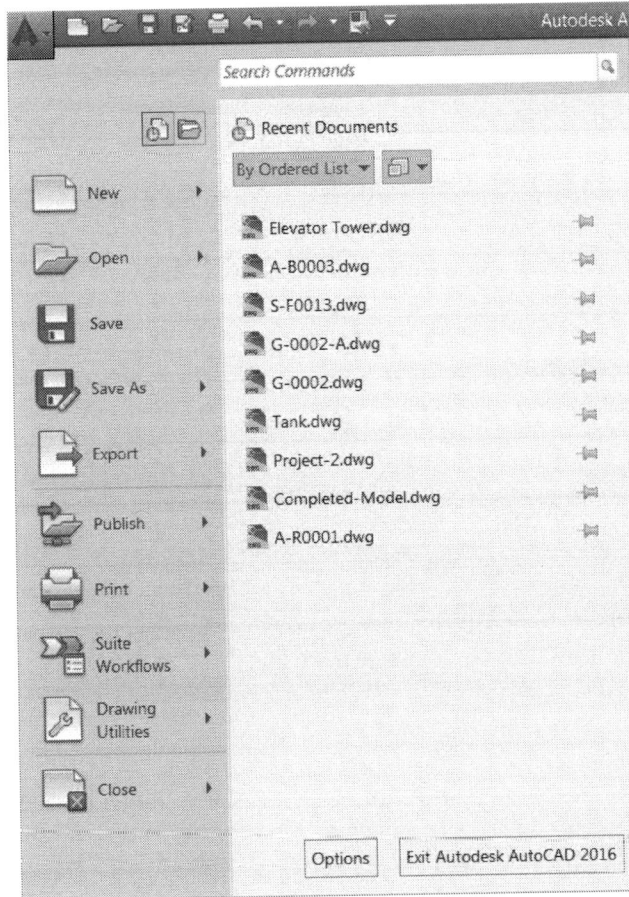

Figure 4 The application menu

Ribbon Tabs

These are a collection of ribbon panels that logically group Autodesk Advance Steel and AutoCAD tool buttons.

Ribbon Panels

These are logical groups of Autodesk Advance Steel and AutoCAD tool buttons. Some ribbon panels have a down arrow at the bottom that is used to expand the panel and show additional buttons.

ViewCube

This is a visualization tool used to display the static views of the model from various directions. You can click on one of the faces or the edges of the ViewCube to display the model from that particular direction. You can also click on one of the corners of the ViewCube to display the 3D view of the model from that direction.

Advance Steel Tool Palette

This palette comprises of eight tabs on the left side with each tab showing a number of tool buttons on the right. These tools are used in conjunction with various Autodesk Advance Steel tools available on the ribbon tabs and panels. You can customize this palette to better suit your needs.

Command Line

This is where all the prompt sequences and options are displayed when you invoke an AutoCAD or Autodesk Advance Steel tool.

Model/Layout Tabs

These tabs are used to switch between the model space or layouts.

Drawing Window

This is the area where the design is displayed and manipulated.

Status Bar

This bar runs through the bottom of the Autodesk Advance Steel window. It hosts various buttons that are used to toggle on or off certain settings such as grids display, object snaps, polar tracking, and so on.

Navigation Bar

Various drawing display tools available in Autodesk Advance Steel are located on this toolbar.

OTHER AUTODESK ADVANCE STEEL PALETTES

In addition to the default tool palette displayed in the Autodesk Advance Steel window, there are a few more palettes that you use on regular basis. These are discussed next.

Connection Vault

This palette provides you various connection types to join structural members together. This palette comprises of four sections. Various types of connections categories are organized in the toolbar format on the left side of this palette, as shown in Figure 5. When you pick a tool button for a connection category, all the available connection types for that category are displayed in the middle section of this palette. The upper-right section shows the preview of the connection type. The lower right section shows the selection order for creating this connection, the sections that can be selected, and also a brief description of the selection connection type.

Tip: While creating connections, it is a good idea to review the selection order, description, and options that are displayed in the right section of the **Connection Vault** window.

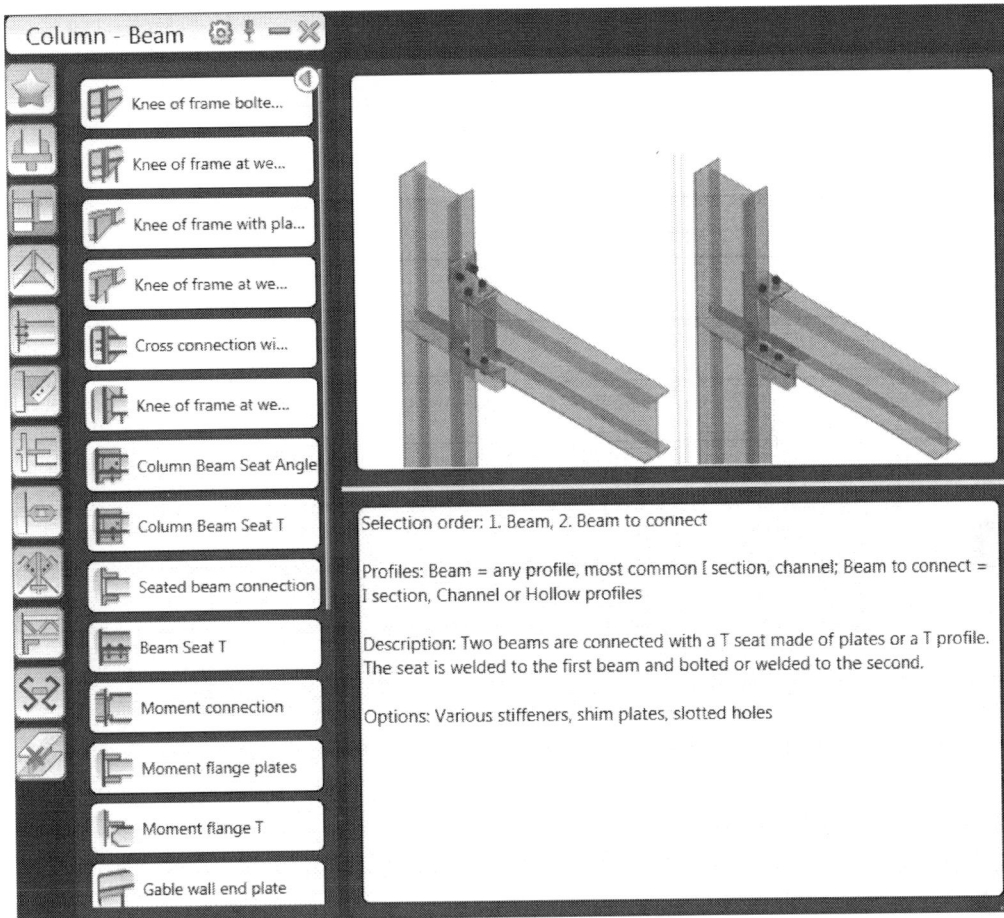

Figure 5 The Connection Vault

Document Manager

The **Document Manager** dialog box manages all the output generated from the Autodesk Advance Steel model, such as detailed drawings, BOMs, Parts Lists, and DSTV/NC files, as shown in Figure 6. This dialog box also allows you to preview, open, revise, delete, and print these outputs.

Advance Steel Management Tools

The **Advance Steel Management Tools** dialog box is used to change the default settings of various structural elements in the project. It can also be used to select units of the project or convert the database of the structural elements. Figure 7 shows the **Home** page of this dialog box. To change any of these settings, select that option from the **Home** page; a new tab will be displayed at the bottom showing the options that you can change. Once you have changed the settings, click the **Apply** button; the tab will be automatically closed and you will return to the **Home** page.

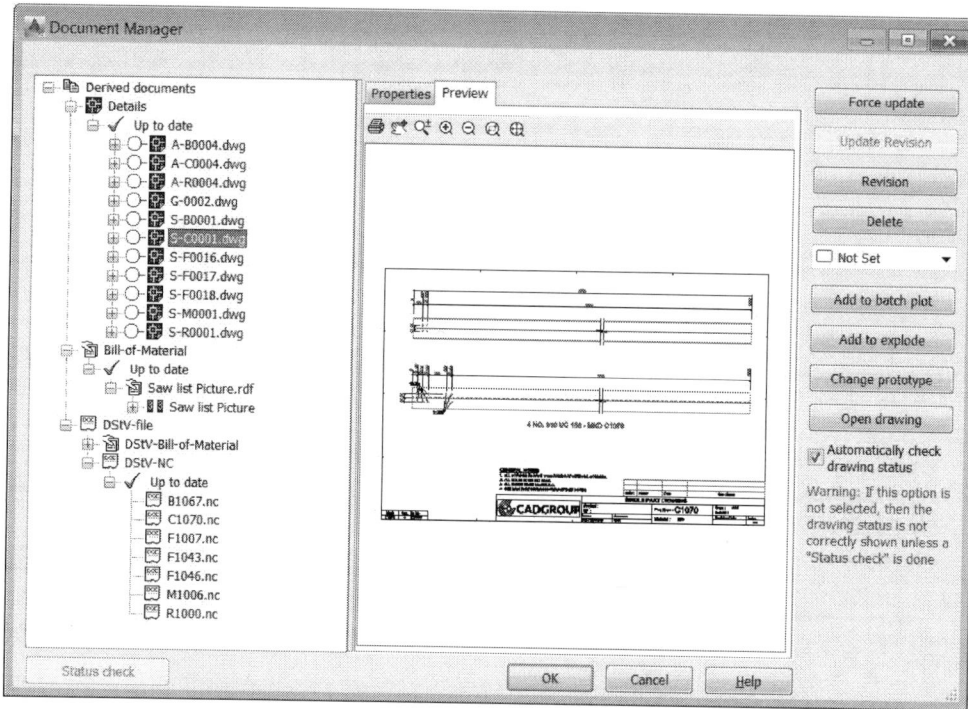

Figure 6 *The **Document Manager** dialog box*

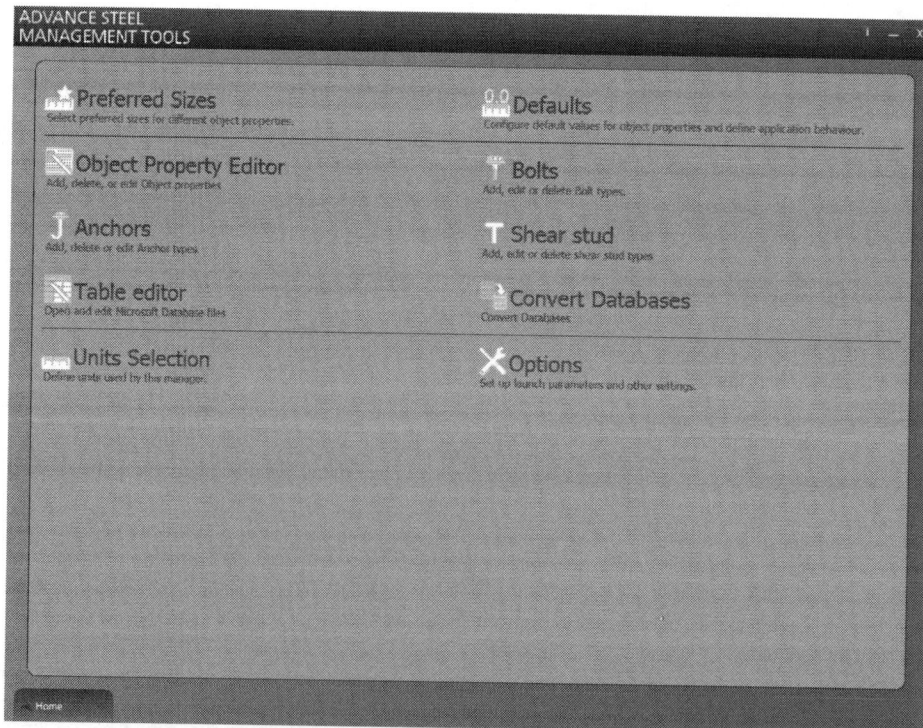

Figure 7 *The **Advance Steel Management Tools** dialog box*

UNDERSTANDING THE FOLDER STRUCTURE

To manage various output files generated from the Autodesk Advance Steel model, there is a folder structure automatically created once you start to generate any type of output from the model. It is extremely important for you to understand the folder structure for successful delivery of a project. The following example explains this process.

1. You have created a folder with the name **ABC123** and saved the current Autodesk Advance Steel project file in that folder with the name **MyProject123.dwg**.

2. You generate any output file such as detailed drawings, lists, or NC files, a folder is created inside the **ABC123** folder with the name **MyProject123**. Inside this folder, there are subfolders that are automatically created for various types of output files, as shown in Figure 8.

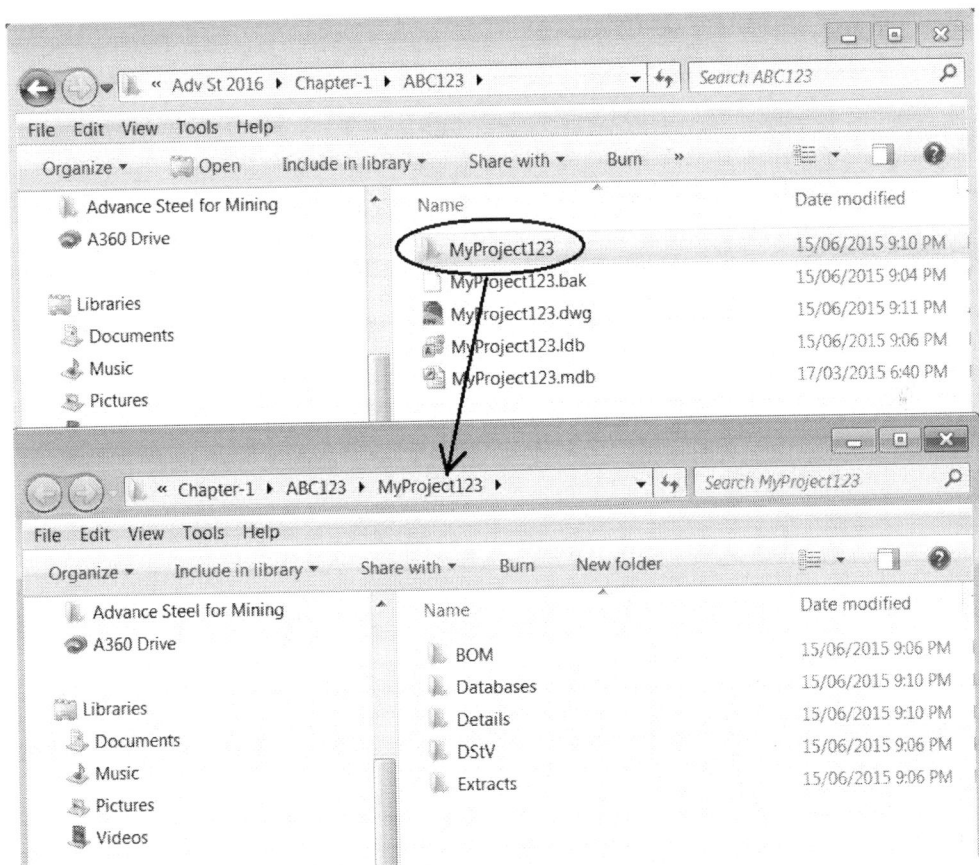

Figure 8 *Folder structure automatically created for output files*

THE USER COORDINATE SYSTEM (UCS)

The location and orientation of various three dimensional (3D) objects created in Advance Steel are dependent on the location and orientation of the User Coordinate System (UCS). Because Advance Steel runs on the AutoCAD platform, the AutoCAD UCS tool is used to locate and orient the plane to create 3D objects. To make it easier for you to change the UCS, the **Advance Steel Tool Palette** provides the **UCS** tab that provides buttons to define some important types of UCSs. This tab also provides tools to rotate UCS to around the X, Y, or Z axis, as shown in Figure 9.

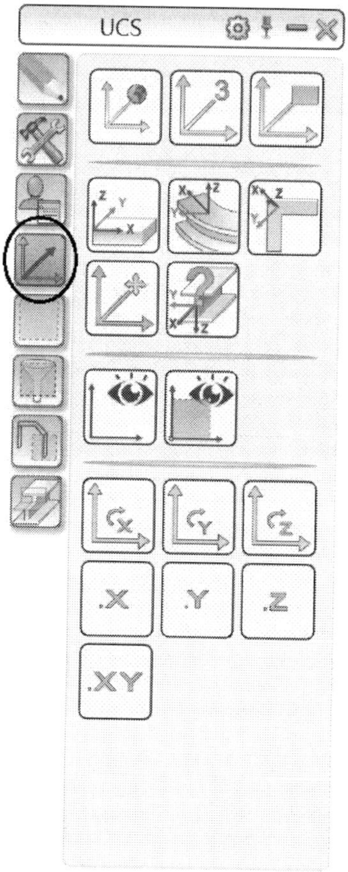

Figure 9 The *UCS* tab of the *Advance Steel Tool Palette*

Tip: The tool buttons of the Advance Steel objects that are dependent on the UCS orientation will display an extended line of text in the tooltip informing you about the UCS on which that object will be created.

CHANGING VIEWPOINTS AND VISUAL STYLES

While modeling in Advance Steel, you will be required to change the viewpoint and visual style of the model display at various times. To make it convenient for you to do this, the option to change viewpoints and visual styles are available at the top left corner of the drawing window. The **View Controls** option is used to change the viewpoint. When you click this option, the list of available viewpoints is displayed, as shown in Figure 10. You can click on the viewpoint that you want to activate. The options at the bottom of this list are used to set the current view to parallel or perspective.

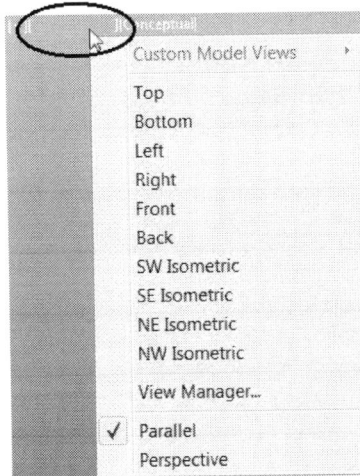

Figure 10 Changing the viewpoint

Similarly, to change the visual styles, you can use the **Visual Style Controls** option. When you click on this option, the list of various visual styles is displayed for you to select from, as shown in Figure 11. Clicking the **Visual Styles Manager** at the bottom of this list displays the **Visual Styles Manager** palette that allows you to create your own custom visual styles or edit the default styles.

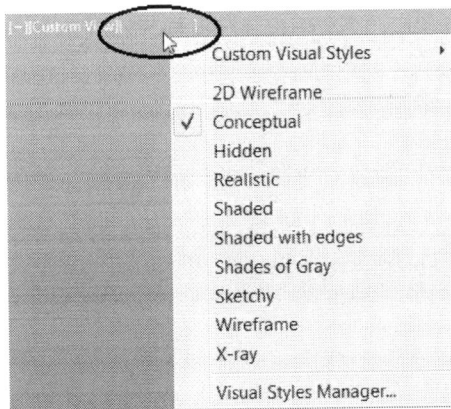

Figure 11 Changing the visual style

Note: *You can also click on one of the faces of the ViewCube to change the viewpoint.*

You can also change the visual styles from **View** *ribbon tab >* **Visual Styles** *ribbon panel >* **Visual Styles** *flyout.*

NAVIGATING THROUGH THE MODEL

Advance Steel provides various navigation tools to help you navigate through the model. These tools are available on the **Navigation Bar** on the right side of the drawing window. Because these tools are standard AutoCAD tools, this book considers that you are familiar with these tools. As a result, they are not discussed in detail in this book.

What I do

*I generally do not use the **Steering Wheel** or **Orbit** tools. I use the wheel button of the mouse to navigate around the model. To zoom in or out, simply scroll the wheel button of the mouse forward or back. Double-click the wheel button of the mouse to zoom to the extents of the model. To pan, press and hold the wheel button and drag the mouse. To orbit, press and hold down the SHIFT key and the wheel button of the mouse and drag.*

INVOKING THE ADVANCE STEEL TOOLS

Most of the Advance Steel tools are invoked using the buttons available on the ribbon panels. Depending on what type of tool you want to invoke, you will have to switch between various ribbon tabs to go to the required ribbon panel. It is important to remember that most of the Advance Steel tools are macros written inside some other tool. As a result, unlike AutoCAD tools, pressing the ENTER key or the SPACEBAR will not invoke the Advance Steel tools.

Sections of the Command Prompt Sequence

For some tools, the prompt sequences that are displayed comprise of multiple sections, separated by a comma (,) or a full stop (.). For example, the following is the prompt sequence that is displayed while creating a building grid:

Please define two diagonal points for grid, origin:_

In this prompt sequence, the section of the prompt sequence before the comma informs you that this type of grid is created by defining two diagonal points for the grids. The second section of the prompt sequence after the comma prompts you to specify the origin point of the grid.

Similarly, the following is the prompt sequence that is displayed while creating a single axis grid:

Please define end points of the grid line. Start point:_

In this prompt sequence, the section of the prompt sequence before the full stop informs you that this type of grid is created by defining endpoints for the grid axis. The second section of the prompt sequence after the full stop prompts you to specify the start point of the grid axis.

Tool Button Flyouts

Some of the buttons display an arrow at the bottom or on the right. These arrows are called flyouts and when clicked, show additional related tool buttons. Figure 12 shows the **Rolled I section** flyout on the **Home** ribbon tab > **Objects** ribbon panel. Similarly, Figure 13 shows the **Concrete beam** flyout on the same ribbon panel.

*Figure 12 The **Rolled I section** flyout*

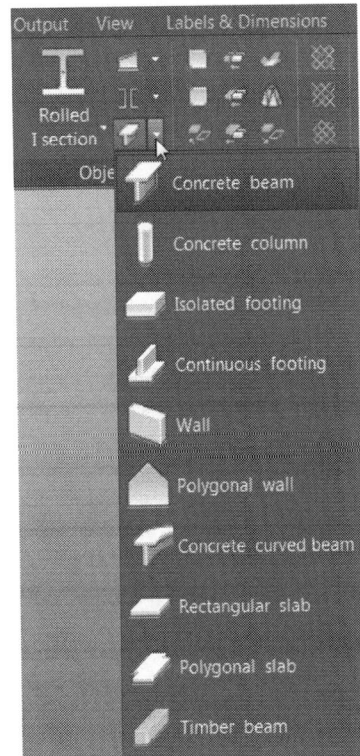

*Figure 13 The **Concrete beam** flyout*

Note: The tool buttons in the flyouts are "sticky buttons". This means that the button that you click last will become the current button in the flyout and will be displayed in the ribbon panel.

HOME RIBBON TAB VS OTHER RIBBON TABS

Most of the basic Advance Steel tools can be invoked from various ribbon panels available on the **Home** ribbon tab. However, if you need to invoke advanced tools, you will have to switch to the **Objects**, **Extended Modeling**, or **Output** ribbon tab. If you are familiar with customizing work-space in AutoCAD, you can add additional tool buttons on the panels available on the **Home** ribbon tab as well.

LOCATING POINTS IN THE DRAWING WINDOW

While creating Advance Steel objects, you will have to specify their locations in the drawing window. This can be done by snapping to the key points of existing objects using object snaps or by specifying the coordinates of the objects. These two methods are discussed next.

Locating Points using Object Snaps

All the object snaps of AutoCAD can also be used to locate Advance Steel objects. However, in addition to using the standard AutoCAD object snaps, there are two Advance Steel specific object snaps are also available for you to use. These are discussed next.

Flange Middle

This object snap is used to snap to the mid point of the flange face of a structural steel member.

Grid Intersection

As the name suggests, this option lets you snap to the intersection point of the grids.

Locating Points by Specifying Coordinates

Whenever you are creating an Advance Steel object, the prompt sequences will display dynamic input boxes to type in the X and Y coordinates of that object. You can use the TAB key on the keyboard to switch between the two dynamic input boxes. Once you have typed the required coordinates, you will have to press the ENTER key for Advance Steel to accept the input.

SELECTING OBJECTS

While creating the structure model, you will need to select various objects in the drawing window. All the object selection modes from AutoCAD also with in Advance Steel. However, pick object, window selection, and crossing selection are the most widely used selection methods in Advance Steel. These three methods are briefly discussed next.

Pick Object

This method allows you to click on an object to select it. The clicked object will be selected and highlighted in the drawing window.

Window Selection

This method is used to select multiple objects and is activated by clicking in the blank area of the drawing window and moving the cursor to the right. On doing so, a Blue box with continuous lines is created. This box is called a Window and only the objects that are fully inside this box will be selected. Note that pressing and holding down the left mouse button and dragging the mouse to the right creates a Window Lasso, which is a freeform shape for selecting the objects that are fully inside the shape.

Crossing Selection

This method is used to select multiple objects and is activated by clicking in the blank area of the drawing window and moving the cursor to the left. On doing so, a Green box with dashed lines is created. This box is called a Crossing and all the objects that are partially or fully inside this box will be selected. Note that pressing and holding down the left mouse button and dragging the mouse to the left creates a Crossing Lasso, which is a freeform shape for selecting the objects that are partially or fully inside the shape.

> **Note**: *The colors mentioned above are based on the default AutoCAD colors. These colors can be changed using the Options dialog box.*

What I do

While teaching AutoCAD or Advance Steel, I tell everyone that these are one of the most user-friendly software in the world. Anything that this software wants you to do, it will be displayed in the prompt sequences. Simply read the prompt sequence to understand what you need to do. Getting into the habit of reading prompt sequences will make the process of learning and using this software really easy.

GRIP EDITING

Most of the Advance Steel objects can be edited using grips, which are the Blue boxes that are displayed on various objects when you select them without invoking any tool. When you hover the cursor over the grip, a triad will be displayed. This triad is used to edit the Advance Steel structural member. Remember that the editing operation that can be performed will depend on the type of object selected. For example, if you drag the axis of the triad along the length of the beam, its length will be changed. However, if you drag it along the other two directions, it will be stretched. Figure 14 shows the length of a beam member being changed using the grip displayed at the top of this beam. In this case, the triad is moved along the beam to change its length. Figure 15 shows the orientation of the beam being changed using the same grip. In this case, triad is being dragged along one of the horizontal directions.

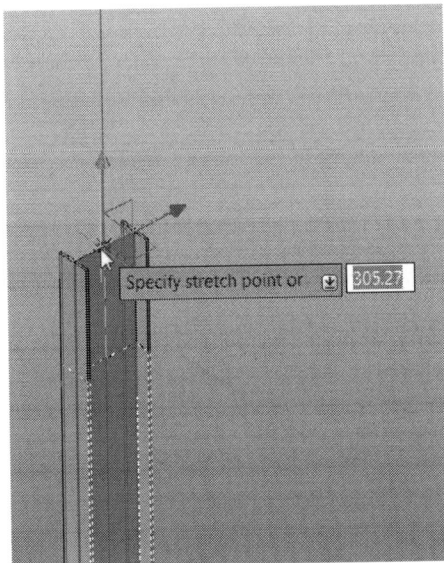

Figure 14 Editing the length of the beam using the grip displayed at the top end

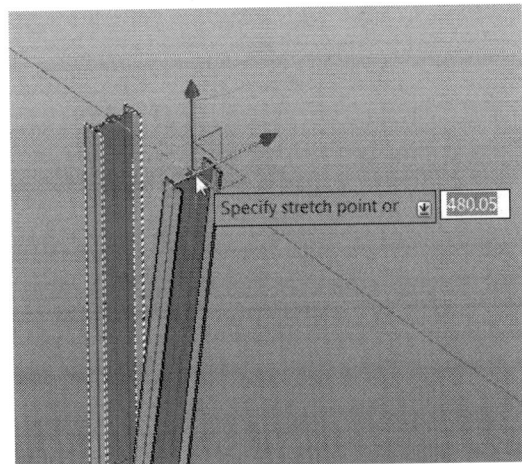

Figure 15 Editing the orientation of the beam using the grip displayed at the top end

USE OF LAYERS IN ADVANCE STEEL

Layers play an important role in the Advance Steel drawing. The default template you start contains predefined layers and as you start creating advance steel objects, they are automatically placed on their respective layers. The advantage is that you can control the visibility of any Advance Steel object using its layers. You can customize the color, linetypes, and lineweights of these layers by editing the template that you use. Figure 16 shows the **Layer Properties Manager** showing the layers of a default **ASTemplate.dwt** file.

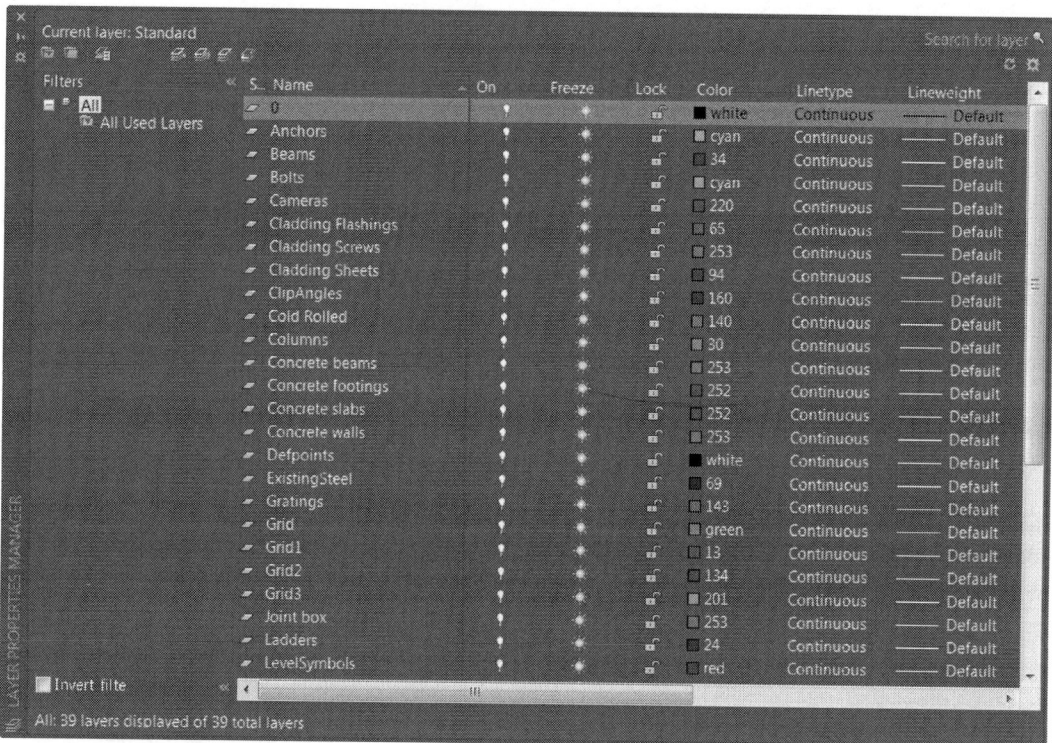

Figure 16 *The default layers available in the **ASTemplate.dwt** file*

*Tip: The assignment of the layers can be controlled using the **Advance Steel Management Tools** dialog box > **Defaults** > panel > **Layer assignments** > **General** option. Generally, the CAD Manager of a company configures the layer settings so that the entire team can work with the same default settings.*

CREATING AND EDITING AUXILIARY OBJECTS

In Advance Steel, certain objects such as grids and level symbols are referred to as auxiliary objects. The reason is that these objects are used as references while creating structural model. In this section, you will learn about creating and editing these auxiliary objects.

Creating and Editing Grids

Grids are used as references to locate the structural members of the Advance Steel model.

There are five tools to create grids, which are available on the **Objects** ribbon tab > **Grid** ribbon panel, as shown in Figure 17. The same panel also shows tools to edit the grid axes. All the methods of creating and editing the grids are discussed next.

Figure 17   The Objects ribbon tab > Grid ribbon panel with the grid command buttons

Creating Building Grid

Home Ribbon Tab > Objects Ribbon Panel > Building Grid
Objects Ribbon Tab > Grid Ribbon Panel > Building Grid

This tool is used to create a 4X4 grid by specifying two opposite corners. These two corners are used to define the spacing between the grid axes. Remember that the grid will be created on the XY plane of the current UCS. When you invoke this tool, the following prompt sequence will be displayed:

Please define two diagonal points for grid, origin:_

In this prompt sequence, you are being prompted to specify the origin of the grid, which is the lower left corner of the grid. You can specify the coordinates of the origin point by typing the values in the prompt sequence or you can specify the location on the screen by picking a point in the drawing window. Once you specify the first point, the following prompt sequence will be displayed:

Second point:_<59' 11/16",39' 4 7/16">:_

As displayed in the prompt sequence, you can either specify the second point, or press ENTER to accept the default value of the second point. Once you specify the second point, a 4x4 grid is created. Figure 18 shows a default building grid created from 0,0 coordinates to the default second point coordinates.

What I do

*Advance Steel has a special object snap option called **GRID Intersection Point**. I make sure that this option is selected in the object snaps whenever I am working with grids. This allows me to place structural members at the grid intersection point.*

Figure 18 A default 4x4 building grid

Editing Building Grids

As shown in Figure 18, the building grid is a 4x4 grid. Also, by default, numbers are used to label the grids along both X and Y directions. All this can be edited by double-clicking on the grid axes. It is important to mention here that the grid axes are edited independently along both directions. Therefore, you need to double-click on the grid axes along that direction that you want to change. On doing so, the **Advance Steel - Grid lines, parallel** dialog box is displayed in the **Total** tab. This tab is used to edit the overall dimension of grids along the selected direction. Also, the **Label type** list in the **Automatic label** area on the right side of this dialog box can be used to change the labelling of the grids along the selected direction.

*Note: There is no **OK** or **Apply** button in the Advance Steel dialog boxes. All the changes you make in the dialog boxes are displayed live in the drawing window. Sometimes, some changes may need to be refreshed, which can be done by simply clicking in another edit box or another field in the same dialog box.*

To change the number of grid axes along the selected direction, click the **Sequence** tab on the left of this dialog box. Now, in the **Definition** area on the right of this dialog box, enter the number of grid axes required in the **Number** edit box. Figure 19 shows the **Advance Steel - Grid lines, parallel** dialog box in the **Sequence** tab after changing the number of grid axes along the selected direction to 9.

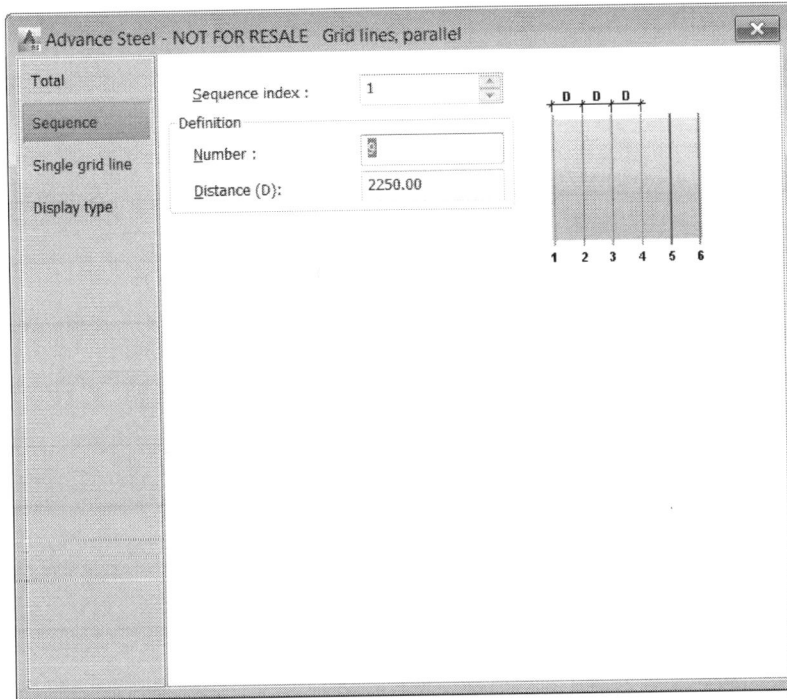

*Figure 19 The **Sequence** tab of the **Advance Steel - Grid lines, parallel** dialog box*

Creating Single Axis Grid

Objects Ribbon Tab > Grid Ribbon Panel > Single axis

This method is used to create a single grid axis along the specified direction. When you invoke this tool, the following prompt sequence is displayed:

Please define end points of the grid line. Start point:_

In this prompt sequence, specify the start point of the grid axis. On doing so, you will be prompted to specify the endpoint of the grid axis. You can specify the endpoint using the polar tracking or by typing the length and angle of the grid axis. As soon as you specify the endpoint, the **Advance Steel - Grid lines, parallel** dialog box is displayed in which you can change the length and label type of the grid axis.

Tip: Whenever you are prompted to specify a point in Advance Steel, you can specify it as relative X and Y coordinate values or length and angle values. To specify the values as relative X and Y coordinates, type a value and it will automatically be taken as the X coordinate value in the first dynamic input box. Next, press Comma (,) and the second dynamic input box is automatically changed to the Y coordinate input box. To specify the value as length and angle, type the length value in the first dynamic input box and then press the TAB key; the first value is automatically considered as the length and the second dynamic input box now allows you to enter the angle value.

Creating Grid with Four Lines

Objects Ribbon Tab > Grid Ribbon Panel > Grid with 4 axes

This method is used to create four grid axes along a specified direction. When you invoke this tool, the following prompt sequence is displayed:

Please define end points of the grid line. Start point:_

In this prompt sequence, specify the start point of the grid axis. On doing so, you will be prompted to specify the end point of the grid axis. You can specify the endpoint using the polar tracking or by typing the length and angle of the grid axis. As soon as you specify the endpoint, the following prompt sequence is displayed:

Direction and length of the group:_

In this prompt sequence, you need to specify the overall spacing between the four grid axes. You can use the polar tracking to specify the spacing. On doing so, the **Advance Steel - Grid lines, parallel** dialog box is displayed in which you can change the length and spacing of the grid axes in the **Total** tab. You can also change the number of grid axes in the **Sequence** tab.

Creating Grid with Groups by Distance

Objects Ribbon Tab > Grid Ribbon Panel > Circular grid with single axis

This method is used to create a group of grids defined by distance and direction values. When you invoke this tool, the following prompt sequence is displayed:

Please define end points of the grid line. Start point:_

In this prompt sequence, specify the start point of the grid group. On doing so, you will be prompted to specify the end point of the grid axis. You can specify the endpoint using the polar tracking or by typing the length and angle of the grid axis. As soon as you specify the endpoint, the following prompt sequence is displayed:

Direction of the group

In this prompt, move the cursor along the direction in which you want to add additional grid axes in the group. On doing so, the following prompted sequence will be displayed:

Distance between grid lines (0") :

Once you specify the distance, this prompt will be repeated again, but it will be if you want to create multiple indices in the grid group with different values. If you do not want to do that, press ENTER to accept the default value in the group. On doing so, the **Advance Steel - Grid lines parallel** dialog box will be displayed. The options in this dialog box are similar to those discussed in the previous section.

Creating Circular Grid with Single Axis

Objects Ribbon Tab > Grid Ribbon Panel > Circular grid with single axis

This method is used to create a single axis circular grid by specifying the start point, endpoint, and a point on the curve. The point on the curve is basically used to specify the radius of the grid curve. On specifying the point on the curve, the **Advance Steel - Curved grid line** dialog box is displayed. In this dialog box, you can edit the radius of the grid axis and also edit the label type, as shown in Figure 20.

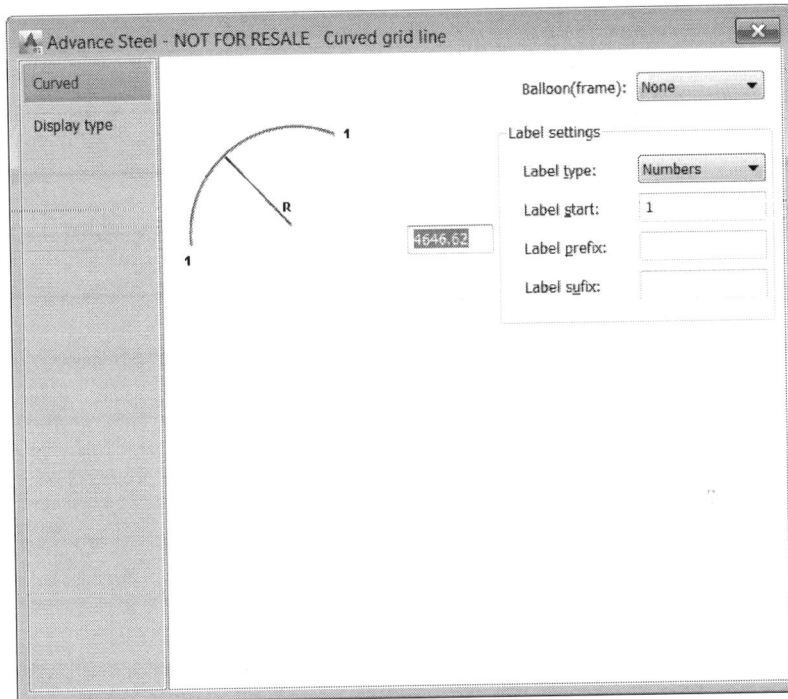

*Figure 20 The **Curved** tab of the **Advance Steel - Curved grid lines** dialog box*

Deleting Grid Axes

Objects Ribbon Tab > Grid Ribbon Panel > Delete axes

This tool is used to delete one or more selected lines of an existing grid. Note that you can only delete one grid axis along the selected direction at a time. When you invoke this tool, the following prompt sequence is displayed:

Please select the grid line to be deleted:

Select the grid axis that you want to delete; you will be informed that 1 axis has been found. If required, select a grid axis along the other direction as well; you will be informed that 1 more axis is found and the total is now 2. After selecting the required grid axes, press the ENTER key; the selected lines will be deleted. Figure 21 shows a 4X6 grid and Figure 22 shows the same grid after deleting grid axis number 3 along the Y direction.

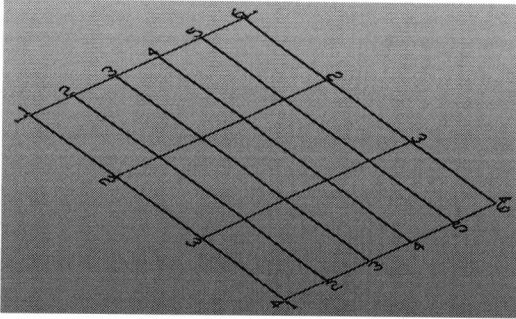

Figure 21 A 4X6 grid

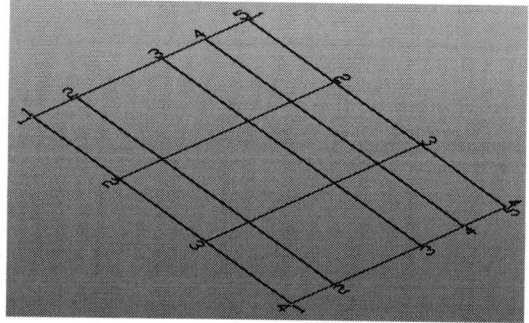

Figure 22 The same grid after deleting an line

Adding Grid Axes

Objects Ribbon Tab > Grid Ribbon Panel > Add axes

This tool is used to add one or more lines to an existing grid. Note that the grid axes will be added along the direction that you select. When you invoke this tool, the following prompt sequence is displayed:

Please select the grid axis after which the group is to be inserted:

In this prompt, select the grid axis after which you want to add additional lines. After selecting the grid axis, press ENTER; you will be prompted to specify how many grid axes do you want to add. Specify the value and press ENTER. You will then be prompted to specify the distance between the grid axes. Once you specify the distance, the grid axes will be added and the rest of the grid axes will shift to maintain the distance between the original grid axes.

Trimming Grid Axes

Objects Ribbon Tab > Grid Ribbon Panel > Trim axes

This tool is used to trim one or more grid axes using one or more boundary objects. These boundary objects are used as cutting edges to trim the grid axes along one or both directions. When you invoke this tool, you will be prompted to select the boundary object. Select one or more objects to be used as the boundary objects and then press ENTER. Next, you will be prompted to select grid axis that is to be cut. Select one or more grid axes along one or both direction, depending upon the boundary objects that will be used for trimming. Once you have selected all the grid axes to be trimmed, press ENTER to finish the tool and view the results.

Extending Grid Axes

| Objects Ribbon Tab > Grid Ribbon Panel > Extend axes |

This tool is used to extend one or more grid axes using one or more boundary objects. These boundary objects are used to extend the grid axes along one or both directions. When you invoke this tool, you will be prompted to select the boundary object. Select one or more objects to be used as the boundary objects and then press ENTER. Next, you will be prompted to select the grid axis that is to be extended. Select one or more grid axes along one or both direction, depending upon the boundary objects that will be used for extending. Once you have selected all the grid axes to be extended, press ENTER to finish the tool and view the results.

Figure 23 shows a 6X6 grid and a circle to be used for trimming and extending the grid axes. Figure 24 shows the same grid after trimming and extending the grid axes.

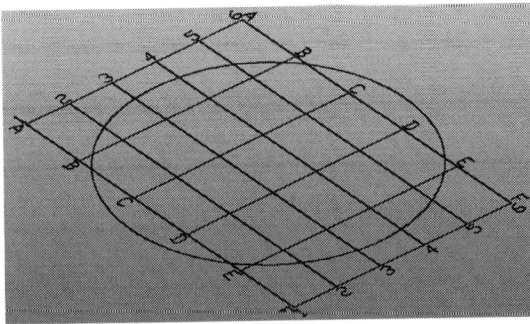

Figure 23 A 6x6 grid with a circle to be used as the boundary for trimming and extending

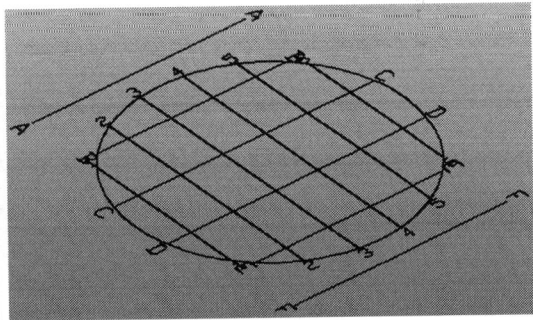

Figure 24 The grid after trimming and extending some grid axes using the circle as the boundary

Tip: In Figure 24, you can use the **Delete axes** tool to one by one delete grid axes A and F.

Note: You will learn about level symbols, which is the other type of auxiliary object, in later chapters.

INSERTING CONCRETE OBJECTS

Concrete objects form the foundation of most of the structural designs. To facilitate this, Advance Steel provides a number of tools to insert concrete objects. The tools to insert these concrete objects are available on the **Home** ribbon tab > **Objects** ribbon panel > **Concrete beam** flyout, as shown in Figure 25. These tools are also available on the **Objects** ribbon tab > **Other objects** ribbon panel.

These tools are discussed next.

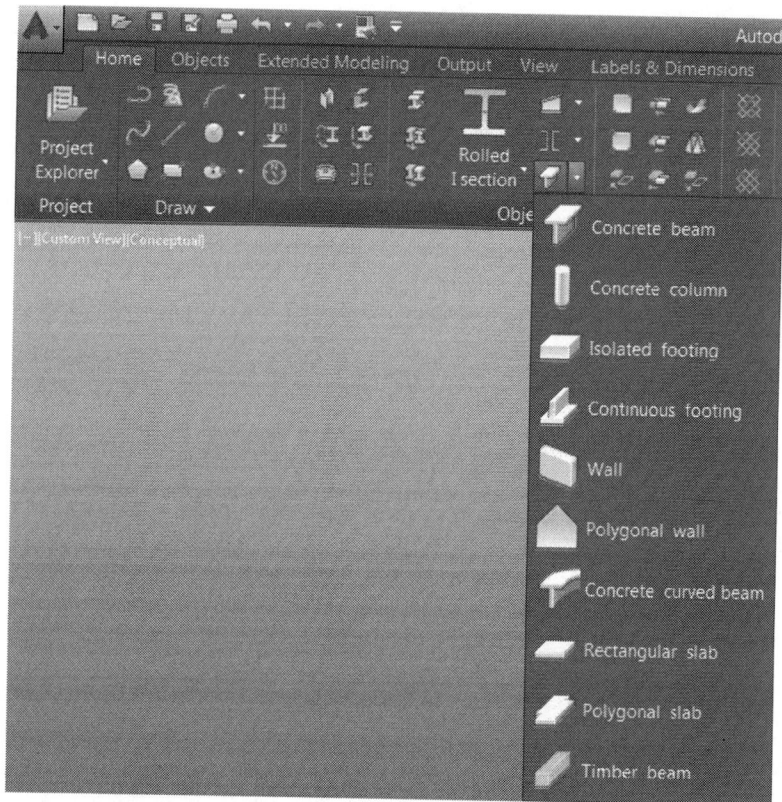

Figure 25 *Various commands to insert concrete objects*

Inserting Concrete Beam

Home Ribbon Tab > Objects Ribbon Panel > Concrete beam
Objects Ribbon Tab > Other objects Ribbon Panel > Concrete beam
---

The **Concrete beam** tool is used to insert a concrete beam. When you invoke this tool, you will be prompted to specify the start point and the endpoint of the system axis. System axis is located on the Advance Steel members and is used to locate them in the drawing window. The two points can be specified by entering the values or by specifying object snap points. As soon as you specify the endpoint of the system axis, the **Advance Steel - Beam [1]** dialog box is displayed. The numeric value in the dialog box name will keep incrementing as you add additional beams. Various tabs of this dialog box and their options are discussed next.

Section & Material Tab

The **Section & Material** tab shown in Figure 26 is the default tab that is active in the dialog box when you specify the second point of the system axis. This tab is used to specify the section type and material of the beam that you have placed. These options are discussed next.

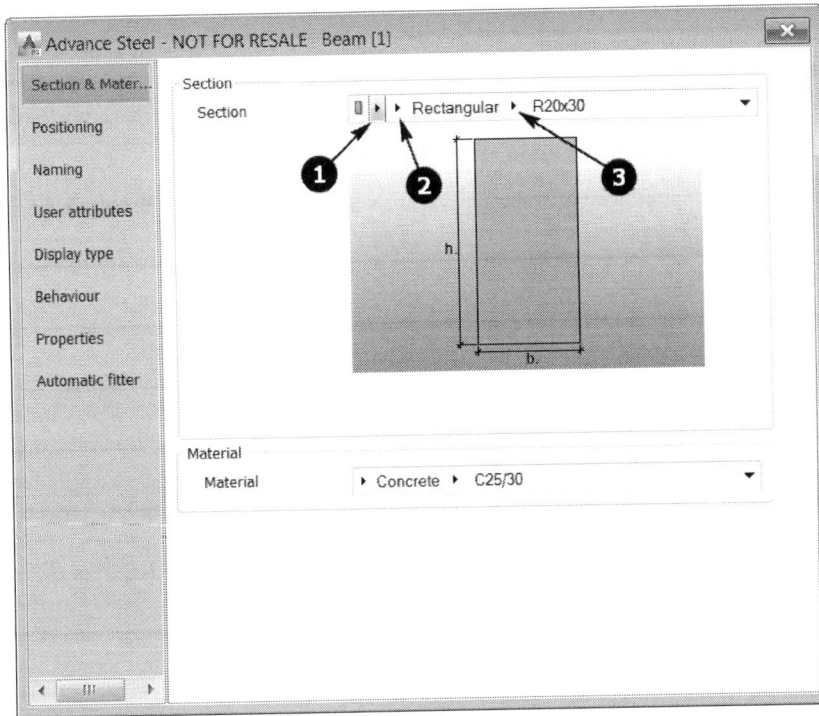

Figure 26 The Section & Material tab of the Advance Steel - Beam [1] dialog box

Section Area

This area provides the options to specify the **Section Class**, **Section Type**, and **Section Size** of the concrete beam you are inserting. These three settings are configured using the three flyouts marked as **1**, **2**, and **3** respectively in Figure 26. The beam class can be **Concrete Precast** or **Standard Profiles**, as shown in Figure 27.

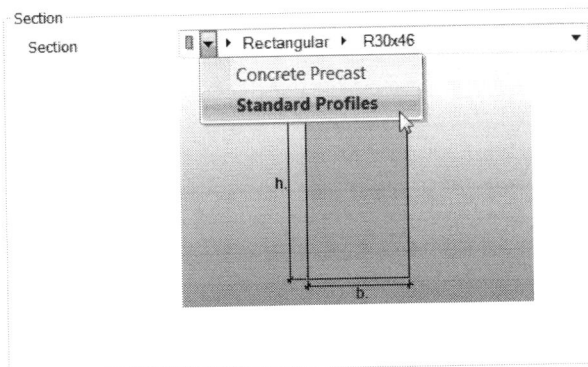

Figure 27 Selecting the concrete beam class

The beam **Section Type** options will be dependent on the beam **Section Class** selected. Figure 28 shows the section type for the **Standard Profiles** class and Figure 29 shows the section type for the **Concrete Precast** class.

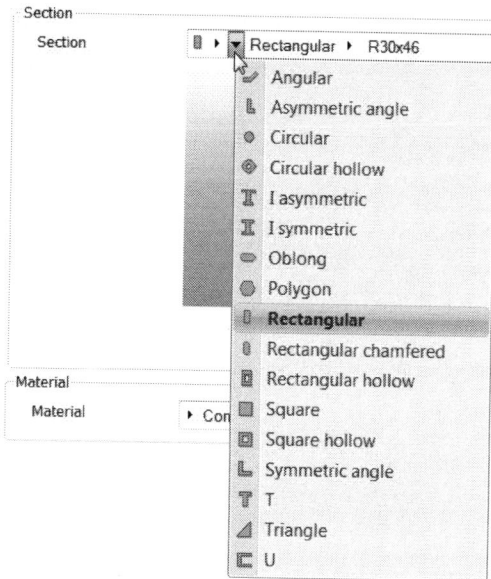

Figure 28 Selecting the beam section type for the Standard Profiles class

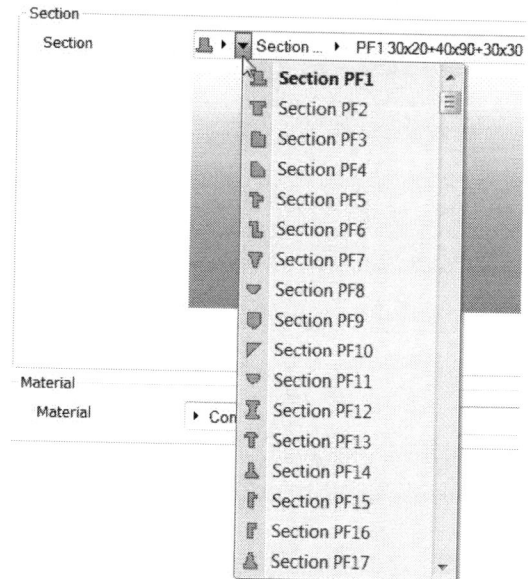

Figure 29 Selecting the beam section type for the Concrete Precast class

Based on the **Section Type** option selected, the **Section Size** flyout will show you the available sizes that you can select from.

Material Area

This area provides the options to specify the material and grade. The first flyout is used to specify the material type and the second flyout is used to specify the material grade.

Positioning Tab

The **Positioning** tab shown in Figure 30 is used to specify the location and orientation of the concrete beam with respect to the system axis points you defined to insert the beam. These options are discussed next.

Offset Area

This area provides the options to specify the location of the beam in relation to the system axis. You can select the radio buttons to define the system axis at the four corner points, midpoints of the four edges, or even the center point of the beam. For asymmetrical members shapes such as asymmetric angles or i-asymmetric member, you can also select the radio button to define the system axis at the center of gravity of the concrete beam. This radio button is marked by arrow **(1)** in Figure 30. The **X** and **Y** edit boxes are used to define offset values along the X and Y directions from the system axis.

Angle Area

This area is used to rotate the concrete beam to a standard angle by clicking its radio button or by entering a custom value in the **Angle** edit box.

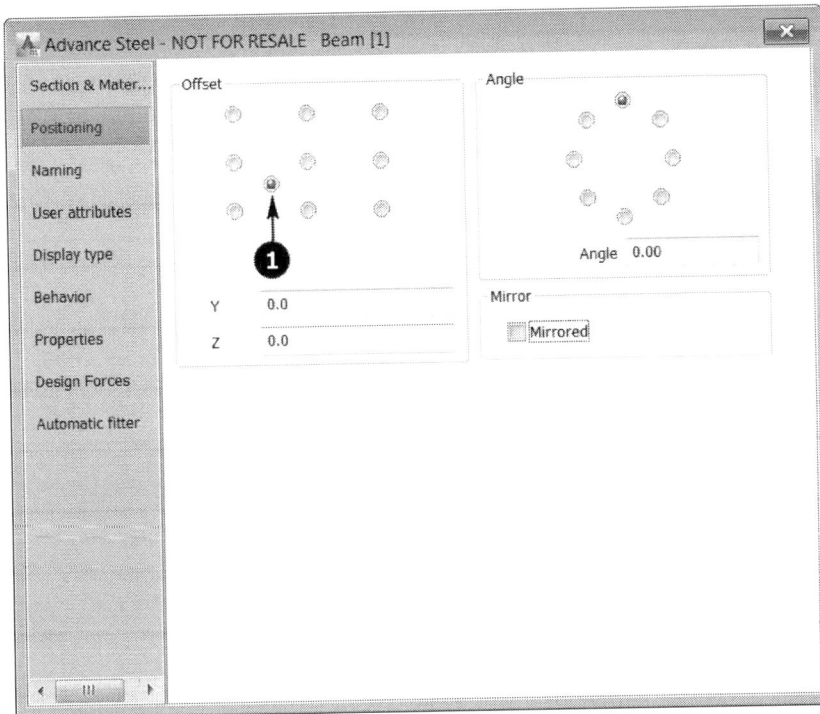

Figure 30 The **Positioning** *tab of the* **Advance Steel - Beam [1]** *dialog box*

Mirror Area
Selecting the **Mirrored** tick box in this area allows you to mirror the concrete beam around its vertical axis.

Naming Tab
The **Naming** tab shown in Figure 31 is used to specify the naming and Bill of Material (BOM) information for the concrete beam you inserted. Normally in Advance Steel, you will use the automatic number tools. However, if you want to manually number certain items, you can use the naming options in this tab. These options are discussed next.

Naming Area
This area provides the options to specify the single part mark and the prefix for single part mark to be assigned to the concrete beam.

Level
This list allows you to select the level at which the beam is inserted. By default, it will display **Structure 1** as the value if there are no levels defined.

Model Role
Model role controls how the structural member is documented, dimensioned, and labeled in the detail drawings. As a result, it is extremely important to specify the model roles correctly. You can select the **Model Role** tick box and then select the role type of the inserted concrete beam from this list.

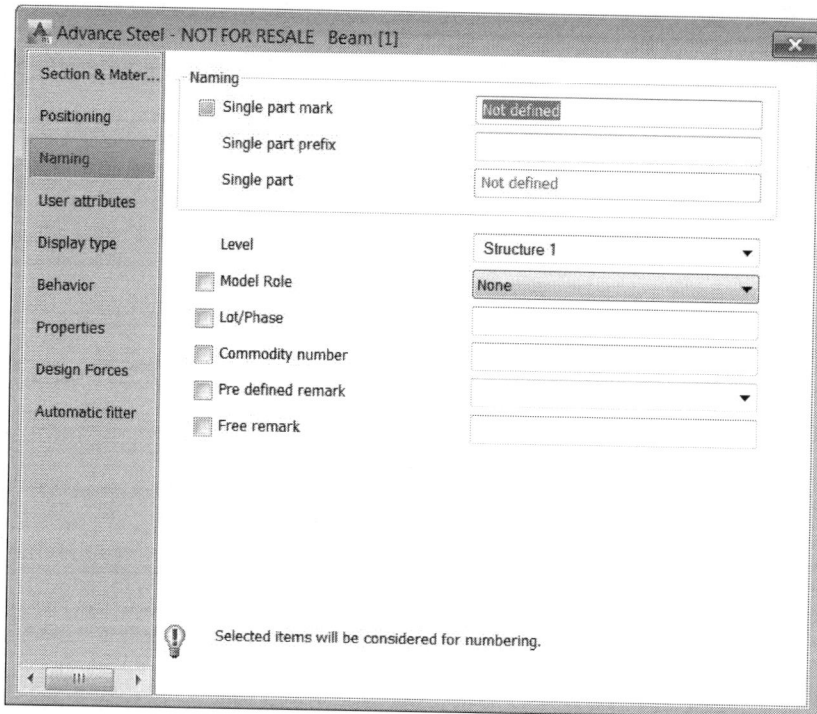

*Figure 31 The **Naming** tab of the **Advance Steel - Beam [1]** dialog box*

Lot/Phase
This option allows you to assign the lot or phase to the concrete beam.

Commodity number
This option allows you to assign the commodity number to the concrete beam.

Pre defined remark
Advance Steel allows you to save specific remarks about the structural members in the **Notes** table of the **AstorBase.mdb** file. Those remarks can be selected from this list. By default, the **AstorBase.mdb** file is located in the **ProgramData > Autodesk >Advance Steel 2016 > Steel\Data** folder.

*Tip: By default, the **Notes** table of the **AstorBase.mdb** file has four entries. You can add more entries, if required. Once you add or edit entries in this database file, you will have to invoke the **Update defaults** button from the **Home** ribbon tab > **Settings** ribbon panel. The new or edited remarks will then be displayed in the **Pre defined remark** list.*

Free remark
This option allows you to add any other remark related to the concrete beam.

User attributes Tab

The **User attributes** tab shown in Figure 32 is used to specify additional user-defined attributes, which can be displayed in the titleblock while generating the detail drawings.

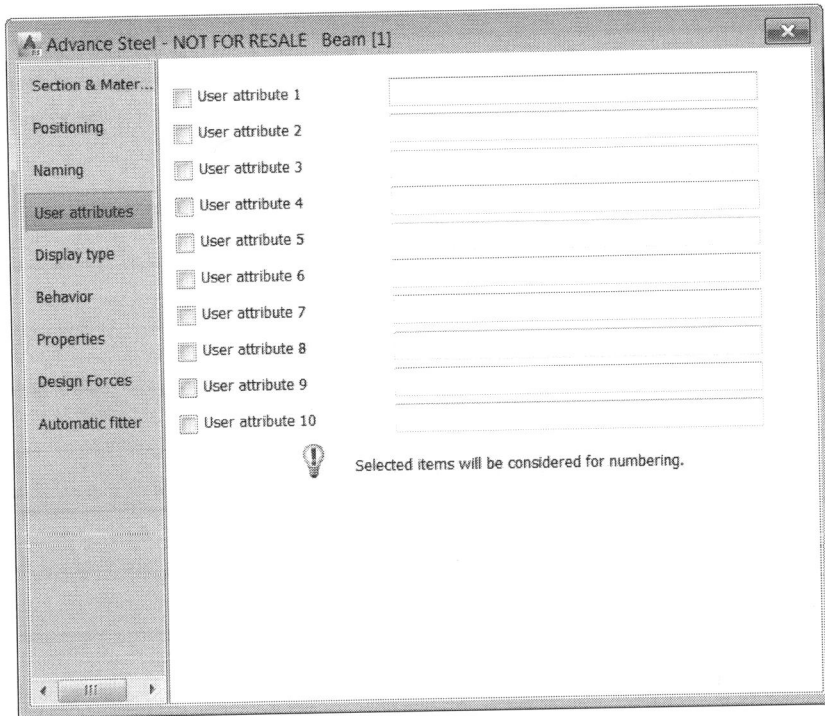

*Figure 32 The **User attribute** tab of the **Advance Steel - Beam [1]** dialog box*

Display type Tab

The **Display type** tab shown in Figure 33 is used to control the representation of the concrete member in the drawing window. Various display types are discussed next.

Off

If this radio button is selected, the display of the concrete beam will be turned off.

> **Tip**: To turn on the visibility of the hidden members, you can use the **All Visible** tool on the **Advance Steel Tool Palette > Quick views** tab.

Standard

This is the default representation type of the inserted beam. In this representation, the beam and the system axis is displayed. This representation is recommended for large structural models.

Features

This representation type is used when you have created some cut features on the concrete beam. With this representation, the edges of the object used to create the cut feature will also be displayed.

Exact

This radio button is used to display the exact representation of the concrete beam.

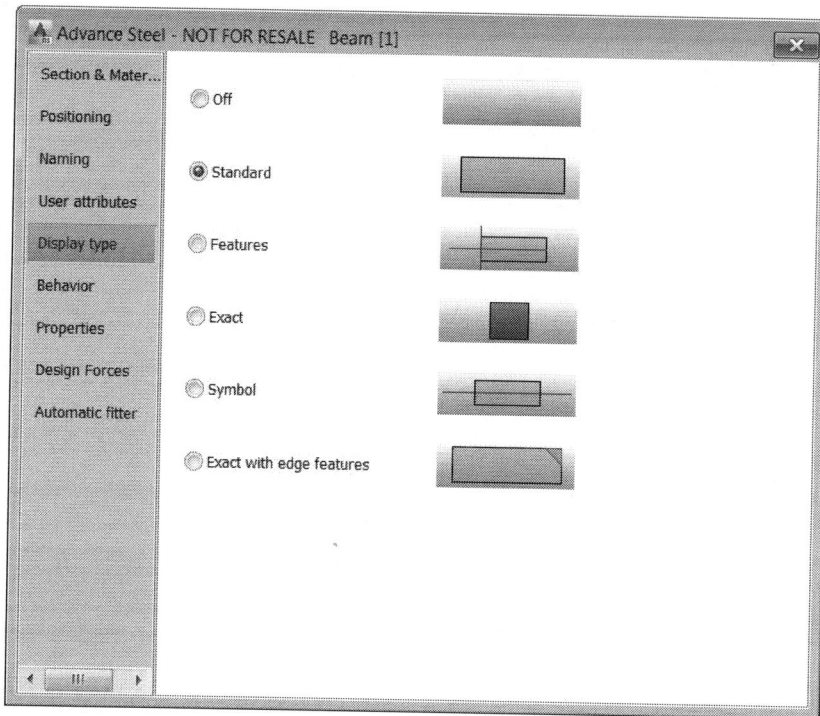

*Figure 33 The **Display type** tab of the **Advance Steel - Beam [1]** dialog box*

Symbol
If this radio button is selected, the display of the concrete beam will be turned off and only the system axis will be displayed.

Exact with edge features
This radio button is used to display the exact representation of the beam, along with all the edge features, such as fillets or chamfers.

Behavior Tab
The **Behavior** tab shown in Figure 34 is used to define how this concrete beam will be used in other Advance Steel operations. These options are discussed next.

used for numbering
If this tick box is selected, the concrete beam will be used for single part numbering.

used for Lists
If this tick box is selected, the concrete beam will be used in various parts lists and Bill of Materials (BOM).

explicit quantity
This edit box is used to specify the quantity of the concrete beams required. By default, the

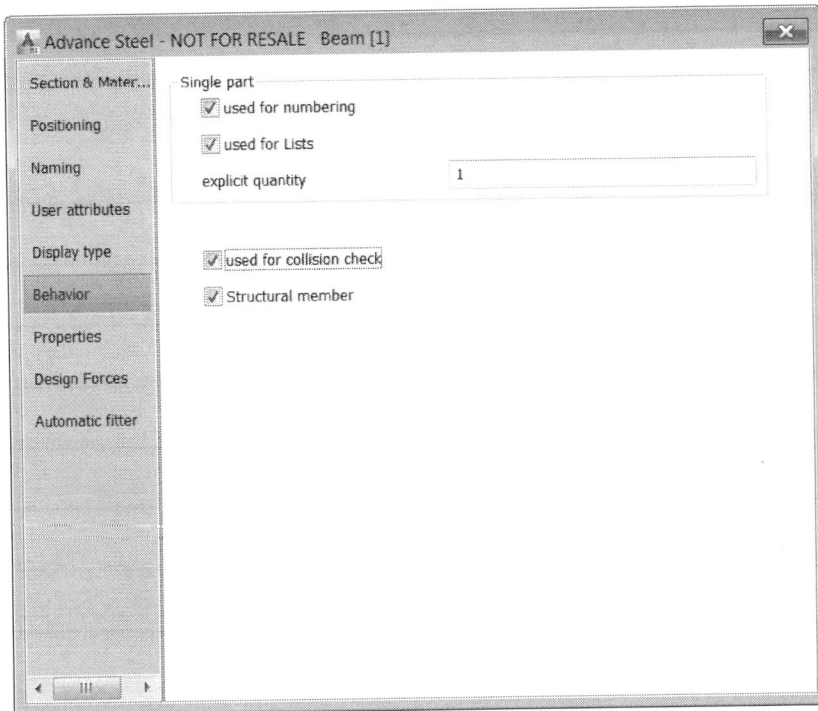

*Figure 34 The **Behavior** tab of the **Advance Steel - Beam [1]** dialog box*

value is set to 1. You can increase this value if you want some extra number of beams to be displayed in the BOM.

used for collision check
Advance Steel provides a tool to check the collision of various structural members. Selecting this tick box will include the concrete beam as part of the collision check process.

Structural member
If this tick box is selected, the concrete beam will be treated as a structural member in the current model.

Properties Tab
The **Properties** tab displays various physical, geometric, and section properties of the concrete beam, as shown in Figure 35.

Design Forces Tab
The **Design Forces** tab shown in Figure 36 is used to specify the values for the design forces to be considered for the concrete beam. These values can be specified based on the **LRFD** standard or the **ASD** standard. Selecting the **Automatic values** tick box allows Advance Steel to automatically define these values.

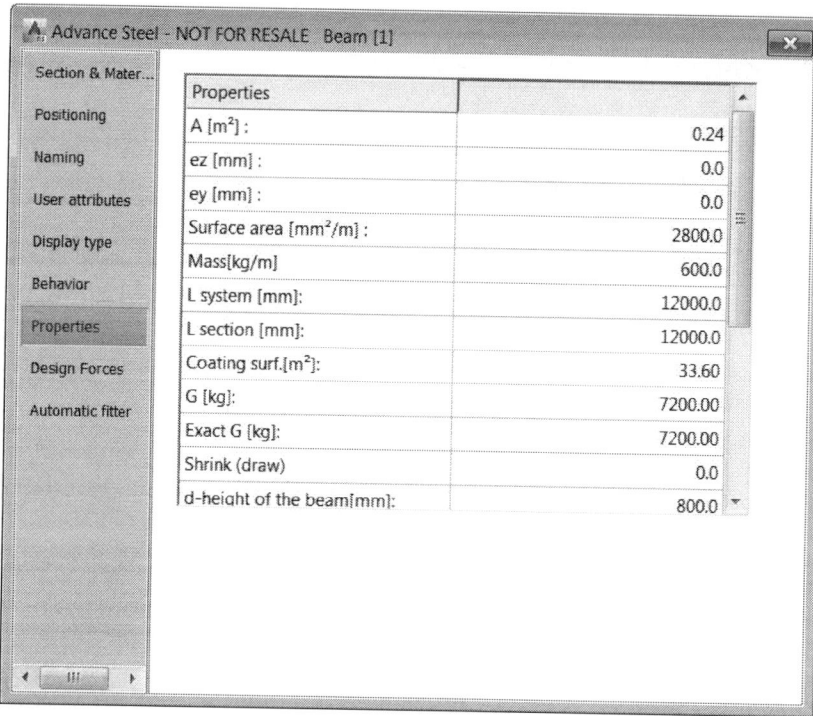

Figure 35 The **Properties** tab of the **Advance Steel - Beam [1]** dialog box

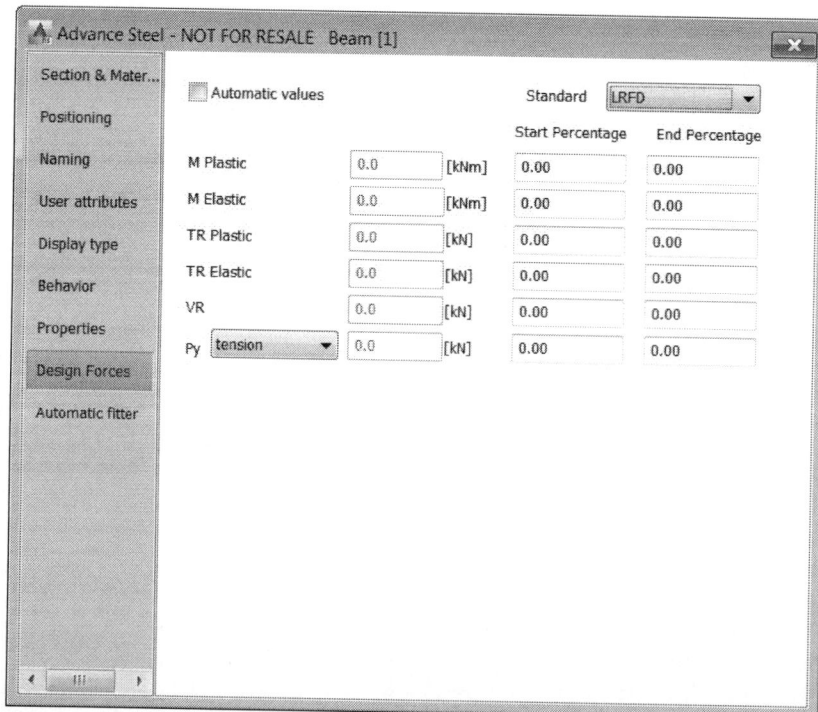

Figure 36 The **Design Forces** tab of the **Advance Steel - Beam [1]** dialog box

Automatic fitter Tab

The options available on the **Automatic fitter** tab, shown in Figure 37, are discussed next.

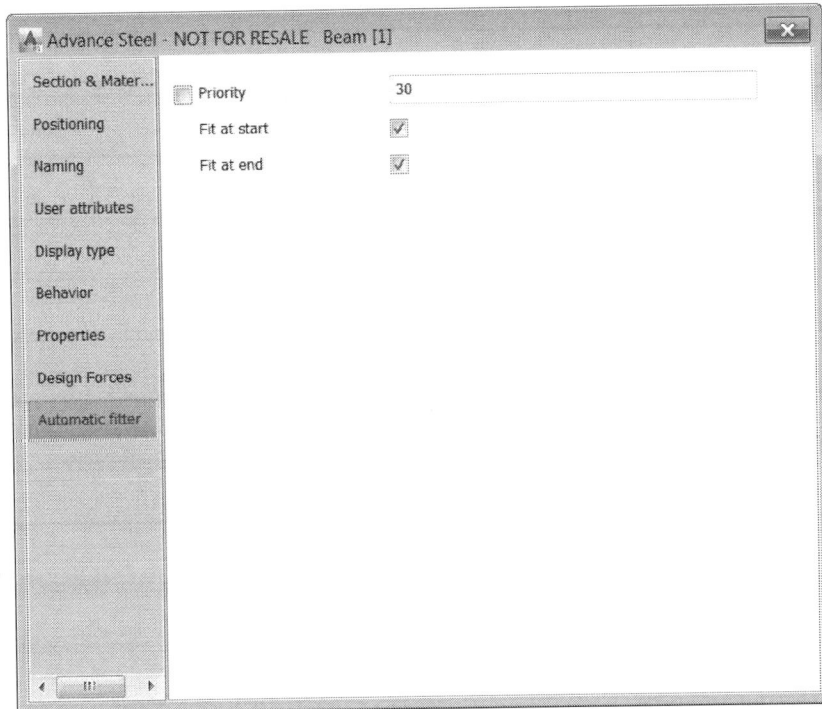

*Figure 37 The **Automatic fitter** tab of the **Advance Steel - Beam [1]** dialog box*

Priority

This option is extremely important at the intersection of beams. By default, if there are two beams intersecting, the first beam gets the priority over the second, which allows the first beam to remain a continuous element, as shown in Figure 38. If you want the second beam to be the continuous element, increase its priority to be a value more than **30**, which is the default value for all the beams. On doing so, the second beam will be the continuous element, as shown in Figure 39.

Fit at start/Fit at end

These options work in conjunction with the **Priority** option and are used to control how the beam with the priority will appear at the corner intersection of another beam. By default, these tick boxes are selected. As a result, the beam with the priority is extended at the corner. In Figure 40, beam 1 has the priority and because it is visible, it is automatically extended. Figure 41 shows the same beam with the **Fit at end** tick box cleared.

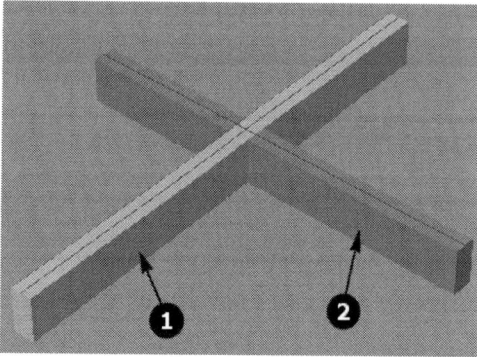

Figure 38 The default priority of beams allows the first beam to be a continuous element

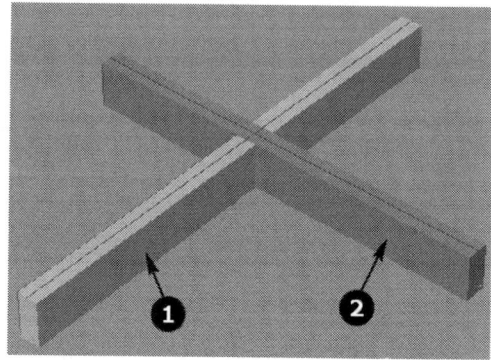

Figure 39 The priority of beam 2 increased more than beam 1

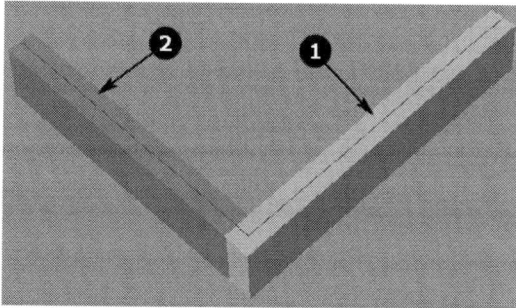

*Figure 40 Beam 1 with the **Fit at end** tick box selected*

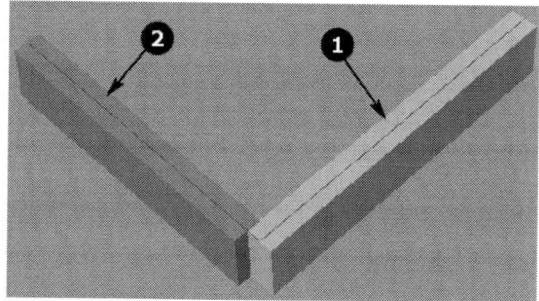

*Figure 41 Beam 1 with the **Fit at end** tick box cleared*

Inserting Concrete Column

Home Ribbon Tab > Objects Ribbon Panel > Concrete beam > Concrete column
Objects Ribbon Tab > Other objects Ribbon Panel > Concrete column

The **Concrete column** tool is used to insert a concrete column. When you invoke this tool, you will be prompted to locate the start point of the system axis, which is the location of the bottom face of the column. On specifying that point, the **Advance Steel - Beam [X]** dialog box will be displayed, where the number X will depend on how many concrete elements you have placed in the past. All the tabs and the options in this dialog box are similar to the ones discussed for the **Concrete beam** tool.

Inserting Isolated Footing

Home Ribbon Tab > Objects Ribbon Panel > Concrete beam > Isolated footing
Objects Ribbon Tab > Other objects Ribbon Panel > Isolated footing

The **Isolated footing** tool is used to insert an isolated footing. When you invoke this tool, you will be prompted to specify the position of the footing. On specifying that point, the

Advance Steel - Isolated Footing [X] dialog box will be displayed with the **Shape & Material** tab active, as shown in Figure 42. In the dialog box, the number X will depend on how many concrete elements you have placed in the past. The options in the **Shape & Material** tab are discussed next.

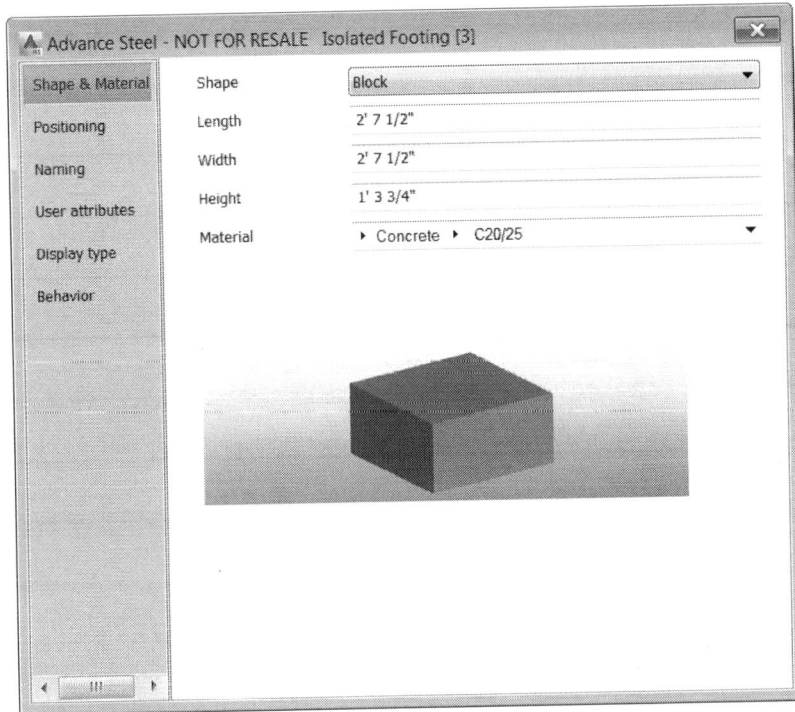

*Figure 42 The **Shape & Material** tab of the **Advance Steel - Isolated Footing** dialog box*

Shape
This list allows you to select the shape of the concrete footing. The available shapes are **Block** or **Cylinder**.

Length/Radius
This is where you specify the length or radius of the footing, depending on the shape selected from the **Shape** list.

Width
This is where you specify the width of the block-shaped footing.

Height
This is where you specify the height of the block or cylinder-shaped footing.

Material
This list allows you to specify the material and the grade of material.

The remaining tabs and options in the **Advance Steel - Isolated Footing [X]** dialog box are similar to those discussed in the **Concrete beam** tool.

Inserting Continuous Footing

Home Ribbon Tab > Objects Ribbon Panel > Concrete beam > Continuous footing
Objects Ribbon Tab > Other objects Ribbon Panel > Continuous footing

The **Continuous footing** tool is used to insert continuous by specifying the start point and endpoint of the system axis. On specifying the two points, the **Advance Steel - Beam [X]** dialog box will be displayed, where the number X will depend on how many concrete elements you have placed in the past. All the tabs and the options in this dialog box are similar to the ones discussed for the **Concrete beam** tool.

Inserting Wall

Home Ribbon Tab > Objects Ribbon Panel > Concrete beam > Wall
Objects Ribbon Tab > Other objects Ribbon Panel > Wall

This tool is used to insert continuous wall segments standing vertically on the XY plane of the current User Coordinate System (UCS). When you invoke this tool, you will be prompted to specify the first point of the wall. On specifying this point, you will then be prompted to specify the next point. If you want to insert only a single wall, you can press the ENTER key after specifying the second point. However, if you want to insert multiple wall segments, you can keep specifying the endpoints of the wall segments in the previous prompts. Remember that although you are inserting multiple wall segments, they will still be treated as individual walls.

You can also use the **Close** option in the command line to close the wall segments into a closed loop. The tool line options also let you undo the previous segment or change the UCS for the next segment. Once you have specified the last point, press the ENTER key; the **Advance Steel - Wall [X]** dialog box will be displayed in the **Shape & Material** tab, as shown in Figure 43. In this dialog box, the number X will depend on how many concrete elements you have placed in the past. The options in the **Shape & Material** tab are discussed next.

Length
If you have placed a single wall segment, this edit box will show the length of that segment. However, if you have inserted multiple wall segments with varying lengths, this edit box will be blank. In this case, if you enter a value in this edit box, all the wall segments created using this **Wall** tool will be changed to that value.

Height
This edit box is used to specify the height of the wall.

Convert to polygon
Selecting this tick box will convert the wall into a polygonal element. As a result, you will be able

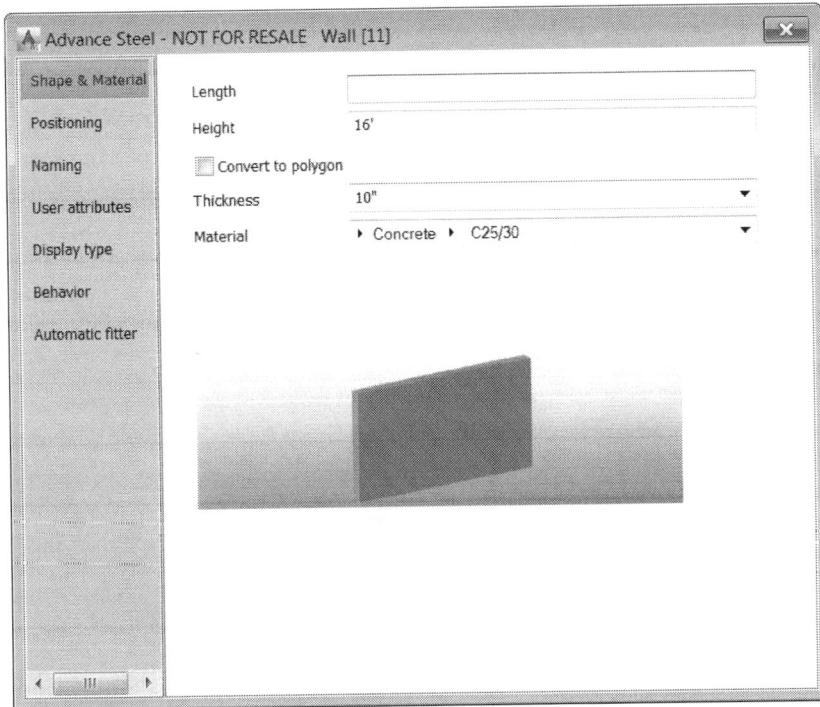

Figure 43 The **Shape & Material** *tab of the* **Advance Steel - Wall** *dialog box*

to drag the top endpoints of the wall independent of the bottom endpoints. However, if this tick box is not selected, dragging one of the endpoints of the wall will change its overall length.

Thickness

This edit box is used to specify the thickness of the wall.

Material

This list allows you to specify the material and the grade of material.

The remaining tabs and options in the **Advance Steel - Wall [X]** dialog box are similar to those discussed in the **Concrete beam** tool.

Inserting Polygonal Wall

Home Ribbon Tab > Objects Ribbon Panel > Concrete beam > Polygonal wall
Objects Ribbon Tab > Other objects Ribbon Panel > Polygonal wall

This tool is used to insert a polygonal wall lying horizontally on the XY plane of the current User Coordinate System (UCS). When you invoke this tool, you will be prompted to specify the points that define the contour of the wall. Once you have specified all the required points, press the ENTER key; the **Advance Steel - Wall [X]** dialog box will be displayed. The options in this dialog box are similar to those discussed in the **Wall** tool.

Inserting Curved Concrete Beam

> **Home Ribbon Tab > Objects Ribbon Panel > Concrete beam >**
> **Concrete curved beam**
> **Objects Ribbon Tab > Other objects Ribbon Panel > Concrete curved beam**

This tool is used to insert a curved concrete beam by specifying the start point, endpoint, and a point on the curve. Once you have specified all three points, the **Advance Steel - Beam [X]** dialog box will be displayed. The options in this dialog box are similar to those discussed in the **Concrete beam** tool, with the exception that in this case, the option to change the radius of the curved beam is available on the **Positioning** tab. Figure 44 shows an I-symmetric concrete curved beam.

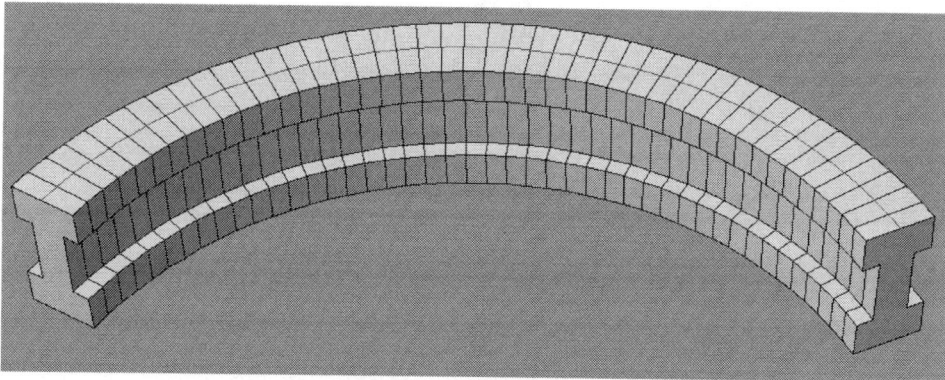

Figure 44 An I-symmetric shape concrete curved beam

Inserting Rectangular Slab

> **Home Ribbon Tab > Objects Ribbon Panel > Concrete beam > Rectangular slab**
> **Objects Ribbon Tab > Other objects Ribbon Panel > Rectangular slab**

This tool is used to insert a rectangular slab by specifying two opposite corners of the slab. As soon as you specify the second point, the **Advance Steel - Slab [X]** dialog box is displayed, which is similar to the dialog box that is displayed for the **Isolated footing** tool. You can change the length and width value of the slab in the **Shape & Material** tab. The remaining options in this dialog box are similar to those discussed in the previous tools.

Tip: To use the second corner of the slab for defining its length and width, enter the length and width values separated by a comma (,) when you are prompted to specify the endpoint of the slab's diagonal line.

Inserting Polygonal Slab

Home Ribbon Tab > Objects Ribbon Panel > Concrete beam > Polygonal slab
Objects Ribbon Tab > Other objects Ribbon Panel > Polygonal slab

---

This tool is used to insert a polygonal slab. When you invoke this tool, you will be prompted to specify the points that define the contour of the slab. Once you have specified all the required points, press the ENTER key; the **Advance Steel - Slab [X]** dialog box will be displayed. The options in this dialog box are similar to those discussed in the previous tool.

EDITING CONCRETE SECTIONS

At various times during a project, you need to change the type, size, orientation, and other properties of the sections you have inserted. This can be done by simply double-clicking on the section you want to edit. If you want to change multiple sections, select all of them and then right-click to display the shortcut menu. From that menu, select **Advance Properties**; the **Advance Steel - Beam [X]** dialog box will be displayed. The changes made in this dialog box will be reflected in all the beams.

Hands-on Tutorial (STRUC)	In this tutorial, you will complete the following tasks: 1. Start a new Advance Steel project and configure its settings. 2. Create the grid, as shown in Figures 45 and 46. 3. Insert the concrete slab using the dimensions shown in Figure 46. 4. Insert concrete columns on the slab using the dimensions shown in Figure 46. 5. Insert isolated footings using the dimensions shown in Figure 46. 6. Change the UCS to the top of the columns and place another concrete slab at the top of the columns.

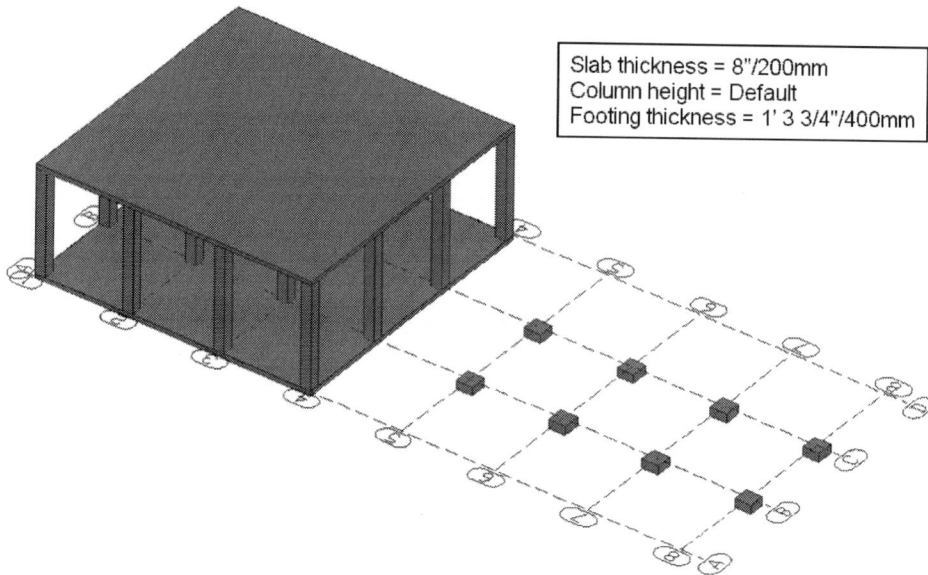

Slab thickness = 8"/200mm
Column height = Default
Footing thickness = 1' 3 3/4"/400mm

Figure 45 *The grid and concrete layout for the tutorial*

Section 1: Starting a New Advance Steel Project and Configuring its Settings

In this section, you will start a new Advance Steel project using the default template. You will then setup the author information in **Management Tools** and then finally use the **Project settings** tool to configure the settings of this project.

1. Start Advance Steel 2016 from **Start > All Programs > Autodesk > Advance Steel 2016 > Advance Steel 2016** or from the shortcut icon on the desktop of your computer.

When Advance Steel is loaded, it automatically starts a default template selected based on the country settings you selected while installing this program.

2. From the **Home** ribbon tab > **Settings** ribbon panel, click **Management Tools** to display the **Management Tools** window.

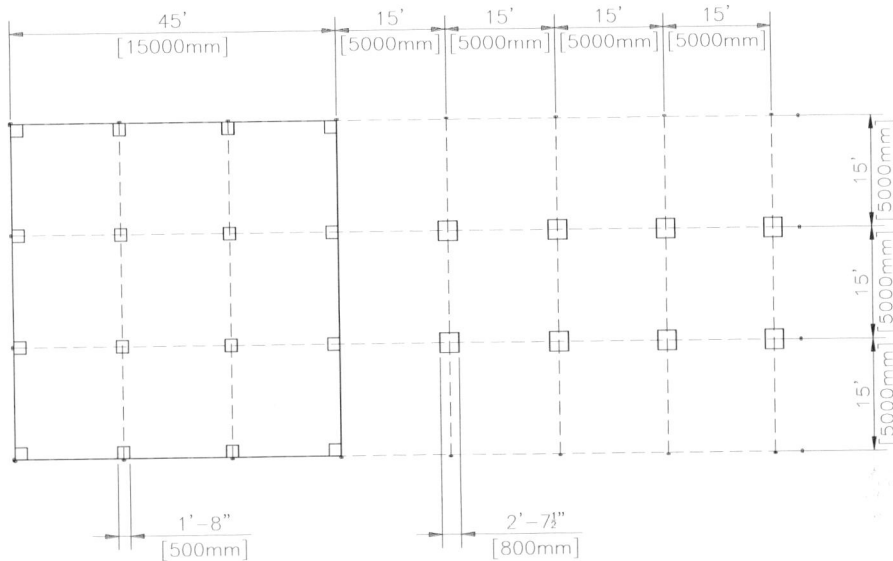

Figure 46 *Dimensions required for the tutorial*

3. Click **Options** in this window to display the **Options** tab.

4. From the right of the **Current author** list, click **Create new author**; the **Create new author** dialog box is displayed.

5. Enter your initials in the **Author key** edit box, as shown in Figure 47.

6. Enter your full name in the **Author name** edit box, as shown in Figure 47.

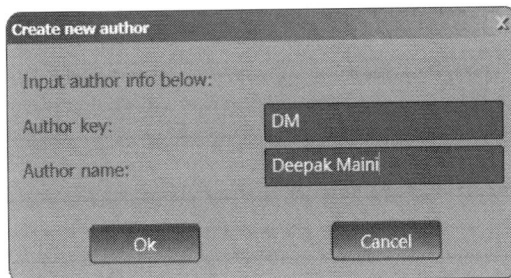

Figure 47 *The **Create new author** dialog box*

7. Click **OK** in the dialog box; the new author you created is listed as the current author.

8. Close the **Management Tools** window by clicking the **X** button on the top right corner of this window.

9. From the same ribbon panel, click **Project settings**; the **Advance Steel - Project data** dialog box is displayed with the **Project Info 1** tab active.

10. In the **Project** edit box, type **Up and Running**.

11. In the **Project No** edit box, type **STRUCC01**.

12. Click the **Length unit** tab from the left pane of the dialog box and make sure the units in the right pane are configured according to your needs.

13. Click **OK** in the dialog box.

Section 2: Changing the Visual Style

In this section, you will change the visual style to **Conceptual**. You will also modify this style to make sure the occluded edges are not displayed.

1. From the top left of the drawing window, click **Visual Style Controls** and select **Conceptual**, as shown in Figure 48.

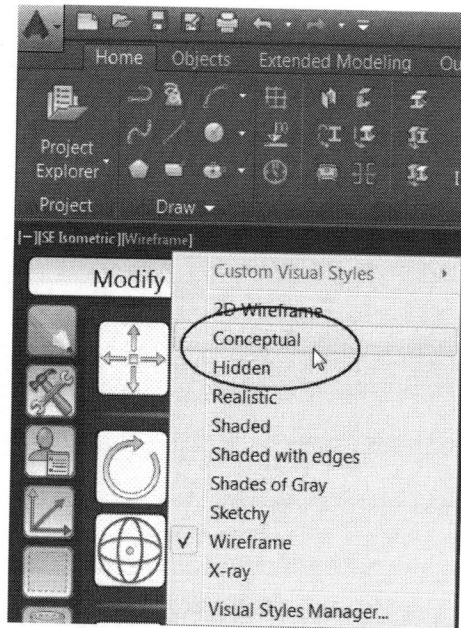

Figure 48 Changing the visual style

Next, you will edit this visual style so that the occluded edges are not displayed in the model.

2. Click on **Visual Style Controls** again and then from the bottom of the list, select **Visual Styles Manager**; the **VISUAL STYLES MANAGER** window is displayed.

3. From the **Available Visual Styles in Drawing** area at the top of this window, click **Conceptual** to display the settings related to this style.

4. Scroll down in this window to the **Occluded Edges** area.

5. Click on the right of the **Show** option and select **No** from the list.

6. Close the **VISUAL STYLE MANAGER** window.

What I do

*I generally use **Conceptual** visual style while creating 3D models in Advance Steel. Therefore, to avoid changing this style in every project, I open my default template and edit the **Conceptual** visual style in it. I also make this style the current style in the template. As a result, every new project that I start using that template, it automatically has my preferred visual style settings.*

Section 3: Creating Grids

As seen in Figures 45 and 46, the grid needed for this project is not a standard 4X4 grid. As a result, you will use the **Building Grid** tool to create this grid and then modify it.

1. From the **Status Bar** at the bottom of the command line, make sure the **Dynamic Input** button is turned on (Blue color).

2. From the **Home** ribbon tab > **Objects** ribbon panel, invoke the **Building Grid** tool; you are prompted to define two diagonal points for grid.

   In the following steps, you will enter numeric values. As discussed earlier in this book, you are provided the values in **Imperial** as well as **Metric** units. You can select the relevant dimension values for your project. Remember that the Metric units are defined with **mm** as the suffix. However, while entering these values in the software, you do not need to enter the units.

3. Type **20',20'** or **1000,1000** as the first corner point of the grid and press ENTER; you are prompted to specify the second point.

4. Type **105',45'** or **35000,15000** and press ENTER; a 4X4 grid is created, as shown in Figure 49.

   You will now modify this grid. It is important to note that the horizontal and vertical grids are created as separate sets. Therefore, they need to be modified separately.

5. Double-click on one of the four vertical grid lines that are parallel to the Y axis; the **Advance Steel - Axis, parallel** (Imperial) or **Grid lines, parallel** (Metric) dialog box is displayed.

   *Tip: The Advance Steel dialog boxes do not have the **OK** or **Apply** buttons. All the values you enter in the dialog boxes are live updated in the drawing window.*

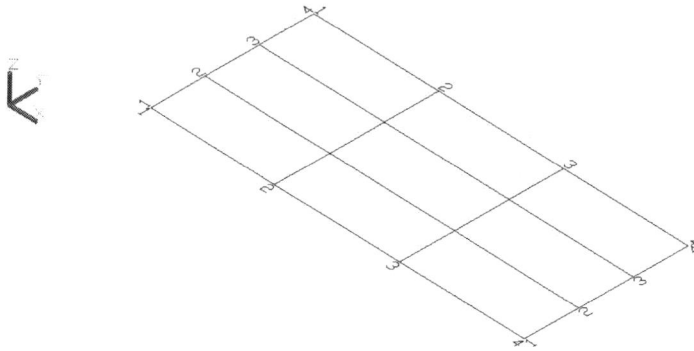

Figure 49 The default 4X4 grid

6. From the **Balloon(Frame)** list on the right, select **Edging** to add a frame around the grid numbers.

    Next, you will activate the **Group** tab (Imperial) or **Sequence** tab (Metric) in the dialog box to change the number of grid lines.

7. From the left pane of the dialog box, invoke the **Group** tab (Imperial) or **Sequence** tab (Metric); the four vertical grid lines are highlighted in Red in the drawing window.

8. In the **Definition** area > **Number** edit box, type **8**, as shown in Figure 50 and press ENTER; the number of vertical grid lines are increased to 8 in the drawing window.

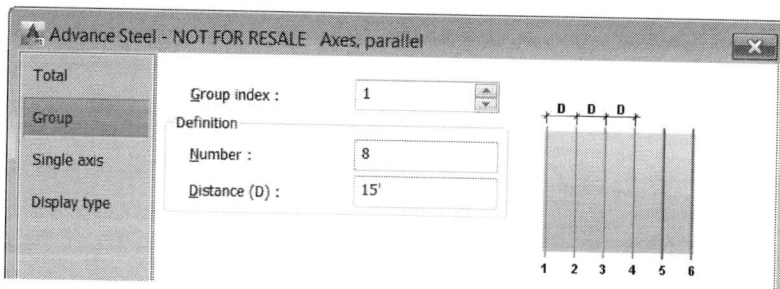

Figure 50 Changing the grid settings

9. Close the dialog box.

    Next, you will modify the horizontal grid lines and change their labels to capital letters.

10. Double-click on one of the four horizontal grid lines that are parallel to the X axis; the **Advance Steel - Axis, parallel** (Imperial) or **Grid lines, parallel** (Metric) dialog box is displayed.

11. From the **Balloon(Frame)** list on the right, select **Edging** to add a frame around the grid numbers.

12. From the **Label type** list, select **Capital letters**; the labels on the horizontal grid lines are changed in the drawing window.

13. Close the dialog box.

Notice how the grid labels at the four corner grid points are overlapping. If you want to avoid it, you can select the grip points of the grid and drag the labels away so they do not overlap. But, that will mean when you generate a drawing, such as an anchor plan, at a later stage, that extra distance between the grid labels will also be dimensioned. However, you can delete that extra dimension in the drawing.

14. Click on the grid label **A** at the **8A** grid intersection point; the Blue grip points are displayed on the grid line.

15. Use the grip point to drag the grid label to the right using 0-degree polar tracking through a distance of **5'** or **1500**.

16. Similarly, drag the grid label **8** at the **8D** grid intersection point **5'** or **1500** along the 90-degree polar tracking. The grid, after making this change, is shown in Figure 51.

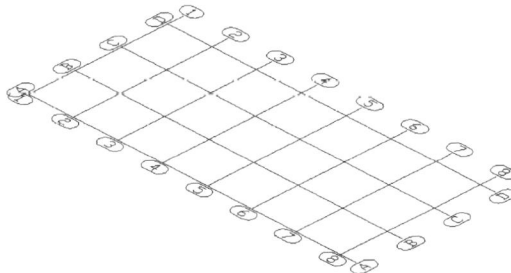

Figure 51 The 8X4 grid created for the project

Section 4: Inserting Concrete Slab

In this section, you will insert the concrete slab using the **Rectangular slab** tool. This slab will be inserted between grid points **1A** and **4D**. To snap to these grid intersection points, you will need to make sure the **GRID Intersection Points** option is selected in object snaps.

1. Make sure the **GRID Intersection Points** object snap is turned on in the **Object Snap** settings on the **Status Bar**.

2. From the **Objects** ribbon tab > **Other objects** ribbon panel, invoke the **Rectangular slab** tool; you are prompted to specify the start point of the slab's diagonal line.

3. Select the **1A** grid point as the start point of the slab's diagonal line; you are prompted to specify the end point of the slab's diagonal line.

4. Select the **4D** grid intersection point; the concrete slab is created and the **Advance Steel - Slab [1]** dialog box is displayed.

5. Click the **Positioning** tab from the left pane and make sure the second button is selected in the **Justification** area.

6. Click the **Naming** tab from the left pane.

7. From the **Model Role** list, select **Slab**.

8. Close the dialog box. Figure 52 shows the model, after inserting the slab.

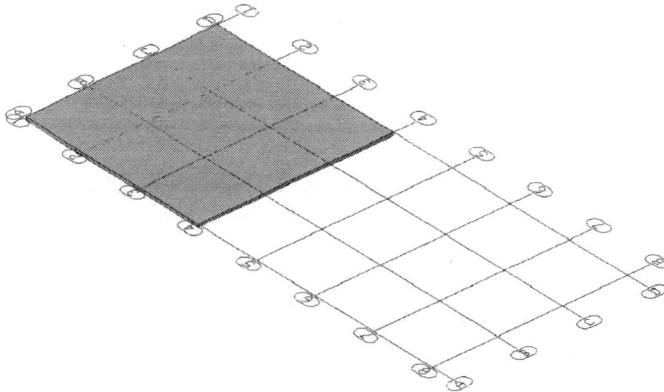

Figure 52 Concrete slab inserted using the grid points

Section 5: Inserting and Copying Concrete Columns

In this section, you will insert concrete columns using the **Concrete column** tool. You will then use the **Copy** tool from the **Advance Steel Tool Palette** to copy this column at all the required locations.

1. From the **Objects** ribbon tab > **Other objects** ribbon panel, invoke the **Concrete column** tool; you are prompted to locate the start point of the system axis.

2. Select the **1A** grid intersection point; a concrete column is inserted and the **Advance Steel - Beam [2]** dialog box is displayed with the **Section & Material** tab active.

    By default, a rectangular concrete column is inserted. You need to change the shape to square column.

3. From the **Section** list, click the second flyout and select **Square**, as shown in Figure 53; the size is automatically set to the available column size.

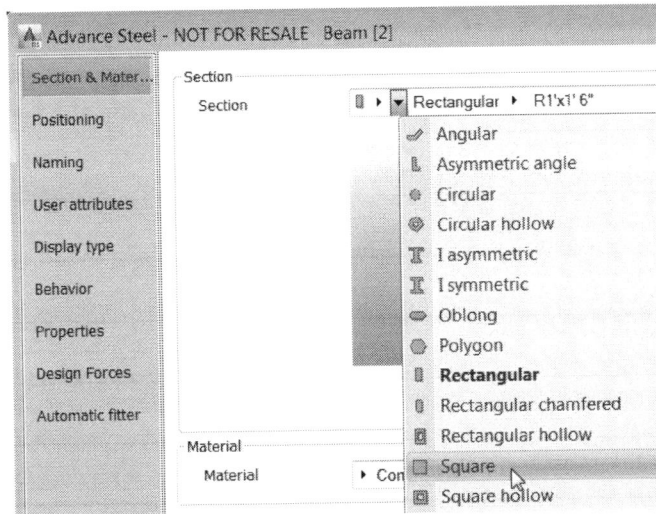

*Figure 53 Selecting the **Square** section type*

By default, the center of the bottom face of the column is placed at the selected grid point. However, in this case, you need to align the grid point to the lower left corner of the column. This will be done using the **Positioning** tab of the dialog box.

4. From the left pane of the dialog box, activate the **Positioning** tab.

5. From the **Offset** area, select the radio button at the lower right corner and close the dialog box; the lower left corner of the column is aligned with the grid point.

    Next, you will copy this column at grid intersection points 2A, 3A, and 4A.

6. From the **Modify** tab of the **Advance Steel Tool Palette**, invoke the **Copy** tool; you are prompted to select the object to copy.

7. Select the column you placed earlier in this section and then press ENTER to accept the selection; you are prompted to specify the base point.

    It is better to turn off all the object snaps other than the **GRID Intersection Points** so that you can easily select the required grid points.

8. Turn off all object snaps other than the **GRID Intersection Points** from the **Object Snap** settings on the **Status Bar**.

9. Now, select the **1A** grid intersection point as the base point; you are prompted to select the second point.

10. Select the **2A**, **3A**, and **4A** grid intersection points to place three more copies of columns.

11. Press the ESC key to exit the **Copy** tool.

Notice that the alignment of the columns is not right. You will now fix these alignments.

12. Select the columns placed at **2A** and **3A** grid points.

13. Right-click in the blank area of the drawing window and select **Advance Properties** from the shortcut menu; the **Advance Steel - Beam [X]** dialog box is displayed.

14. Select the **Positioning** tab.

15. From the **Offset** area, select the bottom center radio button; the columns are aligned correctly with the grid points.

16. Close the dialog box.

17. Double-click on the column placed at the **4A** grid intersection point; the **Advance Steel - Beam [X]** dialog box is displayed.

18. From the **Offset** area, select the bottom left radio button; the column is aligned correctly with the grid point.

19. Close the dialog box.

Next, you need to copy all these four columns to the grid axes B, C, and D. But it is recommended to first hide the slab by turning its layer off.

20. Turn off the **Concrete slab** layer.

21. Invoke the **Copy** tool and then using bottom midpoint of the fourth column shown in Figure 54 as the base point, copy the four columns to the grid intersection points shown in Figure 54.

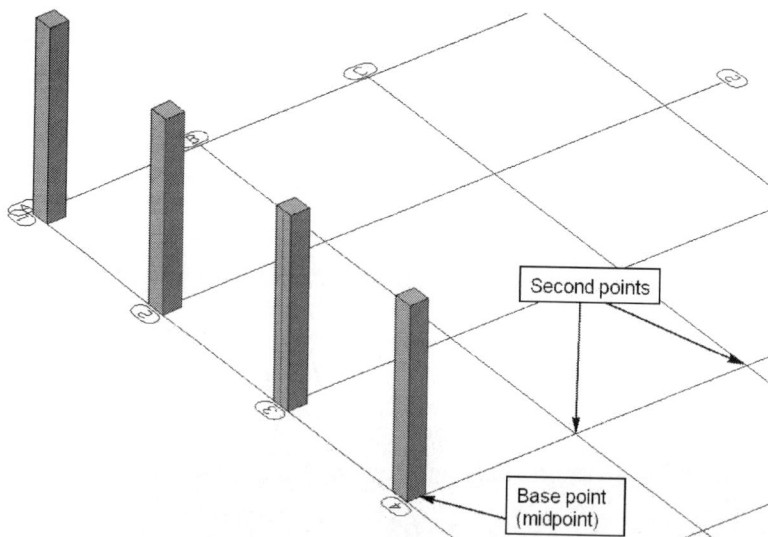

Figure 54 Copying the columns

22. Similarly, copy the four columns on grid axis **A** to grid axis **D** using the top right endpoint of the bottom face of the fourth column as the base point. Figure 55 shows the model after copying all the columns.

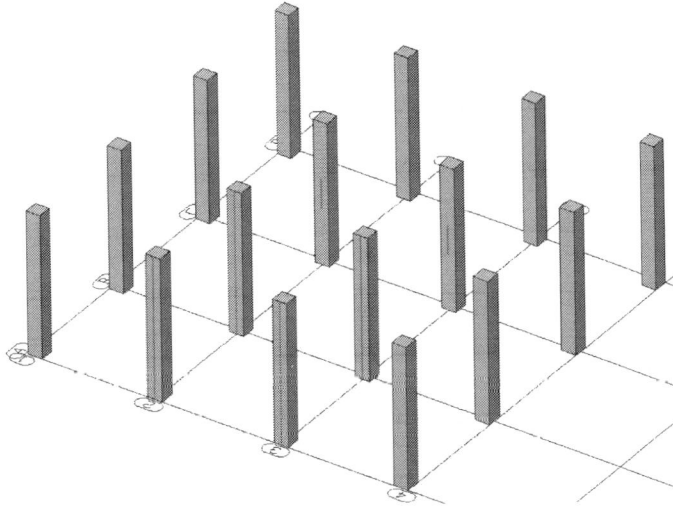

Figure 55 *After copying all the columns*

Section 6: Inserting and Arraying Isolated Footings

In this section, you will insert one instance of the isolated footing shown in Figures 45 and 46 and then array it to create the remaining instances.

1. From the **Objects** ribbon tab > **Other objects** ribbon panel, invoke the **Isolated footing** tool; you are prompted to specify the position of the footing.

2. Select the **5B** grid intersection point; the footing is placed there and the **Advance Steel - Isolated Footing [X]** dialog box is displayed.

3. Close the dialog box by clicking the **[X]** button on the top right of this dialog box.

   Next, you will array the isolated footing to create the remaining instances.

4. Using the **Array** tool, create a disassociative rectangular array of the isolated footings with 2 rows and 4 columns. The spacing values between the rows and columns are 15'/5000mm.

5. Turn on the visibility of the **Concrete slab** layer. Figure 56 shows the model after creating and arraying the isolated footings.

Section 7: Inserting Concrete Slab on Top of Columns

In this section, you will insert a concrete slab on top of the concrete columns. To be able to do that, you first need to create a UCS on the top face of one of the column.

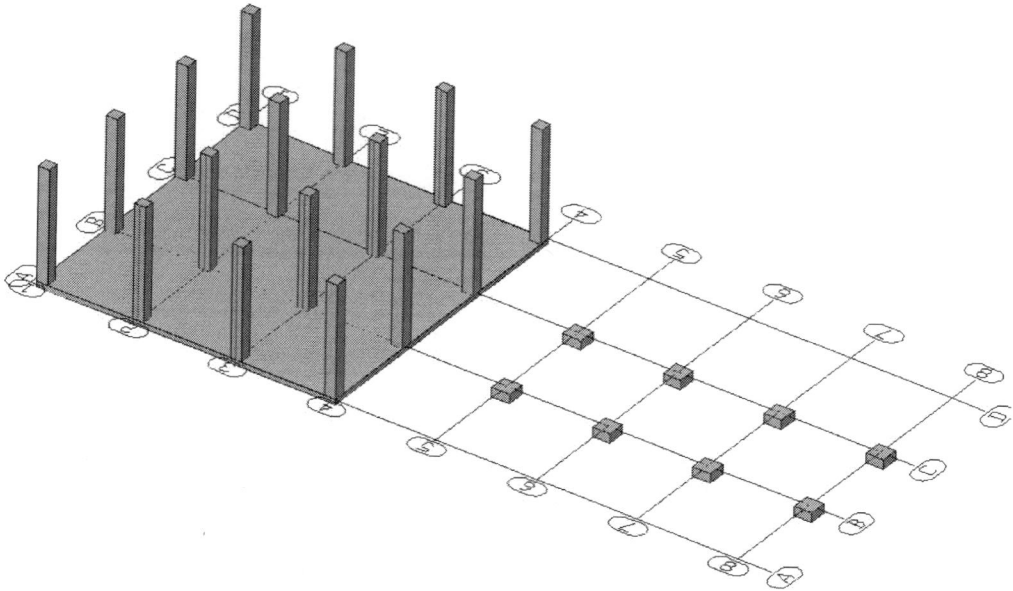

Figure 56 The model after inserting and arraying the isolated footings

1. From the **Advance Steel Tool Palette > UCS** tab > **UCS 3 points** tool, create a UCS at the top face of one of the columns.

2. From the **Objects** ribbon tab > **Other objects** ribbon panel, invoke the **Rectangular slab** tool; you are prompted to specify the start point of the slab's diagonal line.

3. Select the endpoint marked as **1** in Figure 57 as the start point of the slab; you are prompted to specify the end point of the slab's diagonal line.

4. Select endpoint marked as **2** in Figure 57 as the endpoint of the slab; the concrete slab is created and the **Advance Steel - Slab [X]** dialog box is displayed.

5. Click the **Positioning** tab from the left pane.

6. Click the first button in the **Justification** area to place the slab on top of the columns.

7. Click the **Naming** tab from the left pane.

8. From the **Model Role** list, select **Slab**.

9. Close the dialog box. Figure 58 shows the drawing window after inserting the slab.

10. Change the UCS back to world UCS.

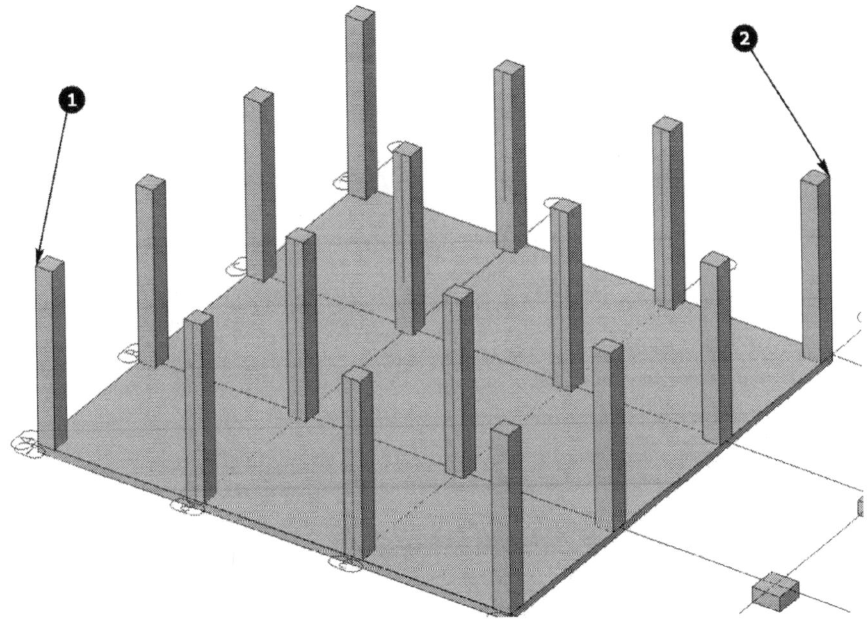

Figure 57 *The start and end points of the slab*

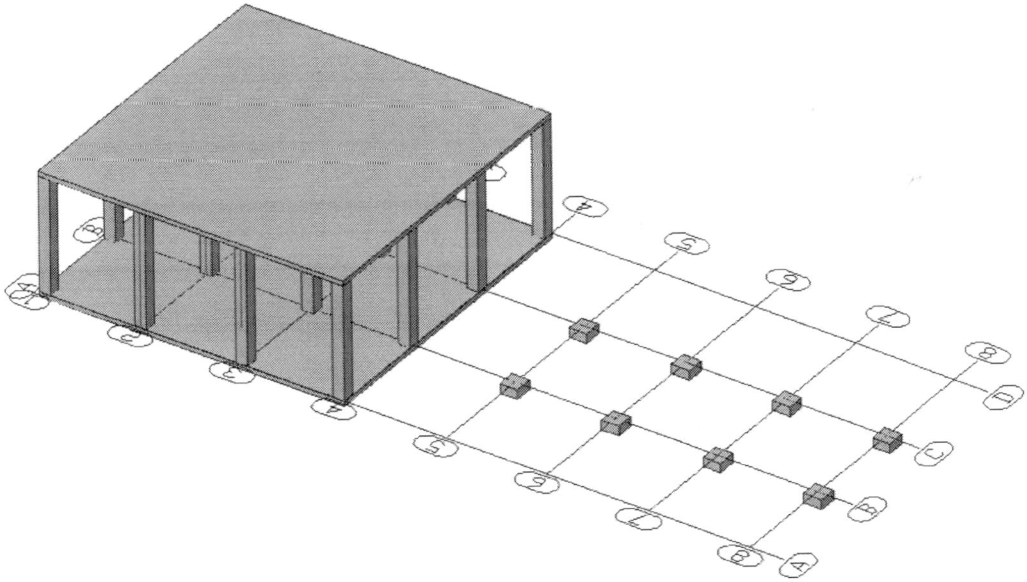

Figure 58 *Completed model*

Next, you will save this file in the folder where you copied the tutorial files of this book. By default, the **C01** folder has a completed file, for your reference.

11. Save the file with the name **c01-Struc.dwg** in the **Tutorial Files > C01 > Struc** folder.

Hands-on Tutorial (BIM)	In this tutorial, you will complete the following tasks: 1. Start a new Advance Steel project and configure its settings. 2. Create the grid, as shown in Figures 59 and 60. 3. Insert the concrete slab using the dimensions shown in Figure 60. 4. Insert isolated footings using the dimensions shown in Figure 60. 5. Insert concrete columns on the slab using the dimensions shown in Figure 60.

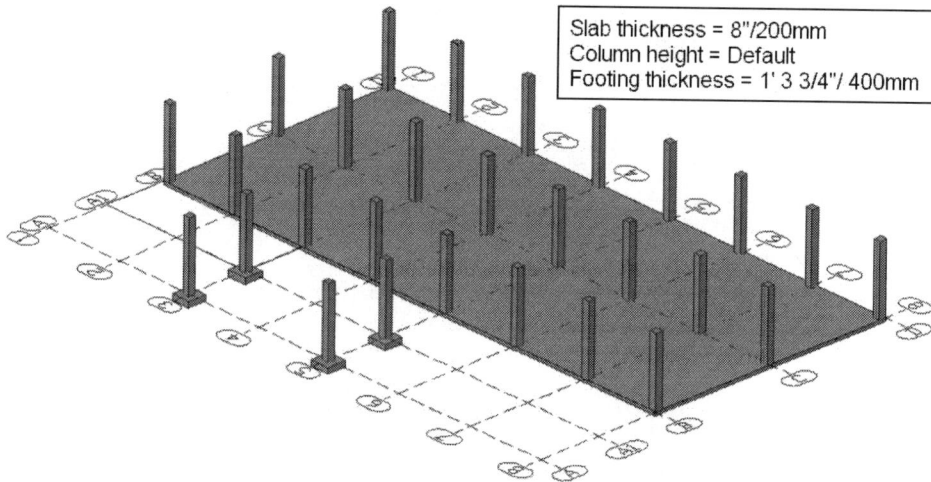

Slab thickness = 8"/200mm
Column height = Default
Footing thickness = 1' 3 3/4"/ 400mm

Figure 59 *The grid and concrete layout for the tutorial*

Section 1: Starting a New Advance Steel Project and Configuring its Settings

In this section, you will start a new Advance Steel project using the default template. You will then setup the author information in **Management Tools** and then finally use the **Project settings** tool to configure the settings of this project.

1. Start Advance Steel 2016 from **Start > All Programs > Autodesk > Advance Steel 2016 > Advance Steel 2016** or from the shortcut icon on the desktop of your computer.

   When Advance Steel is loaded, it automatically starts a default template selected based on the country settings you selected while installing this program.

2. From the **Home** ribbon tab > **Settings** ribbon panel, click **Management Tools** to display the **Management Tools** window.

3. Click **Options** in this window to display the **Options** tab.

4. From the right of the **Current author** list, click **Create new author**; the **Create new author** dialog box is displayed.

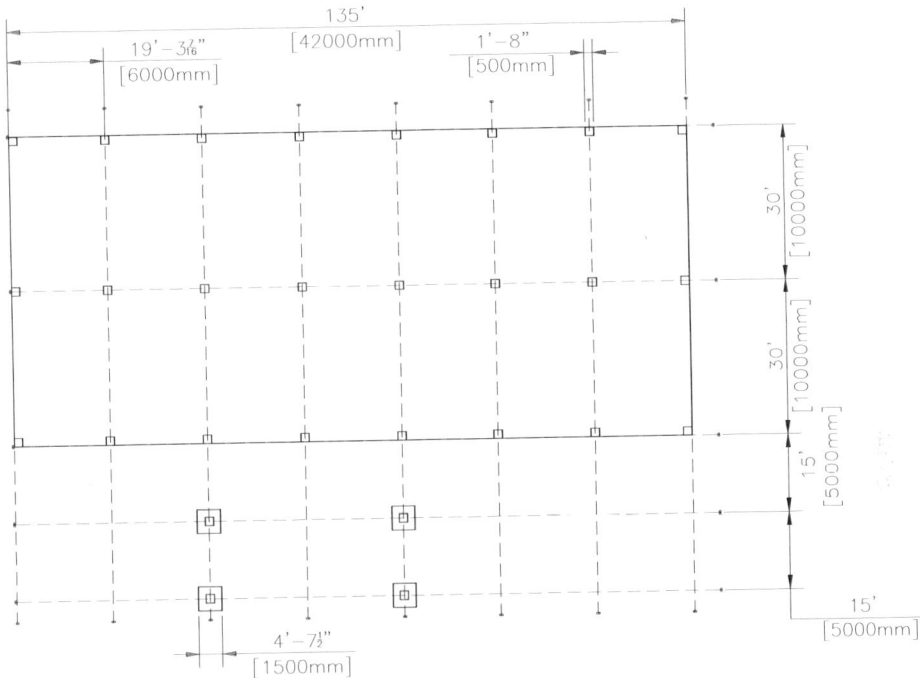

Figure 60 *Dimensions required for the tutorial*

5. Enter your initials in the **Author key** edit box, as shown in Figure 61.

6. Enter your full name in the **Author name** edit box, as shown in Figure 61.

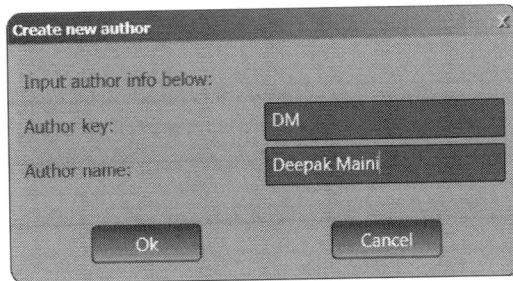

Figure 61 *The* ***Create new author*** *dialog box*

7. Click **OK** in the dialog box; the new author you created is listed as the current author.

8. Close the **Management Tools** window by clicking the **X** button on the top right corner of this window.

9. From the same ribbon panel, click **Project settings**; the **Advance Steel - Project data** dialog box is displayed with the **Project Info 1** tab active.

10. In the **Project** edit box, type **Up and Running**.

11. In the **Project No** edit box, type **BIMC01**.

12. Click the **Length unit** tab from the left pane of the dialog box and make sure the units in the right pane are configured according to your needs.

13. Click **OK** in the dialog box.

Section 2: Changing the Visual Style

In this section, you will change the visual style to **Conceptual**. You will also modify this style to make sure the occluded edges are not displayed.

1. From the top left of the drawing window, click **Visual Style Controls** and select **Conceptual**, as shown in Figure 62.

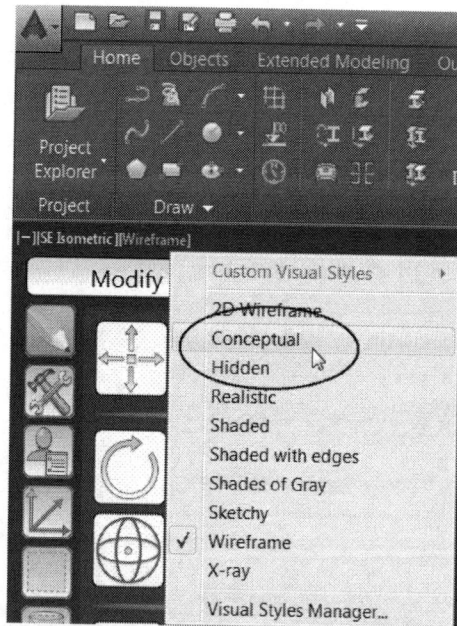

Figure 62 *Changing the visual style*

Next, you will edit this visual style so that the occluded edges are not displayed in the model.

2. Click on **Visual Style Controls** again and then from the bottom of the list, select **Visual Styles Manager**; the **VISUAL STYLES MANAGER** window is displayed.

3. From the **Available Visual Styles in Drawing** area at the top of this window, click **Conceptual** to display the settings related to this style.

4. Scroll down in this window to the **Occluded Edges** area.

5. Click on the right of the **Show** option and select **No** from the list.

6. Close the **VISUAL STYLE MANAGER** window.

What I do

*I generally use **Conceptual** visual style while creating 3D models in Advance Steel. Therefore, to avoid changing this style in every project, I open my default template and edit the **Conceptual** visual style in it. I also make this style the current style in the template. As a result, every new project that I start using that template, it automatically has my preferred visual style settings.*

Section 3: Creating Grids

As seen in Figures 59 and 60, the grid needed for this project is not a standard 4X4 grid. As a result, you will use the **Building Grid** tool to create this grid and then modify it.

1. From the **Status Bar** at the bottom of the command line, make sure the **Dynamic Input** button is turned on (Blue color).

2. From the **Home** ribbon tab > **Objects** ribbon panel, invoke the **Building Grid** tool; you are prompted to define two diagonal points for grid.

   In the following steps, you will enter numeric values. As discussed earlier in this book, you are provided the values in **Imperial** as well as **Metric** units. You can select the relevant dimension values for your project. Remember that the Metric units are defined with **mm** as the suffix. However, while entering these values in the software, you do not need to enter the units.

3. Type **20',20'** or **1000,1000** as the first corner point of the grid and press ENTER; you are prompted to specify the second point.

4. Type **135',95'** or **42000,30000** and press ENTER; a 4X4 grid is created, as shown in Figure 63.

   You will now modify this grid. It is important to note that the horizontal and vertical grids are created as separate sets. Therefore, they need to be modified separately.

5. Double-click on one of the four vertical grid lines that are parallel to the Y axis; the **Advance Steel - Axis, parallel** (Imperial) or **Grid lines, parallel** (Metric) dialog box is displayed.

*Tip: The Advance Steel dialog boxes do not have the **OK** or **Apply** buttons. All the values you enter in the dialog boxes are updated live in the drawing window.*

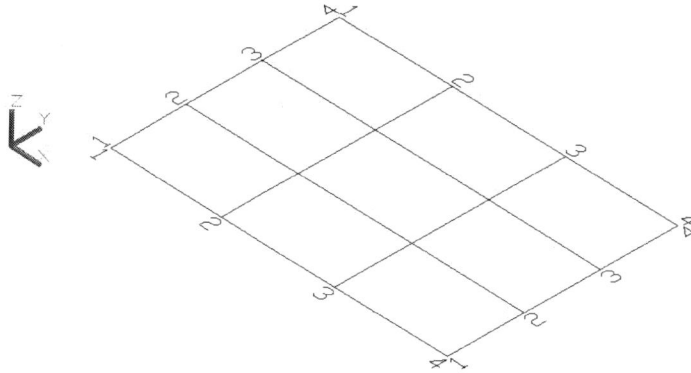

Figure 63 *The default 4X4 grid*

6.   From the **Balloon(Frame)** list on the right, select **Edging** to add a frame around the grid numbers.

Next, you will activate the **Group** tab (Imperial) or **Sequence** tab (Metric) in the dialog box to change the number of grid lines.

7.   From the left pane of the dialog box, invoke the **Group** tab (Imperial) or **Sequence** tab (Metric); the four vertical grid lines are highlighted in Red in the drawing window.

8.   In the **Definition** area > **Number** edit box, type **8**, as shown in Figure 64, and press ENTER; the number of vertical grid lines are increased to 8 in the drawing window.

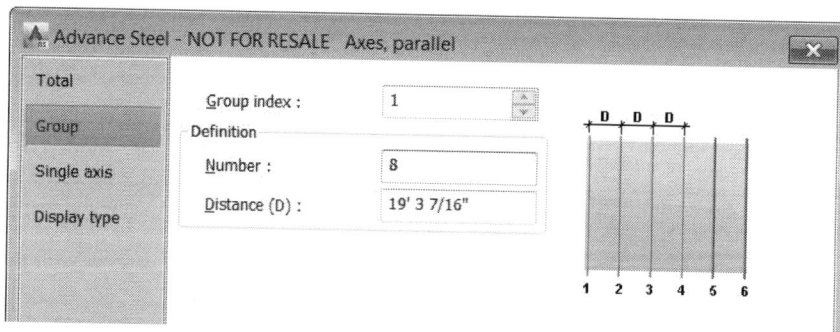

Figure 64 *Changing the grid settings*

9.   Close the dialog box.

You will now modify the horizontal grid lines and change their labels to capital letters.

10. Double-click on one of the four horizontal grid lines that are parallel to the X axis; the **Advance Steel - Axis, parallel** (Imperial) or **Grid lines, parallel** (Metric) dialog box is displayed.

11. From the **Balloon(Frame)** list on the right, select **Edging** to add a frame around the grid numbers.

12. From the **Label type** list, select **Capital letters**; the labels on the horizontal grid lines are changed in the drawing window.

    Next, you need to add grid axis A1 between A and B grid axes. This is done using the options in the **Single axis** (Imperial) or **Single grid line** (Metric) tab of the dialog box.

13. From the left pane of the dialog box, click the **Single axis** (Imperial) or **Single grid line** (Metric) tab; grid axis A is highlighted in the drawing window.

14. In the right pane of the dialog box, select the **Side 2** tick box; the **Main axes name** tick box is activated.

15. Select the **Main axes name** (Imperial) or **Use main grid label** (Metric) tick box.

16. Delete the content of the **Prefix** edit box so that it is blank.

17. Enter **1** in the **Suffix** edit box.

18. In the **Side 2 distance** edit box, enter **15'** or **5000**; a new grid axis with the name A1 is added between axes A and B.

19. Close the dialog box by clicking the **X** button on the top right of this dialog box.

    Notice how the grid labels at the four corner grid points are overlapping. If you want to avoid it, you can select the grip points of the grid and drag the labels away so they do not overlap. But, that will mean when you generate a drawing, such as an anchor plan, at a later stage, that extra distance between the grid labels will also be dimensioned. However, you can delete that extra dimension in the drawing.

20. Click on the grid label **A** at the **8A** grid intersection point; the Blue grip points are displayed on the grid line.

21. Use the grip point to drag the grid label to the right using 0-degree polar tracking through a distance of **5'** or **1500**.

22. Similarly, drag the grid label **8** at the **8D** grid intersection point **5'** or **1500** along the 90-degree polar tracking. The grid, after making this change, is shown in Figure 65.

Section 4: Inserting Concrete Slab

In this section, you will insert the concrete slab using the **Rectangular slab** tool. This slab will be inserted between grid points 1B and 8D. To snap to these grid intersection points, you will need to make sure the **GRID Intersection Points** option is selected in object snaps.

1. Make sure the **GRID Intersection Points** object snap is turned on in the **Object Snap** settings on the **Status Bar**.

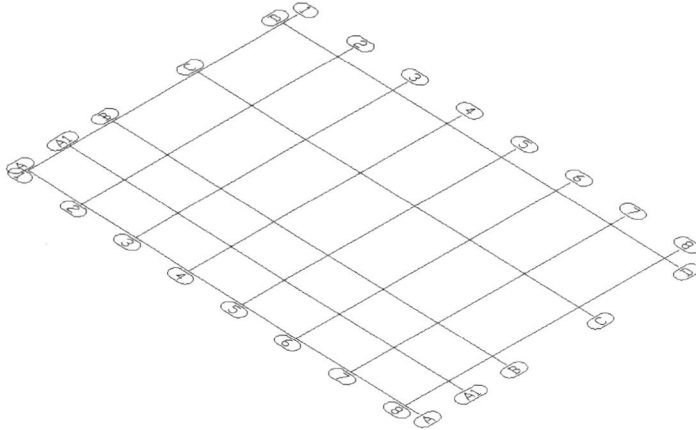

Figure 65 *The 8X5 grid created for the project*

2. From the **Objects** ribbon tab > **Other objects** ribbon panel, invoke the **Rectangular slab** tool; you are prompted to specify the start point of the slab's diagonal line.

3. Select the **1B** grid intersection point as the start point of the slab's diagonal line; you are prompted to specify the end point of the slab's diagonal line.

4. Select the **8D** grid intersection point; the concrete slab is created and the **Advance Steel - Slab [1]** dialog box is displayed.

5. Click the **Positioning** tab from the left pane and make sure the second button is selected in the **Justification** area.

6. Click the **Naming** tab from the left pane.

7. From the **Model Role** list, select **Slab**.

8. Close the dialog box. Figure 66 shows the model window after inserting the slab.

Section 5: Inserting and Copying Concrete Columns

In this section, you will insert the concrete columns using the **Concrete column** tool. You will then use the **Copy** tool from the **Advance Steel Tool Palette** to copy this column at all the required locations.

1. From the **Objects** ribbon tab > **Other objects** ribbon panel, invoke the **Concrete column** tool; you are prompted to locate the start point of the system axis.

2. Select the **1B** grid point; a concrete column is inserted and the **Advance Steel - Beam [2]** dialog box is displayed with the **Section & Material** tab active.

By default, a rectangular concrete column is inserted. You need to change the shape to square column.

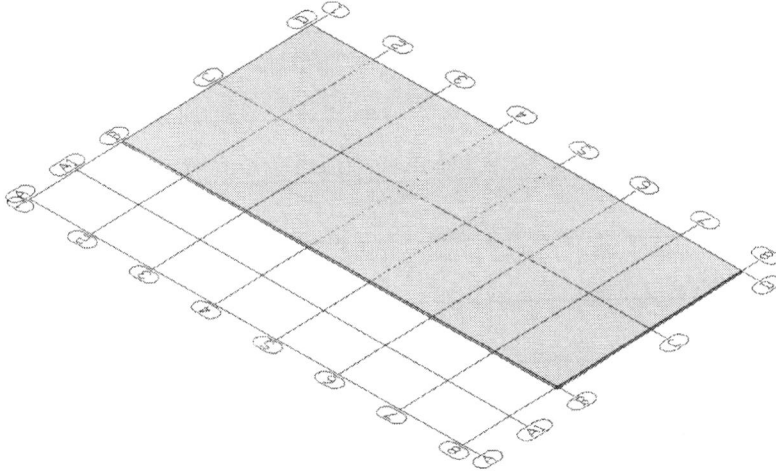

Figure 66 Concrete slab inserted using the grid points

3. From the **Section** list, click the second flyout and select **Square**, as shown in Figure 67; the size is automatically set to the available column size.

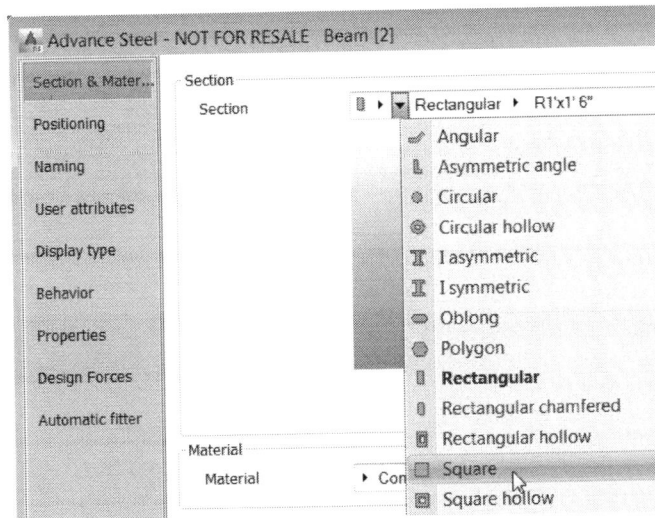

*Figure 67 Selecting the **Square** section type*

By default, the center of the bottom face of the column is placed at the selected grid point. However, in this case, you need to align the grid point to the lower left corner of the column. This will be done using the **Positioning** tab of the dialog box.

4. From the left pane of the dialog box, select the **Positioning** tab.

5. From the **Offset** area, select the radio button at the lower right corner; the lower left corner of the column is aligned with the grid point.

Next, you will copy this column at grid intersection points 2B to 8B.

6. From the **Modify** tab of the **Advance Steel Tool Palette**, invoke the **Copy** tool; you are prompted to select the object to copy.

7. Select the column you placed earlier in this section and then press ENTER to accept the selection; you are prompted to specify the base point.

8. Select the midpoint shown in Figure 68 as the base point of copy; you are prompted to select the second point.

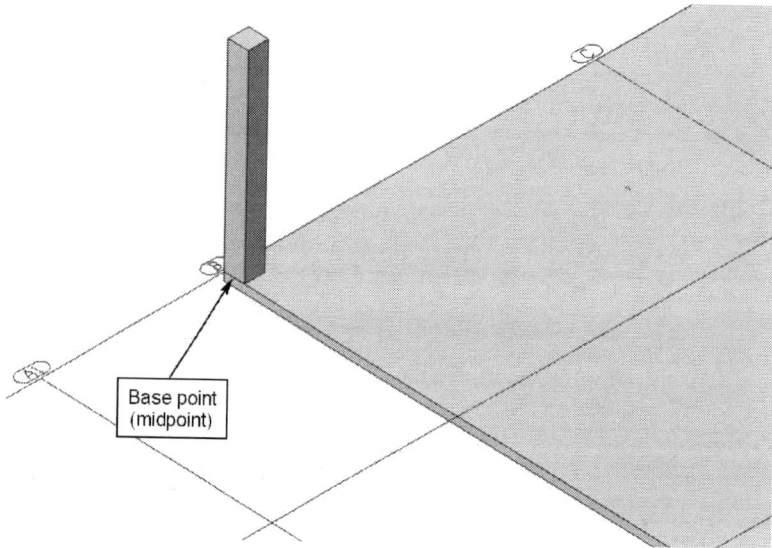

Figure 68 *Selecting the base point for copy*

9. Select the grid intersection points **2B** through to **8B** to place the columns along grid axis B.

10. Press the ESC key to exit the **Copy** tool.

    Notice that the alignment of the last column is not right. You will now fix this alignment.

11. Double-click on the column placed at **8B** grid point; the **Advance Steel - Beam [X]** dialog box is displayed.

12. Select the **Positioning** tab.

13. From the **Offset** area, select the bottom center radio button; the column is aligned correctly with the grid point.

    Next, you need to copy all these columns to grid axes C and D. But it is recommended to first hide the slab by turning its layer off.

14. Turn off the **Concrete slab** layer.

15. Invoke the **Copy** tool.

16. Using the bottom midpoint highlighted as **1** in Figure 69 as the base point, copy the columns to the grid intersection point **8C** also marked as **1** in the same figure.

17. Similarly, copy the previously selected columns using the endpoint marked as **2** in Figure 69 as the base point to the grid intersection point **8D**, also marked as **2** in the same figure.

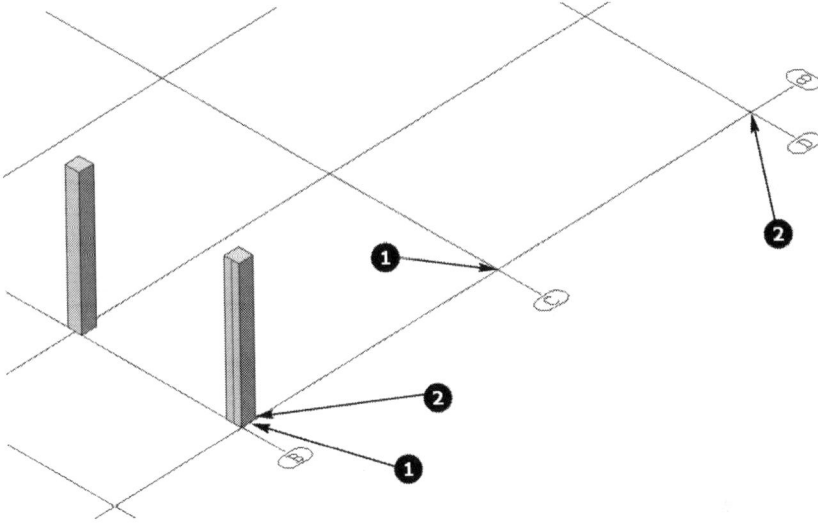

Figure 69 Selecting the points for copy

18. Similarly, copy the columns at grid intersection points **3A**, **3A1**, **5A**, and **5A1**. Figure 70 shows the model after copying all the columns.

Section 6: Inserting and Arraying Isolated Footings

In this section, you will insert one instance of the isolated footing shown in Figures 59 and 60 and then array it to create the remaining instances.

1. From the **Objects** ribbon tab > **Other objects** ribbon panel, invoke the **Isolated footing** tool; you are prompted to specify the position of the footing.

2. Select the **3A** grid intersection point; the footing is placed there and the **Advance Steel - Isolated Footing [X]** dialog box is displayed.

3. In the **Length** and **Width** edit boxes, change the values to **4' 7 1/2"** or **1500**.

4. Close the dialog box by clicking the **[X]** button on the top right of this dialog box.

Next, you will array the isolated footing to create the remaining instances.

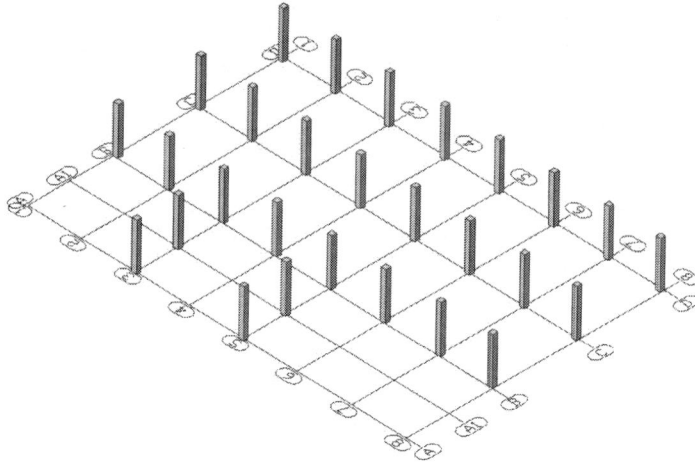

Figure 70   *Model after copying all the columns*

5. Using the **Array** tool, create a disassociative rectangular array of the isolated footings with 2 rows and 2 columns. The spacing between the rows is 15'/5000mm and the spacing between the columns is 38'-6 7/8"/12000mm.

6. Turn on the visibility of the **Concrete slab** layer. Figure 71 shows the completed model after creating and arraying the isolated footings.

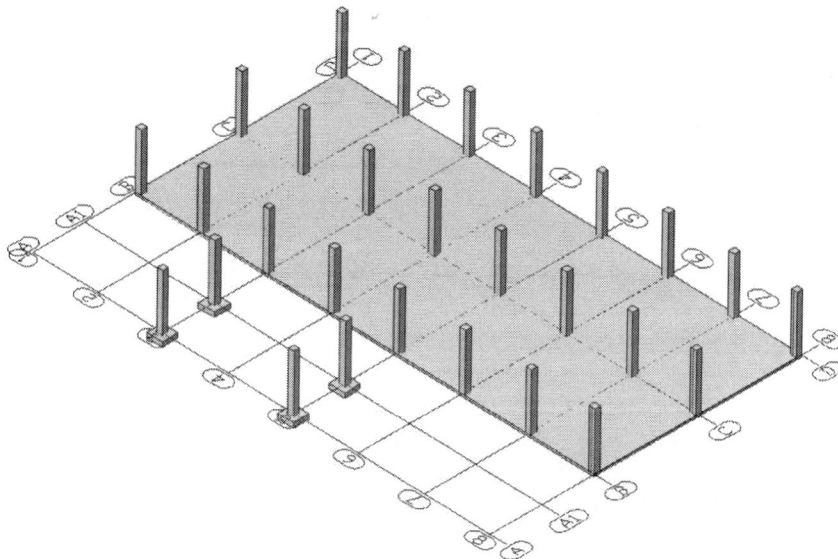

Figure 71   *The completed model for the tutorial*

Next, you will save this file in the folder where you copied the tutorial files of this book.

7. Save the file with the name **c01-BIM.dwg** in the **Tutorial Files > C01 > BIM** folder.

Skill Evaluation

Evaluate your skills to see how many questions you can answer correctly. The answers to these questions are given at the end of the book.

1. Advance Steel has a special object snap option called **GRID Intersection Point**. (True/False)

2. You can double-click on the grid axes to edit them. (True/False)

3. Concrete beams once inserted cannot be changed. (True/False)

4. Model role controls how the structural member is documented, dimensioned, and labeled in the detail drawings. (True/False)

5. By default, if there are two concrete beams intersecting, the second beam gets the priority over the first. (True/False)

6. Which tool is used to create a 4X4 grid by specifying two opposite corners?

    (A) **Building Grid**   (B) **Grid with groups by distance**
    (C) **Grid with 4 axes**   (D) **4x4 Grid**

7. Which of the following two shapes are available while creating isolated footings?

    (A) Cylinder   (B) Ellipse
    (C) Block   (D) Square

8. Which tool is used to insert a curved concrete beam?

    (A) **Beam**   (B) **Curved beam**
    (C) **Concrete curved beam**   (D) **Concrete beam**

9. Which tool is used to insert a continuous footings?

    (A) **Footing**   (B) **Isolated footing**
    (C) **Liner footing**   (D) **Continuous footing**

10. What is the default display type of the concrete member in the drawing window?

    (A) **Standard**   (B) **Features**
    (C) **Exact**   (D) **Off**

Class Test Questions

Answer the following questions:

1. Explain briefly the process of creating a circular grid with a single axis.

2. How will you add an additional grid axis to an existing grid?

3. Explain briefly how will you create a concrete slab on top of a column?

4. While creating a building grid, how may points do you need to specify?

5. Which snap point allows you to snap to the grid intersection point?

Chapter 2 – Inserting and Editing Structural Sections

The objectives of this chapter are to:

√   *Explain the process of inserting straight structural sections*
√   *Edit the properties of straight structural sections*
√   *Explain the process of inserting continuous straight beams*
√   *Explain the process of inserting curved structural sections*
√   *Edit the properties of curved structural sections*
√   *Explain the process of inserting structural sections on lines or polylines*
√   *Explain how to split structural sections*
√   *Explain how to merge various structural sections*

INSERTING STRAIGHT STRUCTURAL SECTIONS

Advance Steel provides a number of tools to insert straight structural members. Most of these tools are available on the **Home** ribbon tab > **Objects** ribbon panel > **Rolled I Section** flyout, as shown in Figure 1. In this section, you will learn about those tools.

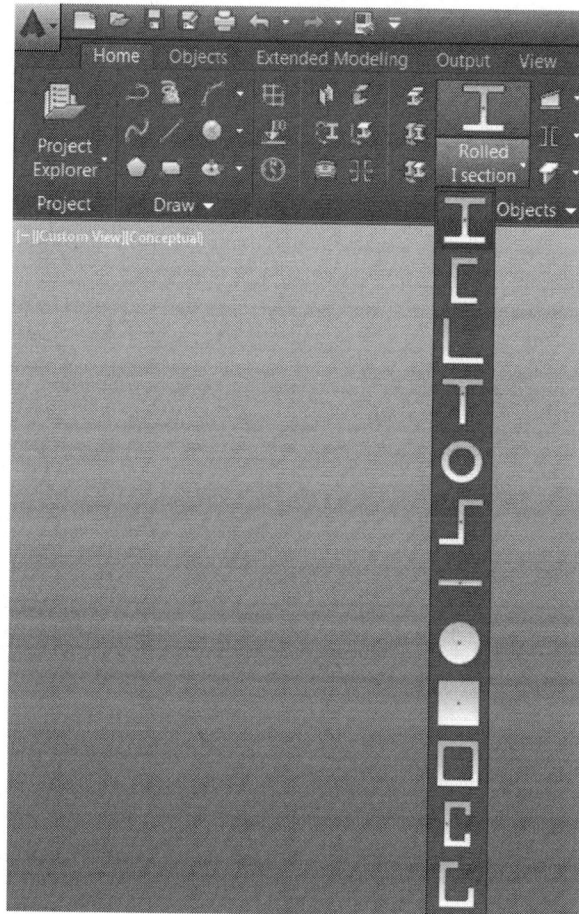

*Figure 1   The **Rolled I Section** flyout displaying various tools*

Inserting Rolled I-sections

Home Ribbon Tab > Objects Ribbon Panel > Rolled I Section
Objects Ribbon Tab > Beams Ribbon Panel > Rolled I Section

This tool is used to insert single or multiple rolled I-sections by specifying the start point and the endpoint of the sections. When you invoke this tool, you will be prompted to specify the start point of the system axis. Once you specify the start point, you will then be prompted to specify the endpoint of the system axis. On specifying the endpoint, the rolled I-section will be inserted but you will again be prompted to specify the start point of the system axis of the next section. You can continue inserting multiple rolled I-sections by specifying the start and end points of the system axes. Once you have inserted all the required sections, press the ENTER key; the **Advance Steel - Beam [X]** dialog box will be displayed in the **Section & Material** tab, as shown in Figure 2. In this dialog box name, the number X will depend on how many beams have already been inserted. The options available in various tabs of this dialog box are discussed next.

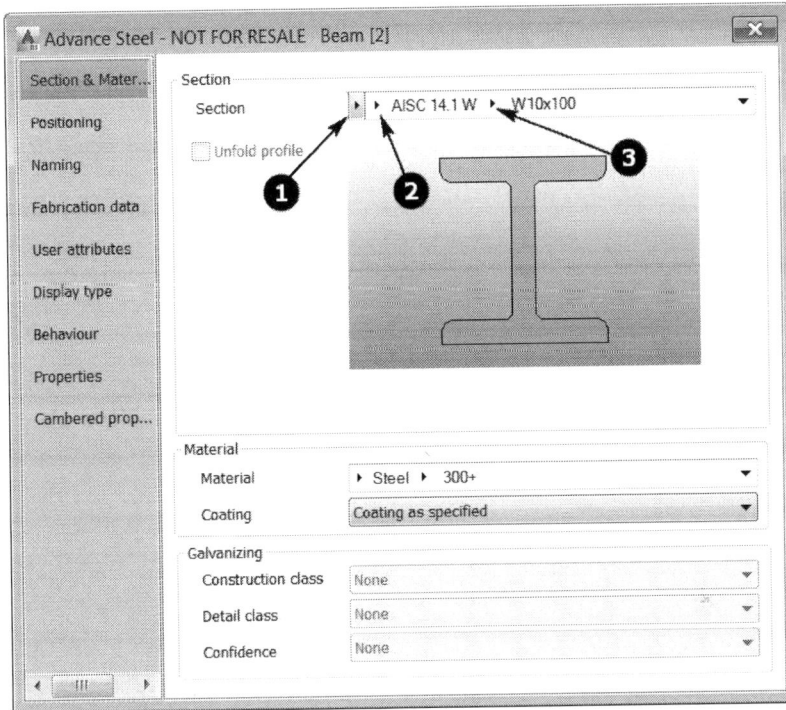

Figure 2 The Section & Material tab of the Advance Steel - Beam [X] dialog box

Section & Material Tab

The **Section & Material** tab shown in Figure 2 is used to specify the section type, material, coating, and galvanizing options of the beams that you have inserted. These options are discussed next.

Section Area

This area provides the options to select the **Section Class**, **Section Type**, and **Section Size**. These three settings are configured using the three flyouts labeled as **1**, **2**, and **3** respectively in Figure 2. Figure 3 shows some of the **Section Class** options available in the list.

The **Section Type** options will be dependent on the **Section Class** selected. Figure 4 shows various section types for an **I Section** class.

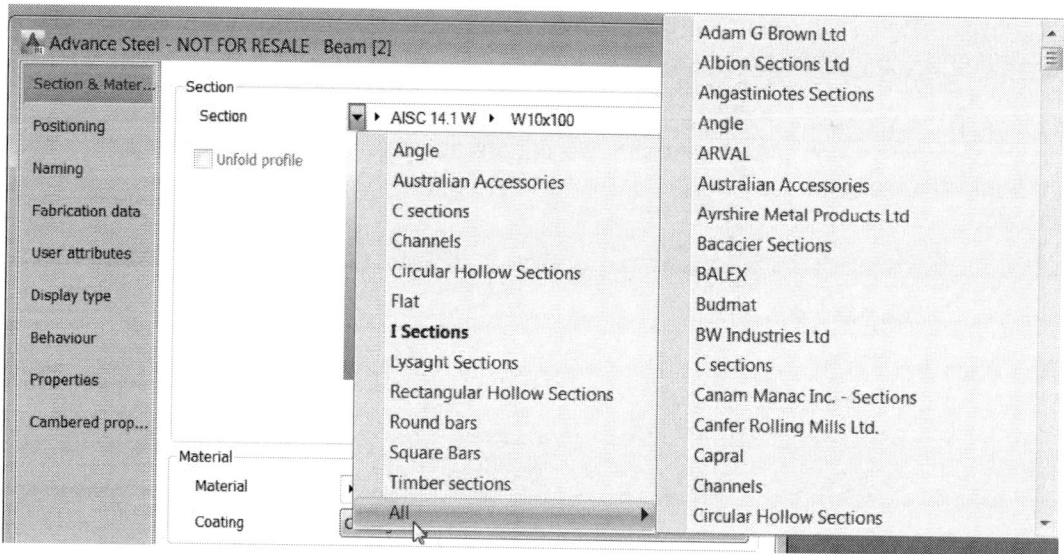

Figure 3 The beam **Section Class** *options*

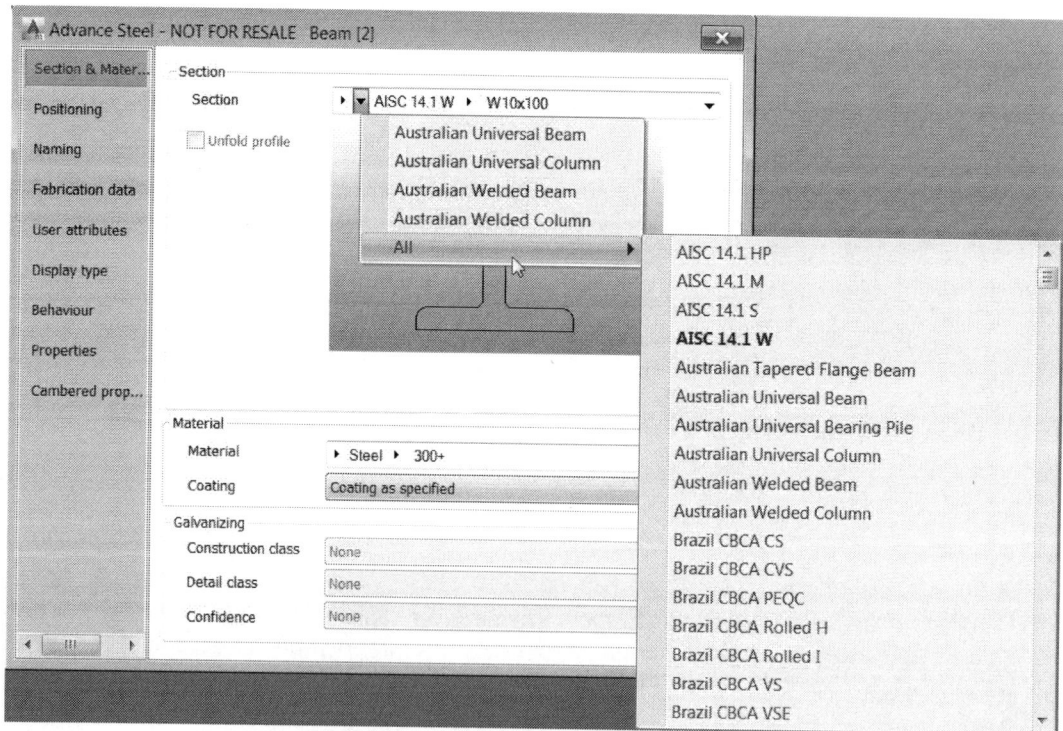

Figure 4 The beam **Section Type** *options*

The **Section Size** flyout shows you all the available sizes based on the **Section Type** option selected.

Unfold Profile
This tick box is only active for rectangular, square, or circular hollow sections or cold rolled profiles. If you select this tick box, you will be able to display the flattened profile of the section in the detail views.

Material
This area provides the options to specify the material and grade. The first flyout is used to specify the material type and the second flyout is used to specify the material grade.

Coating
This list allows you to select the coating required for the sections you inserted.

Galvanizing
This area provides the options to specify the galvanizing information for the sections.

Positioning Tab
The **Positioning** tab shown in Figure 5 is used to specify the location and orientation of the structural section with respect to the placement points. These options are discussed next.

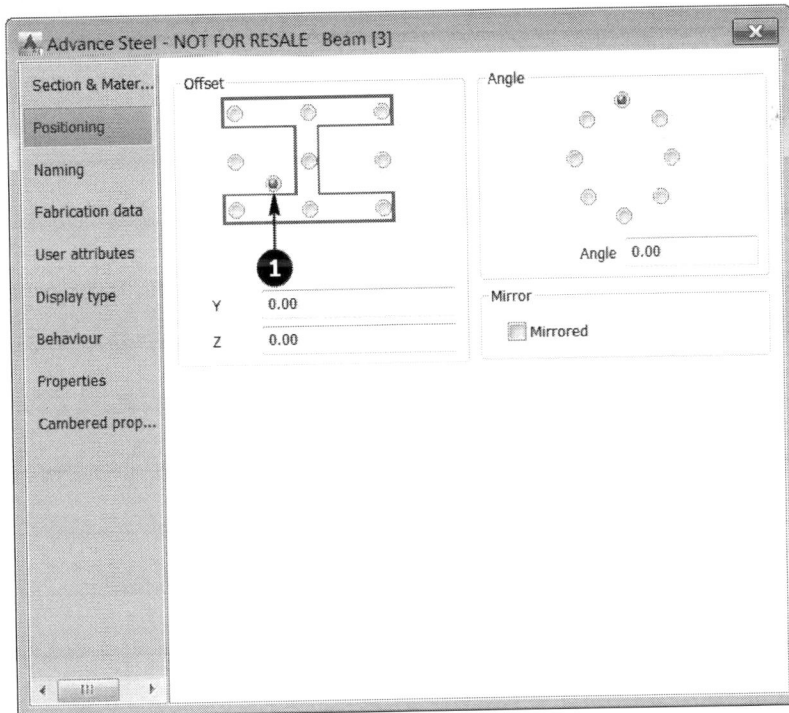

Figure 5 *The* ***Positioning*** *tab of the* ***Advance Steel - Beam [X]*** *dialog box*

Offset Area

This area provides the options to specify the location of the beam in relation to the system axis. You can select the radio buttons to define the system axis at the four corner points, midpoints of the four edges, or the center point of the beam. For asymmetrical section shapes such as unequal sections, you can also select the radio button to define the system axis at the center of gravity of the structural section. This radio button is marked by arrow **(1)** in Figure 5. The **X** and **Y** edit boxes are used to define offset values along the X and Y directions from system axis.

Angle Area

This area is used to rotate the structural section to a standard angle by clicking its radio button or by entering a custom value in the **Angle** edit box.

Mirror Area

Selecting the **Mirrored** tick box in this area allows you to mirror the structural section around its vertical axis.

Naming Tab

The **Naming** tab shown in Figure 6 is used to specify the naming and Bill of Material (BOM) information for the structural section you inserted. These options are discussed next.

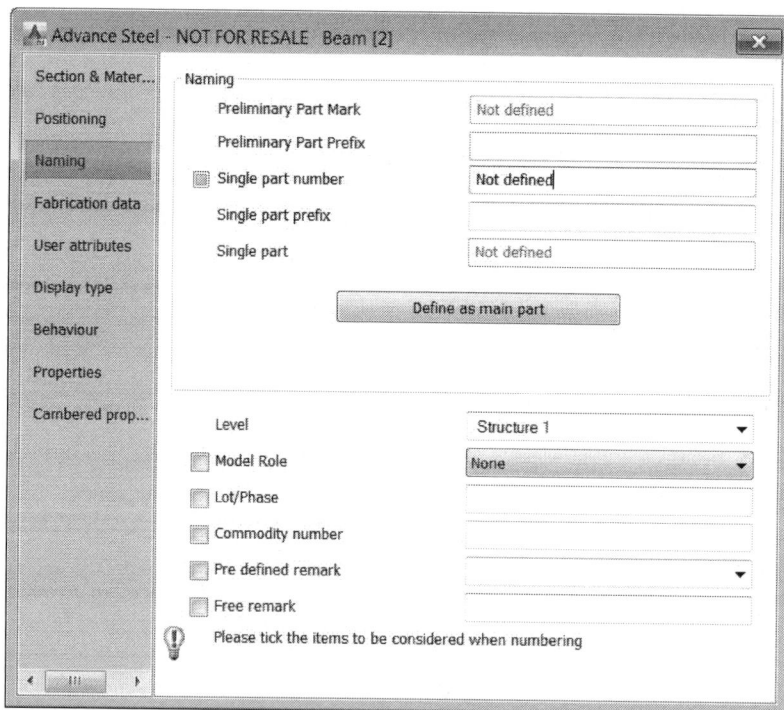

*Figure 6 The **Naming** tab of the **Advance Steel - Beam [X]** dialog box*

Naming Area

This area provides the options to specify the preliminary part mark or single part numbers and the prefixes for the same. Clicking the **Define as main part** button allows you to define the inserted beams as the main part of the assembly. As a result, three more edit boxes will be displayed that will allow you to enter the assembly numbering options.

Level
This list allows you to specify the project explorer structure to which the beam will be linked. By default, it will display **Structure 1**.

Model Role
Model role controls how the structural section is numbered, documented, dimensioned, and labeled in the detail drawings. It also helps in improving selection through selection filters. As a result, it is extremely important to specify the model roles correctly. You can select the **Model Role** tick box and then select the role type of the inserted structural section beam from this list.

Lot/Phase
This option allows you to assign the lot or phase to the structural section.

Commodity number
This option allows you to assign the commodity number to the structural section.

Pre defined remark
Advance Steel allows you to save specific remarks about the structural sections in the **Notes** table of the **AstorBase.mdb** file. Those remarks can be selected from this list. By default, the **AstorBase.mdb** file is located in the **ProgramData > Autodesk >Advance Steel 2016 \Steel\Data** folder.

*Tip: By default, the **Notes** table of the **AstorBase.mdb** file has four entries. You can add more entries, if required. Once you add or edit entries in this database file, you will have to invoke the **Update defaults** button from the **Home** ribbon tab > **Settings** ribbon panel. The new or edited remarks will then be displayed in the **Pre defined remark** list.*

Free remark
This option allows you to add any other remark related to the concrete beam.

Fabrication data Tab
The options in the **Fabrication data** tab, shown in Figure 7, are used to specify the information related to the fabrication of the structural section and also to support the fabrication software. These options include setting the assembly properties when the inserted structural section is a part of an assembly and the object properties. You can also set the assembly approval status from the **Approval status** list.

User attributes Tab
The **User attributes** tab shown in Figure 8 is used to specify additional user-defined attributes for the inserted structural section that can later be displayed in the titleblocks while generating the fabrication drawings. The user-defined attributes that have their tick boxes selected will be considered while numbering.

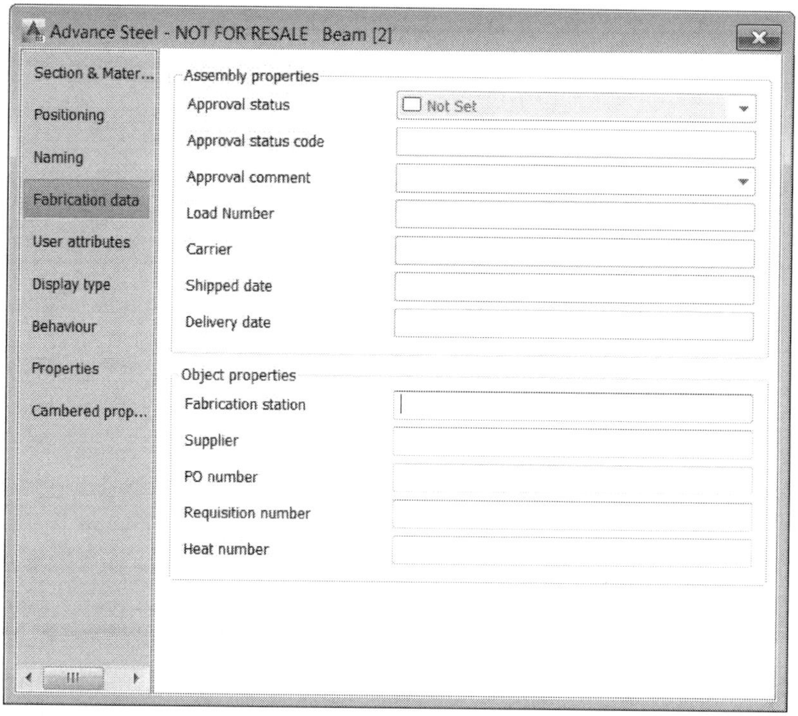

Figure 7 *The **Fabrication data** tab of the **Advance Steel - Beam [X]** dialog box*

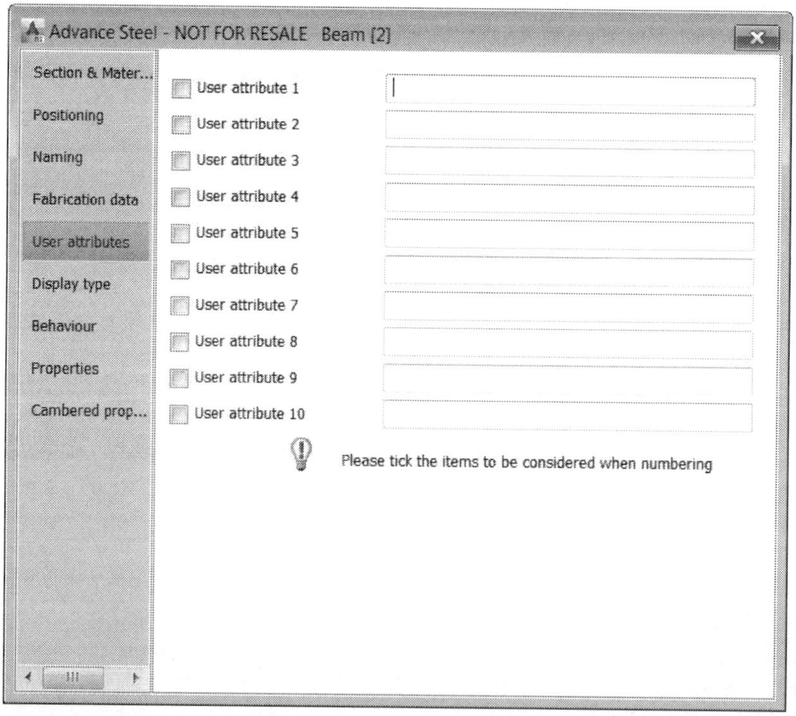

Figure 8 *The **User attribute** tab of the **Advance Steel - Beam [X]** dialog box*

Display type Tab

The **Display type** tab shown in Figure 9 is used to control the representation of the structural section in the drawing window. Various display types are discussed next.

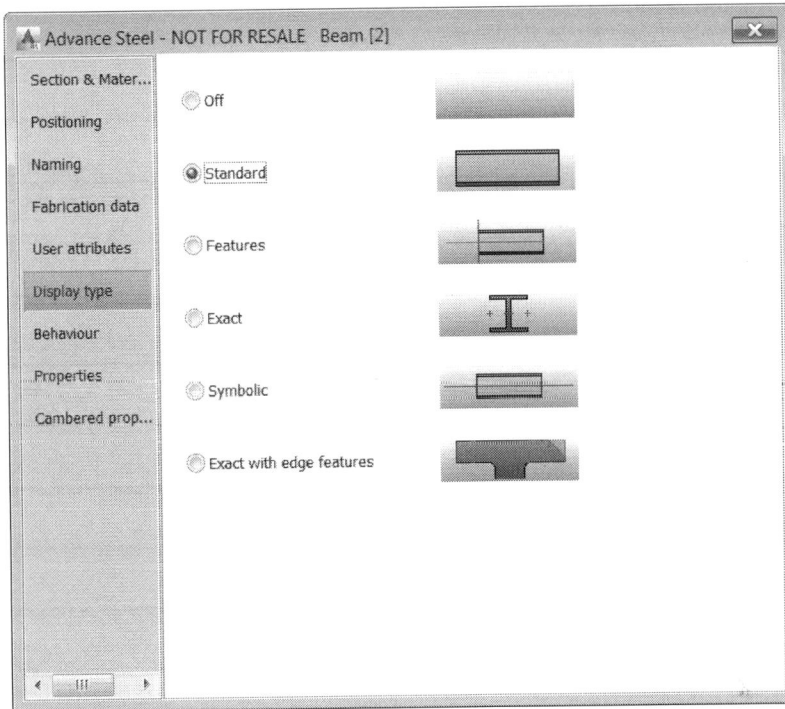

Figure 9 The Display type tab of the Advance Steel - Beam [X] dialog box

Off

If this radio button is selected, the display of the structural section will be turned off.

*Tip: To turn on the visibility of the hidden members, you can use the **All Visible** tool on the **Advance Steel Tool Palette > Quick views** tab.*

Standard

This is the default representation type of the structural section. In this representation, the structural section and the system axis is displayed. This representation is recommended for large structural models.

Features

This representation type is used when you have created some cut features on the structural section and you wish to display the cut elements for editing or deleting the feature. With this representation, the edges of the object used to create the cut feature will also be displayed.

Exact

This radio button is used to display the exact representation of the structural section.

Symbolic

If this radio button is selected, the display of the structural section will be turned off and only the system axis will be displayed.

Exact with edge features

This radio button is used to display the exact representation of the structural section, along with all the edge features, such as fillets or chamfers.

Behaviour Tab

The **Behaviour** tab shown in Figure 10 is used to define how the structural section will be used in other Advance Steel operations. These options are discussed next.

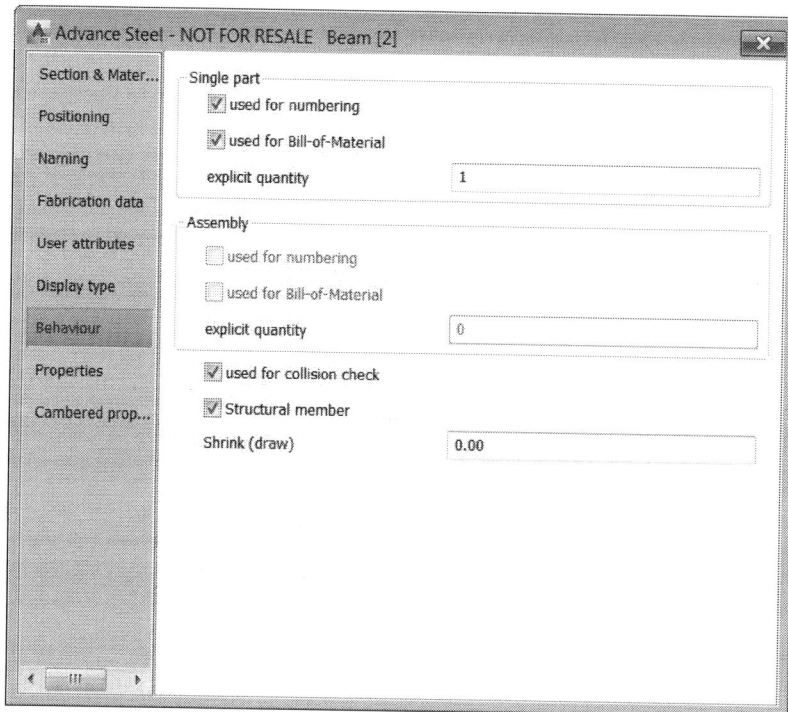

*Figure 10 The **Behaviour** tab of the **Advance Steel - Beam [X]** dialog box*

used for numbering

If this tick box is selected, the structural section will be used for single part numbering.

used for Bill-of-Material

If this tick box is selected, the structural section will be used in various parts lists and Bills of Material (BOMs).

explicit quantity

This edit box is used to specify the quantity of the structural section required. By default, the value is set to 1. If required, you can increase this value if you want some extra beams to be displayed in the BOM.

Note: The **Assembly** options will be displayed only when the inserted structural section is a part of an assembly.

used for collision check
Advance Steel provides a tool to check the collision of various structural sections and other parts. Selecting this tick box will include the inserted beam as part of the collision check process.

Structural member
If this tick box is selected, the inserted section will be treated as a structural member in the current model.

Shrink (draw)
This edit box is used to define the change in the length of the structural section caused due to deformation during the assembly process. The value in this edit box is entered in the current linear units. Entering a positive value reduces the length and entering a negative value increases the length. The change in lengths will be reflected in the shop drawings, numbering, BOMs, and NC files.

Properties Tab
The **Properties** tab displays various physical, geometric, and section properties of the section, as shown in Figure 11.

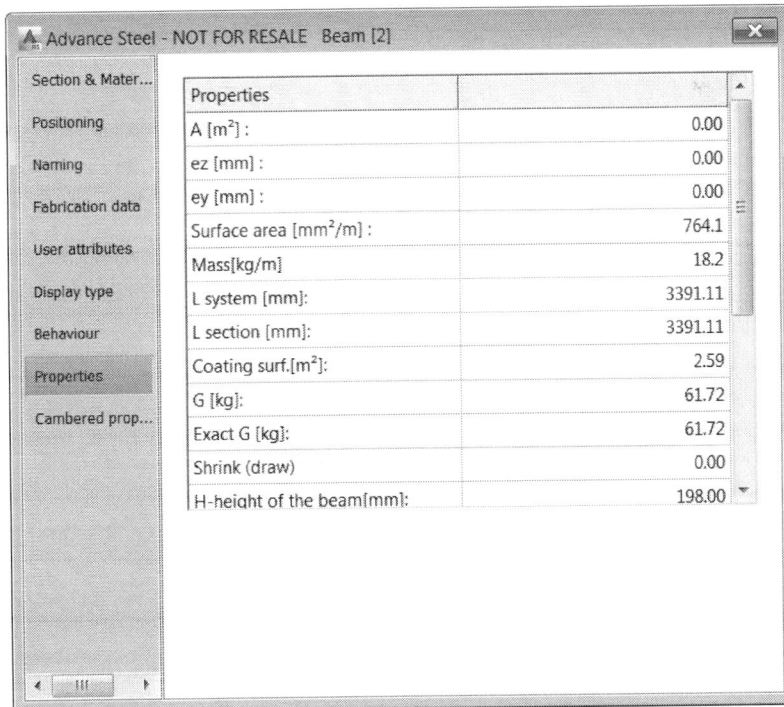

Properties	
A [m²] :	0.00
ez [mm] :	0.00
ey [mm] :	0.00
Surface area [mm²/m] :	764.1
Mass[kg/m]	18.2
L system [mm]:	3391.11
L section [mm]:	3391.11
Coating surf.[m²]:	2.59
G [kg]:	61.72
Exact G [kg]:	61.72
Shrink (draw)	0.00
H-height of the beam[mm]:	198.00

*Figure 11 The **Properties** tab of the **Advance Steel - Beam [X]** dialog box*

Cambered properties Tab

Cambered structural sections are the ones that are fabricated as curved sections but become straight when assembled because of their own weight or due to external loads. The **Cambered properties** tab shown in Figure 12 allows you to define the inserted structural section as a cambered section. These options are discussed next.

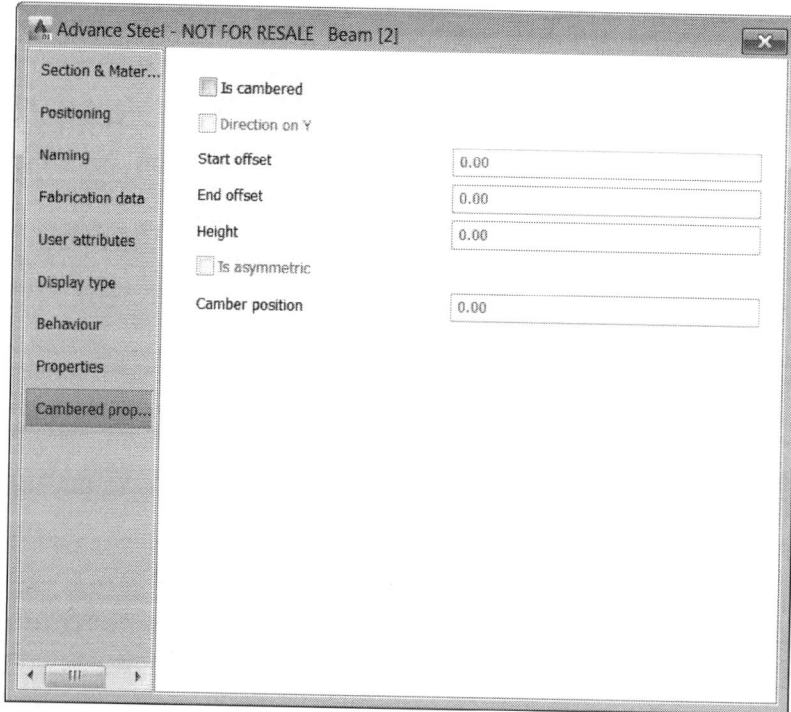

*Figure 12 The **Cambered properties** tab of the **Advance Steel - Beam [X]** dialog box*

Is cambered

Select this tick box if the structural section you inserted is cambered. Note that the remaining options in this tab will only activate if this tick box is selected.

> *Note: It is important to mention here that the cambered sections will appear straight in the modeling environment. The cambering will only be shown in the drawings.*

Direction on Y

This tick box is selected if the structural member is cambered along its Y direction. If this tick box is not selected, the beam will be cambered along its X direction.

Start offset

This edit box is used to define the offset distance of the point where the cambering will begin from the start point of the structural section.

End offset
This edit box is used to define the offset distance of the point where the cambering will finish to the endpoint of the structural section.

Is asymmetric
If this tick box is selected, the structural section will not be symmetrically cambered. In this case, the camber position will have to be defined in the edit box that will be activated after you select this tick box.

Camber position
As mentioned above, if the structural section is not symmetrically cambered, this edit is used to define the camber position.

Inserting Channel Sections

Home Ribbon Tab > Objects Ribbon Panel > Rolled I Section > Channel section
Objects Ribbon Tab > Beams Ribbon Panel > Rolled I Section > Channel section

This tool is used to insert single or multiple channel sections by specifying the start point and the endpoint of the sections. When you invoke this tool, you will be prompted to specify the start point of the system axis. Once you specify the start point, you will then be prompted to specify the endpoint of the system axis. On specifying the endpoint, the channel section will be inserted, but you will again be prompted to specify the start point of the system axis of the next section. You can continue inserting multiple channel sections by specifying the start and endpoints of the system axes. Once you have inserted all the required sections, press the ENTER key; the **Advance Steel - Beam [X]** dialog box will be displayed. In this dialog box name, the number X will depend on how many beams have already been inserted. The options available in various tabs of this dialog box are the same as those discussed in the **Rolled I section** tool.

Inserting Angle Sections

Home Ribbon Tab > Objects Ribbon Panel > Rolled I Section > Angle section
Objects Ribbon Tab > Beams Ribbon Panel > Rolled I Section > Angle section

This tool is used to insert single or multiple angle sections by specifying the start point and the endpoint of the sections. The process of inserting the angle sections is exactly the same as that discussed for the **Rolled I section** tool.

Inserting T Sections

Home Ribbon Tab > Objects Ribbon Panel > Rolled I Section > T sections
Objects Ribbon Tab > Beams Ribbon Panel > Rolled I Section > T sections

This tool is used to insert single or multiple T sections by specifying the start point and the endpoint of the sections. The process of inserting the T sections is exactly the same as that discussed for the **Rolled I section** tool.

Inserting Circular Hollow Sections (CHS)

> **Home Ribbon Tab > Objects Ribbon Panel > Rolled I Section > Circular hollow sections**
> **Objects Ribbon Tab > Beams Ribbon Panel > Rolled I Section > Circular hollow sections**

This tool is used to insert single or multiple circular hollow sections (CHS) by specifying the start point and the endpoint of the sections. The process of inserting these sections is the same as that discussed for the **Rolled I section** tool.

Inserting Z-Steel Sections

> **Home Ribbon Tab > Objects Ribbon Panel > Rolled I Section > Z-steel**
> **Objects Ribbon Tab > Beams Ribbon Panel > Rolled I Section > Z-steel**

This tool is used to insert single or multiple Z-steel sections by specifying the start point and the endpoint of the sections. The process of inserting these sections is the same as that discussed for the **Rolled I section** tool.

Inserting Flat Sections

> **Home Ribbon Tab > Objects Ribbon Panel > Rolled I Section > Flat**
> **Objects Ribbon Tab > Beams Ribbon Panel > Rolled I Section > Flat**

This tool is used to insert single or multiple flat sections by specifying the start point and the endpoint of the sections. The process of inserting these sections is the same as that discussed for the **Rolled I section** tool.

Inserting Round Bar Sections

> **Home Ribbon Tab > Objects Ribbon Panel > Rolled I Section > Round bar**
> **Objects Ribbon Tab > Beams Ribbon Panel > Rolled I Section > Round bar**

This tool is used to insert single or multiple round bar sections by specifying the start point and the endpoint of the sections. The process of inserting these sections is the same as that discussed for the **Rolled I section** tool.

Inserting Square/Rectangular Hollow Sections (SHS/RHS)

> **Home Ribbon Tab > Objects Ribbon Panel > Rolled I Section > Square/Rectangular Hollow sections**
> **Objects Ribbon Tab > Beams Ribbon Panel > Rolled I Section > Square/Rectangular Hollow sections**

This tool is used to insert single or multiple square or rectangular hollow sections (SHS/RHS) by specifying the start point and the endpoint of the sections. The process of inserting these sections is the same as that discussed for the **Rolled I section** tool.

Inserting Cold Rolled Sections

Home Ribbon Tab > Objects Ribbon Panel > Rolled I Section > Cold rolled sections
Objects Ribbon Tab > Beams Ribbon Panel > Rolled I Section > Cold rolled sections

This tool is used to insert single or multiple square or rectangular hollow sections (SHS/RHS) by specifying the start point and the endpoint of the sections. The process of inserting these sections is the same as that discussed for the **Rolled I section** tool.

Inserting Other Sections

Home Ribbon Tab > Objects Ribbon Panel > Rolled I Section > Other sections
Objects Ribbon Tab > Beams Ribbon Panel > Rolled I Section > Other sections

This tool is used to insert other nonstandard sections by specifying the start point and the endpoint of the sections. The process of inserting these sections is the same as that discussed for the **Rolled I section** tool.

INSERTING CURVED STRUCTURAL SECTIONS

Home Ribbon Tab > Objects Ribbon Panel > Curved beam
Objects Ribbon Tab > Beams Ribbon Panel > Curved beam

The **Curved beam** tool allows you to insert a curved structural section by specifying three points. The first point is the start point of the section, the second point is the endpoint of the section and the third point is a point on the curve. On specifying the third point, the **Advance Steel - Beam [X]** dialog box will be displayed. The options in most tabs of this dialog box are similar to those discussed for the **Rolled I section** tool, except for the options under the **Linear dimension properties** on the **Position** tab shown in Figure 13. These options are discussed next.

Radius

This edit box allows you to enter the radius of the curved section. While creating the section, you can click any point as the third point and then change the radius value using this edit box to an accurate value.

Tolerance

This edit box allows you to enter the tolerance to define the number of bends in the curved section. Smaller the tolerance value, more the number of bends, resulting in more accurate curved section.

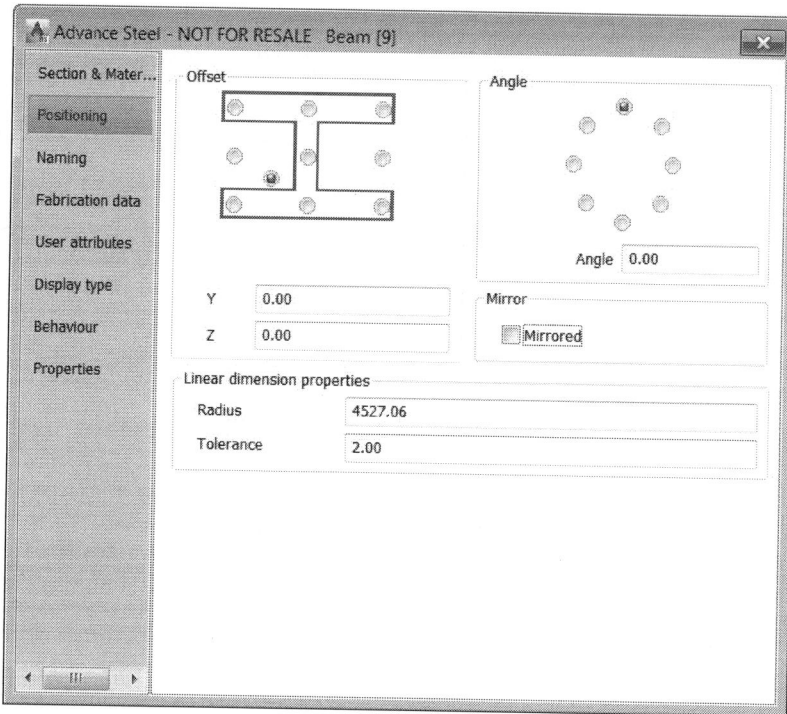

Figure 13 *The **Position** tab showing the options to change the radius and tolerance of a curved section*

INSERTING STRUCTURAL SECTIONS ON LINES, ARCS, OR POLYLINES

Advance Steel allows you to insert structural sections on existing lines, arcs, or polylines. The methods to insert these types of sections are discussed next.

Inserting Structural Sections on Lines or Arcs

Home Ribbon Tab > Objects Ribbon Panel > Beam, from line
Objects Ribbon Tab > Beams Ribbon Panel > Beam, from line

The **Beam, from line** tool allows you to insert structural sections on one or more selected lines or arcs. When you invoke this tool, you will be prompted to select lines or arcs. Once you have selected all the required lines or arcs, press the ENTER key; you will be prompted to specify whether or not you want to delete the selected lines or arcs. Once you specify this option, the **Advance Steel - Beam [X]** dialog box will be displayed. The options in various tabs of this dialog box are similar to those discussed in the earlier sections. Note that each section inserted on the selected lines or arcs will be individual sections and will be listed separately in the BOMs. You will also be able to double-click on any of the sections to change the properties of that particular section. Figure 14 shows 2D lines and arcs and Figure 15 shows the structural members inserted using those lines and arcs.

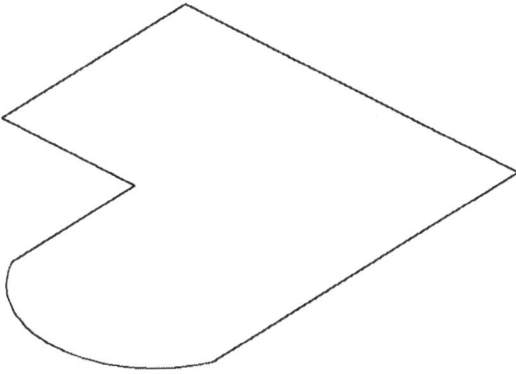

Figure 14 Lines and arcs

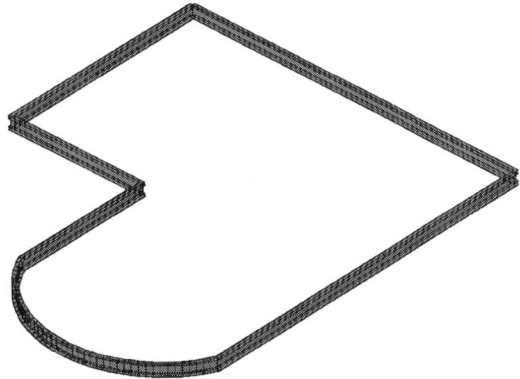

Figure 15 After inserting structural sections

What I do

*A number of times structural engineers export analytical model from a structural engineering program in the form of a line diagram. In a case like this, I use the **Beam, from line** tool to insert structural sections of the desired specifications on those lines. It is a quick way of creating a structural model using the engineers analytical diagram.*

Inserting Structural Sections on Polylines

Home Ribbon Tab > Objects Ribbon Panel > Beam, polyline
Objects Ribbon Tab > Beams Ribbon Panel > Beam, polyline

If the structural model you are creating consists of beams that are bent into the required shape, you can use the **Beam, polyline** tool. When you invoke this tool, you will be prompted to specify the start point or use the **Polyline** option to select an existing polyline. You can select a 2D or a 3D polyline to convert into bent structural sections. Once you have selected the required polylines, press the ENTER key; you will be prompted to specify whether or not you want to delete the polylines. After you specify the option for this prompt, the **Advance Steel - Beam [X]** dialog box will be displayed. The options in various tabs of this dialog box are similar to those discussed in the previous sections. Remember that all the structural sections inserted using a polyline are treated as a single beam. Figure 16 shows a polyline with two line segments and an arc tangent to those segments. Figure 17 shows a bent structural section inserted using the polyline.

It is important to mention here that the corner treatments of the structural sections placed on lines will be different than those placed on polylines. This is because the structural sections placed on lines are treated as individual sections and polylines are treated as a single bent section. Figure 18 shows the corner treatments of sections inserted on two lines labeled as **1** and **2** and Figure 19 shows the corner treatment of a section inserted on a polyline labeled as **1**.

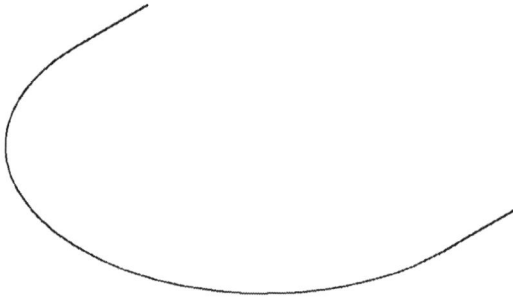

Figure 16 *Polyline with lines and arc segments*

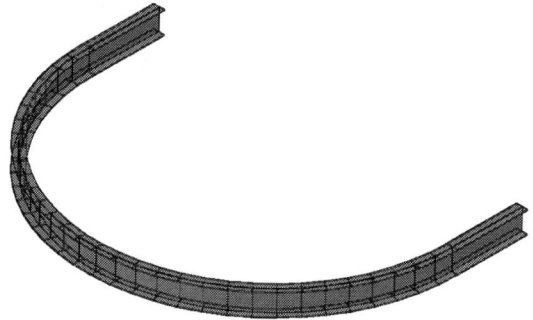

Figure 17 *After inserting the structural section*

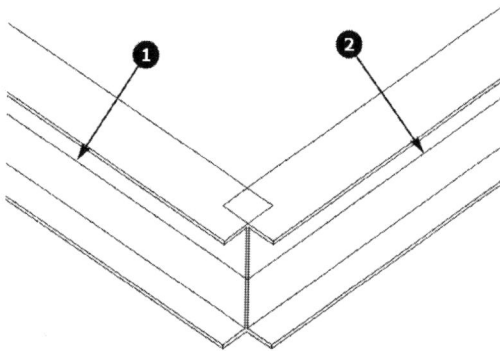

Figure 18 *Corner treatment of structural sections inserted using lines*

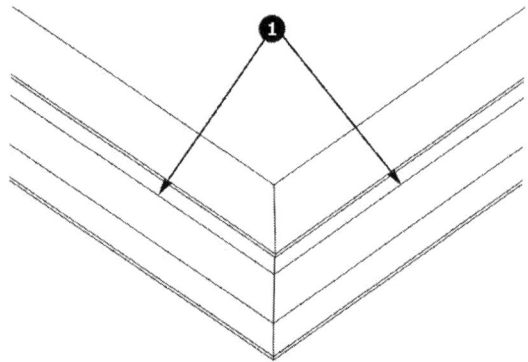

Figure 19 *Corner treatment of structural sections inserted using a polyline*

INSERTING COLUMNS

Home Ribbon Tab > Objects Ribbon Panel > Column
Objects Ribbon Tab > Beams Ribbon Panel > Column

Advance Steel provides you with the **Column** tool that is used to insert single or multiple columns of a default length. The columns will be inserted standing vertically on the XY plane of the World Coordinate System (WCS). When you invoke this tool, you will be prompted to specify the start point of the system axis. When you specify the point, a column will be placed at that location and you will again be prompted to specify the start point of the system axis. After placing all the required columns, press the ENTER key; the **Advance Steel - Beam [X]** dialog box will be displayed. The options in various tabs of this dialog box are similar to those discussed in the earlier sections.

Tip: The structural columns placed using the **Column** tool will automatically have their model role defined as **Column** in the **Naming** tab of the **Advance Steel - Beam [X]** dialog box.

INSERTING CONTINUOUS STRUCTURAL SECTIONS

```
Objects Ribbon Tab > Beams Ribbon Panel > Continuous Beam
```

The **Continuous Beam** tool allows you to insert multiple end-connected structural sections. When you invoke this tool, you will be prompted to specify the start point of the system axis. On specifying the start point, you will be prompted to specify the endpoint of the system axis. This prompt will keep repeating, allowing you to insert multiple end-connected sections. Once you have inserted all the required sections, press the ENTER key; the **Advance Steel - Beam [X]** dialog box will be displayed. The options in various tabs of this dialog box are the same as those discussed in the earlier sections.

Remember that each section inserted using the **Continuous Beam** tool is treated as an individual section. As a result, there is no corner treatment defined at the corners of the end-connected sections.

EDITING INSERTED STRUCTURAL SECTIONS

At various times during a project, you need to change the type, size, orientation, and other properties of the sections you have inserted. This can be done by simply double-clicking on the section you want to edit. If you want to change multiple sections, select all of them and then right-click to display the shortcut menu. From that menu, select **Advance Properties**; the **Advance Steel - Beam [X]** dialog box will be displayed. The changes made in this dialog box will be reflected in all the beams.

SPLITTING AND MERGING STRUCTURAL SECTIONS

A number of structural designs will require you to split or merge structural sections, once you have placed them. The following sections explain these two editing operations.

Splitting Structural Sections

```
Home Ribbon Tab > Objects Ribbon Panel > Split beams
Objects Ribbon Tab > Beams Ribbon Panel > Split beams
```

The **Split beams** tool allows you to split a structural section into two. While splitting, you can also decide to insert a gap between the two resulting beams. When you invoke this tool, you will be prompted to select objects. Select all the structural sections you want to split and then press the ENTER key. On doing so, you will be prompted to select the split point or define the gap. Select the split points on all the selected structural sections. Alternatively, you can first specify the split gap and then select the split points. After selecting all the split points, press the ENTER key. All the selected structural sections will be split at the points you defined. Figure 20 shows a beam placed on top of two columns and Figure 21 shows the same beam split with a gap value.

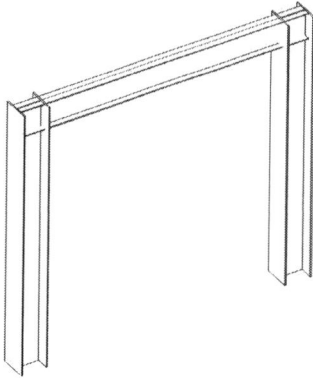

Figure 20 *A beam placed on top of two columns*

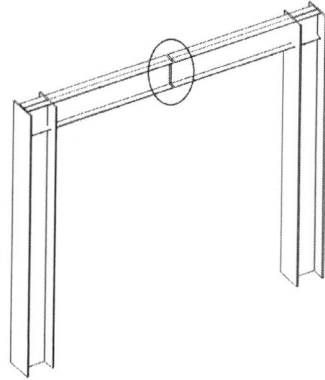

Figure 21 *After splitting the beam with a gap*

Merging Structural Sections

Home Ribbon Tab > Objects Ribbon Panel > Merge beams
Objects Ribbon Tab > Beams Ribbon Panel > Merge beams

The **Merge beams** tool allows you to merge two coplanar structural sections into a single section. The beams can be collinear or at an angle to each other. However, the only criteria to remember is that the structural sections should be in the same plane. If the sections are at an angle to each other, they will be converted into a single bent section. Figure 22 shows two coplanar sections at an angle to each other before merging and Figure 23 shows a bent section created after merging.

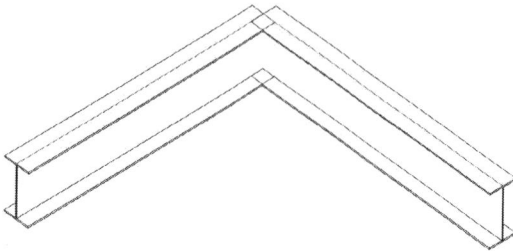

Figure 22 *Two structural sections before merging*

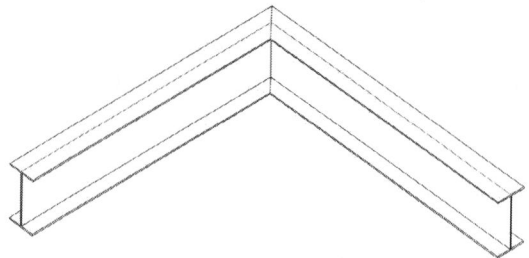

Figure 23 *A single section created after merging*

Matching Properties of Sections

Tools Ribbon Tab > Properties Ribbon Panel > Match Properties

Advance Steel allows you to match properties of various sections using the AutoCAD **Match Properties** tool. Similar to AutoCAD, when you invoke this tool in Advance Steel, you will be prompted to select the source object. At this prompt, you can also select the **Settings** option to display the **Advance Steel - Set match properties** dialog box. This dialog box is similar to those discussed earlier in this chapter. You can deselect any property you do not want to match with the destination section by clearing its tick box.

What I do

The **Match Properties** tool is used quite often in Advance Steel. As a result, I add this tool to the **Quick Access** toolbar. This allows me to easily invoke this tool whenever required without the need of switching to the **Tools** ribbon tab.

Hands-on Tutorial (STRUC)	In this tutorial, you will complete the following tasks: 1. Open the **C02-Struc.dwg** file. 2. Insert the four columns, as shown in Figure 24. 3. Insert the perimeter beams at level 1 using the dimensions shown in Figure 25. 4. Insert the filter beams at level 1 using the dimensions shown in Figure 25. 5. Insert the perimeter beams at level 2 using the dimensions shown in Figure 26. 6. Insert the filter beams at level 2 using the dimensions shown in Figure 26.

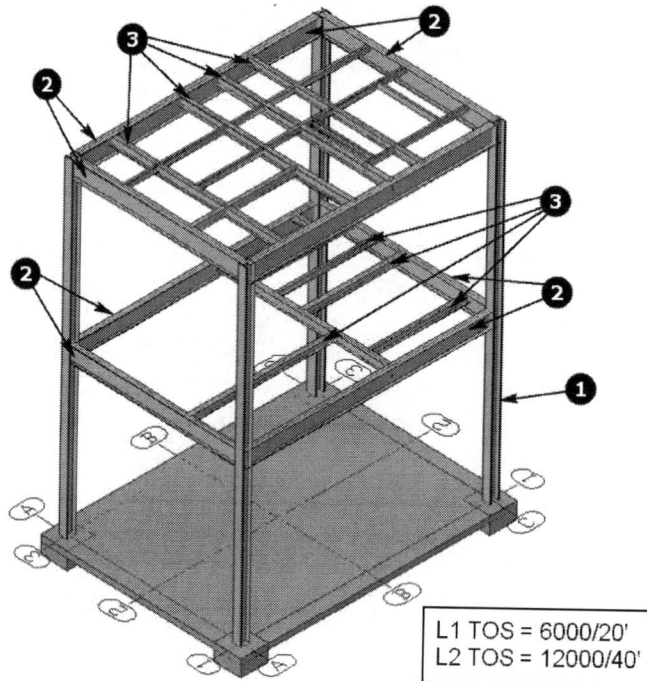

L1 TOS = 6000/20'
L2 TOS = 12000/40'

Figure 24 Two levels of the Elevator Tower for the Structural tutorial

Column Size

The sections used in this tutorial are **Australian Standard (AS)** for Metric or **AISC** standard for Imperial. However, you are free to use your country standard sections. The column size is given below. The level 1 and level 2 section sizes are given on the following pages:

Sections 1 (Columns): AISC 14.1 W > W12x210 / Australian Universal Column > 310 UC 158

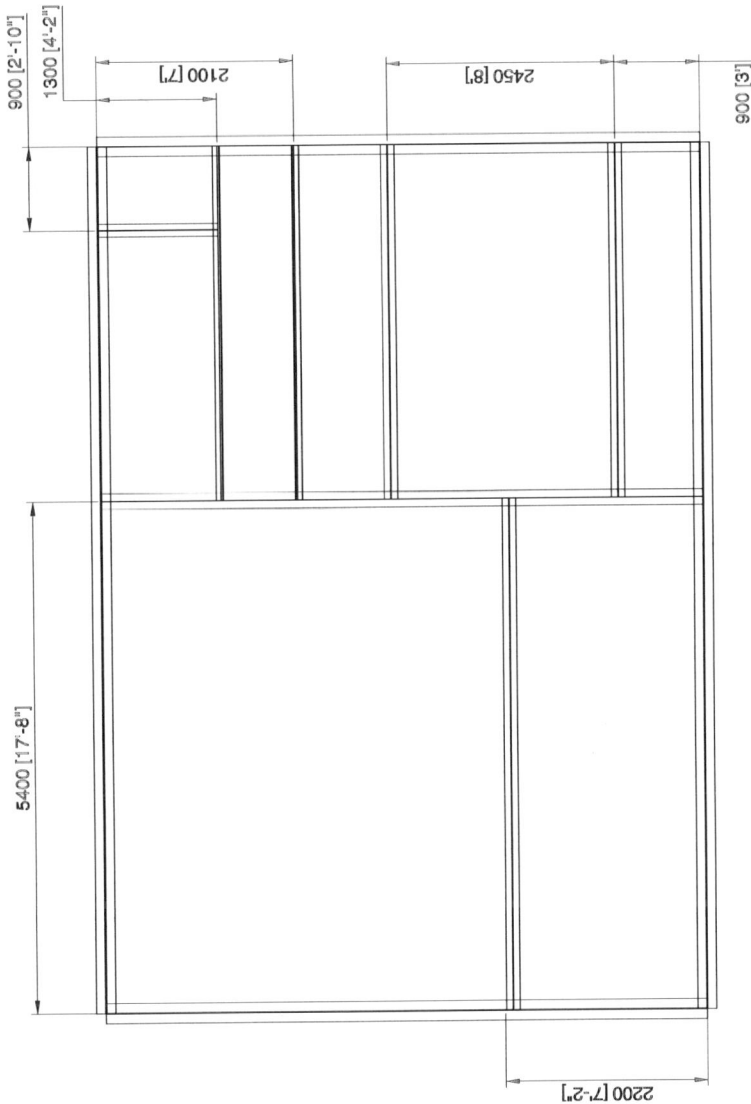

Figure 25 Dimensions for level 1

Section Sizes for Level 1

The sizes of the sections used on level 1 are listed below:

Sections 2 (Perimeter Beams): AISC 14.1 S > S24X121 / Australian Universal Beam > 530 UB 82.0
Sections 3 (Filter Beams): AISC 14.1 S > S12X50 / Australian Universal Beam > 250 UB 31.4
Remaining Sections on Level 1: AISC 14.1 C Channels > C9X20 / Australian Parallel Flange Channel > 200 PFC

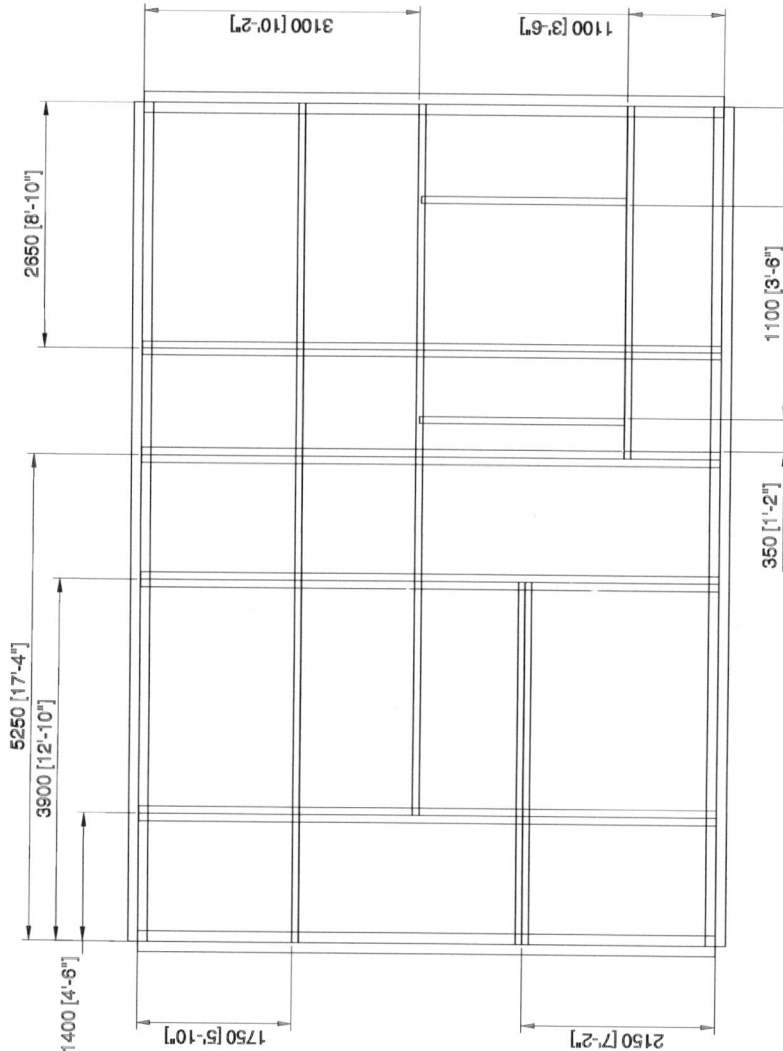

Figure 26 Dimensions for level 2

Section Sizes for Level 2

The sizes of the sections used on level 2 are listed below:

Sections 2 (Perimeter Beams): AISC 14.1 S > S24X121 / Australian Universal Beam > 530 UB 82.0

Sections 3 (Filter Beams): AISC 14.1 S > S12X50 / Australian Universal Beam > 250 UB 31.4

Remaining Sections on Level 2: AISC 14.1 C Channels > C9X20 / Australian Parallel Flange Channel > 200 PFC

Section 1: Opening the File and Inserting Columns

In this section, you will open the **C02-Struc-Imperial.dwg** file or the **C02-Struc-Metric.dwg** file and then add one of the four columns. You will then copy this column to create three additional copies of it at the grid intersection points.

1. From the **C02 > Struc** folder, open the **C02-Struc-Imperial.dwg** file or the **C02-Struc-Metric.dwg** file, depending on which units you want to use.

   Notice that this file already has the four isolated footings, slab, and grid created. You will now use the **Rolled I section** tool to insert a column at the grid intersection point **A1**.

   > **Note:** *The reason you are not using the **Column** tool to insert the column is that this tool does not prompt for the height of the column while inserting it. As a result, you will need to grip edit the height of the column after inserting. However, if you still want to use the **Column** tool, you can edit the default height of the column using the **Management Tools > Beams > General > Default height for column** edit box.*

2. From the **Home** ribbon tab > **Objects** ribbon panel, invoke the **Rolled I section** tool; you are prompted to locate the start point of the system axis.

3. Select the **A1** grid intersection point and then move the cursor up so that you are tracking along the Z direction.

   As shown in Figure 24, the total height to the Top of Steel on level 2 is 40' or 12000mm. In this tutorial, you will insert a single column of that height.

4. With the cursor tracking along the Z direction, enter **40'** or **12000** as the value; you are again prompted to specify the start point of the system axis.

5. Press ENTER; the **Advance Steel - Beam [X]** dialog box is displayed with the **Section & Material** tab active.

   Notice that in the first flyout of the **Section** list, **I Sections** is automatically selected. You now need to select the section type and section size from the next two flyouts.

6. From the second flyout in the **Section** list, select **AISC 14.1 W** or **Australian Universal Column** as the section type. Alternatively, you can also select your own country standard section type.

7. From the third flyout in the **Section** list, select **W12x210** or **310 UC 158** as the section size.

8. From the left pane of the dialog box, select the **Naming** tab.

9. From the **Model Role** list, select **Column** as the model role for this section.

10. Close the dialog box.

    Next, you will create the remaining three instances of this column using the **Copy** tool.

11. Using the **Advance Steel Tool Palette** > **Modify** tab > **Copy** tool, copy the column to the grid intersection points **A3**, **C1**, and **C3**. The model after inserting the columns is shown in Figure 27.

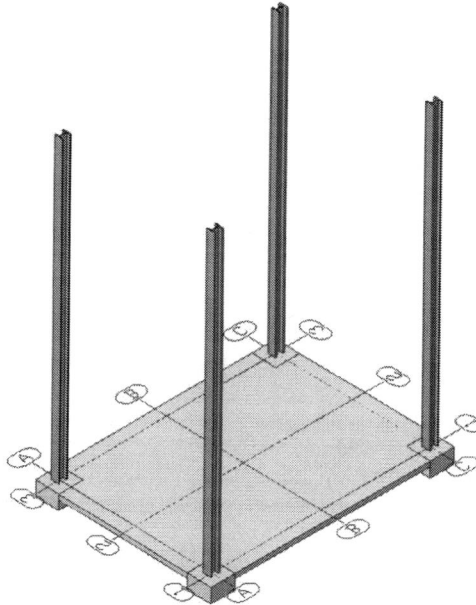

Figure 27 The model after inserting the four columns

Section 2: Inserting Sections at Level 1

Before you start inserting the perimeter and filter beams, it is recommended to change the display type of the four columns so that it is easier for you to insert sections at level 1.

1. Select the four columns and then right-click in the blank area of the drawing window; the shortcut menu is displayed.

2. Select **Advance Properties** from the shortcut menu; the **Advance Steel - Beam [X]** dialog box is displayed.

3. From the left pane of the dialog box, select the **Display type** tab.

4. From the right pane, select **Symbol** (Imperial) or **Symbolic** (Metric); the four columns are displayed as system axes with the actual section displayed partially at the midpoint.

5. Close the dialog box.

   As mentioned earlier, the four beams to be inserted on the outside are **AISC 14.1 S > S24X121** or **Australian Universal Beam > 530 UB 82.0**. You will now use the **Continuous Beam** tool to insert these beams.

6. From the **Objects** ribbon tab > **Beams** ribbon panel, invoke the **Continuous Beam** tool; you are prompted to locate the start point of the system axis.

7. Select the midpoint of one of the columns; you are prompted to specify the endpoint of the system axis.

8. Select the midpoint of the next column; a beam is inserted between the two midpoints and you are again prompted to specify the endpoint of the system axis.

9. Similarly, select the midpoints of the remaining columns and then of the first column again to insert four beams.

10. After selecting the last midpoint, press the ENTER key; the **Advance Steel - Beam [X]** dialog box is displayed.

    The first flyout in the **Section** list automatically shows **I Sections** selected. You will now select the section type and size from the second and third flyouts.

11. From the second flyout of the **Section** list, select **AISC 14.1 S** or **Australian Universal Beam** as the section type.

12. From the third flyout of the **Section** list, select **S24X121** or **530 UB 82.0** as the section size.

13. From the left pane of the dialog box, select the **Positioning** tab.

14. From the right pane, in the **Offset** area, select the top center radio button; all the four beams are aligned such that their system axes are at the top face.

15. From the **Naming** tab > **Model Role** list, select **Beam** as the model role.

16. Close the dialog box. The model after inserting beams at level 1 is shown in Figure 28.

    Next, you need to insert the remaining sections on level 1. It is recommended that you isolate the perimeter beams of this level to make it easier for you to insert the remaining sections on this level.

17. With no tool active, select the four perimeter beams inserted in the above steps.

18. Right-click in the blank area of the drawing window and select **Isolate > Isolate Objects** from the shortcut menu; all objects other than the selected perimeter beams are turned off.

19. Change the current view to the plan view. Also, right-click on the ViewCube and then select **Parallel** from the shortcut menu to change the current camera to parallel projection.

    Before you start inserting the sections, it is important to make sure only the **Node** and **Intersection** object snaps are selected. Also, you need to make sure the **Object Snap Tracking** is turned on.

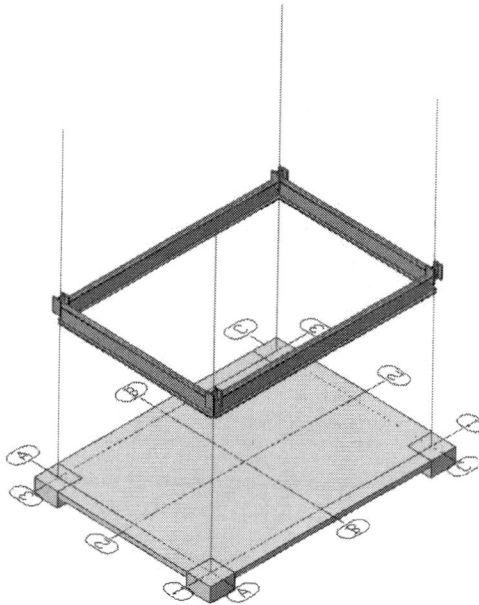

Figure 28 *The model after inserting beams at level 1*

20. From the **Object Snap** settings, make sure only the **Node** and **Intersection** options are selected.

21. From the Status Bar, make sure the **Object Snap Tracking** is turned on.

22. From the **Objects** ribbon tab > **Beams** ribbon panel, invoke the **Rolled I section** tool; you are prompted to locate the start point of the system axis.

23. Snap to the **Node** point at the top left corner of the perimeter beams and then move the cursor to the right to track the 0-degree angle, as shown in Figure 29.

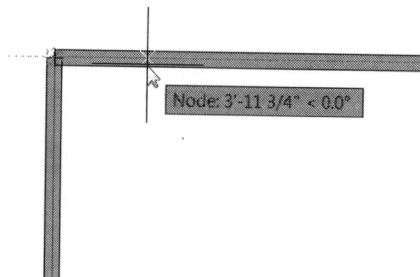

Node: 3'-11 3/4" < 0.0°

Figure 29 *Tracking the 0-degree angle*

24. Once the tracking line is displayed, enter **17'8"** or **5400** as the value; a new beam starts from that point.

25. Move the cursor down to track the 270-degrees angle.

26. With the tracking line displayed, move the cursor to the system axis of the perimeter beam at the bottom. Click to end the beam when the **Intersection** snap point is displayed, as shown in Figure 30; a new beam is inserted and you are again prompted to locate the start point of the system axis.

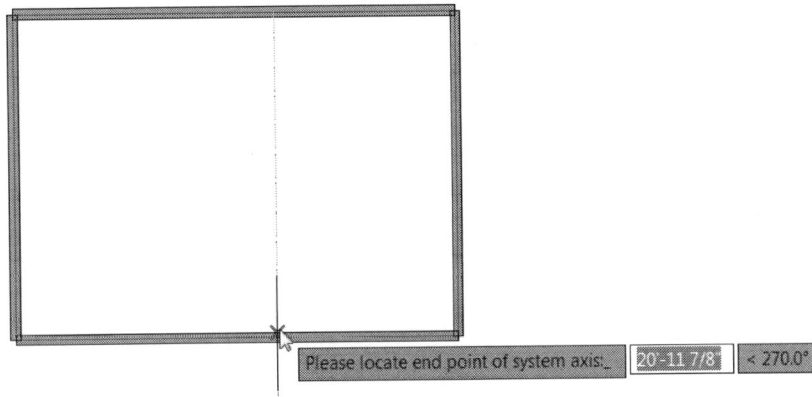

Please locate end point of system axis:_ 20'-11 7/8" < 270.0°

Figure 30 Specifying the endpoint of the first filter beam

Ideally, you can insert all the required sections and then edit the properties of all of them together. However, in this case, you will edit the size of the first filter beam first so that there is no gap between this section and other sections that will connect to this one.

27. Press ENTER; the **Advance Steel - Beam [X]** dialog box is displayed.

28. From the second flyout in the **Section** list, select **AISC 14.1 S** or **Australian Universal Beam** as the section type.

29. From the third flyout in the **Section** list, select **S 12X50** or **250 UB 31.4** as the section size.

30. From the left pane of the dialog box, select the **Positioning** tab.

31. From the right pane, in the **Offset** area, select the top center radio button; the beam is aligned such that its system axis is at the top face.

32. From the left pane of the dialog box, select the **Naming** tab.

33. In the right pane, from the **Model Role** list, select **Beam** and then close the dialog box.

34. Invoke the **Rolled I section** tool again.

35. Snap to the **Node** point at the lower left corner of the perimeter beams and then move the cursor up to track the 90-degrees angle.

36. Once the tracking line is displayed, enter **7'2"** or **2200** as the value; a new beam starts from that point.

37. Move the cursor to the right to track the 0-degree angle.

38. With the tracking line displayed, move the cursor to the last beam you inserted. Click to end the beam when the **Intersection** snap point is displayed, as shown in Figure 31; a new beam is inserted and you are again prompted to locate the start point of the system axis.

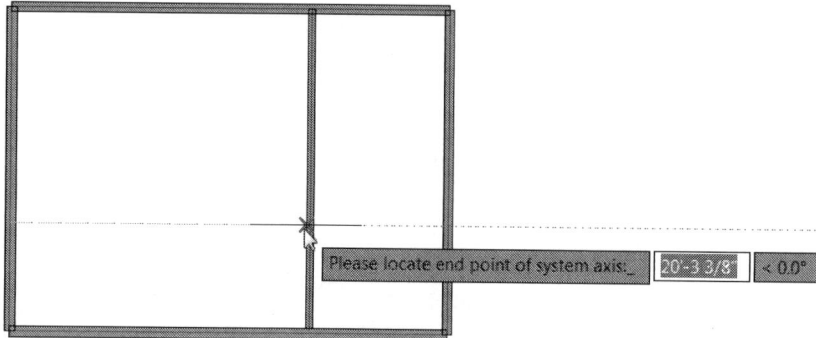

Figure 31 Specifying the endpoint of the second filter beam

39. Snap to the **Node** point at the lower right corner of the perimeter beams and then move the cursor up to track the 90-degrees angle.

40. Once the tracking line is displayed, enter **3'** or **900** as the value; a new beam starts from that point.

41. Move the cursor to the left to track the 180-degrees angle.

42. With the tracking line displayed, move the cursor to the vertical beam you inserted in earlier steps. Click to end the beam when the **Intersection** snap point is displayed; a new beam is inserted and you are again prompted to locate the start point of the system axis.

43. Snap to the right **Node** point of the last beam you inserted and then move the cursor up to track the 90-degrees angle.

44. Once the tracking line is displayed, enter **8'** or **2450** as the value; a new beam starts from that point.

45. Move the cursor to the left to track the 180-degrees angle.

46. With the tracking line displayed, move the cursor to the vertical beam you inserted in earlier steps. Click to end the beam when the **Intersection** snap point is displayed; a new beam is inserted and you are again prompted to locate the start point of the system axis.

47. Press ENTER; the **Advance Steel - Beam [X]** dialog box is displayed.

    The section sizes and positioning are automatically defined based on the last section you placed. So you do not need to change anything.

48. From the left pane of the dialog box, select the **Naming** tab.

49. In the right pane, from the **Model Role** list, select **Beam** and then close the dialog box.

    Next, you need to insert the two channels.

50. From the **Object** ribbon tab > **Beams** ribbon panel > **Rolled I section** flyout, invoke the **Channel section** tool; you are prompted to locate the start point of the system axis.

51. Snap to the **Node** point at the top right corner of the perimeter beams and then move the cursor down to track the 270-degrees angle.

52. Once the tracking line is displayed, enter **4'2"** or **1300** as the value; a new beam starts from that point.

53. Move the cursor to the left to track the 180-degrees angle.

54. With the tracking line displayed, move the cursor to the right edge of the vertical beam. Click to end the beam when the **Intersection** snap point is displayed, as shown in Figure 32; a new beam is inserted and you are again prompted to locate the start point of the system axis.

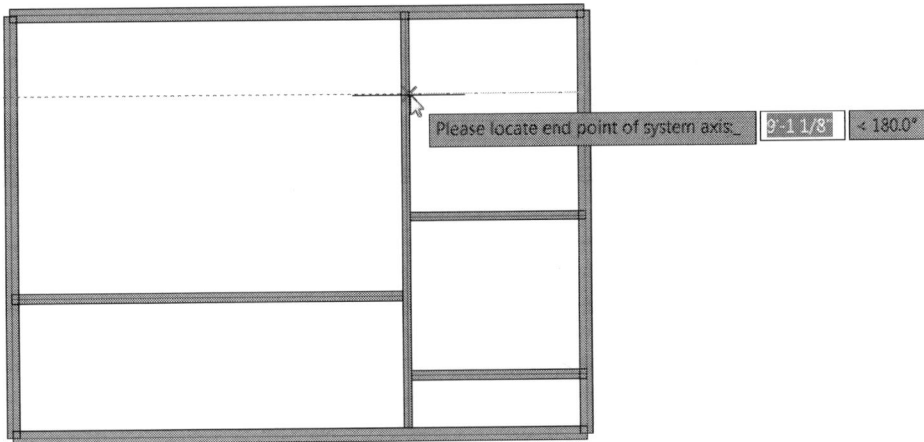

Figure 32 Specifying the endpoint of the channel

55. Snap to the **Node** point at the top right corner of the perimeter beams and then move the cursor down to track the 270-degrees angle.

56. Once the tracking line is displayed, enter **7'** or **2100** as the value; a new beam starts from that point.

57. Move the cursor to the left to track the 180-degrees angle.

58. With the tracking line displayed, move the cursor to the right edge of the vertical beam. Click to end the beam when the **Intersection** snap point is displayed; a new beam is inserted and you are again prompted to locate the start point of the system axis.

    The last beam to be inserted is an I-section at an offset of 2'10" or 900 from the top right corner of the perimeter beams. However, you can insert that beam as a channel and then use the **Match Properties** tool to make it the same as the I-sections inserted earlier.

59. Snap to the **Node** point at the top right corner of the perimeter beams and then move the cursor to the left to track the 180-degrees angle.

60. Once the tracking line is displayed, enter **2'10"** or **900** as the value; a new beam starts from that point.

61. Move the cursor to down to track the 270-degrees angle.

62. With the tracking line displayed, move the cursor to the top edge of the first channel you inserted. Click to end the beam when the **Intersection** snap point is displayed; a new beam is inserted and you are again prompted to locate the start point of the system axis.

63. Press the ENTER key; the **Advance Steel - Beam [X]** dialog box is displayed.

    Notice that the first flyout in the **Section** list automatically selects **Channel** because of the tool that was used to insert these sections.

64. From the second flyout in the **Section** list, select **AISC 14.1 C Channels** or **Australian Parallel Flange Channel** as the section type.

65. From the third flyout in the **Section** list, select **C9X20** or **200 PFC** as the section size.

66. From the left pane of the dialog box, select the **Positioning** tab.

67. From the right pane, in the **Offset** area, select the top center radio button; all the channels are aligned such that their system axes are at the top face.

68. From the left pane of the dialog box, select the **Naming** tab.

69. In the right pane, from the **Model Role** list, select **Beam**.

70. Close the dialog box.

    Next, you need to change the orientation of the second channel you placed. It needs to be inserted as a mirror of the current orientation.

71. Double-click on the second channel you placed; the **Advance Steel - Beam [X]** dialog box is displayed.

72. From the left pane of the dialog box, select the **Positioning** tab.

73. From the right pane, select the **Mirrored** tick box from the **Mirror** area; the channel is oriented correctly.

   Lastly, you need to change the third channel you placed to I-section similar to the ones you placed earlier.

74. Invoke the **Match Properties** tool; you are prompted to select the source object.

75. Select one of the I-sections you placed as filter beams; you are prompted to select the destination object.

76. Select the third channel you inserted; it is changed to an I-section. Notice that it still maintains its positioning.

77. Press ENTER to terminate the tool. The model after inserting all the sections on level 1 is shown in Figure 33.

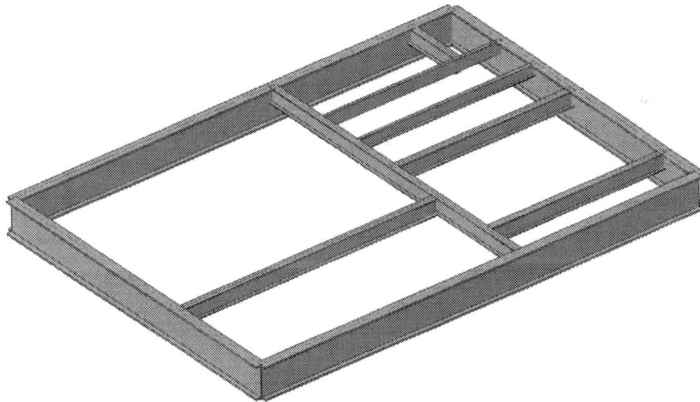

Figure 33 The model after inserting all the sections at level 1

Section 3: Inserting Sections at Level 2

Before you start inserting sections at level 2, you first need to end object isolation so you can see all the items in the model.

1. With no tool active, right-click in the blank area of the drawing window and select **Isolate > End Object Isolation** from the shortcut menu; all the items in the model are redisplayed in the drawing window.

Because the perimeter beams at level 2 are the same as those on level 1, you can copy them using the **Copy** tool.

2. Using the **Copy** tool and the midpoint of one of the columns as the base point, copy the four perimeter beams to the top **Node** point of the same column. This will copy the perimeter beams from level 1 to level 2. Figure 34 shows the model after copying the perimeter beams from level 1 to level 2.

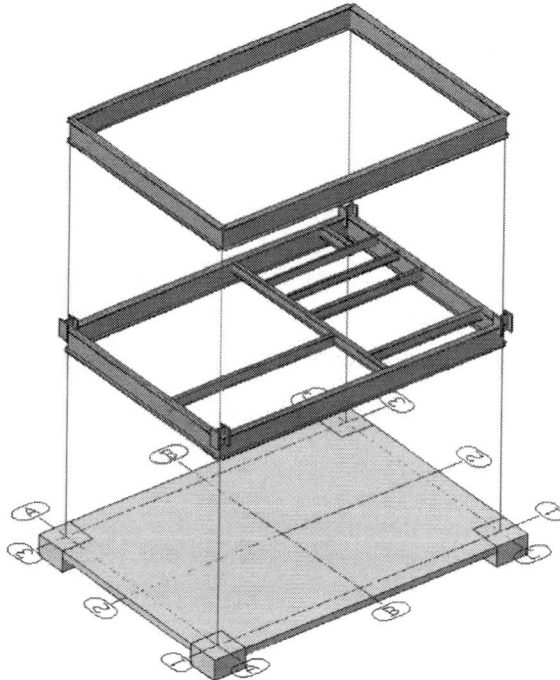

Figure 34 The model after copying the perimeter beams from level 1 to level 2

It is better to isolate the top perimeter beams before you start inserting the filter beams.

3. Select the four perimeter beams at level 2 and isolate them.

4. Change the current view to plan view with parallel projection.

5. From the **Objects** ribbon tab > **Beams** ribbon panel, invoke the **Rolled I section** tool; you are prompted to locate the start point of the system axis.

6. Snap to the **Node** point at the top right corner of the perimeter beams and then move the cursor to the left to track the 180-degrees angle.

7. Once the tracking line is displayed, enter **8'10"** or **2650** as the value; a new beam starts from that point.

8. Move the cursor down to track the 270-degrees angle.

9. With the tracking line displayed, move the cursor to the inner edge of the perimeter beam at the bottom. Click to end the beam when the **Intersection** snap point is displayed, as shown in Figure 35; a new beam is inserted and you are again prompted to locate the start point of the system axis.

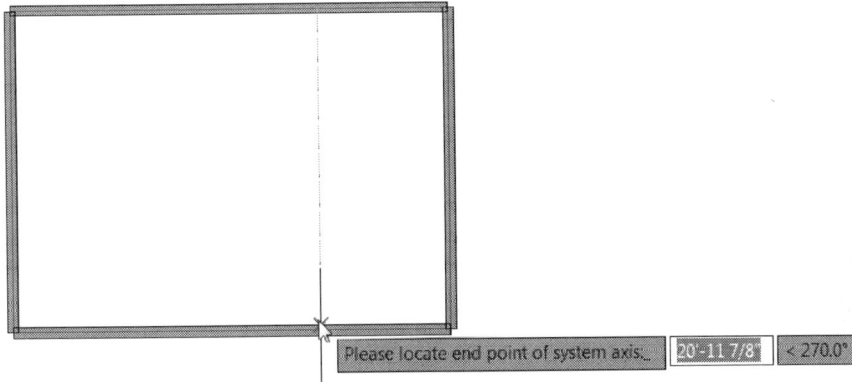

Figure 35 Specifying the endpoint of the first filter beam at level 2

10. Snap to the **Node** point at the top left corner of the perimeter beams and then move the cursor to the right to track the 0-degree angle.

11. Once the tracking line is displayed, enter **4'6"** or **1400** as the value; a new beam starts from that point.

12. Move the cursor down to track the 270-degrees angle.

13. With the tracking line displayed, move the cursor to the inner edge of the perimeter beam at the bottom. Click to end the beam when the **Intersection** snap point is displayed; a new beam is inserted and you are again prompted to locate the start point of the system axis.

14. Snap to the top left **Node** point of the perimeter beams and then move the cursor to the right to track the 0-degree angle.

15. Once the tracking line is displayed, enter **12'10"** or **3900** as the value; a new beam starts from that point.

16. Move the cursor down to track the 270-degrees angle.

17. With the tracking line displayed, move the cursor to the inner edge of the perimeter beam at the bottom. Click to end the beam when the **Intersection** snap point is displayed, as shown in Figure 36; a new beam is inserted and you are again prompted to locate the start point of the system axis.

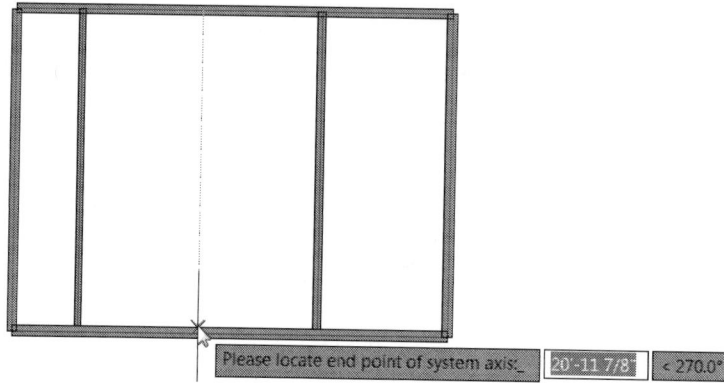

Figure 36 *Specifying the endpoint of the filter beam at level 2*

18. Snap to the top left **Node** point of the perimeter beams and then move the cursor to the right to track the 0-degree angle.

19. Once the tracking line is displayed, enter **17'4"** or **5250** as the value; a new beam starts from that point.

20. Move the cursor down to track the 270-degrees angle.

21. With the tracking line displayed, move the cursor to the inner edge of the perimeter beam at the bottom. Click to end the beam when the **Intersection** snap point is displayed; a new beam is inserted and you are again prompted to locate the start point of the system axis.

22. Snap to the bottom left **Node** point of the perimeter beams and then move the cursor up to track the 90-degrees angle.

23. Once the tracking line is displayed, enter **7'2"** or **2150** as the value; a new beam starts from that point.

24. Move the cursor to the right to track the 0-degree angle.

25. With the tracking line displayed, move the cursor to the **Intersection** snap point on the left edge of the third filter beam, as shown in Figure 37, and then click to end the beam; a new beam is inserted and you are again prompted to locate the start point of the system axis.

26. Press ENTER; the **Advance Steel - Beam [X]** dialog box is displayed.

27. From the second flyout in the **Section** list, select **AISC 14.1 S** or **Australian Universal Beam** as the section type, if not already selected.

28. From the third flyout in the **Section** list, select **S 12X50** or **250 UB 31.4** as the section size, if not already selected.

29. From the left pane of the dialog box, select the **Positioning** tab.

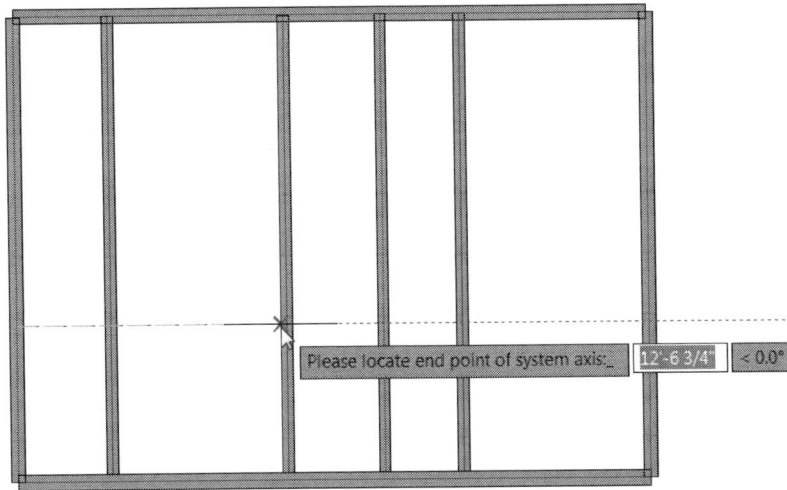

Figure 37 Specifying the endpoint of the horizontal filter beam

30. From the right pane, in the **Offset** area, select the top center radio button; all the beams are aligned such that their system axes are at the top face.

31. From the left pane of the dialog box, select the **Naming** tab.

32. In the right pane, from the **Model Role** list, select **Beam**.

33. Close the dialog box.

    Next, you need to insert the channels.

34. From the **Object** ribbon tab > **Beams** ribbon panel > **Rolled I section** flyout, invoke the **Channel section** tool; you are prompted to locate the start point of the system axis.

35. Snap to the **Node** point at the top left corner of the perimeter beams and then move the cursor down to track the 270-degrees angle.

36. Once the tracking line is displayed, enter **5'10"** or **1750** as the value; a new beam starts from that point.

37. Move the cursor to the right to track the 0-degree angle.

38. With the tracking line displayed, move the cursor to the inside edge of the right perimeter beam. Click to end the beam when the **Intersection** snap point is displayed, as shown in Figure 38; a new beam is inserted and you are again prompted to locate the start point of the system axis.

39. Snap to the **Node** point at the top right corner of the perimeter beams and then move the cursor down to track the 270-degrees angle.

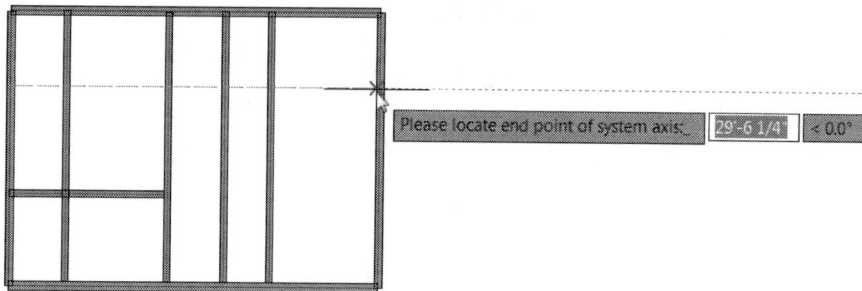

Figure 38 Specifying the endpoint of the channel

40. Once the tracking line is displayed, enter **10'2"** or **3100** as the value; a new beam starts from that point.

41. Move the cursor to the left to track the 180-degrees angle.

42. With the tracking line displayed, move the cursor to the right edge of the first filter beam from the left. Click to end the beam when the **Intersection** snap point is displayed, as shown in Figure 39; a new beam is inserted and you are again prompted to locate the start point of the system axis.

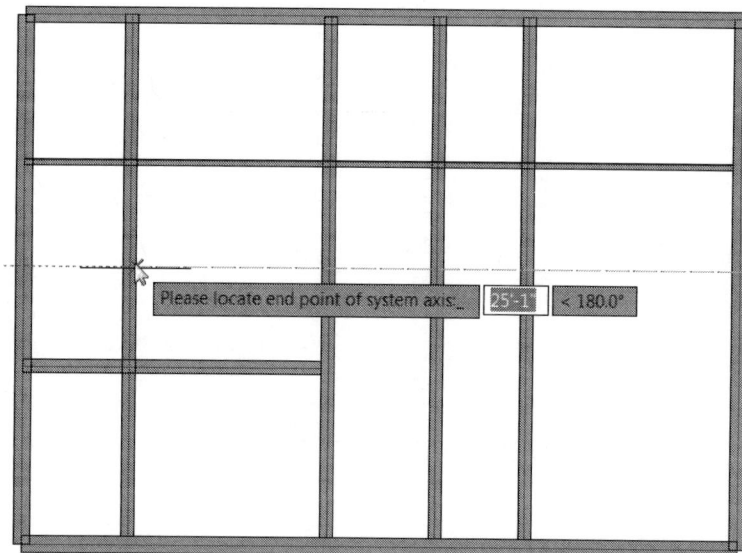

Figure 39 Specifying the endpoint of the second channel

43. Snap to the **Node** point at the bottom right corner of the perimeter beams and then move the cursor up to track the 90-degrees angle.

44. Once the tracking line is displayed, enter **3'6"** or **1100** as the value; a new beam starts from that point.

45. Move the cursor to the left to track the 180-degrees angle.

46. With the tracking line displayed, move the cursor to the right edge of the second filter beam from the right. Click to end the beam when the **Intersection** snap point is displayed, as shown in Figure 40; a new beam is inserted and you are again prompted to locate the start point of the system axis.

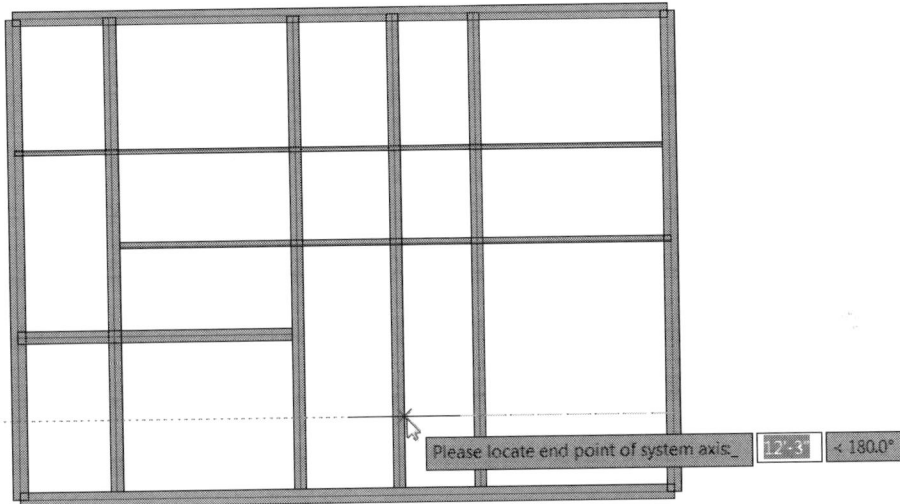

Figure 40 Specifying the endpoint of the channel

47. Similarly, insert the remaining channels. When you finish inserting all the channels, press the ENTER key; the **Advance Steel - Beam [X]** dialog box is displayed.

    Notice that the first flyout in the **Section** list automatically selects **Channel** because of the tool that was used to insert these sections.

48. From the second flyout in the **Section** list, select **AISC 14.1 C Channels** or **Australian Parallel Flange Channel** as the section type, if not already selected.

49. From the third flyout in the **Section** list, select **C9X20** or **200 PFC** as the section size.

50. From the left pane of the dialog box, select the **Positioning** tab.

51. From the right pane, in the **Offset** area, select the top center radio button, if not already selected; all the beams are aligned such that their system axes are at the top face.

52. From the left pane of the dialog box, select the **Naming** tab.

53. In the right pane, from the **Model Role** list, select **Beam**.

54. Close the dialog box. The model after inserting all the I-sections and channels is displayed in Figure 41.

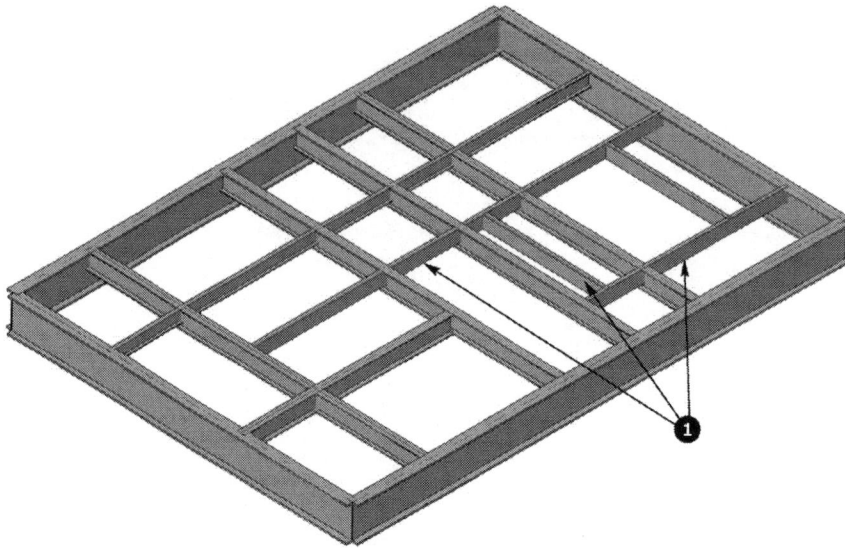

Figure 41 *The model after inserting all the sections at level 2*

In Figure 41, you need to mirror the orientation of the channels labeled as **1**.

55. Select the three channels labeled as **1** in Figure 41.

56. Right-click in the blank area of the graphics window and select **Advance Properties** from the shortcut menu; the **Advance Steel - Beam [X]** dialog box is displayed.

57. From the left pane of the dialog box, select the **Positioning** tab.

58. From the right pane, select the **Mirrored** tick box from the **Mirror** area; the beam is oriented correctly.

59. Close the dialog box.

Note: On mirroring the orientation of the channels, they will move slightly by a value around 9/16" or 14mm. To make this tutorial simple and easier to work with, you will ignore this change.

Section 4: Splitting Sections

The beams and channels you inserted in the previous section need to be split at the points where they intersect with the other sections. This is done using the **Split beams** tool.

Note: The **Split beams** tool allows you to split multiple sections together. However, you need to be careful while splitting multiple sections as all the selected sections will be split at every single split point.

1. From the **Objects** ribbon tab > **Beams** ribbon panel, invoke the **Split beams** tool; you are prompted to select objects.

2. Select the four vertical I-sections labeled as **1** in Figure 42 and then press ENTER; you are prompted to select the split point or define the gap.

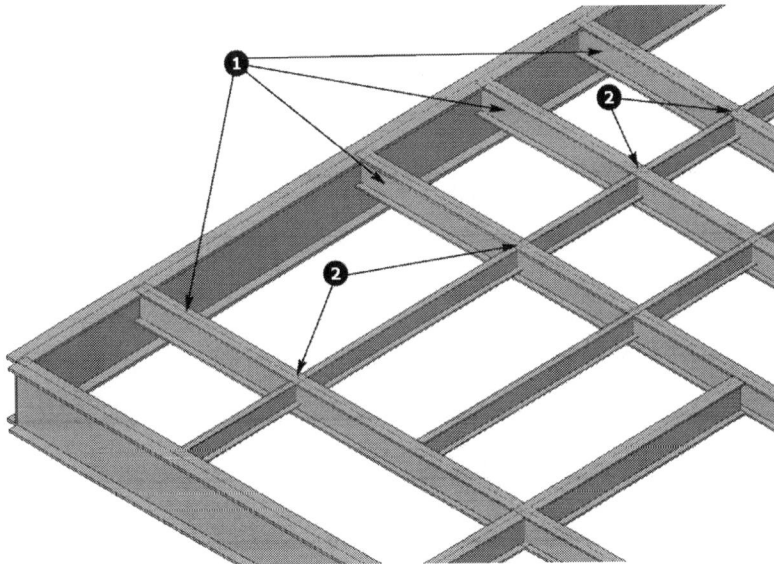

Figure 42 Selecting the sections to split and the split point

3. Select any one of the intersection points labeled as **2** in Figure 42 and then press ENTER.

   Notice that although you selected only one of the points labeled as **2**, all the selected I-sections are still split along a plane at that point.

4. Similarly, use the **Split beams** tool to split the remaining I-sections and channels.

Section 5: Restoring the Visibility of all Objects

In this section, you will end the object isolation and also turn the display type of the columns back to standard display.

1. With no tool active, right-click in the blank area of the drawing window and select **Isolate > End Object Isolation** from the shortcut menu; the visibility of all the objects is restored.

2. Select the symbol lines of the four columns.

3. Right-click in the blank area of the drawing window and select **Advance Properties** from the shortcut menu; the **Advance Steel - Beam [X]** dialog box is displayed.

4. From the **Display type** tab, select **Standard**; the visibility of all the columns is restored. The completed model of the tutorial is shown in Figure 43. In this figure, the projection type is set to perspective.

5. Save the file and then close it.

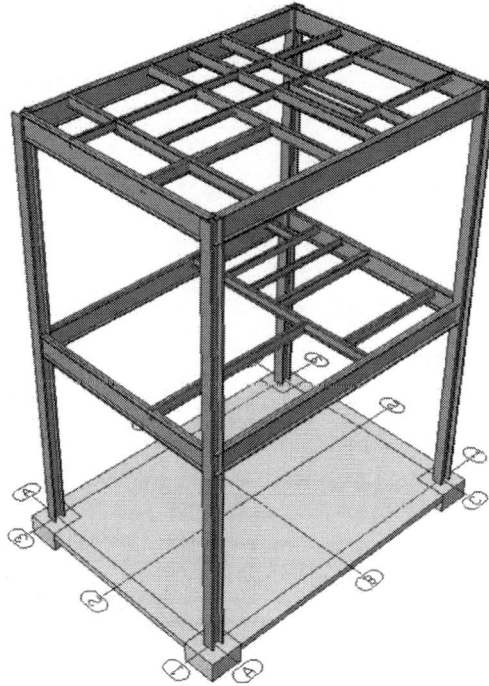

Figure 43 *The perspective view of the completed model*

Hands-on Tutorial (BIM)	*In this tutorial, you will complete the following tasks:*
	1. *Open the **C02-BIM.dwg** file.*
	2. *Insert the columns, as shown in Figure 44.*
	3. *Insert the perimeter beams at level 1 using the dimensions shown in Figure 45.*
	4. *Insert the filter beams at level 1 using the dimensions shown in Figure 45.*
	5. *Copy the perimeter and filter beams to level 2.*

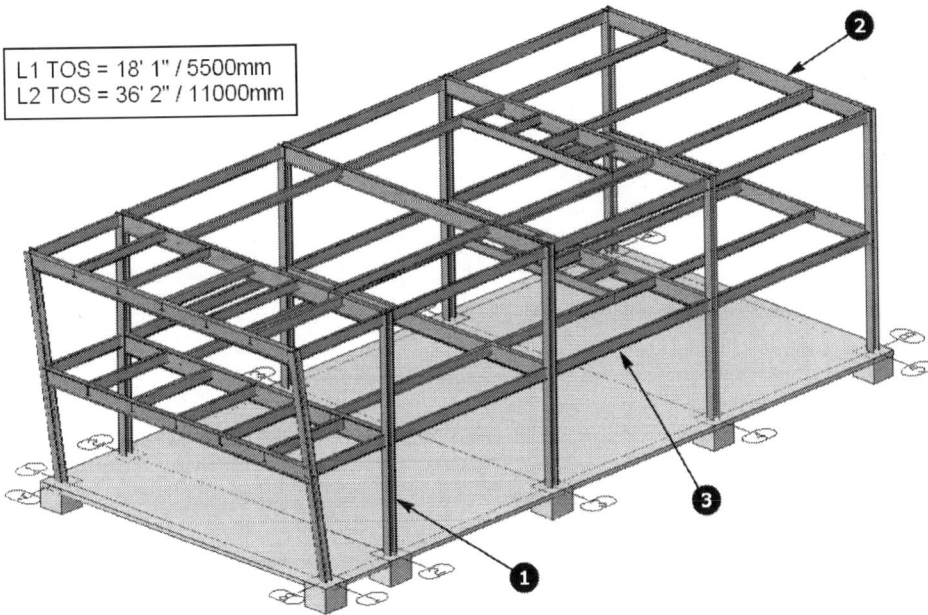

Figure 44 *Two levels of the Zone A of the hospital structure for the BIM tutorial (Image courtesy Autodesk, Inc., © 2014)*

Section Sizes

The sections used in this tutorial are **Australian Standard (AS)** for Metric or **AISC** standard for Imperial. However, you are free to use your country standard sections. The section sizes are given below. The level 1 and level 2 section sizes are given on the following page:

Sections 1 (Columns): AISC 14.1 W > W14x145 / Australian Welded Column > 400 WC 328
Sections 2 (Main Perimeter Beams): AISC 14.1 W > W30x124 / Australian Welded Beam > 800 WB 122
Sections 3 (Other Perimeter Beams and Filter Beams): AISC 14.1 W > W21x57 / Australian Universal Beam > 530 UB 92.4

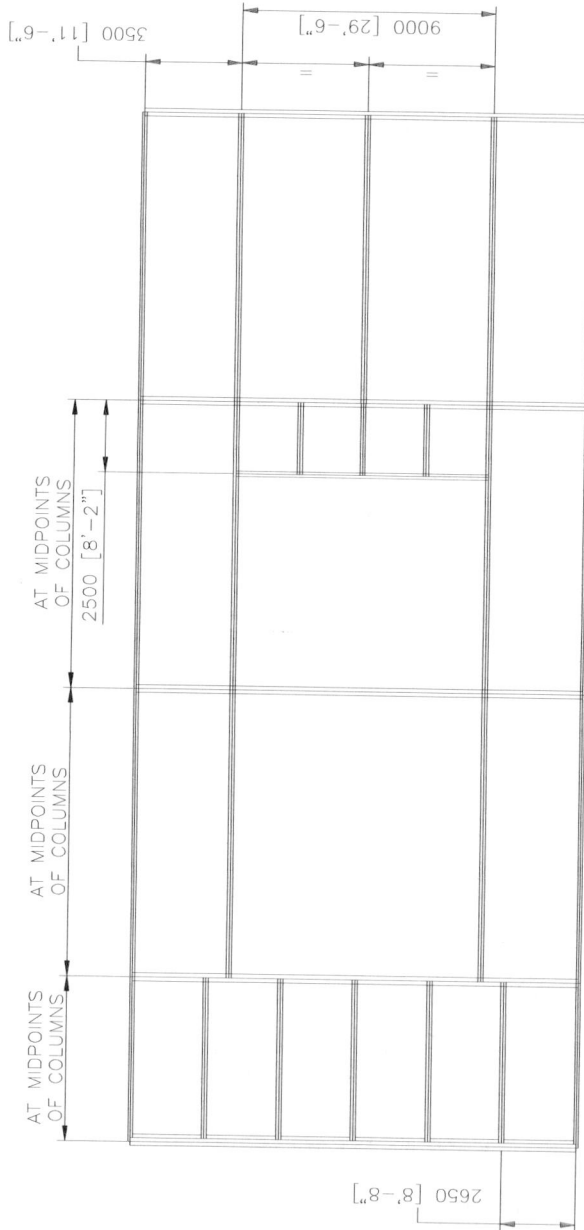

Figure 45 Dimensions for level 1 and level 2

Section 1: Opening the File and Inserting Columns

In this section, you will open the **C02-BIM-Imperial.dwg** file or the **C02-BIM-Imperial.dwg** file and then add one of the front columns. You will then modify this column using its top grip point and then create a copy of it at the other side of the concrete slab.

1. From the **C02 > BIM** folder, open the **C02-BIM-Imperial.dwg** file or the **C02-BIM-Metric.dwg** file, depending on which units you want to use.

    Notice that this file already has the isolated footings, slab, and grid created. You will now use the **Rolled I section** tool to insert a column at the grid intersection point **B1**.

    > **Note**: *The reason you are not using the **Column** tool to insert the column is that this tool does not prompt for the height of the column while inserting it. As a result, you will need to grip edit the height of the column after inserting. However, if you still want to use the **Column** tool, you can edit the default height of the column using the **Management Tools > Beams > General > Default height for column** edit box.*

2. From the **Home** ribbon tab > **Objects** ribbon panel, invoke the **Rolled I section** tool; you are prompted to locate the start point of the system axis.

3. Select the **B1** grid intersection point and then move the cursor up so that you are tracking along the Z direction.

    As shown in Figure 44, the total height to the Top of Steel on level 2 is 36'2"/11000mm. In this tutorial, you will insert a single column of that height.

4. With the cursor tracking along the Z direction, enter **36'2"/11000** as the value; you are again prompted to specify the start point of the system axis.

5. Press ENTER; the **Advance Steel - Beam [X]** dialog box is displayed with the **Section & Material** tab active.

    Notice that in the first flyout of the **Section** list, **I Sections** is automatically selected. You now need to select the section type and section size from the next two flyouts.

6. From the second flyout in the **Section** list, select **AISC 14.1 W** or **Australian Welded Column** as the section type. Alternatively, you can also select your own country standard section type.

7. From the third flyout in the **Section** list, select **W14x145** or **400 WC 328** as the section size.

8. From the left pane of the dialog box, select the **Positioning** tab.

9. From the **Offset** area, make sure the center radio button is selected.

10. From the left pane of the dialog box, select the **Naming** tab.

11. From the **Model Role** list, select **Column** as the model role for this section.

12. Close the dialog box.

    Next, you will edit this column using its top grip point and move that point 5' or 1500mm along the 180-degrees tracking direction.

13. Select the column you just placed; it is highlighted and its grip points are displayed.

14. Click on the top grip point and move the cursor along the -X direction to track the 180-degrees angle.

15. Once the tracking line is displayed, type **5'** or **1500** and then press the ENTER key; the angle of the column is modified.

16. Press the ESC key two times to exit the grip editing mode.

17. Using the **Copy** tool, copy this column to the **A1** grid intersection point. The model after copying the column is shown in Figure 46.

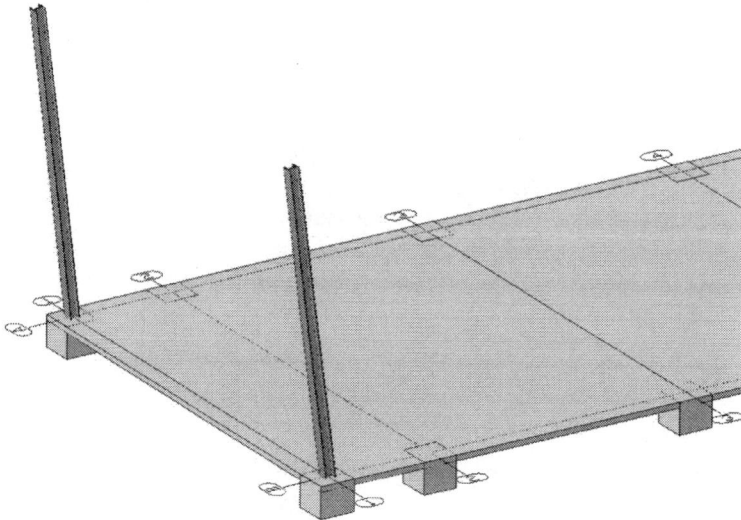

Figure 46 The model after inserting the two columns

Next, you will insert the straight columns. You need to insert just one column and then you can use the **Array** tool to create the remaining instances of the columns.

18. From the **Home** ribbon tab > **Objects** ribbon panel, invoke the **Rolled I section** tool; you are prompted to locate the start point of the system axis.

19. Select the **B2** grid intersection point and then move the cursor up so that you are tracking along the Z direction.

20. With the cursor tracking along the Z direction, enter **36'2"** or **11000** as the value; you are again prompted to specify the start point of the system axis.

21. Press ENTER; the **Advance Steel - Beam [X]** dialog box is displayed with the **Section & Material** tab active.

    Advance Steel remembers the last section that you inserted and inserts the same section next time. Therefore, in your case, the required section type and size is automatically selected from the **Section** list. If not, you can use the following steps to insert the section.

22. From the second flyout in the **Section** list, select **AISC 14.1 W** or **Australian Welded Column** as the section type. Alternatively, you can also select your own country standard section type.

23. From the third flyout in the **Section** list, select **W14x145** or **400 WC 328** as the section size.

24. From the left pane of the dialog box, select the **Naming** tab.

25. From the **Model Role** list, select **Column** as the model role for this section.

26. Close the dialog box.

    Next, you will array this column.

27. Using the **Array** tool, create a rectangular array of the column you just inserted. Make sure you clear the **Associative** button to make the array disassociative. You need to create four columns and two rows. Use the grid intersection points to specify the spacing between the rows and columns. The model after inserting all the columns is shown in Figure 47.

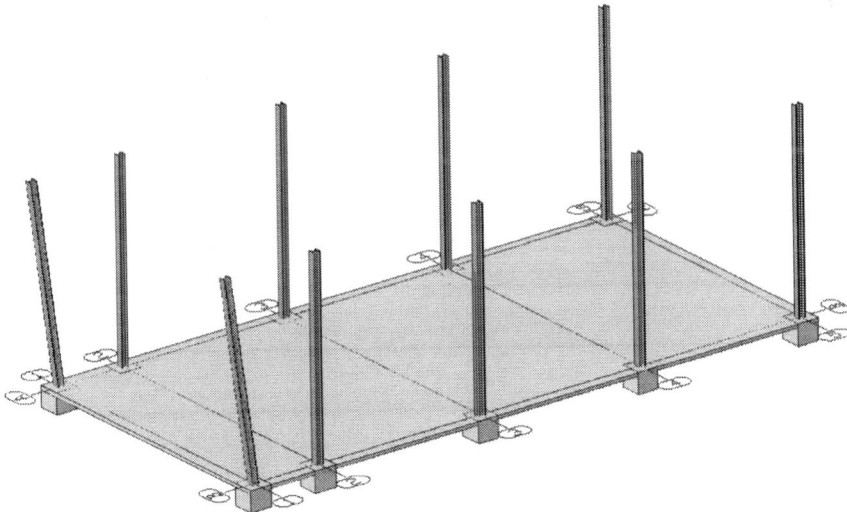

Figure 47 *The model after inserting all the columns*

Section 2: Inserting Sections at Level 1

Before you start inserting the perimeter and filter beams, it is recommended to change the display type of the columns so that it is easier for you to insert sections at level 1.

1. Select all the columns and then right-click in the blank area of the drawing window; the shortcut menu is displayed.

2. Select **Advance Properties** from the shortcut menu; the **Advance Steel - Beam [X]** dialog box is displayed.

3. From the left pane of the dialog box, select the **Display type** tab.

4. From the right pane, select **Symbol** (Imperial) or **Symbolic** (Metric); the four columns are displayed as system axes with the actual section displayed partially at the midpoint.

5. Close the dialog box.

   As mentioned earlier, the main perimeter beams to be inserted on the outside are **AISC 14.1 W > W30x124** or **Australian Welded Beam > 800 WB 122**. You will use the **Continuous Beam** tool to insert the four perimeter beams and then change the properties of two of the beams.

6. From the **Objects** ribbon tab > **Beams** ribbon panel, invoke the **Continuous Beam** tool; you are prompted to locate the start point of the system axis.

7. Select the midpoint of the column at the **B1** grid intersection point; you are prompted to specify the endpoint of the system axis.

8. Select the midpoint of the column at the **A1** grid intersection point; you are prompted to specify the endpoint of the system axis.

9. Select the midpoint of the column at the **A5** grid intersection point; you are prompted to specify the endpoint of the system axis.

10. Select the midpoint of the column at the **B5** grid intersection point; you are prompted to specify the endpoint of the system axis.

11. Select the midpoint of the column at the **B1** grid intersection point; you are prompted to specify the endpoint of the system axis.

12. Press the ENTER key; the **Advance Steel - Beam [X]** dialog box is displayed.

    The first flyout in the **Section** list automatically shows **I Sections** selected. You will now select the section type and size from the second and third flyouts.

13. From the second flyout of the **Section** list, select **AISC 14.1 W** or **Australian Welded Beam**.

14. From the third flyout of the **Section** list, select **W30x124** or **800 WB 122** as the section size.

15. From the left pane of the dialog box, select the **Positioning** tab.

16. From the right pane, in the **Offset** area, select the top center radio button; all the beams are aligned such that their system axes are in line with the midpoints of the columns.

17. From the **Naming** tab > **Model Role** list, select **Beam** as the model role.

18. Close the dialog box. The model, after inserting the four beams, is shown in Figure 48.

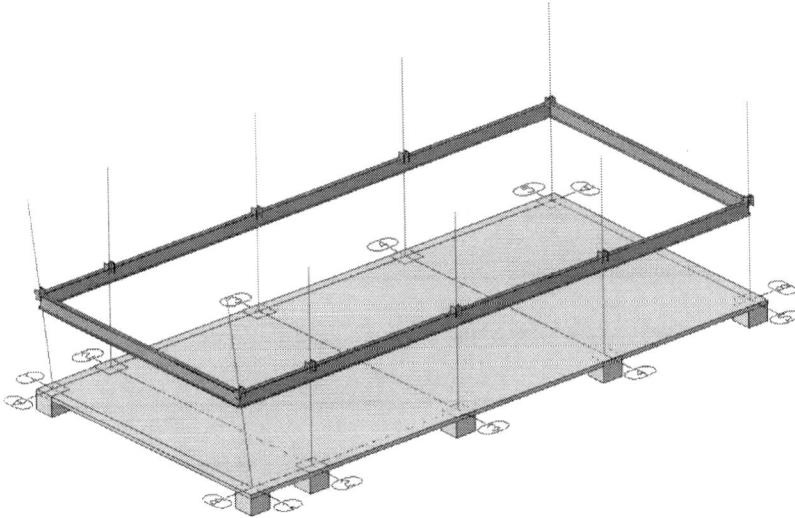

Figure 48 The model after inserting the perimeter beams at level 1

Next, you need to change the sizes of the two longer beams.

19. Select the two longer beams.

20. Right-click in the blank area of the drawing window and select **Advance Properties** from the shortcut menu; the **Advance Steel - Beam [X]** dialog box is displayed.

21. Select **AISC 14.1 W > W21x57 / Australian Universal Beam > 530 UB 92.4** as the beam size.

22. Close the dialog box.

Next, you will insert a beam between the midpoints of the columns at the **A2** and **B2** grid intersection points. You will then copy this beam to create the remaining two instances.

23. Invoke the **Rolled I section** tool; you are prompted to locate the start point of the system axis.

24. Select the midpoint of the column at the **B2** grid intersection point; you are prompted to specify the endpoint of the system axis.

25. Select the midpoint of the column at the **A2** grid intersection point; you are prompted to specify the endpoint of the system axis.

26. Press the ENTER key; the **Advance Steel - Beam [X]** dialog box is displayed.

27. From the **Section** list, select **AISC 14.1 W > W30x124** or **Australian Welded Beam > 800 WB 122**.

28. From the left pane of the dialog box, select the **Positioning** tab.

29. From the right pane, in the **Offset** area, select the top center radio button; the beam is aligned such that its system axis is aligned with the midpoint of the columns.

30. From the **Naming** tab > **Model Role** list, select **Beam** as the model role.

31. Close the dialog box.

32. Copy this beam to the midpoints of the columns at **B3** and **B4** grid intersection points. The model after inserting and copying all the beams is shown in Figure 49.

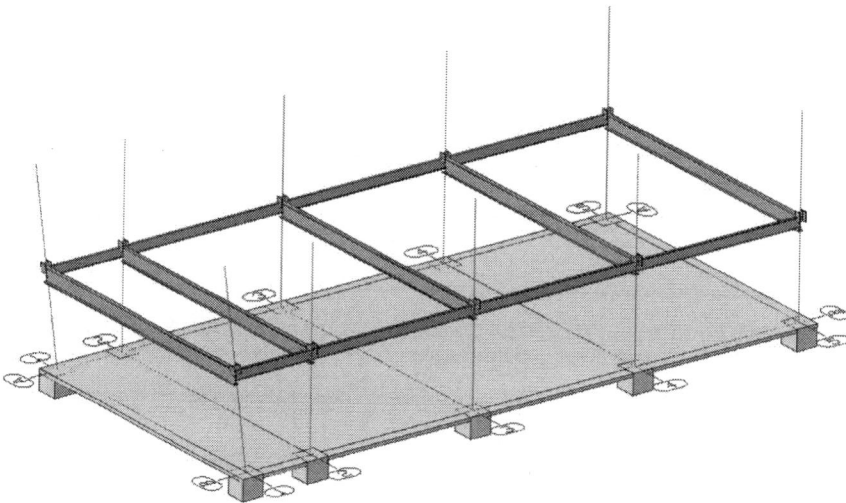

Figure 49 *The model after inserting and copying the beams*

Next, you need to insert the remaining filter beams. To make it easier for you to insert those beams, it is better to isolate the beams at level 1.

33. Select all the beams at level 1.

34. Right-click in the blank area of the drawing window and select **Isolate > Isolate Objects** from the shortcut menu; the visibility of all the unselected objects will be turned off.

35. Change the view to Top view with parallel projection.

Before proceeding any further, it is important to make sure only the **Node** and **Intersect** object snap points are turned on.

36. From the **Object Snap** settings, make sure only the **Node** and **Intersect** object snap options are turned on.

37. Invoke the **Rolled I section** tool; you are prompted to locate the start point of the system axis.

38. Snap to the top right **Node** point of the perimeter beam and then move the cursor down to track the 270-degrees angle, as shown in Figure 50.

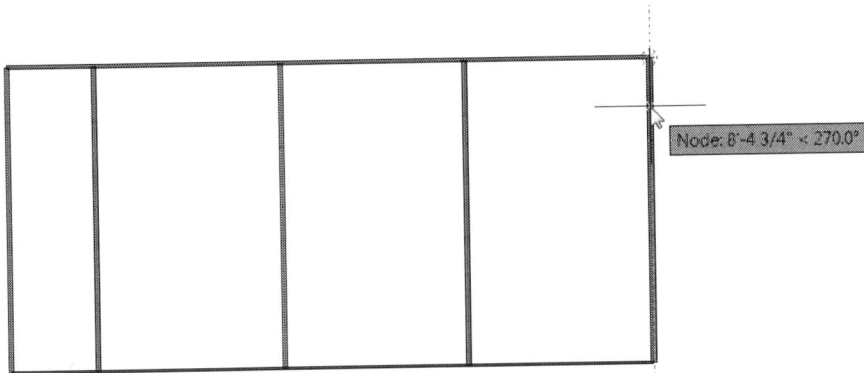

Figure 50 *Tracking the Node point to start the beam*

39. Once the tracking line is displayed, type **11'6"** or **3500** and press ENTER; a new beam starts from this point.

40. Move the cursor horizontally towards the left to track the 180-degrees angle.

41. With the tracking line displayed, snap to the **Intersection** point of the first filter beam on the left side, as shown in Figure 51; you are again prompted to locate the start point of the system axis.

42. Snap to the bottom right **Node** point and then move the cursor up to track the 90-degrees angle.

43. Once the tracking line is displayed, type **11'6"** or **3500** and press ENTER; a new beam starts from this point.

44. Move the cursor horizontally towards the left to track the 180-degrees angle.

45. With the tracking line displayed, snap to the **Intersection** point of the first filter beam on the left side, as shown in Figure 51; you are again prompted to locate the start point of the system axis.

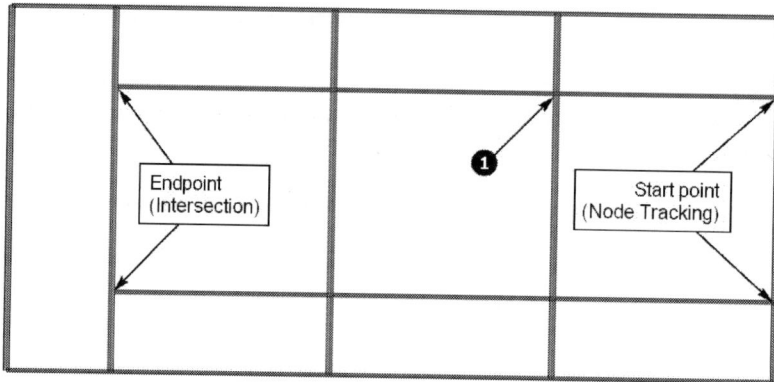

Figure 51   The start points and endpoints of the beams

The next beam will be inserted by tracking the intersection of the two beams labeled as **1** in Figure 51.

46. Snap to the **Intersection** point labeled as **1** in Figure 51 above and move the cursor to the left to track the 180-degrees angle.

47. Once the tracking line is displayed, type **8'2"** or **2500** and press the ENTER key; a new beam starts from this point.

48. Move the cursor vertically down to track the 270-degrees angle.

49. With the tracking line displayed, snap to the **Intersection** point of the first filter beam below, as shown in Figure 52; you are again prompted to locate the start point of the system axis.

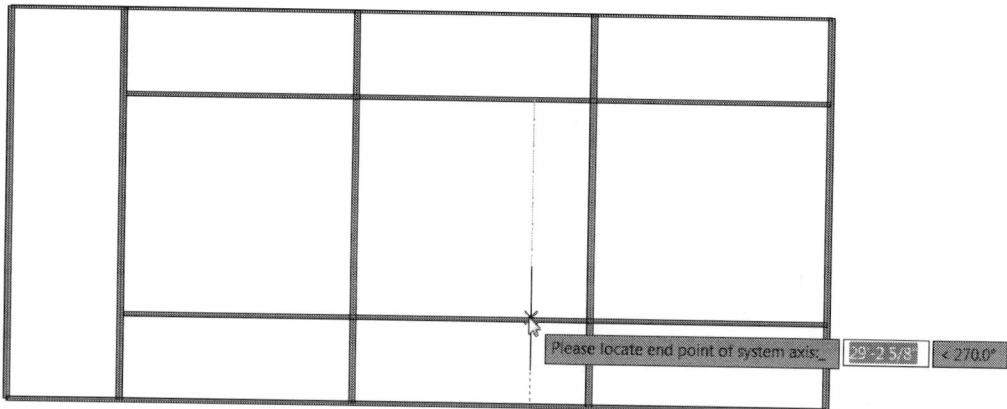

Figure 52   Specifying the endpoint of the beam

The next beam needs to be inserted by tracking the top **Node** point of the beam you just inserted. As shown in Figure 45, the total distance between the top filter beam and the bottom filter beam is 29'6" or 9000mm. Therefore, you need to track down half of this value to specify the start point of the new beam.

50. Snap to the top **Node** point of the beam you just inserted and then move the cursor down to track the 270-degrees angle.

51. Once the tracking line is displayed, type **14'9"** or **4500** and press ENTER; a new beam starts from this point.

52. Move the cursor horizontally towards the right to track the 0-degree angle.

53. With the tracking line displayed, snap to the **Intersection** point of the right perimeter beam; you are again prompted to locate the start point of the system axis.

    The next beam will be inserted using the same **Node** point that you used for the previous beam. However, this time you need to track down 7'4 1/2" or 2250mm.

54. Snap to the **Node** point labeled as **1** in Figure 53 and then move the cursor down to track the 270-degrees angle.

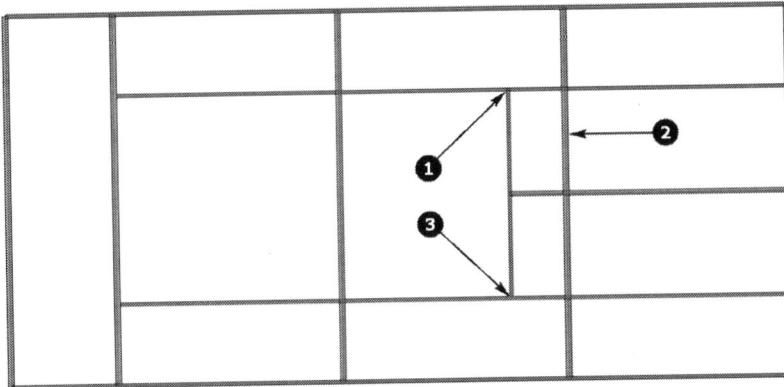

Figure 53 Locating the start and end points of the beams

55. Once the tracking line is displayed, type **7'4.5"** or **2250** and press ENTER; a new beam starts from this point.

56. Move the cursor horizontally towards the right to track the 0-degree angle.

57. With the tracking line displayed, snap to the **Intersection** point of the beam labeled as **2** in Figure 53; you are again prompted to locate the start point of the system axis.

58. Snap to the **Node** point labeled as **3** in Figure 53 and then move the cursor up to track the 90-degrees angle.

59. Once the tracking line is displayed, type **7'4.5"** or **2250** and press ENTER; a new beam starts from this point.

60. Move the cursor horizontally towards the right to track the 0-degree angle.

61. With the tracking line displayed, snap to the **Intersection** snap point of the beam labeled as **2** in Figure 53; you are again prompted to locate the start point of the system axis.

    Finally, you will place the beam by tracking up from the lower left **Node** point.

62. Snap to the lower left **Node** point and then move the cursor up to track the 90-degrees angle.

63. Once the tracking line is displayed, type **8'8"** or **2650** and press ENTER; a new beam starts from this point.

64. Move the cursor horizontally towards the right to track the 0-degree angle.

65. With the tracking line displayed, snap to the **Intersection** point of the beam on the right, as shown in Figure 54; you are again prompted to locate the start point of the system axis.

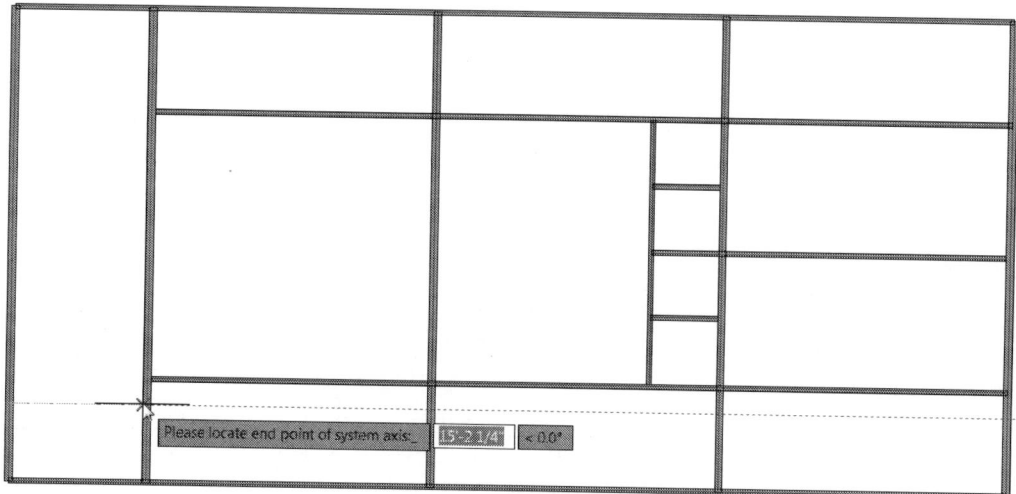

Figure 54 Specifying the endpoint of the beam

66. Press the ENTER key; the **Advance Steel - Beam [X]** dialog box is displayed.

67. From the **Section** list, select **AISC 14.1 W > W21x57** or **Australian Universal Beam > 530 UB 92.4**.

68. From the left pane of the dialog box, select the **Positioning** tab.

69. From the right pane, in the **Offset** area, select the top center radio button; the beam is aligned such that its system axis is aligned with the midpoint of the columns.

70. From the **Naming** tab > **Model Role** list, select **Beam** as the model role.

71. Close the dialog box.

72. Change the view to a 3D view with parallel projection. The model, after inserting all the beams, is shown in Figure 55.

    Next, you need to create four more copies of the last beam.

73. Using the **Copy** tool, create four more copies of the last beam with the spacing of **8'8"** or **2650**. The model after copying the beams is shown in Figure 56.

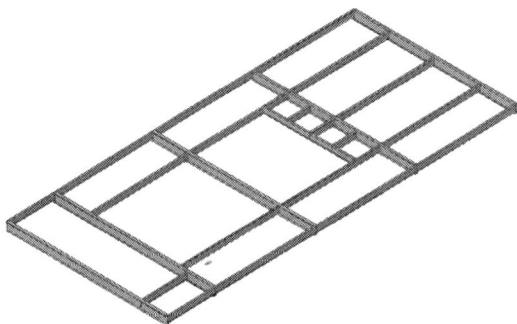

Figure 55 The 3D view of the model after inserting all the beams

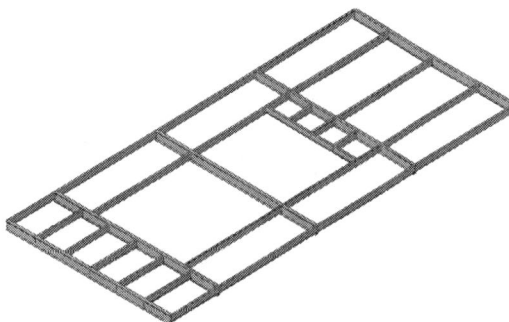

Figure 56 The model after copying the beams

Section 3: Splitting Beams

Some of the filter beams inserted in the previous section need to be split in order to place joints in later chapters. In Advance Steel, the beams are split using the **Split beams** tool.

1. From the **Home** ribbon tab > **Objects** ribbon panel, invoke the **Split beams** tool; you are prompted to select the beams to be split.

2. Select all the beams labeled as **1** in Figure 57 and then press the ENTER key; you are prompted to select split point or define the gap.

3. Select the **Intersection** snap points labeled as **2** in Figure 57 and then press the ENTER key; the five selected beams are split at the specified points.

Section 4: Inserting Sections at Level 2

The sections at level 2 are exactly similar to the ones at level 1, except at the front where the length of the beams is different. Therefore, you will copy the beams from level 1 to level 2 and then modify the length of the beams at the front. But first, you need to end the object isolation and turn the column visibility to standard.

1. Right-click in the blank area of the drawing window and select **Isolate > End Object Isolation** from the shortcut menu; the visibility of all the hidden objects is restored.

    Notice that the columns are still displayed as system axes. You will now restore the standard display of all the columns.

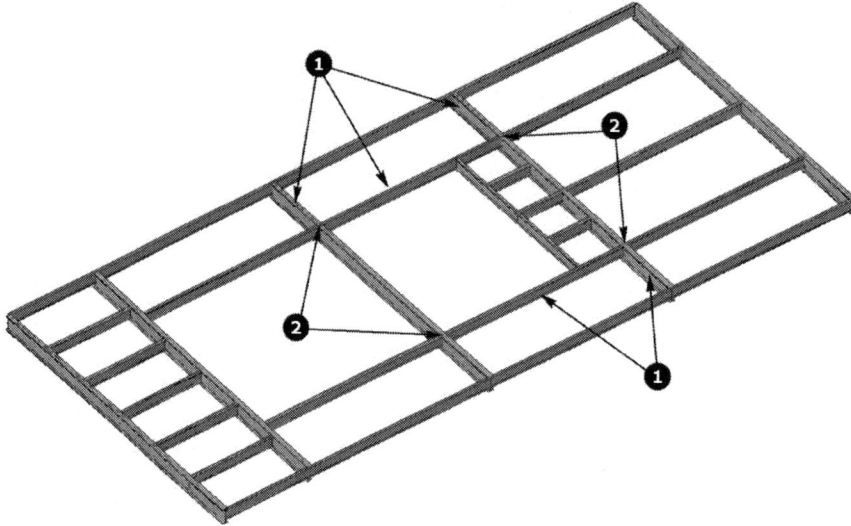

Figure 57 *The beams to be split and the split points*

2. Select the system axes representing all the columns.

3. Right-click in the blank area of the drawing window and select **Advance Properties** from the shortcut menu; the **Advance Steel - Beam [X]** dialog box is displayed.

4. Activate the **Display type** tab and select the **Standard** radio button; the standard display of the columns is restored.

5. Close the dialog box. The model after restoring the visibility is shown in Figure 58.

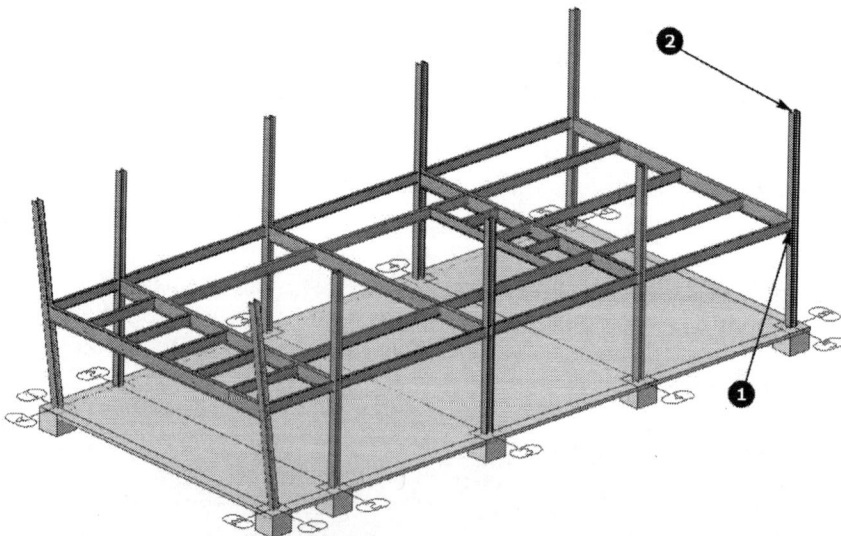

Figure 58 *The model after restoring the visibility of all the objects*

Next, you will copy all the beams at level 1 to level 2.

6. Using the **Copy** tool, copy all the beams from level 1 to level 2. Use the **Midpoint** labeled as **1** in Figure 58 as the base point and the **Node** point labeled as **2** as the second point.

    Once you copy the beams, you will notice that the front beam does not align with the inclined columns at the front, as shown in Figure 59.

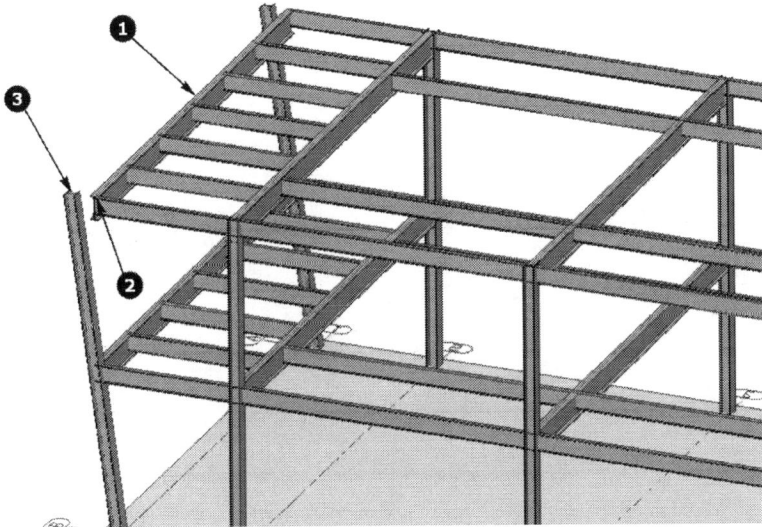

Figure 59 The beam to be selected to move and the reference points for the move

7. Using the **Move** tool, move the beam labeled **1** in Figure 59 with the **Node** point labeled **2** as the base point to the **Node** point labeled **3**.

    Next, you need to extend the seven beams at the front to the perimeter beam you moved in the previous step. This is done using the **Advance Trim/Extend** tool available on the **Tools** tab of the **Advance Steel** tool palette.

8. If the **Advance Steel** tool palette is not displayed in the drawing window, invoke it using the **Home** ribbon tab > **Extended Modeling** ribbon panel > **Advance Steel Tool Palette** tool.

9. From the left of this tool palette, invoke the **Tools** tab.

10. Scroll down in this tab and invoke the **Advance Trim/Extend** tool; you are prompted to select the operation mode.

11. Select **Extend** in the previous prompt sequence; you are prompted to select an option.

12. Select **System** from the previous prompt sequence; you are prompted to select the boundary objects.

13. Select the perimeter beam that you moved earlier in this section; you are again prompted to select objects.

14. Press the ENTER key; you are prompted to select members to be extended.

15. One by one select all the seven beams at the front and then press ENTER; all the beams are extended to the front beam.

16. Press the ENTER key to terminate the tool. The completed model is shown in Figure 60.

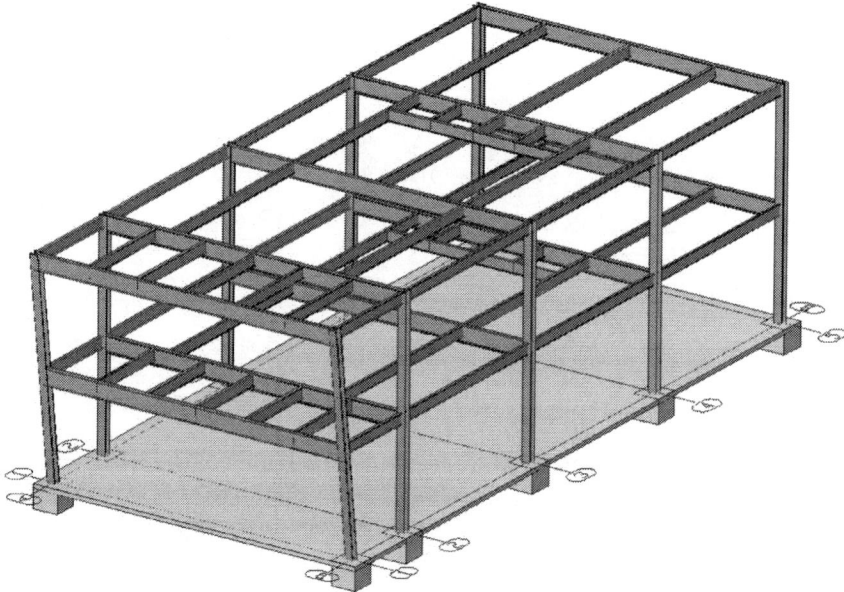

Figure 60 *The completed model of the tutorial*

17. Save the file and then close it.

Skill Evaluation

Evaluate your skills to see how many questions you can answer correctly. The answers to these questions are given at the end of the book.

1. Advance Steel only allows you to insert straight structural sections. (True/False)

2. While positioning the sections, you can also place them at their center of gravity. (True/False)

3. Symbolic display is the default display of structural sections. (True/False)

4. In Advance Steel, you can match properties of two sections. (True/False)

5. In Advance Steel, you can insert continuous sections. (True/False)

6. Which tool is used to insert a curved section?

   (A) **Curved beam**   (B) **Curved section**
   (C) **Arc beam**   (D) **Arc section**

7. Which tool is used to match properties of one section to the other?

   (A) **Sections**   (B) **Properties**
   (C) **Select similar**   (D) **Match Properties**

8. Which tab of the **Advance Steel - Beam [X]** dialog box is used to display various physical, geometric, and section properties of the section?

   (A) **Properties**   (B) **Physical**
   (C) **Sections**   (D) **Physical Properties**

9. Which tool is used to insert columns in Advance Steel?

   (A) **Beams/Columns**   (B) **Insert**
   (C) **Beam**   (D) **Column**

10. Which tab provides the option to place the section as a mirrored copy of itself?

    (A) **Positioning**   (B) **Naming**
    (C) **Mirror**   (D) **Orientation**

Class Test Questions

Answer the following questions:

1. Explain briefly the process of changing the display of the sections to system axis.

2. How will you isolate selected sections in the drawing window?

3. Explain briefly how will you edit the size of multiple sections together?

4. Explain briefly the process of splitting beams?

5. What is the process of matching properties of the sections?

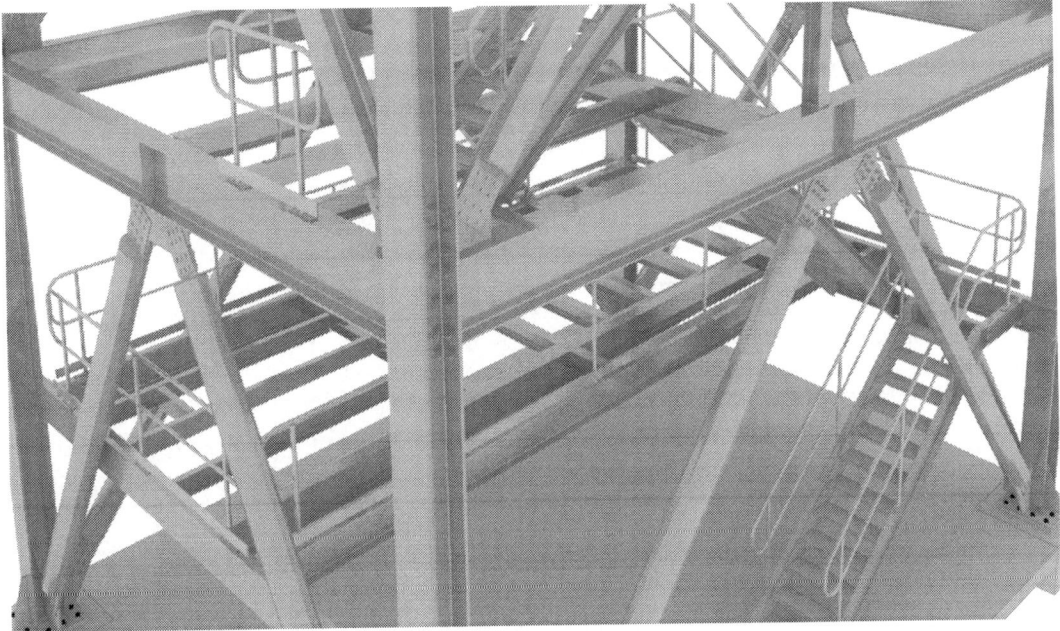

Chapter 3 – Advanced Structural Elements – I

The objectives of this chapter are to:

√   *Explain the process of inserting and editing portal and gable frames*
√   *Explain the process of inserting and editing mono-pitch frames*
√   *Explain the process of inserting and editing purlins*
√   *Explain the process of inserting and editing trusses*

ADVANCED STRUCTURAL ELEMENTS

Advanced structural elements are a combination of various structural sections placed using a single tool. These elements behave as a single unit and can be edited using a single dialog box. Advance Steel provides a number of tools to insert these advanced structural elements. These tools are discussed next.

INSERTING PORTAL/GABLE FRAMES

> **Home Ribbon Tab > Extended Modeling Ribbon Panel > Portal/Gable Frame**
> **Extended Modeling Ribbon Tab > Structural Elements Ribbon Panel >**
> **Portal/Gable Frame**

The **Portal/Gable Frame** tool allows you to insert a portal or a gable frame. Before you learn about creating these frames, it is important for you to understand various components of these frames. Figure 1 shows a portal frame. Various components of this frame are numbered and are explained below.

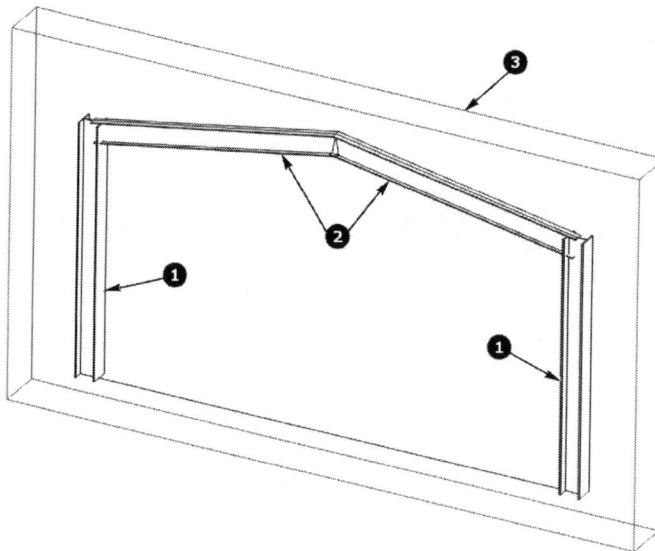

Figure 1 Portal frame and its components

Item 1: Columns
Item 2: Rafters
Item 3: Joint Box (If you delete this, the portal frame will be exploded into individual sections)

Figure 2 shows a gable frame. Various components of this frame are numbered and are explained below.

Item 1: Columns
Item 2: Rafters
Item 3: Gable Post

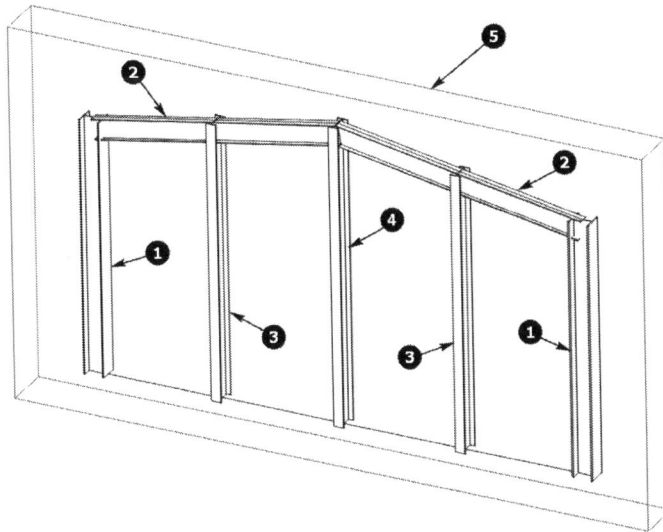

Figure 2 Gable frame and its components

Item 4: Center Gable Post
Item 5: Joint Box (If you delete this, the portal frame will be exploded into individual sections)

When you invoke this tool, you will be prompted to select the base point of the first column. On specifying this point, you will be prompted to select the top point of the second column. In this prompt, you can actually define the bottom endpoint of the second column, which in turn defines the width of the frame. Finally, you will be prompted to enter the point for the apex height or define the angle. The angle option is used to define the angle value between the rafters. Once you have specified this option, the **Advance Steel - Portal frame [X]** dialog box will be displayed. The same dialog box changes to **Advance Steel - Gable frame [X]** dialog box if you change the type of frame in the **Properties** tab to **Gable frame**. The options in **Properties**, **Set out**, **Sections**, and **Position of frame** tabs of both these dialog boxes are the same and are discussed next.

Properties Tab (Portal/Gable Frame)

As mentioned earlier, by default when you specify the three points, the **Advance Steel - Portal frame [X]** dialog box is displayed in the **Properties** tab, as shown in Figure 3. The options in this tab are discussed next.

Type

This list allows you to select the option to create a portal or a gable frame.

Name

This edit box allows you to enter the name of the portal or gable frame.

Tip: There is no **OK** or **Apply** button in the dialog boxes. As you edit the values, the changes are reflected live on the model because the **Automatic** tick box is selected. If this tick box is not selected, you will have to click **Update now** to view the changes.

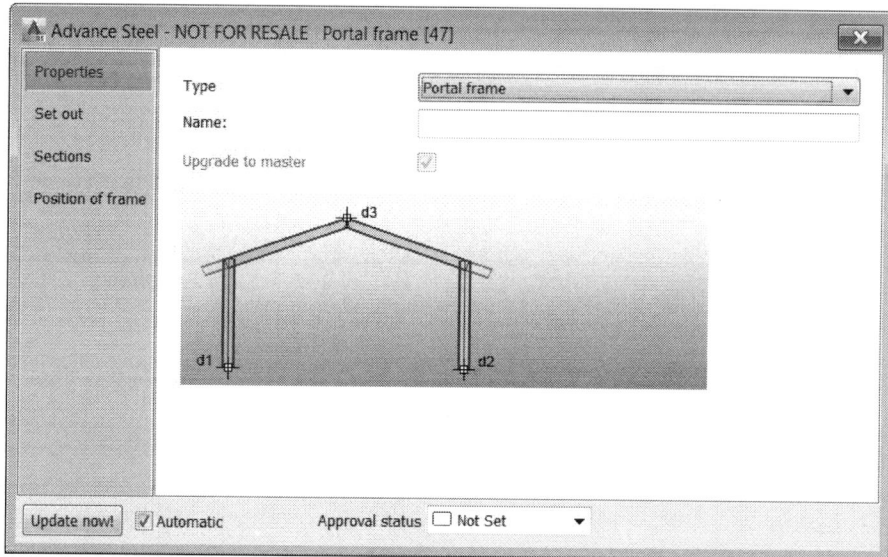

Figure 3 *The* **Properties** *tab of the* **Advance Steel - Portal frame [X]** *dialog box*

Upgrade to master

This tick box is grayed out while creating a new portal or gable frame.

Set out Tab (Portal/Gable Frame)

The **Set out** tab provides options to edit the size of the portal or gable frame, as shown in Figure 4. Notice that all these options are numbered and the preview image in this tab shows each of those numbered values. This makes it easy for you to understand what value you are changing. All these options are discussed next.

Symmetrical roof

By default, this tick box is selected. As a result, the eaves height and roof slope on both sides of the frame will be the same. Clear this tick box if the frame you are creating is not symmetrical.

Span of frame

This edit box is used to define the span of the frame. This value is automatically calculated from the second point you define while creating the frame. If that value was not correct, you can edit it here.

Total height

This value defines the total height of the frame from the bottom of the columns to the apex.

Eaves level 1

This value defines the height of the column 1. If you are creating a symmetrical roof frame, the same value will be used for the column 2 as well.

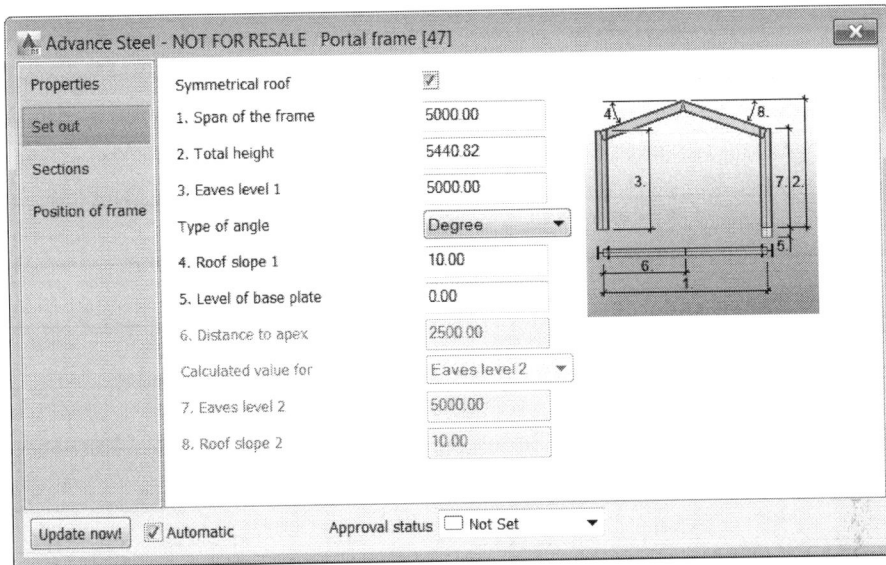

*Figure 4 The **Set out** tab of the **Advance Steel - Portal frame [X]** dialog box*

Type of angle
This list provides the following options:

Degree
This is the default option and is used when you want to define the slope of the roof in degrees.

Percent
This option is used when you want to define the slope of the roof in terms of percentage of the value.

Roof slope 1
This edit box is used to define the slope of the roof on side 1 of the frame. If you are creating a symmetrical roof frame, the same value will be used for side 2 also.

Level of base plate
If the portal or gable frame you are creating is to be mounted on a base plate, you can enter the thickness of that plate here. The two columns will then be shortened by that value at the bottom.

Distance to apex
This option will be active when the roof of the frame you are creating is asymmetric. In that case, this edit box will be used to specify the distance of apex from the first column.

Calculated value for
This option will also be active only when the roof of the frame you are creating is asymmetric. In that case, this list allows you to specify whether the second side of the frame will be defined by entering the height of the second column or the slope of the roof on that side.

Eaves level 2

This edit box is active when the roof of the frame you are creating is asymmetric and is used to define the height of the column 2.

Roof slope 1

This edit box is active when the roof of the frame you are creating is asymmetric and is used to define the slope of the roof on side 1 of the frame.

Sections Tab (Portal/Gable Frame)

The **Sections** tab shown in Figure 5 is used to define the sizes of the columns and rafters of the portal or gable frame. These options are discussed next.

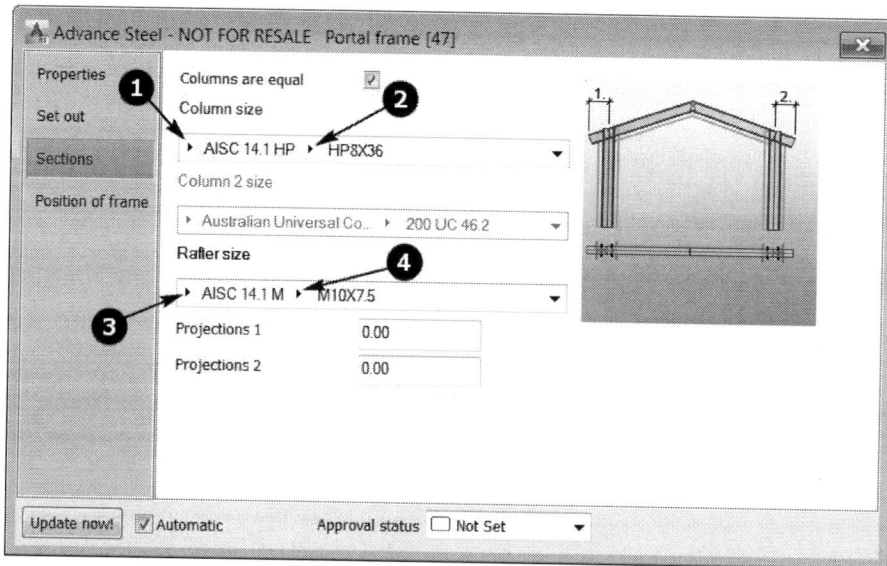

Figure 5 *The **Sections** tab of the **Advance Steel - Portal frame [X]** dialog box*

Columns are equal

This tick box is selected by default and is used to ensure the columns on both sides of the frame are equal.

Column size

This list is used to specify the type and size of the column. The first flyout in this list (marked as **1**) allows you to select the type of the column you want to use. The second flyout (marked as **2**) will show you all the sizes available for the selected column type.

Column 2 size

This list is only active when you clear the **Columns are equal** tick box and is used to specify the type and size of the second column.

Rafter size

This list is used to specify the type and size of the rafter. The first flyout in this list (marked as **3**) allows you to select the type of the rafter you want to use. The second flyout (marked as **2**) will show you all the sizes available for the selected rafter type.

Projections 1

This edit box is used to enter the value by which you want to overhang the rafter beyond the first column.

Projections 2

This edit box is used to enter the value by which you want to overhang the rafter beyond the second column.

Position of frame Tab (Portal/Gable Frame)

This tab is used to define the position of the columns and rafters with respect to the points used to define the frame, as shown in Figure 6. These options are discussed next.

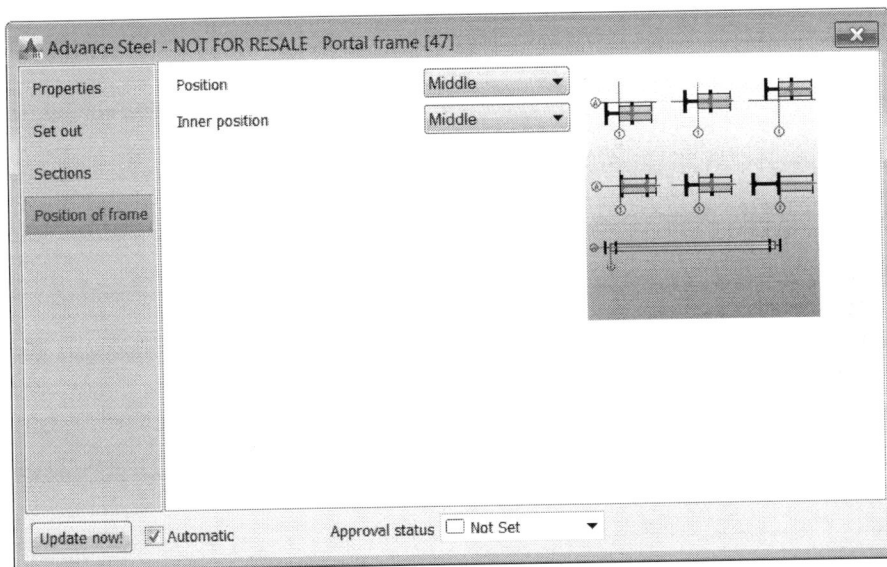

Figure 6 *The* ***Position of frame*** *tab of the* ***Advance Steel - Portal frame [X]*** *dialog box*

Position

This list is used to specify whether the columns and rafters of the frame will be placed on the left, right, or middle of the system axes of the frame.

Inner position

This list is used to specify whether the columns and rafters of the frame will be placed on the inside, outside, or middle of the system axes of the frame.

Gable sections Tab (Gable Frame)

This tab is only available while creating the gable frames. The options in this tab are used to define size of the gable posts, as shown in Figure 7. These options are discussed next.

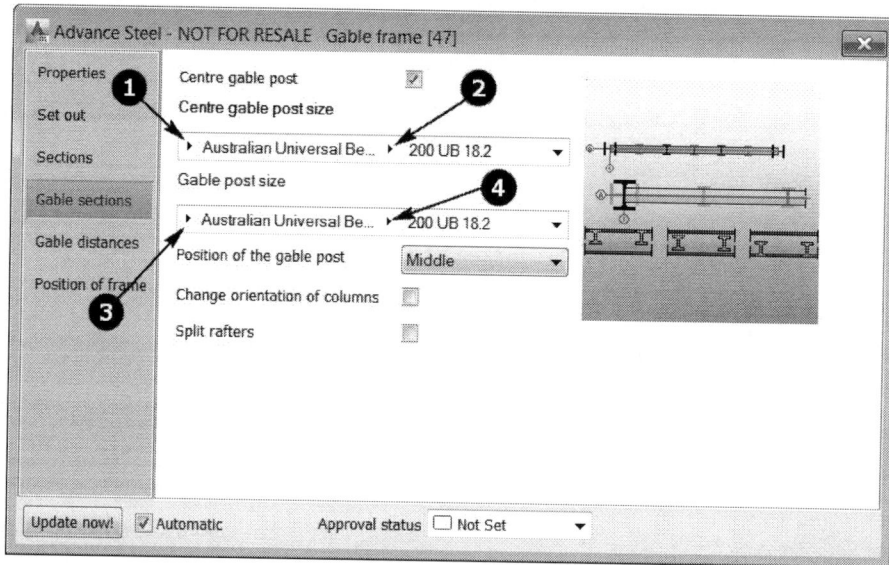

*Figure 7 The **Gable sections** tab of the **Advance Steel - Gable frame [X]** dialog box*

Center gable post

If this tick box is selected, a gable post is also placed at the center of the frame.

Center gable post size

This list allows you to select the type and size of the center gable post. The first flyout in this list (marked as **1**) is used to specify the type of the gable post section. The second flyout (marked as **2**) is used to specify the size of the section.

Gable post size

This list allows you to select the type and size of the rest of the gable posts. The first flyout in this list (marked as **3**) is used to specify the type of the gable post sections. The second flyout (marked as **4**) is used to specify the size of the sections.

Position of the gable post

This list allows you to specify the position of the gable posts. You can select the gable posts to be placed at the middle, inside flush with the columns, inside flush with the frame, outside flush with the column, or outside flush with the frame.

Change orientation of columns

This tick box is selected to rotate the columns by 90-degrees.

Split rafters

This tick box is selected to split the rafters at the point where the system axis of the gable posts intersect the rafters.

Gable distances Tab (Gable Frame)

This tab is only available while creating the gable frames. The options in this tab are used to define size of the gable posts, as shown in Figure 8. These options are discussed next.

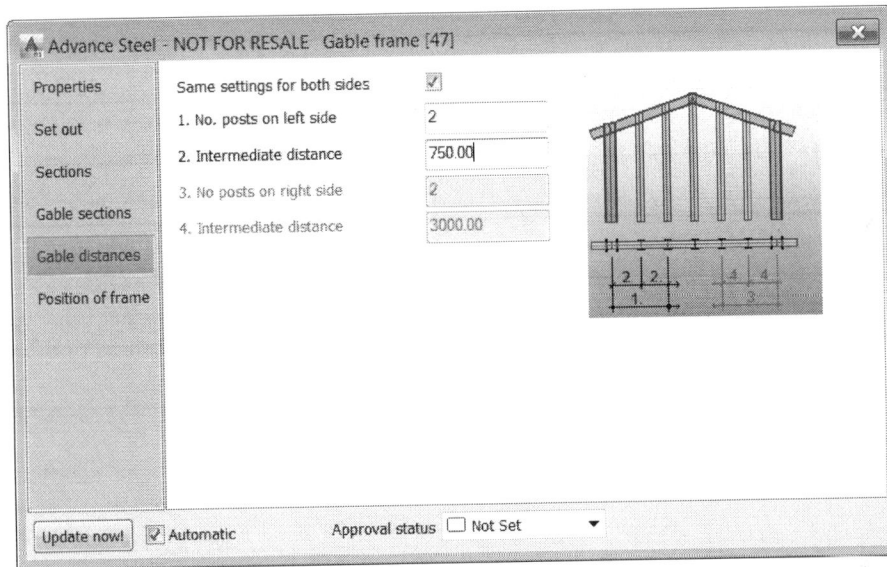

*Figure 8 The **Gable distances** tab of the **Advance Steel - Gable frame [X]** dialog box*

Same settings for both sides

This tick box is selected to specify the same distance settings on both sides of the gable frame.

No. of posts on left side

This edit box is used to specify the number of gable posts required on the left side of the frame.

Intermediate distance

This edit box is used to define the intermediate distance between the gable posts on the left side of the frame.

No. of posts on right side

This edit box is only active if the **Same settings for both sides** tick box is cleared and is used to specify the number of gable posts required on the right side of the frame.

Intermediate distance

This edit box is only active if the **Same settings for both sides** tick box is cleared and is used to define the intermediate distance between the gable posts on the right side of the frame.

INSERTING MONO-PITCH FRAMES

Home Ribbon Tab > Extended Modeling Ribbon Panel > Mono-pitch frame
Extended Modeling Ribbon Tab > Structural Elements Ribbon Panel >
Mono-pitch frame

Mono-pitch frames have a single rafter inserted between the two columns. This rafter can be a horizontal section, as shown in Figure 9, or at an angle, as shown in Figure 10.

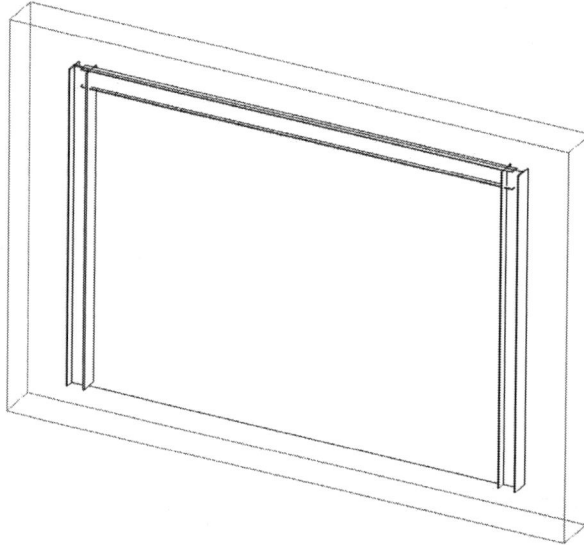

Figure 9 *Mono-pitch frame with horizontal rafter*

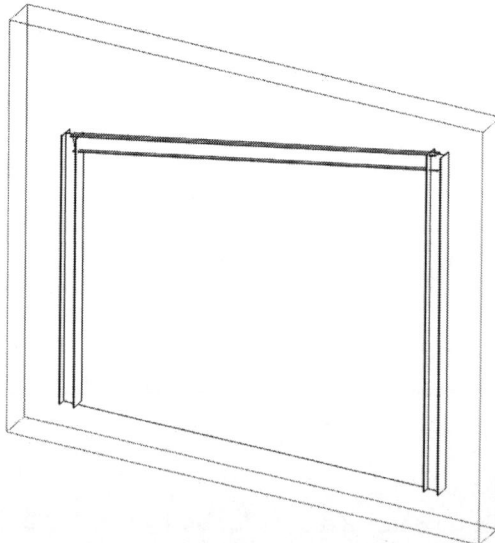

Figure 10 *Mono-pitch frame with the rafter at an angle*

When you invoke this tool, you will be prompted to select the base point of the first column. On specifying this point, you will be prompted to select the top point of the second column. In this prompt, you can actually define the bottom endpoint of the second column, which in turn defines the width of the frame. On specifying this point, the **Advance Steel - Portal frame [X]** dialog box will be displayed. The options in various tabs of this dialog box are the same as those discussed in portal/gable frames dialog boxes.

EDITING PORTAL/GABLE/MONO-PITCH FRAMES

It is important to note that you cannot double-click on any section of a frame to edit the frame. On doing so, the **Advance Steel - Beam** dialog box will be displayed to edit the selected beam, but all the options in that dialog box will be grayed out. To edit a frame, select one of its sections or the joint box. Next, right-click and select **Advance Joint Properties** from the shortcut menu, as shown in Figure 11.

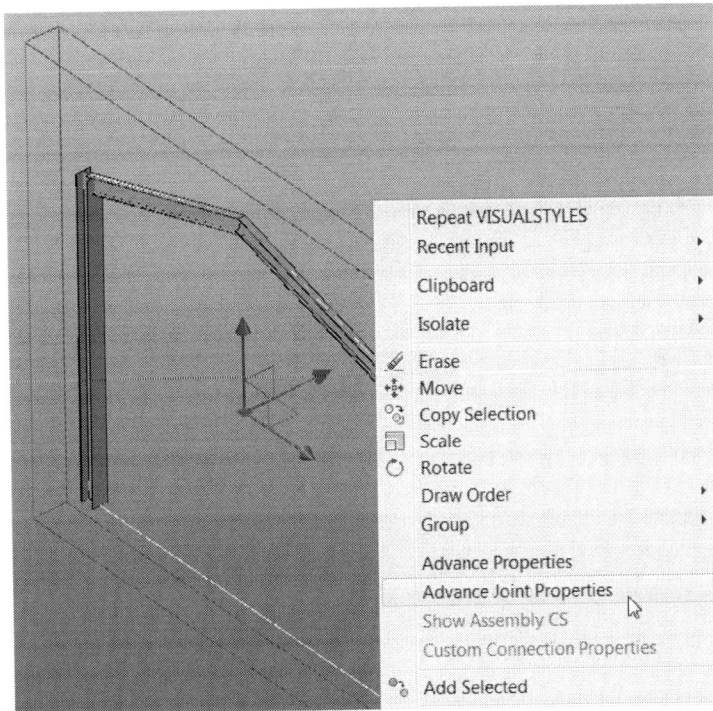

Figure 11 *Editing a frame*

On doing so, the **Advance Steel - Portal/Gable frame [X]** dialog box will be displayed. All the tabs and options in this dialog box are the same as those discussed while inserting these frames.

Note: As mentioned earlier, if you delete the frame joint box, the frame will be exploded into individual sections with no relationship with the other sections in the frame.

All the members of the frames are treated as beams and are placed on a layer called **Beams**.

INSERTING PURLINS

Home Ribbon Tab > Extended Modeling Ribbon Panel > Purlins
Extended Modeling Ribbon Tab > Structural Elements Ribbon Panel > Purlins

Purlins are secondary framing sections placed equi-spaced on top of the selected rafters. These sections are generally used to provide additional support for the roof deck. In Advance Steel, these sections are placed on a layer called **Cold Rolled**. Figure 12 shows four mono-pitch frames with purlins placed on top of their rafters. Note that the purlins will also have a joint box placed around them because all the purlins are treated as a group of sections. In this model, the display of the frame and purlin joint boxes is turned off using layers.

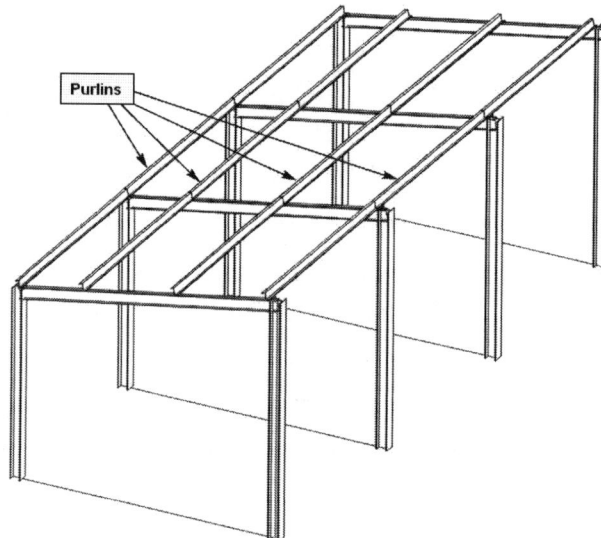

Figure 12 Purlins placed on top of the rafters of mono-pitch frames

When you invoke the **Purlins** tool, you will be prompted to select the supporting beams. In this prompt, select all the rafters on which you want to place the purlins and then press ENTER. Next, you will be prompted to specify if you want to select a reference point. This reference point can then be used to define the location of purlins. After you specify the option in this prompt, you will then be prompted to specify whether or not you want to select additional reference beam. Once you specify the option in this prompt, the **Advance Steel - Purlins [X]** dialog box will be displayed with the **Properties** tab active. The options in this tab are similar to those discussed while inserting frames. The remaining tabs of this dialog box are discussed next.

Sections Tab

The options in the **Sections** tab, shown in Figure 13, are used to specify the information related to purlin sections. These options are discussed next.

Section size

This list is used to specify the type and size of the purlin sections. The first flyout in this list allows

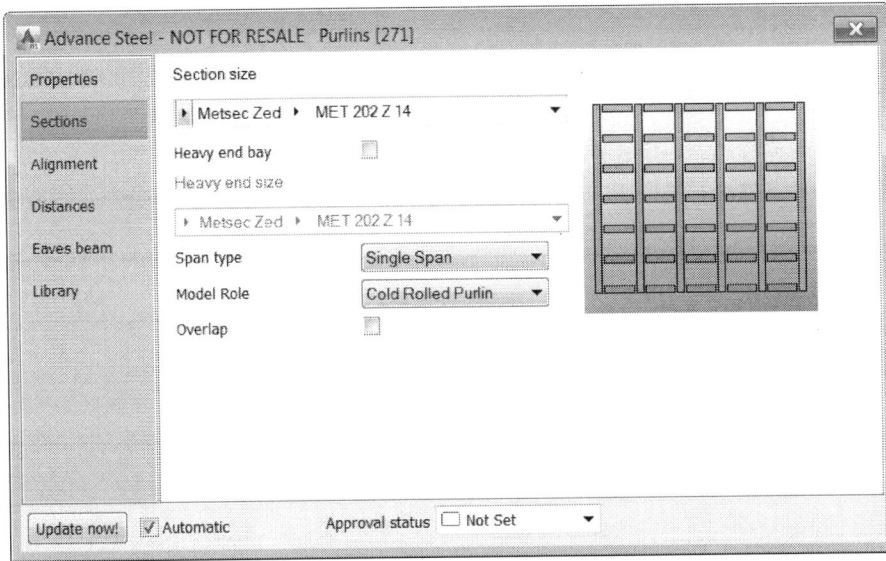

Figure 13 The **Sections** tab of the **Advance Steel - Purlins [X]** dialog box

you to select the type of the section you want to use. The second flyout will show you all the sizes available for the selected section type.

Heavy end bay

This tick box is selected if you want to insert the end purlin sections heavier than the middle sections. Figure 14 shows heavy end purlins sections placed on mono-pitch frames.

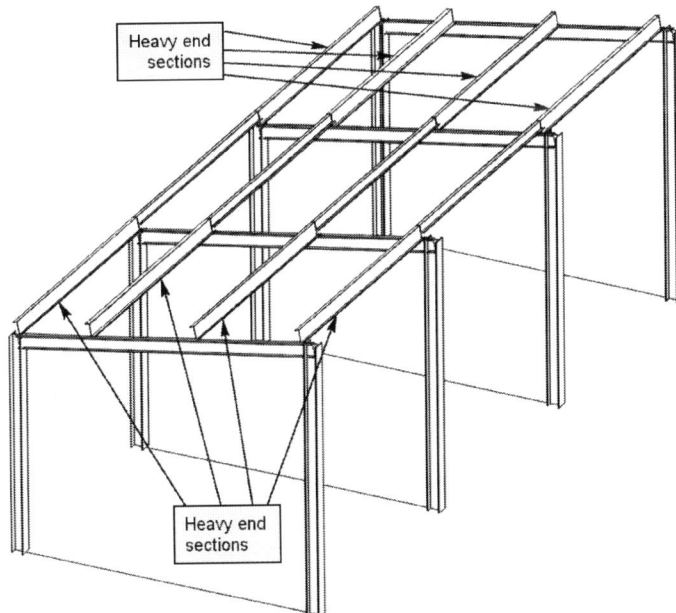

Figure 14 Heavy end purlins sections on top of mono-pitch frames

Heavy end size

This list is used to select the type and size of the heave end purlin sections. This list will only be active if you select the **Heavy end bay** tick box.

Span type

This list allows you to select the span type for purlins. By default, **Single span** is selected from this list. As a result, individual purlin sections are placed between all the selected rafters. You can also select options to place the purlin sections as right or left hand with swaps, right or left end single, or centered.

Model role

This list is used to define the model role of the purlin sections. By default, **Cold Rolled Purlin** is selected as the model role.

Overlap

Select this tick box if you want the purlin sections to overlap.

Alignment Tab

The options in the **Alignment** tab, shown in Figure 15, are used to specify the alignment of the purlin sections. These options are discussed next.

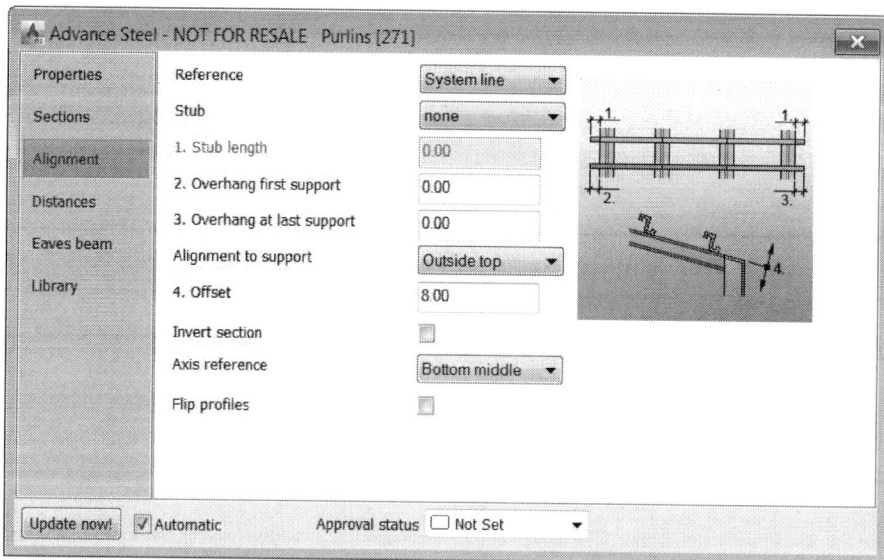

*Figure 15 The **Alignment** tab of the **Advance Steel - Purlins [X]** dialog box*

Reference

This list is used to define the reference for the placement of the purlins. By default, **System line** is selected. As a result, the purlins are aligned to the system axis of the end rafters. If you select the **Outer edge** option, the purlins will be extended to the outer edges of the end rafters.

Stub

This list allows you to specify if you want to insert additional purlin sections extending outside the end rafter sections. By default, the **None** option is selected. As a result, additional sections are not inserted at the ends.

Stub length

This edit box is only active if you are inserting stub sections at one or both ends and is used to define the length of the section. This option is listed as **1** in the preview window on the right.

Overhang first support

This edit box is used to extend the purlins at the first end to create an overhang. This option is listed as **2** in the preview window on the right.

Overhang at last support

This edit box is used to extend the purlins at the other end to create an overhang. This option is listed as **3** in the preview window on the right.

Alignment to support

This list is used to specify the position of the purlins with respect to the rafters selected. By default, the **Outside top** option is selected. As a result, the purlins are placed aligned to the outside top of the rafters. You can also position the purlins to the inside top, outside bottom, inside bottom, or at the center of the rafters.

Offset

This edit box is used to define an offset value between the rafters and purlins. This option is listed as **4** in the preview window on the right.

Invert section

Select this tick box to invert the purlin sections.

Axis reference

This list allows you to select where the system axis reference is located on the purlin sections.

Flip profiles

Select this tick box if you want to reverse the purlin section profiles.

Distances Tab

The options in the **Distances** tab, shown in Figure 16, are used to specify the distances between the purlin sections. These options are discussed next.

Distance reference

This list allows you to specify the location of the first purlin. The remaining purlins will then be defined from the first one using the values defined in the other edit boxes in this tab. The options available in this list are discussed next.

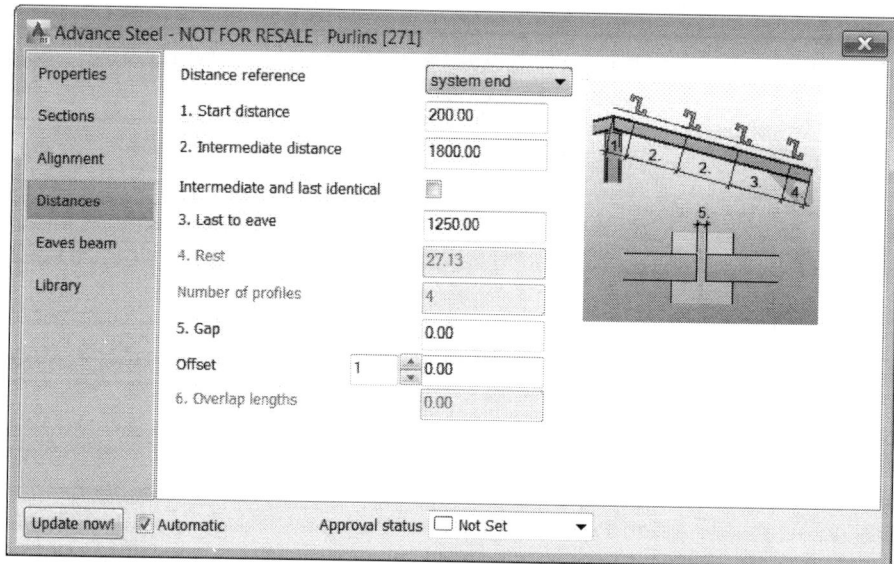

*Figure 16 The **Distances** tab of the **Advance Steel - Purlins [X]** dialog box*

system end
By default, this option is selected. As a result, the first purlin is placed at the end of the system axis of the rafter.

reference point
This option is used to place the first purlin at the reference point defined in the prompt sequence while placing the purlins.

physical end
This option is used to place the first purlin at the physical endpoint of the rafter. The physical end of the rafter is different from the system end if there is an end treatment defined at the end of the rafter, such as rafter shortening due to a connection.

centered on system
This option is used to place the purlins centered around the system axis of the rafters.

centered on physical
This option is used to place the purlins centered around the physical length of the rafters.

Start distance
This edit box allows you to define the start distance of the first purlin from the distance reference selected from the list above. This option is listed as **1** in the preview window on the right.

Intermediate distance
This edit box allows you to define the intermediate distance between the purlins. This option is listed as **2** in the preview window on the right.

Intermediate and last identical
Select this tick box if you want the distance between the second last and the last purlins to be the same as the intermediate distance.

Last to eave
This edit box allows you to enter the distance between the second last and last purlin. This box will only be active if the **Intermediate and last identical** tick box is not selected. This option is listed as **3** in the preview window on the right.

Rest
This edit box shows the distance between the last purlin and the end of the rafter.

Number of profiles
This edit box shows the total number of purlin profiles placed.

Gap
This edit box allows you to enter a gap value between the adjoining sections, as shown in Figure 17.

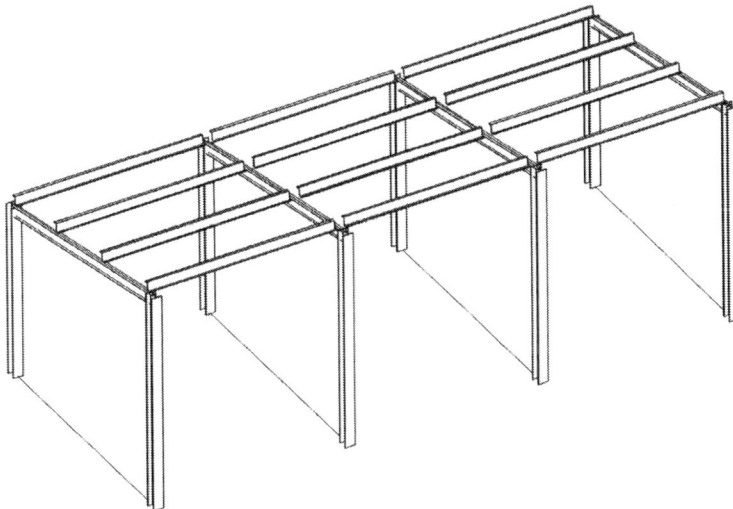

Figure 17 Gap defined between the adjoining purlins

Offset
This option provides a spinner and an edit box. The spinner is used to switch between the purlin sections and the edit box is used to define the offset for the selected purlin sections.

Overlap lengths
This edit box is only active if you have selected the **Overlap** tick box in the **Sections** tab and is used to define the overlap value.

Eaves beam Tab

The options in the **Eaves beam** tab, shown in Figure 18, are used to specify the options related to the section and size of the eaves beam. These options are discussed next.

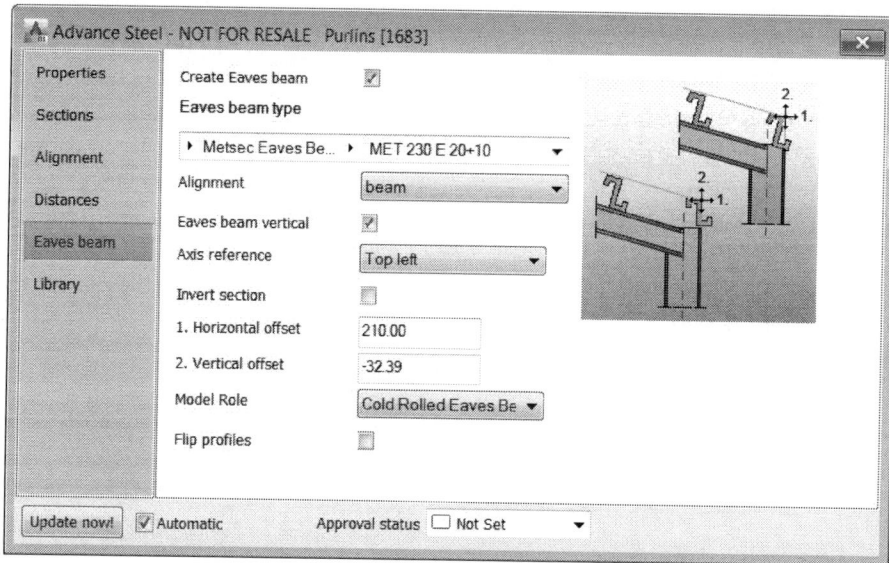

*Figure 18 The **Eaves beam** tab of the **Advance Steel - Purlins [X]** dialog box*

Create Eaves beam

Select this tick box if you want to create an eaves beam as well.

Eaves beam type

This list allows you to select the section type and size for the eaves beam.

Alignment

This list allows you to specify whether the eaves beam is aligned using the **beam** or the **sheeting line** option.

Eaves beam vertical

This tick box is only available if the eaves beam is aligned using the **beam** option. In this case, you can specify whether the eaves beam should be vertical or oriented as the purlins. Figure 19 shows the eaves beam created with this tick box selected and Figure 20 shows the eaves beam with this tick box cleared.

Invert section

Select this tick box to invert the eaves beam section.

Horizontal offset

This edit box is used to define the horizontal offset between the eaves beam and the end of the rafter.

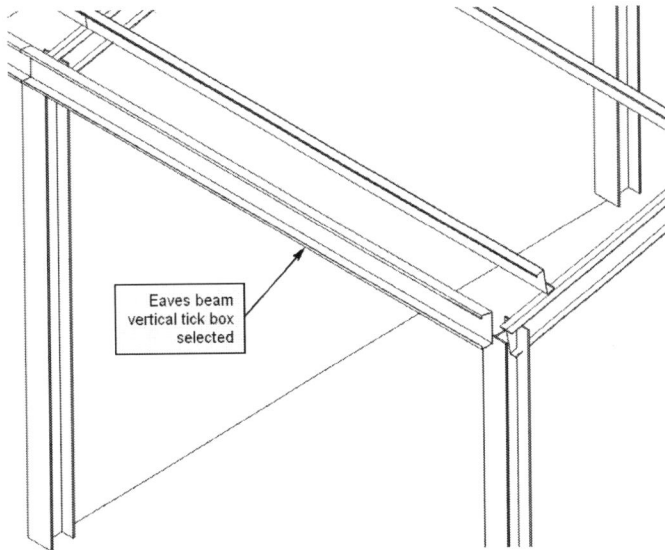

Figure 19 Eaves beam placed with the **Eaves beam vertical** tick box selected

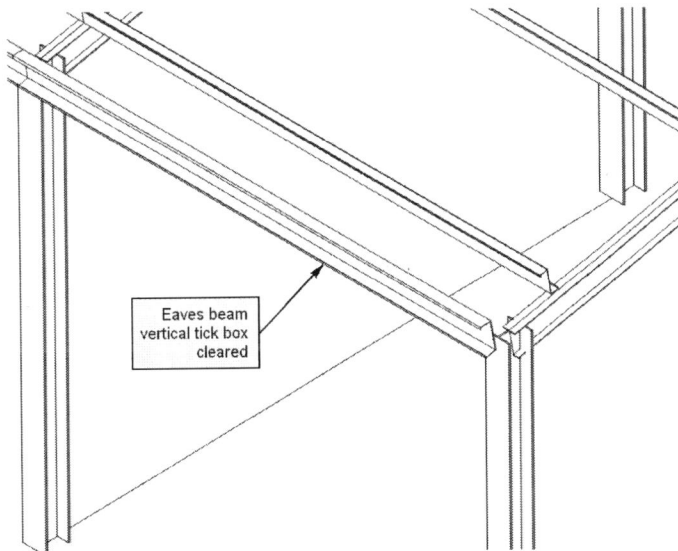

Figure 20 Eaves beam placed with the **Eaves beam vertical** tick box cleared

Vertical offset

This edit box is used to define the vertical offset between the eaves beam and the end of the rafter.

Model role

This list is used to define the model role of the eaves beam. By default, the **Cold Rolled Eaves Beam** is selected.

Flip profiles

Select this tick box to flip the profile of the eaves beam section.

Library Tab

The tab us used to add the defined purlin size and specifications into the library for future use. When you click this tab, the default sizes are listed on the right side, as shown in Figure 21.

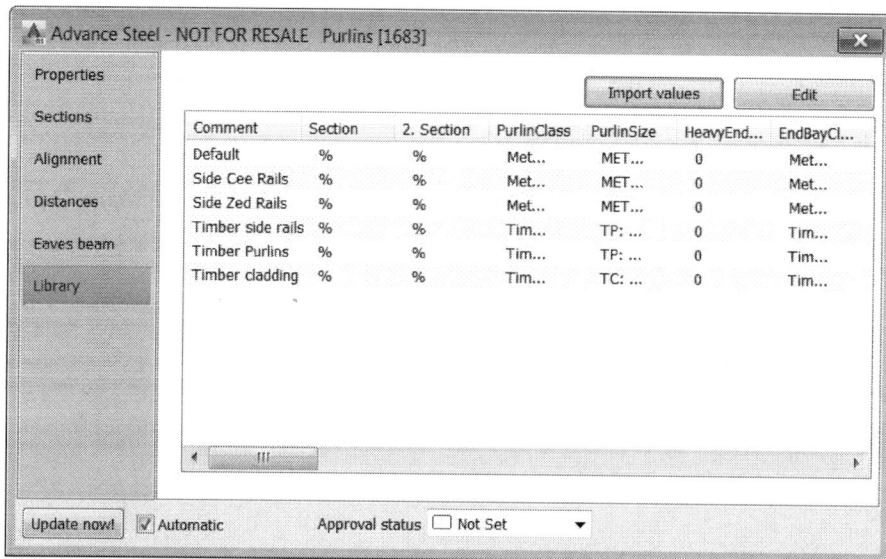

*Figure 21 The **Library** tab of the **Advance Steel - Purlins [X]** dialog box*

To add the sizes you have defined into the library, click the **Import values** button. Next, click the **Edit** button to display the **Table Process** dialog box, as shown in Figure 22.

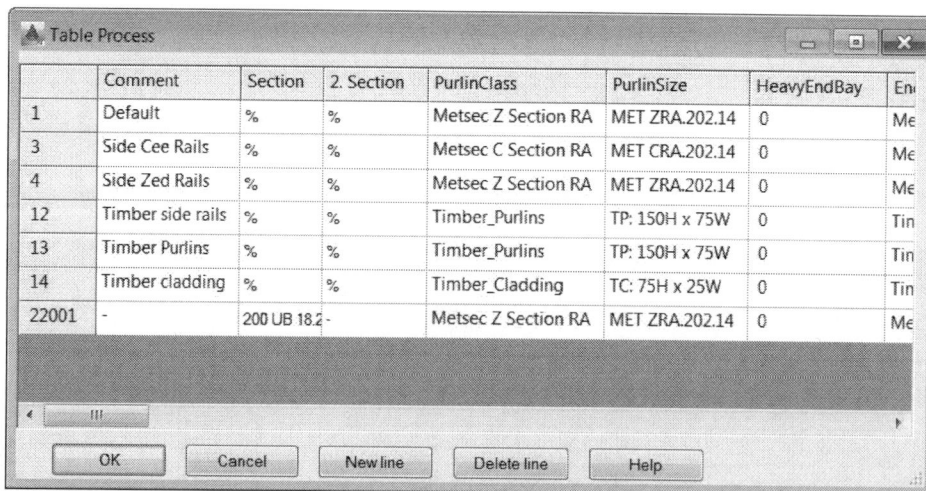

Figure 22 Editing the library values

In this dialog box, the row you have added will be listed as - in the **Comment** column. Click on that field and rename it to whatever name you prefer. Finally, click **OK** to add your purlin settings into the library for future use.

> **Tip**: *Next time you need to insert the purlins with the same specifications, define your selections in the prompt sequences and then click the **Library** tab of the **Advance Steel - Purlins [X]** dialog box. From this tab, click on the row that relates to your specifications and the purlins will be automatically updated with the values from the library.*

INSERTING TRUSSES

> **Home Ribbon Tab > Extended Modeling Ribbon Panel > Truss**
> **Extended Modeling Ribbon Tab > Structural Elements Ribbon Panel > Truss**

Trusses are a group of structural sections placed to support roofs of a building. These are also extensively used in bridge design as they allow effective distribution of loads. In Advance Steel, you can insert various types of trusses. For example, Figure 23 shows a straight truss aligned to the top of two columns and Figure 24 shows a truss with the curved bottom chord.

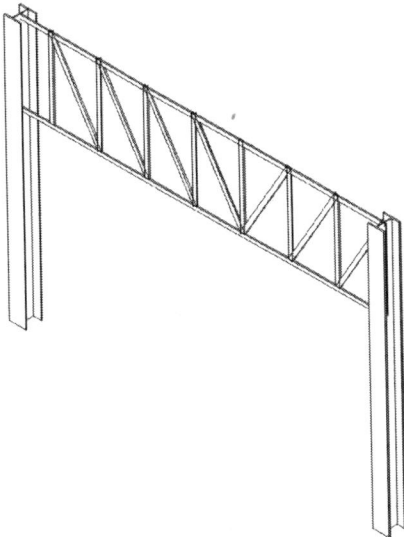

Figure 23 A straight truss placed at the top of two columns

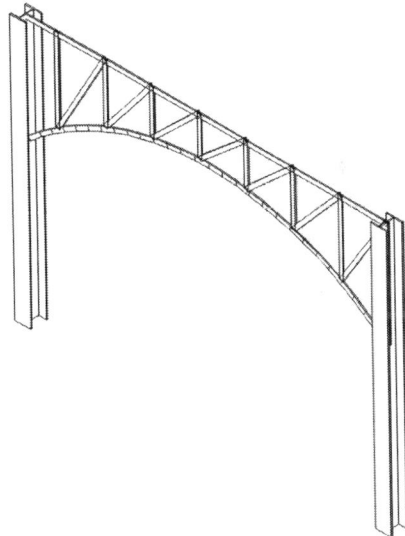

Figure 24 A truss with curved bottom chord placed at the top of two columns

When you invoke the **Truss** tool, you will be prompted to specify the start point and endpoint of the truss. These points are the two endpoints of the bottom chord of the truss. Also, note that the truss direction will depend on the +Z direction of the current UCS. Once you specify these two points, the **Advance Steel - Structural element truss [X]** dialog box will be displayed with the **Properties** tab active. The options in this tab are similar to those discussed while inserting frames. The remaining tabs of this dialog box are discussed next.

Geometry Tab

The options in **Geometry** tab, shown in Figure 25, are used to define the alignment and chord types of the truss. These options are discussed next.

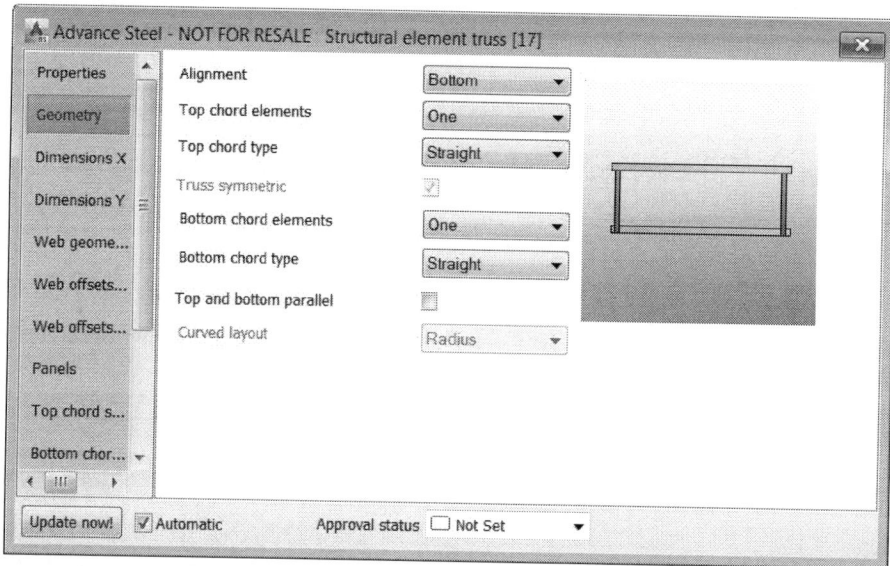

Figure 25 *The* **Geometry** *tab of the* **Advance Steel - Structural element truss [X]** *dialog box*

Alignment

This list is used to specify if the top or bottom of the truss will be aligned to the placement points you define while inserting the truss.

Top chord elements

This list is used to specify the number of the top chord elements. You can decide to have one, two, or three elements in the top chord.

Top chord type

This list is used to specify if the top chord will be straight or curved. Note that if you have defined more than one top chord elements, then you can only define straight top chord type.

Truss symmetric

This tick box is only active if you define two or three top chord elements. In that case, you can select this tick box to create a symmetric truss.

Bottom chord elements

This list is used to specify the number of the bottom chord elements. You can decide to have one element or same as top.

Bottom chord type

This list is used to specify if the bottom chord will be straight or curved. Note that if you have selected **Same as top** from the **Bottom chord elements** list, then you can only define straight bottom chord type.

Top and bottom parallel
Select this tick box to make the top and bottom sections of the truss parallel.

Curved layout
This list is used to specify whether the curves of the top or bottom chords will be defined using their radii or apex distances.

Dimension X Tab
The options in **Dimension X** tab, shown in Figure 26, are discussed next.

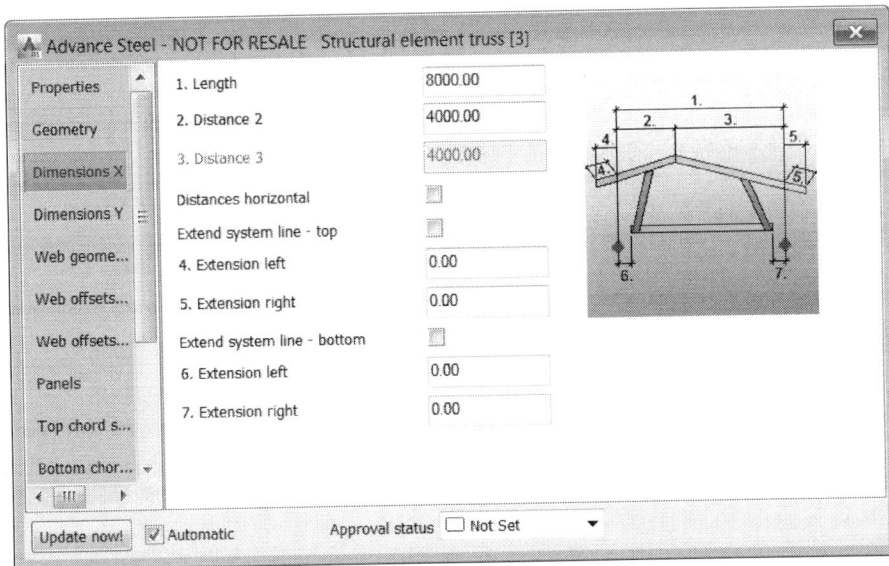

*Figure 26 The **Dimension X** tab of the **Advance Steel - Structural element truss [X]** dialog box*

Length
This edit box is used to define the length of the truss. This value is automatically populated based on the two points you specify while inserting the truss.

Distance 2
This edit box is only active if you select **Two** or **Three** as the value of the **Top chord elements** list on the **Geometry** tab. This value defines the location of the intersection point of the two top chords from the first placement point.

Distance 3
This edit box is only active if you select **Three** as the value of the **Top chord elements** list on the **Geometry** tab. This value defines the length of the middle chord element at the top.

Distances horizontal

This tick box is selected to define the distance values as horizontal.

Extend system line - top

If this tick box is selected and you define the left or right extension of the top chords, the extended chords will be placed above the placement points.

Extension left/Extension right

These edit boxes are used to specify the extension of top chords on the left and right. These values are listed as 4 and 5 in the preview window.

Extend system line - bottom

If this tick box is selected and you define the left or right extension of the bottom chords, the extended chords will be placed above the placement points.

Extension left/Extension right

These edit boxes are used to specify the extension of bottom chords on the left and right. These values are listed as 6 and 7 in the preview window.

Dimension Y Tab

The options in **Dimension Y** tab, shown in Figure 27, are discussed next.

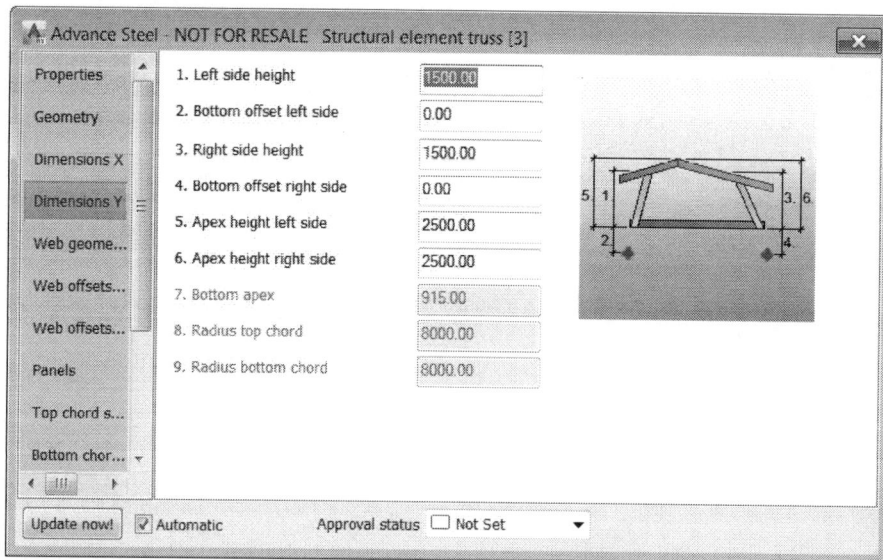

Figure 27 *The **Dimension Y** tab of the **Advance Steel - Structural element truss [X]** dialog box*

Left side height

This edit box is used to specify the height of the truss at the first placement point.

Bottom offset left side

This edit box is used to offset the truss at the first placement point.

Right side height

This edit box is used to specify the height of the truss at the second placement point.

Bottom offset left side

This edit box is used to offset the truss at the second placement point.

Apex height left side

This edit box is only active if you select **Two** or **Three** as the value of the **Top chord elements** list on the **Geometry** tab. This value defines the height of the apex where the first and second top chord elements intersect.

Apex height right side

This edit box is used if you select **Three** as the value of the **Top chord elements** list on the **Geometry** tab. This value defines the height of the apex where the second and third top chord elements intersect.

Bottom apex

This edit box is used to define the height of the apex at the bottom.

Radius top/bottom chord

These edit boxes are used to define the radius of the top and bottom chords.

Web geometry Tab

The options in **Web geometry** tab, shown in Figure 28, are discussed next.

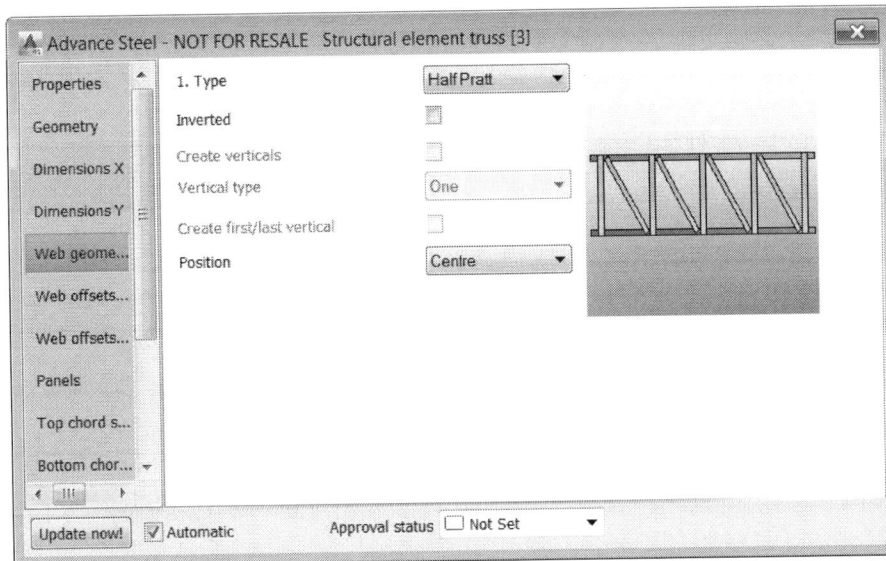

*Figure 28 The **Web geometry** tab of the **Advance Steel - Structural element truss [X]** dialog box*

Type

This list is used to specify the type of truss that will be created. Advance Steel allows you to create **Pratt**, **Half Pratt**, and **Warren** types of trusses. These three types are shown in Figures 29 through 31.

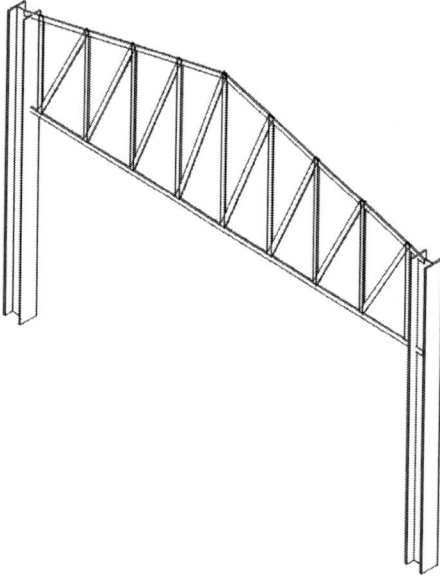

Figure 29 A *Half Pratt* type truss with two chords elements at the top

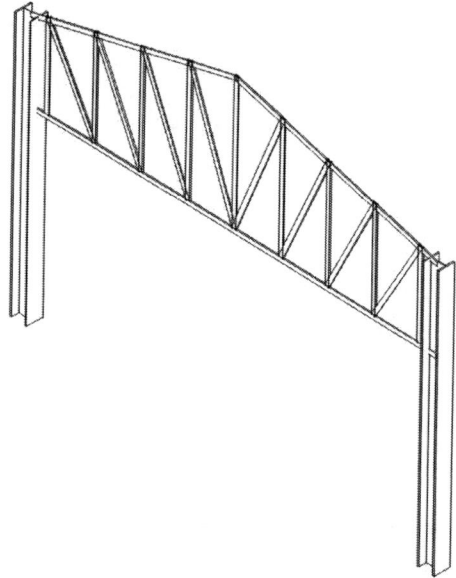

Figure 30 A *Pratt* type truss with two chords elements at the top

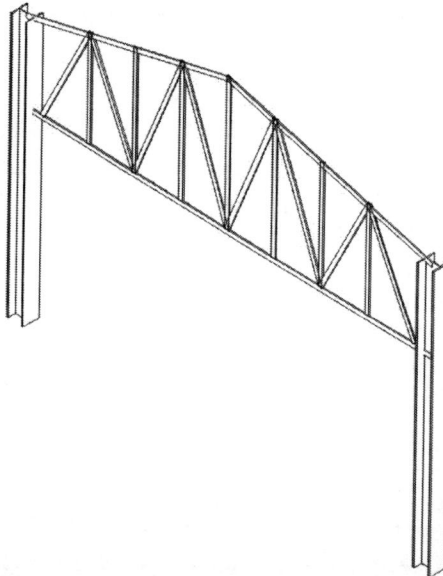

Figure 31 A *Warren* type truss with both vertical types created

Inverted

Select this tick box to invert all the web sections of the truss.

Create verticals

This tick box is only available for the **Warren** type trusses. Selecting this tick box will also insert vertical web sections.

Vertical type

This list is available only for the **Warren** type trusses and is used to specify whether the verticals should be created on the first, second, or both web sections.

Create first/last verticals

This list is also available only for the **Warren** type trusses and when the **Create verticals** tick box is selected. You can use this tick box to specify whether or not the first and last verticals will be created.

Position

This list is used to specify the position of the web sections. You can position them using the **Center**, **Flush side 1**, or **Flush side 2** options.

Web offsets - Top/Web offsets - Bottom Tabs

The options in **Web offsets - Top** and **Web offsets - Bottom** tabs provide options to define whether or not the web sections will be trimmed to match the shapes of the sections to which they are connecting. Figure 32 shows the **Web offsets - Top** tab. The **Web offsets - Bottom** tab also provides similar options, but for the bottom web sections. All these options are discussed next.

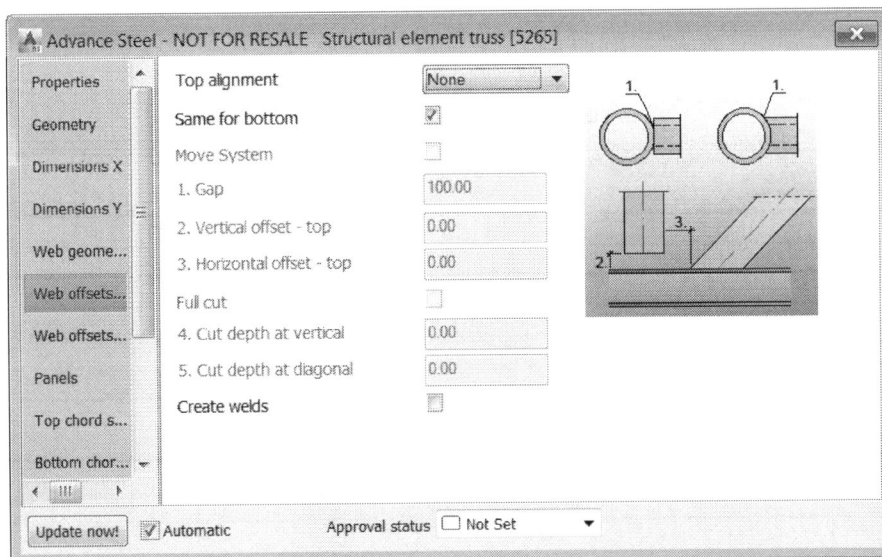

Figure 32 The Web offsets - Top tab of the Advance Steel - Structural element truss [X] dialog box

Top alignment

This list controls the options to cut the web sections. These options are discussed next.

None

If this option is selected, the web sections will not be cut.

Exact cut

If this option is selected, the web sections will be cut to exactly match the shape of the adjacent sections. With this option selected, you will also be able to specify the gap and offset values.

Aligned

If this option is selected, the web sections will be cut and aligned to the adjacent sections.

Same for bottom

Selecting this tick box will apply the top cut settings to the bottom cuts as well.

Move System

Selecting this tick box will move all the web sections in order to reduce the number of cuts required in the section. Figure 33 shows the web sections without this tick box selected and Figure 34 shows the same sections after selecting this tick box.

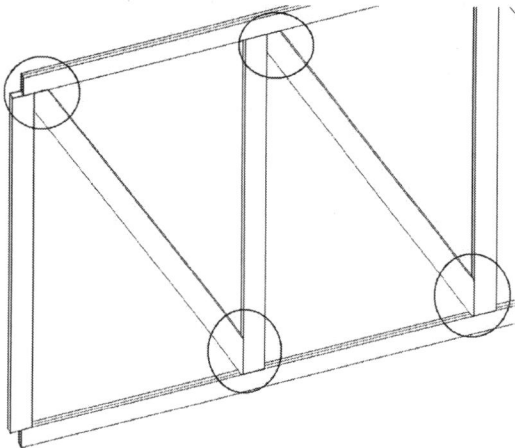

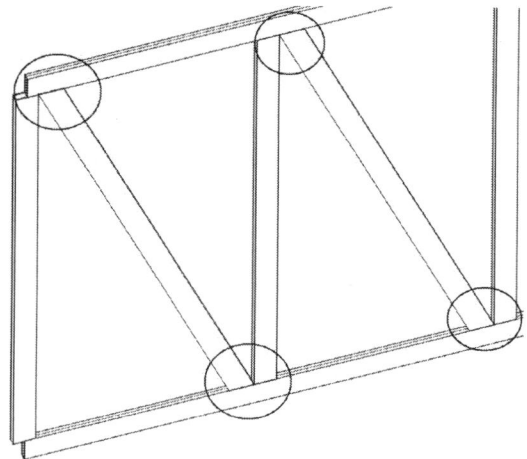

Figure 33 Web sections before moving the systems   *Figure 34 Web sections after moving the systems*

Gap

This edit box is used to define a gap between the web sections and the top and bottom chords. Figure 35 shows web sections with gap value defined and **Move System** tick box cleared. Figure 36 shows the same web sections and same gap value, but the **Move System** tick box selected.

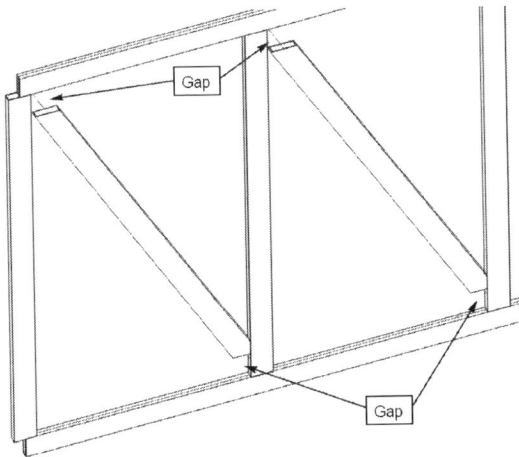

Figure 35 Gap value defined and the **Move System** tick box cleared

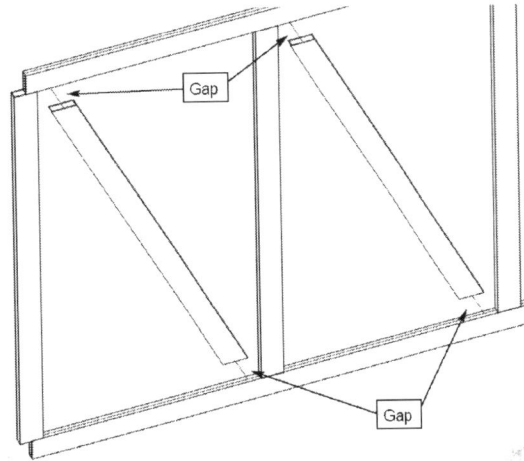

Figure 36 Gap value defined and the **Move System** tick box selected

Vertical offset - top

This edit box is used to define an offset between the vertical web sections and the top chord. If the **Same for bottom** tick box is selected, the same value will also be applied as an offset from the bottom chord. Figure 37 shows the vertical web sections with vertical offset from the top and bottom chords.

Horizontal offset - top

This edit box is used to define an offset at the top between the diagonal web sections and the vertical web sections. If the **Same for bottom** tick box is selected, the same value will also be applied as an offset at the bottom. Figure 38 shows the diagonal web sections with offsets from the vertical web sections.

Full cut

This tick box is only available if you select **Exact cut** from the **Top alignment** list. If this tick box is selected, the web sections will be created with full cuts based on the values defined above. You can clear this tick box to override the cuts at diagonals and verticals using the edit boxes available below this tick box.

Create welds

Select this tick box if you want to create welds between various web sections and top and bottom chords.

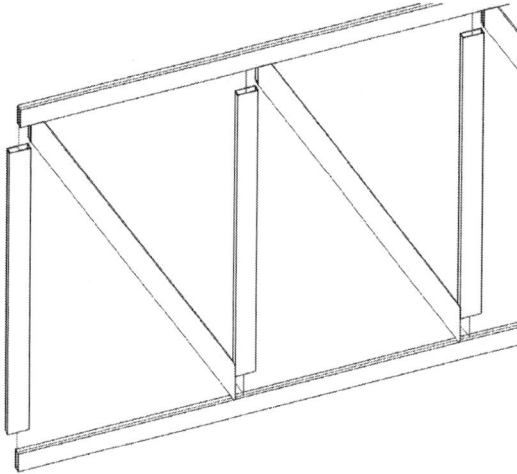

Figure 37 *Vertical web sections with offsets from the top and bottom chords*

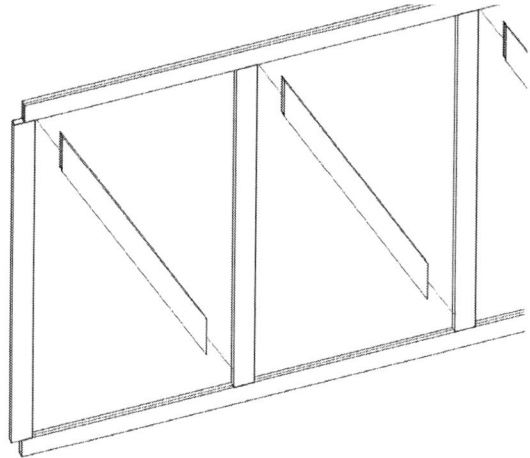

Figure 38 *Diagonal web sections with offsets from the vertical web sections*

Panels Tab

The **Panels** tab, shown in Figure 39, is used to specify the panel layout and the spacing between the panels. These options are discussed next.

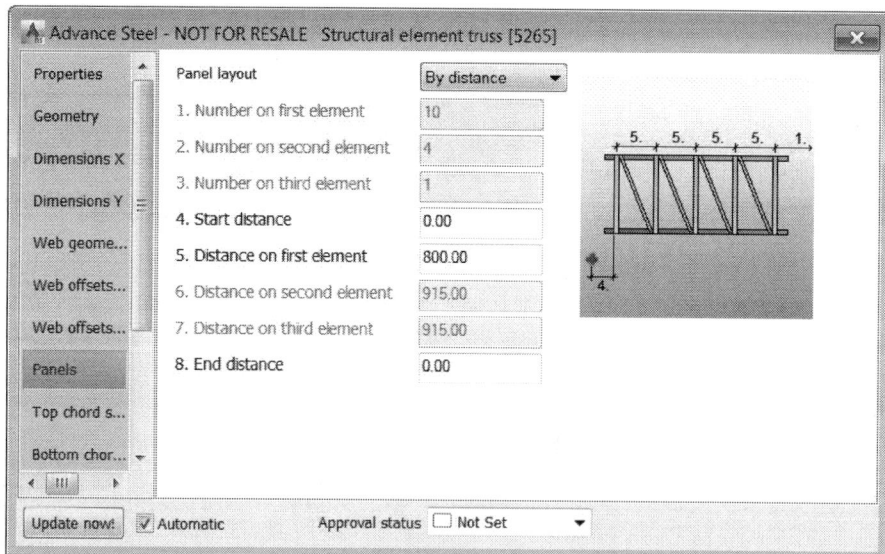

Figure 39 *The **Panels** tab of the **Advance Steel - Structural element truss [X]** dialog box*

Panel layout
This list allows you to select whether you want to specify the panel layout based on the distance between the panels or the total number of panels.

Number on first element
This edit box is only available if you are defining the panel layout by number and is used to specify the number of panels on the first chord element. If you have selected **One** as the option from the **Top chord elements** list on the **Geometry** tab, the value defined in this edit box will be considered as the total number of panels from start to end.

Number on second element
This edit box is only available if you have selected **Two** as the option from the **Top chord elements** list on the **Geometry** tab. The value defined in this edit box will define the number of panels on the second chord element. Figure 40 shows eight panels on the first chord element and four panels on the second.

Number on third element
This edit box is only available if you have selected **Three** as the option from the **Top chord elements** list on the **Geometry** tab. The value defined in this edit box will define the number of panels on the third chord element. Figure 41 shows four panels on the first and third chord elements and two panels on the second.

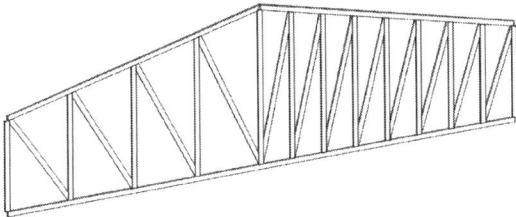

Figure 40 A truss with eight panels on the first chord element and four panels on the second

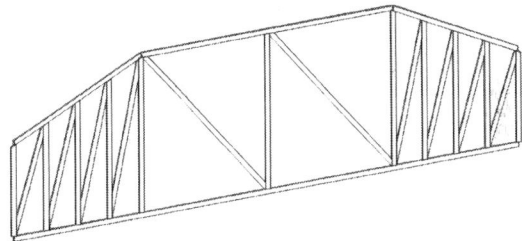

Figure 41 A truss with four panels on the first and third chord element and two panels on the second

Start distance
This edit box is used to specify a gap between the start of the truss and the first panel.

Distance on first element
This edit box is only available if you are defining the panel layout by distance and is used to define the spacing between the panels on the first chord element. If you have selected **One** as the option from the **Top chord elements** list on the **Geometry** tab, all the panels in the truss will be equally spaced using the value defined in this edit box.

Distance on second element

This edit box is only available if you have selected **Two** as the option from the **Top chord elements** list on the **Geometry** tab. The value defined in this edit box will define the spacing between the panels on the second chord element.

Distance on third element

This edit box is only available if you have selected **Three** as the option from the **Top chord elements** list on the **Geometry** tab. The value defined in this edit box will define the spacing between the panels on the third chord element.

Top chord section/Bottom chord section Tabs

The options in **Top chord section** and **Bottom chord section** tabs are used to define the section types and sizes for the top and bottom chords. Figure 42 shows the options in the **Top chord section** tab. The **Bottom chord section** tab provides similar options. These options are discussed next.

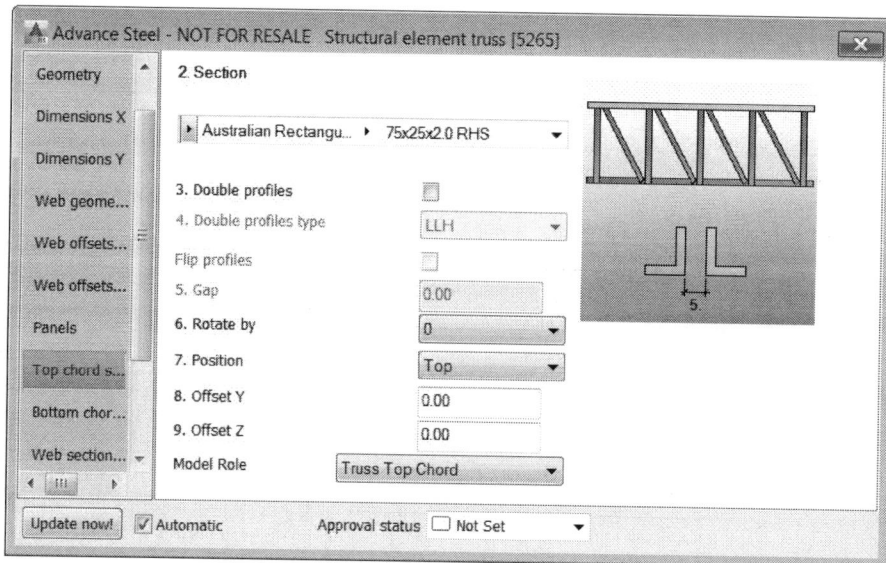

*Figure 42 The **Top chord section** tab of the **Advance Steel - Structural element truss [X]** dialog box*

Section

The two flyouts in this list are used to select the section type and the section size for the top chord. If the **Same as top chord** tick box is selected in the **Bottom chord section** tab, the section type and size selected from this list will also be defined for the bottom chord.

Double profiles

Selecting this tick box will place double profiles for the top and bottom chord.

Double profiles type

This list is generally used when you select unequal angles for the top and bottom chord profiles. Selecting the **LLH** option places the unequal angles such that the longer legs are horizontal. Selecting the **LLV** option will make the longer legs vertical.

Flip profiles

This tick box is only available when you are placing double profiles and is selected to flip the orientation of two chord sections.

Gap

This edit box is only available when you are placing double profiles and is used to specify a gap between the two chord sections.

Rotate by

This list is only available when you are placing a single chord section and is used to rotate that section.

Position

This list is used to specify whether the system axis is defined at the top, bottom, or middle of the chord sections. The system axis will in turn control the alignment of the chord sections with web sections.

Offset Y

This edit box is used to define the offset between the chord section and its system line along the Y direction of the truss.

Offset Z

This edit box is used to define the offset between the chord section and its system line along the Z direction of the truss.

Model Role

This list is used to define the model role of the top truss chord. By default, **Truss Top Chord** is selected from this list. Remember that even though you may have selected the **Same as top chord** tick box is in the **Bottom chord section** tab, the role of the bottom chord is by default set to **Truss Bottom Chord**.

Web section diagonals/Web section verticals Tabs

The options in **Web section diagonals** tab, shown in Figure 43, and **Web section verticals** tab are used to define the section types and sizes for the diagonal and vertical web sections. As you can see, these options are similar to those discussed in the previous tab. Therefore, these options do not need to be explained again.

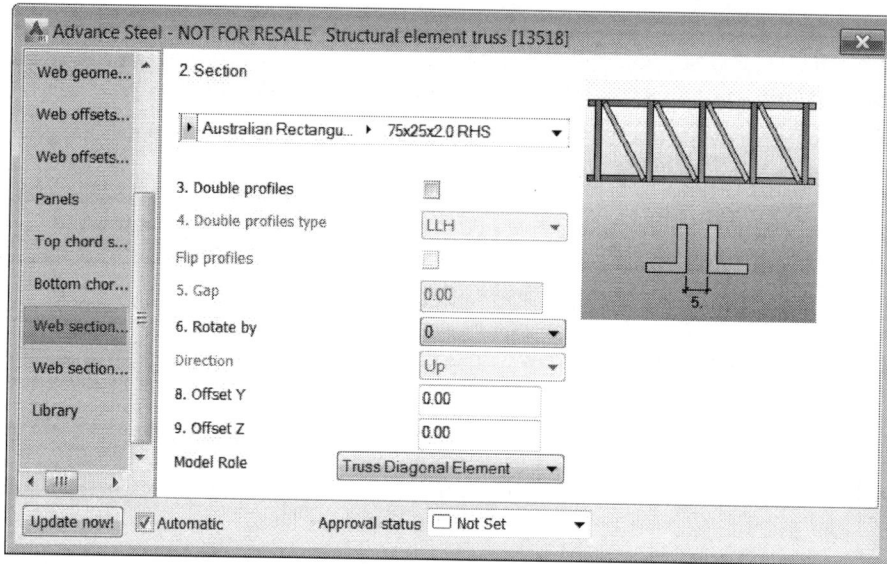

Figure 43 *The* **Top chord section** *tab of the* **Advance Steel - Structural element truss [X]** *dialog box*

Library Tab

The tab is used to add the defined purlin size and specifications into the library for future use. These options are similar to those discussed for purlins.

<table>
<tr><td>

Hands-on Tutorial (STRUC/ BIM)

</td><td>

In this tutorial, you will complete the following tasks:
1. *Open the **C03-Imperial.dwg** or **C03-Metric.dwg** file.*
2. *Insert the gable frame at the front and back, as shown in Figure 44.*
3. *Copy the gable frame and change it into the portal frame.*
4. *Create two additional instances of the portal frame using the **Copy** tool.*
5. *Add purlins to the frame, as shown in Figure 44.*

</td></tr>
</table>

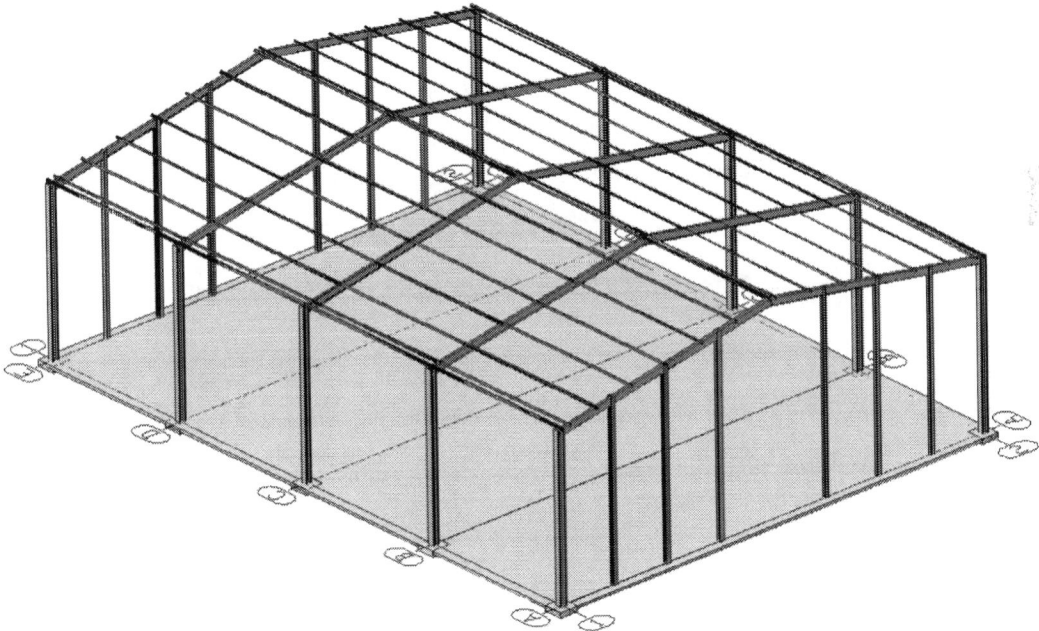

Figure 44   Frame for the tutorial

Frame Sizes

The sections used in this tutorial are **Australian Standard (AS)** for Metric or **AISC** standard for Imperial. However, you are free to use your country standard sections.

Columns: AISC 14.1 W > W14x500 / Australian Universal Column > 310 UC 158
Rafters: AISC 14.1 W > W21x201 / Australian Universal Beam > 460 UB 67.1
Gable Post Size: AISC 14.1 W > W10x60 / Australian Universal Column > 150 UC 37.2
Span of Frame: 120'/36500mm
Total Height: 50'/15250mm
Eaves Level 1: 40'/12000mm
Intermediate Gable Section Distance: 15'/4500mm

Purlin Sizes

The following are the purlin sizes for this tutorial:

Purlin Sections: Ayrshire Zed > AYR Z/15513
Eaves Beam Section: Albion Eaves Beam > ALB EB 20018 - 9
Overhang First and Last Supports: 2'/600mm
Start Distance: 7 7/8"/300mm
Intermediate Distance: 10'/3000mm
Last to Eave: 8'/2500mm
Overlap Length: 2'/600

Section 1: Opening the File and Inserting Gable Frames

In this section, you will open **C03-Imperial.dwg** or **C03-Metric.dwg** file and then add a gable frame at the front. You will then use the **Copy** tool to copy this frame at the back.

1. From the **C03 > Struc-BIM** folder, open the **C03-Imperial.dwg** or **C03-Metric.dwg** file, depending on the units you want to use.

    Notice that this file already has the four isolated footings, slab, and grid created. You will now use the **Portal/Gable Frame** tool to insert a gable frame at the front.

2. From the **Home** ribbon tab > **Extended Modeling** ribbon panel, invoke the **Portal/Gable Frame** tool; you are prompted select the base point for the first column.

3. Select the **A1** grid point as the base point for the first column; you are prompted to select the top point for the second column.

4. Select the **A2** grid point; you are prompted to enter the point for the apex height to define the angle.

5. Move the cursor up so that you are tracking along the +Z direction.

6. Once the tracking line appears, type **50'/15250** as the value and then press ENTER; the **Advance Steel - Portal Frame [X]** dialog box is displayed with the **Properties** tab active.

7. From the **Type** list, select **Gable frame**; the gable sections are added to the frame.

8. Activate the **Set out** tab of the dialog box.

9. In the right pane, enter **40'/12000** as the value in the **Eaves level 1** edit box.

10. Activate the **Sections** tab of the dialog box.

11. From the **Column size** list, select **AISC 14.1 W > W14x500 / Australian Universal Column > 310 UC 158**.

12. From the **Rafter size** list, select **AISC 14.1 W > W21x201 / Australian Universal Beam > 460 UB 67.1**.

13. Activate the **Gable Sections** tab of the dialog box.

14. From the **Gable post size** list, select **AISC 14.1 W > W10x60 / Australian Universal Column > 150 UC 37.2**.

15. Make sure the **Split rafters** tick box is not selected.

16. Activate the **Gable Distances** tab of the dialog box.

17. Enter **3** as the value in the **No. posts on left side** edit box.

18. Enter **15'** or **4500** as the value in the **Intermediate distance** edit box.

19. Close the dialog box. The model, after creating the gable frame, is shown in Figure 45.

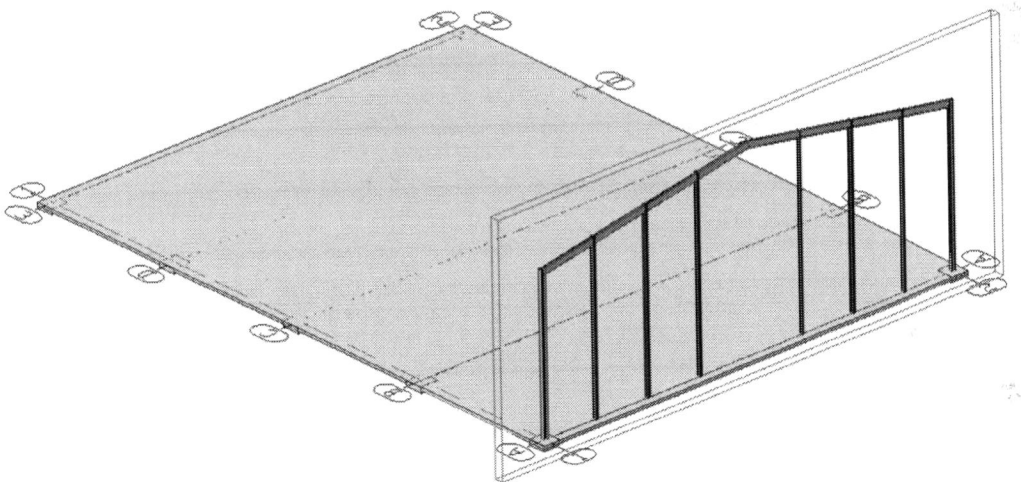

Figure 45 The model after creating the front gable frame

Next, you will copy this frame at the back. Note that to copy the gable frame, you need to select the connection box, displayed in Gray around the gable frame, and not any item of the gable frame.

20. Invoke the **Copy** tool and then select the connection box around the gable frame as the object to copy. Select the **A1** grid intersection point as the base point and **E1** grid intersection point as the second point of copy.

Section 2: Copying and Modifying the Gable Frame to Change it to Portal Frame

In this section, you will copy the front gable frame and then modify it to change it to a portal frame. You will then copy that portal frame and create two more instances.

1. Copy the gable frame from **A1** grid intersection point to **B1** grid intersection point. Remember that you need to select the connection box around the frame as the object to copy.

    You will now convert the gable frame at the **B1** grid intersection point into a portal frame.

2. Double-click on the connection box around the gable frame at the **B1** grid intersection point; the **Advance Steel - Gable Frame [X]** dialog box is displayed with the **Position** tab active.

3. From the **Type** list, select **Portal frame**; the gable posts disappear from the frame.

    Rest of the size of the portal frame is the same as that of the gable frame. So you do not need to change anything else.

4. Close the dialog box.

    You will now copy the portal from to grid intersection points **C1** and **D1**.

5. Copy the gable frame from **B1** grid intersection point to **C1** and **D1** grid intersection points. Remember that you need to select the connection box around the frame as the object to copy.

    To make the drawing window less cluttered, it is better to turn off the visibility of the connection boxes by turning their layer off.

6. Turn off the **Connection boxes** layer. The model after creating all the frames is shown in Figure 46.

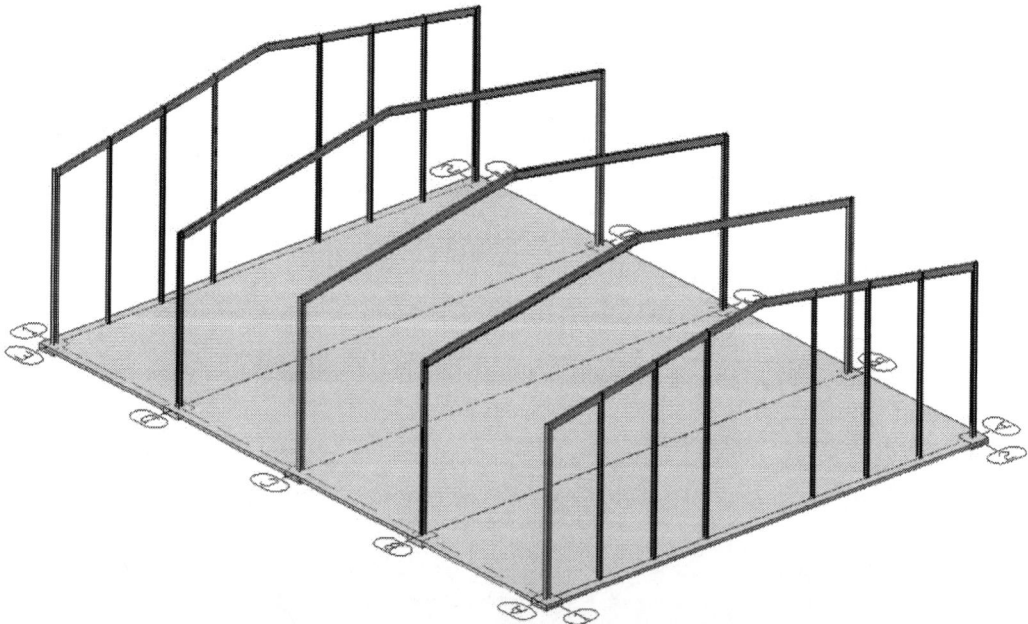

Figure 46 *The model after creating all the frames*

Tip: To edit the portal/gable frame with the **Connection boxes** layer turned off, you need to select a member of the portal/gable frame. Then right-click and select **Advance Joint Properties** from the shortcut menu.

Section 3: Inserting and Modifying Purlins

In this section, you will insert the purlins and then modify them to match the required sizes. You will also add these purlins to the library to be reused on the other side of the rafters.

1. From the **Home** ribbon tab > **Extended Modeling** ribbon panel, invoke the **Purlins** tool; you are prompted select objects.

2. One by one, select all the five rafters on the left side of the frames and then press ENTER; you are prompted to specify if you want to select a reference point.

Note: If you only select the first and last rafters, Advance Steel will insert single continuous sections between the first and last rafters. However, if you select the intermediate rafters as well, the purlin sections are automatically split at each rafter so that you can define overlap values.

3. Press ENTER to accept **No** as the value of the prompt; you are prompted to specify if you want to select an additional beam.

4. Press ENTER to accept **No** as the value of this prompt as well; the **Advance Steel - Purlins** dialog box is displayed.

5. Activate the **Sections** tab of the dialog box.

6. From the **Section size** list, select **Ayrshire Zed > AYR Z/15513**.

7. Select the **Overlap** tick box.

8. Activate the **Alignment** tab of the dialog box.

9. From the **Reference** list, select **Outer edge**, if it is not already selected.

10. In the **Overhang first support** and **Overhang at last support** edit boxes, enter **2'** or **600** as the value.

11. Select the **Invert section** tick box.

12. Activate the **Distances** tab of the dialog box.

13. Enter **7 7/8"** or **300** as the value in the **Start distance** edit box.

14. Enter **10'** or **3000** as the value in the **Intermediate distance** edit box.

15. Enter **8'** or **2500** as the value in the **Last to eave** edit box.

16. Enter **2'** or **600** as the value in the **Overlap lengths** edit box.

17. Activate the **Eaves beam** tab of the dialog box.

18. Select the **Create Eaves beam** tick box.

19. From the **Eaves beam type** list, select **Albion Eaves Beam > ALB EB 20018 - 9**.

20. From the **Alignment** list, select **beam**.

21. Select the **Eaves beam vertical** tick box.

22. From the **Axis reference** list, select **top left**.

23. In the **Horizontal offset** edit box, enter **1' 2 1/2"** or **370**.

24. In the **Vertical offset** edit box, enter **-7"** or **-175**.

25. Select the **Flip profiles** tick box.

    Next, you will add these purlin settings to the library for future use.

26. Activate the **Library** tab of the dialog box.

27. Click the **Save values** button; a new row with the current values is added.

28. Click the **Edit** button to display the **Library** dialog box.

29. Click in the **Comment** field of the newly added row and rename it to **Purlins-Up and Running**.

30. Click **OK** in the **Library** dialog box.

31. Close the **Advance Steel - Purlins [X]** dialog box.

    Next, you need to create the same purlins on the other side of the frames. You will use the settings saved in the library to insert the purlins. However, you will have to invert the profiles on that side.

32. From the **Home** ribbon tab > **Extended Modeling** ribbon panel, invoke the **Purlins** tool again; you are prompted to select objects.

33. One by one, select all the five rafters on the right side of the frames and then press ENTER; you are prompted to specify if you want to select a reference point.

34. Press ENTER to accept **No** as the value of the prompt; you are prompted to specify if you want to select an additional beam.

35. Press ENTER to accept **No** as the value of this prompt as well; the **Advance Steel - Purlins** dialog box is displayed.

36. Activate the **Library** tab of the dialog box.

37. From the list, select **Purlins-Up and Running**; the purlins are modified to match the settings you saved in the library.

    Next, you need to invert the profiles and eaves beam.

38. Activate the **Alignment** tab of the dialog box.

39. Clear the **Invert section** tick box; the purlin sections are inverted.

40. Activate the **Eaves beam** tab of the dialog box.

41. Select the **Invert section** tick box; the eaves beam is inverted.

42. Close the dialog box. The completed model, after inserting the required purlins, is shown in Figure 47.

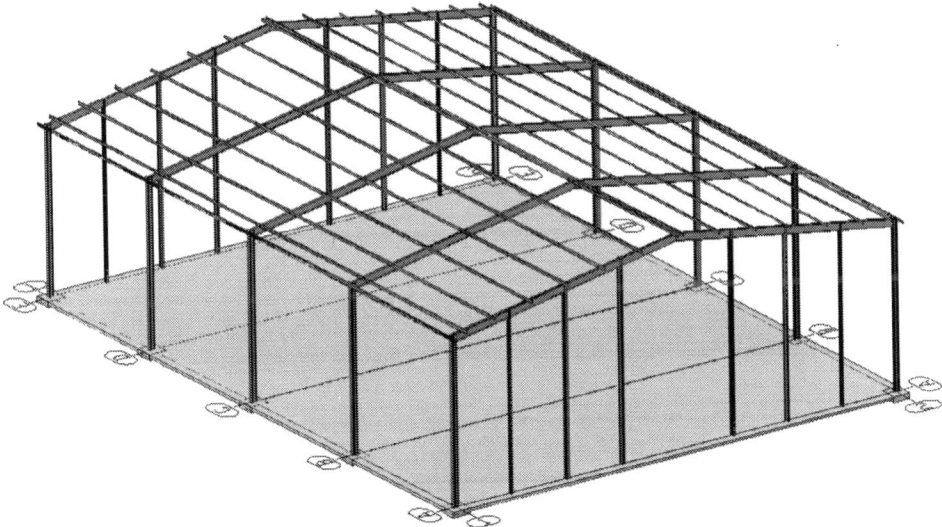

Figure 47 The model, after inserting the purlins

43. Save the file and then close it.

Section 4: Inserting and Modifying Trusses

In this section, you will open the **C03-Trusses-Imperial.dwg** or **C03-Trusses-Metric.dwg** file and then insert the required trusses. You will then copy these trusses to complete the model.

1. From the **C03 > Struc-BIM** folder, open the **C03-Trusses-Imperial.dwg** or **C03-Trusses-Metric.dwg** file, depending on the units you want to use. The model in this file looks similar to the one shown in Figure 48.

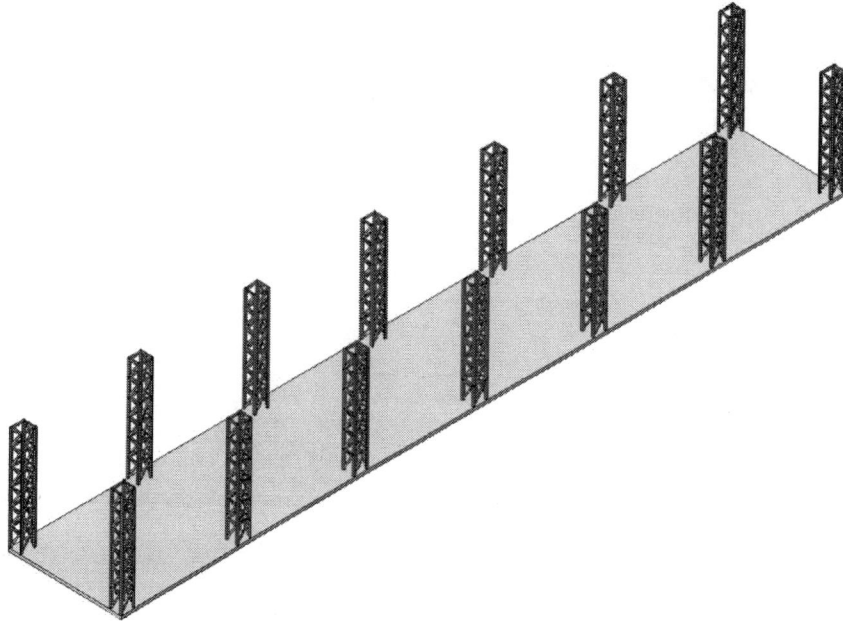

Figure 48 The model to create the trusses

To make it easier for you to insert trusses, you will first isolate the first four brace-supported columns.

2. Change the current view to the top view.

3. Window select the four items encircled in Figure 49.

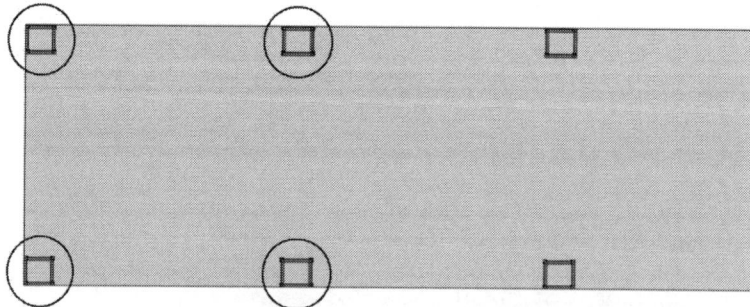

Figure 49 Window select the four items encircled

4. Right-click in the blank area of the drawing window and select **Isolate > Isolate Objects**; the selected objects are isolated and the visibility of the rest of the objects is turned off.

5. Change the current view to Home view with parallel projection.

6. Zoom closer to the visible objects, as shown in Figure 50.

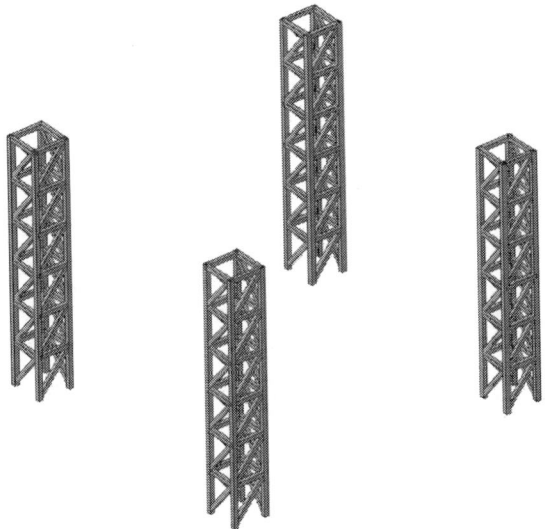

Figure 50 The isolated model to create the trusses

7. From the **Home** ribbon tab > **Extended Modeling** ribbon panel, invoke the **Truss** tool; you are prompted to select the start point.

8. Select the node point labeled as **1** in Figure 51 as the start point of the truss; you are prompted to select the endpoint.

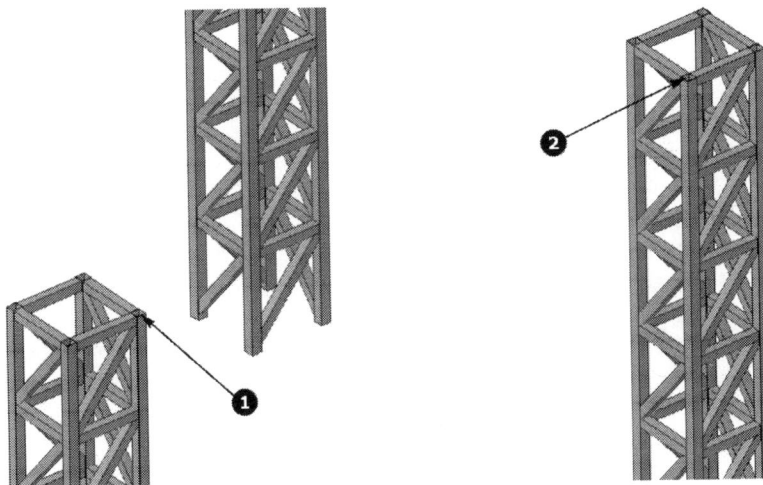

Figure 51 The node points to be selected to insert the truss

9. Select the node point labeled as **2** in Figure 51 as the end point of the truss; the **Advance Steel - Structural element truss [X]** dialog box is displayed.

10. Activate the **Geometry** tab of the dialog box.

11. From the **Alignment** list, select **Top**.

12. Activate the **Dimensions Y** tab of the dialog box.

13. In the **Left side height** and **Right side height** edit boxes, enter **3'** or **850** as the value.

14. In the **Bottom offset left side** and **Bottom offset right side** edit boxes, enter **2'** or **650** as the value.

15. Activate the **Web geometry** tab of the dialog box.

16. From the **Type** list, select **Pratt**; notice the preview of the truss updates to this type in the drawing window.

17. Activate the **Web offsets - Top** tab of the dialog box.

18. From the **Top alignment** list, select **Aligned**; all the web sections are trimmed to match the sections they are connecting to.

19. Make sure the **Same for bottom** tick box is selected.

20. Make sure the values of the **Gap, Vertical offset - top**, and **Horizontal offset - top** edit boxes are set to **0**.

21. Activate the **Panels** tab of the dialog box.

22. From the **Panel layout** list, select **By number**.

23. In the **Number on first element** edit box, enter **8** as the value.

24. In the **Start distance** and **End distance** edit boxes, enter **5"** or **100** as the value.

25. Activate the **Top chord section** tab of the dialog box.

26. From the **Section** list, select **AISC 14.1 HSS rectangular > HSS 3 1/2X2 1/2X3/8** or **Australian Square Hollow Section - CF C350L0 > 75x75x3.0 SHS** as the section type and size.

    By default, the bottom chord section is set the same as the top one. So you do not need to change it.

27. Activate the **Web section diagonals** tab of the dialog box.

28. From the **Section** list, select **AISC 14.1 HSS rectangular > HSS 3 1/2X2X1/4** or **Australian Square Hollow Section - CF C350L0 > 65x65x3.0 SHS** as the section type and size.

29. Activate the **Web section verticals** tab of the dialog box.

30. From the **Section** list, select **AISC 14.1 HSS rectangular > HSS 3 1/2X2X1/4** or **Australian Square Hollow Section - CF C350L0 > 65x65x3.0 SHS** as the section type and size.

Next, you will save this truss in the library for future use.

31. Activate the **Library** tab of the dialog box.

32. Click the **Save values** button; a new row with the current values is added.

33. Click the **Edit** button to display the **Library** dialog box.

34. Click in the **Comment** field of the newly added row and rename it to **Truss-Up and Running**.

35. Click **OK** in the **Library** dialog box.

36. Close the **Advance Steel - Structural element truss [X]** dialog box. The model, after inserting the first truss, is shown in Figure 52.

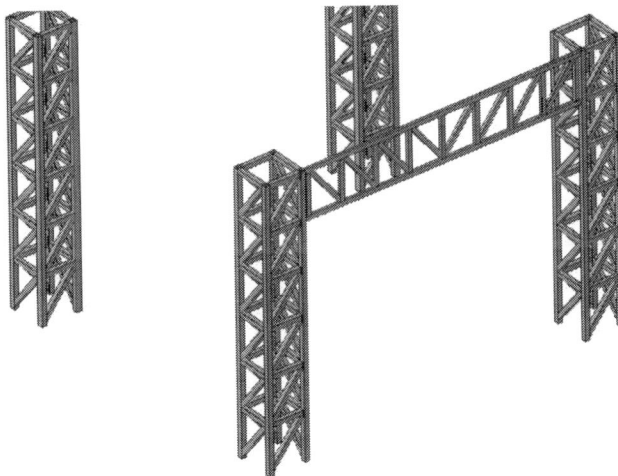

Figure 52 *The model after inserting the first truss*

Next, you will insert another truss using the settings you saved in the truss library.

37. From the **Home** ribbon tab > **Extended Modeling** ribbon panel, invoke the **Truss** tool; you are prompted to select the start point.

38. Select the node point labeled as **1** in Figure 53 as the start point of the truss; you are prompted to select the end point.

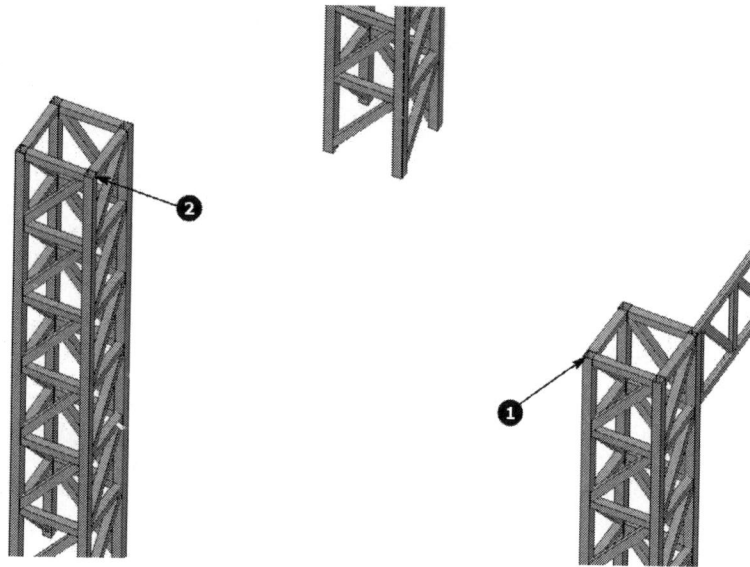

Figure 53   The node points to be selected to insert the truss

39. Select the node point labeled as **2** in Figure 53 as the end point of the truss; the **Advance Steel - Structural element truss [X]** dialog box is displayed.

40. Activate the **Library** tab of the dialog box.

41. Select the **Truss-Up and Running** entry to insert a new truss using the values saved in the library; the truss updates and looks similar to the first truss you inserted.

42. Close the **Advance Steel - Structure element truss [X]** dialog box.

What I do

As seen in the above steps, inserting a structural element with the settings saved in the library saved a lot of time. This is what I generally do whenever I am working on a structural model. Once I enter the required values, I save those settings in the library for future use.

Section 5: Copying Trusses

In this section, you will copy the first and second trusses to the other sides of the brace-supported columns. However, to copy the truss, you cannot select one of the sections of the truss. You need to select the connection box of the truss as the object to copy. By default, in this file, the **Connection boxes** layer is turned off. So you need to first turn this layer on.

1. Turn on the **Connection boxes** layer.

    Notice various other connection boxes are turned on as well. These connection boxes are for the beam trimming. You will learn more about beam trimming in later chapters.

2. Invoke the **Copy** tool and select the connection box labeled as **1** in Figure 54.

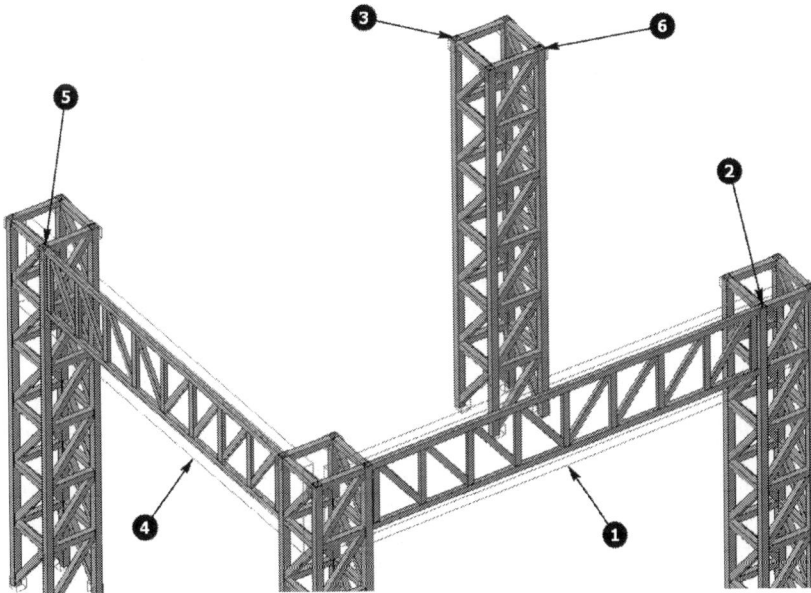

Figure 54 The trusses to be selected and the node points to be used for copying

3. Copy it using the node point labeled as **2** as the base point to the node point labeled as **3** in Figure 54.

4. Similarly, copy the truss labeled as **4** using the node point labeled as **5** as the base point to the node point labeled as **6**. The model, after copying the two trusses, is shown in Figure 55.

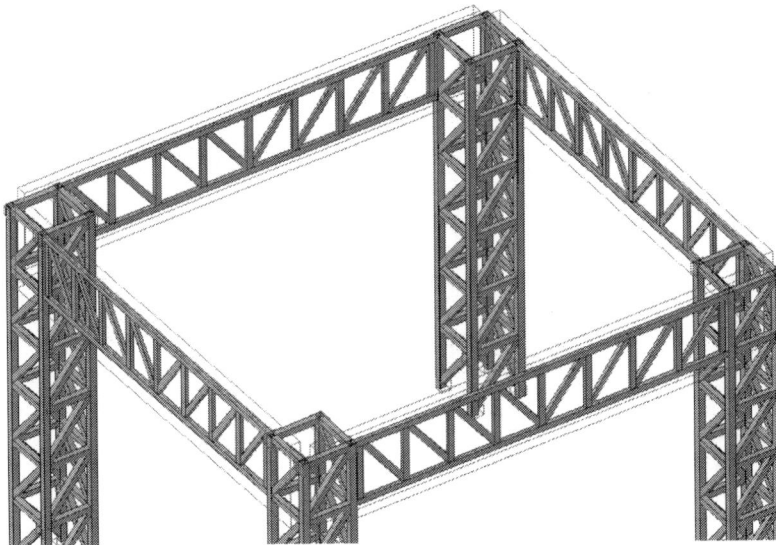

Figure 55 The model after copying the two trusses

Next, you will copy the first, third, and fourth trusses to the remaining places in the model. For this, you will use the **Array** option of the **Copy** tool. But before that, you need to end object isolation in the model.

5. Right-click in the blank area of the drawing window and select **Isolate > End Object Isolation**; the visibility of all the objects in the model is restored.

6. Invoke the **Copy** tool and select the connection boxes of the first, third, and fourth trusses as the objects to copy.

7. Click any point in the blank area of the drawing window as the base point; you are prompted to select the second point or use the **Array** option.

8. In the above prompt, select the **Array** option; you are prompted to enter the number of items to array.

9. Enter **6** as the value and then press ENTER; you are prompted to specify the second point or use the **Fit** option.

10. Move the cursor to track 0-degree angle.

11. Once the tracking line is displayed, enter **22'6"** or **7000** as the value; the required number of trusses are created.

12. Press the ENTER key to end the **Copy** tool. The model, after creating all the trusses, is shown in Figure 56.

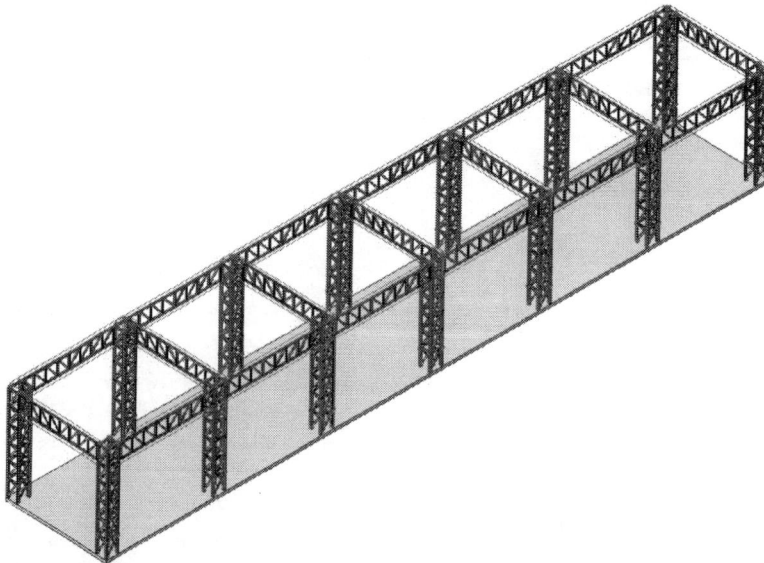

Figure 56 The completed model after inserting all the trusses

13. Save the file and close it.

Skill Evaluation

Evaluate your skills to see how many questions you can answer correctly. The answers to these questions are given at the end of the book.

1. Advance Steel allows you to insert the portal as well as gable frame using the same dialog box. (True/False)

2. You can double-click on any section of the frame to edit the frame. (True/False)

3. Mono frames cannot be inserted in Advance Steel. (True/False)

4. The top and bottom sections of trusses can be straight or curved members. (True/False)

5. Purlins can be inserted with heavier end bay sections. (True/False)

6. Which tool is used to create mono-pitch frames?

   (A) **Portal Frame**   (B) **Gable Frame**
   (C) **Mono Frame**   (D) **Mono-pitch frame**

7. Which tab of the **Advance Steel - Purlins** dialog box is used to define the overlap values for the purlins?

   (A) **Sections**   (B) **Alignment**
   (C) **Properties**   (D) **Distances**

8. Which tool is used to insert trusses in Advance Steel?

   (A) **My Trusses**   (B) **Truss**
   (C) **Trusses**   (D) **Truss Sections**

9. Which tool is used to insert gable frames in Advance Steel?

   (A) **Portal/Gable Frame**   (B) **Gable Frame**
   (C) **Mono-pitch frame**   (D) **Portal Frame**

10. What are the three types of trusses that can be created in Advance Steel?

   (A) **Warren**   (B) **Half Pratt**
   (C) **Prat**   (D) **Circular**

Class Test Questions
Answer the following questions:

1. Explain briefly the process of inserting purlins with the sections split at each rafter.

2. How will you insert a truss with curved top section?

3. Explain briefly how will you create a gable frame.

4. How will you copy a truss?

5. What is the process of editing a portal frame?

Chapter 4 – Inserting the Plates at Beam and Column – Beam Joints

The objectives of this chapter are to:

√ *Familiarize you with the **Connection Vault** palette*
√ *Explain the process of inserting various types of **Plates at Beam** joints*
√ *Explain the process of inserting various types of **Column - Beam** joints*
√ *Explain the process of adding joints to the **Favorites** category*
√ *Explain the process of editing joints*
√ *Explain the process of copying joints*
√ *Explain the process of upgrading a joint to master*

THE CONNECTION VAULT PALETTE

> **Home Ribbon Tab > Extended Modeling Ribbon Panel > Connection vault**
> **Extended Modeling Ribbon Tab > Joint Ribbon Panel > Connection vault**

In Advance Steel, all the joints are inserted using the **Connection Vault** palette. This palette comprises of four main areas, as labeled in Figure 1. These four areas are discussed next.

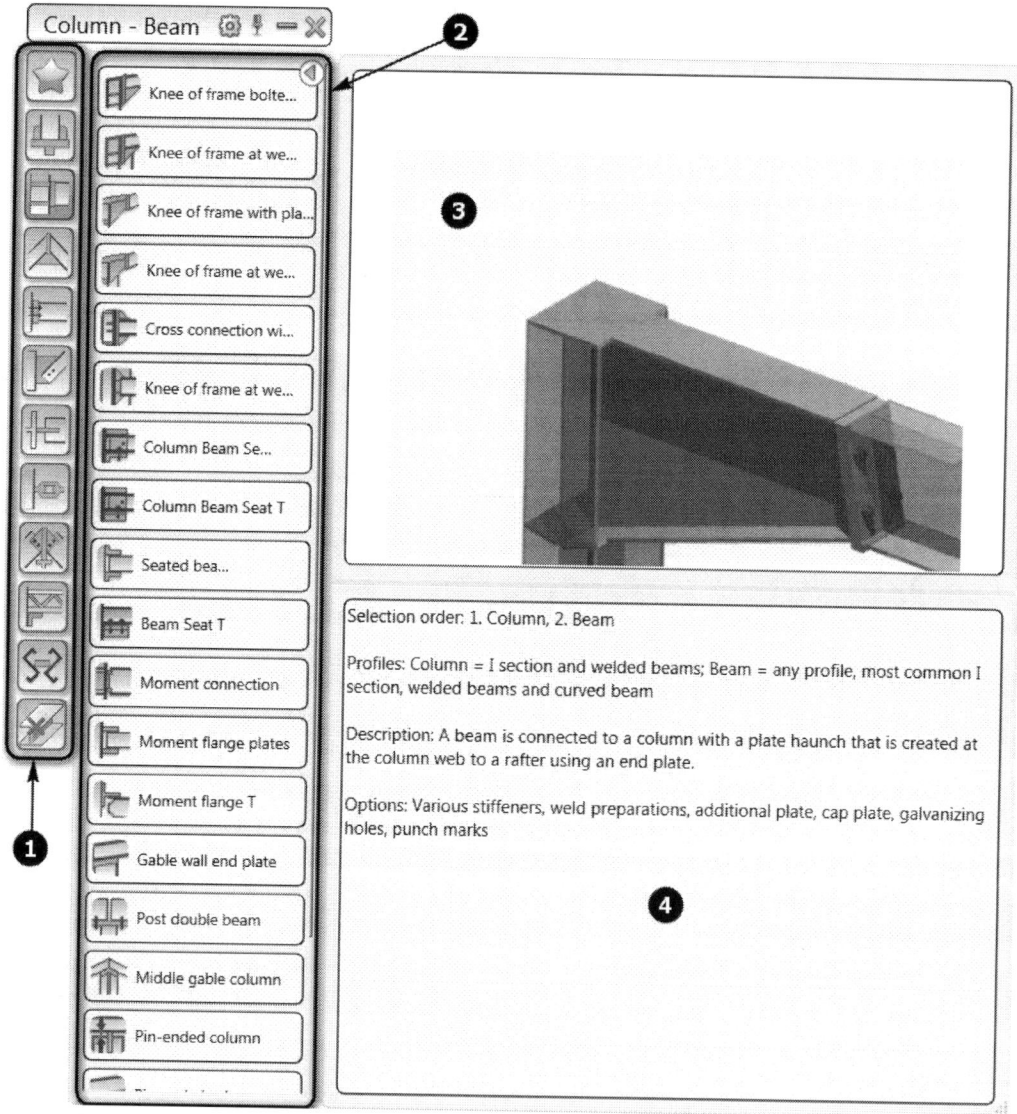

Figure 1 The **Connection Vault** palette

1 = **Joint Categories Area**: This area shows the buttons of various joint categories.

2 = **Joint Type Area**: This area shows various joint types available for the selected joint category.

3 = **Joint Preview Area**: This area shows the preview of the joint type selected.

4 = **Selection and Description Area**: This area shows the selection order for creating the joint and also gives a brief description of the joint. If there are multiple options for the joint type, they will be listed here as well.

INSERTING THE PLATES AT BEAM JOINTS

This category lists various joints that can be added between a structural section and a plate. These joint types are discussed next.

Note: It is important to mention here that most of the joints in this category are similar. Therefore, only certain joints types are discussed in detail.

Base plate Joint

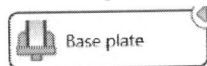

This type of joint is used to insert a base plate at the bottom of a vertical or inclined column. This joint type allows any section type to be used as the column. When you invoke this tool, you will be prompted to select a column. Once you select the column and press the ENTER key, the **Advance Steel - Base plate [X]** dialog box will be displayed with the **Properties** tab active.

Note: The joint dialog boxes that you will see in this chapter are a little different from the ones you have seen in the previous chapters. In the joint dialog boxes, you will find the category buttons in the left pane. When you click on these buttons, various tabs will be displayed that will allow you to configure settings related to that category.

Properties Category

The **Properties** category of the dialog box has three tabs: **Properties**, **Library**, and **Joint design**. The first two tabs are similar to those discussed in the previous chapter. The **Joint design** tab is used to test the joint design you created based on the **AISC** or **EC3** designs. In this tab, click the **Check** button to check if the joint you created passes the design criteria.

Base Plate Category > Base plate layout Tab

When you click on the **Base Plate** category in the left pane of the dialog box, the **Base plate layout** tab is the first tab, as shown in Figure 2. The options available in this tab are discussed next.

Plate thickness

This edit box is used to enter the thickness of the base plate that will be inserted at the bottom of the selected column.

Layout

This list provides you the option to specify the size of the base plate. If you select the **projections** option from this list, the size of the base plate will be defined in terms of its projections from the selected column. Selecting **total** option allows you to manually specify the length and width of the plate. These values are defined in the **Base plate dimensions**

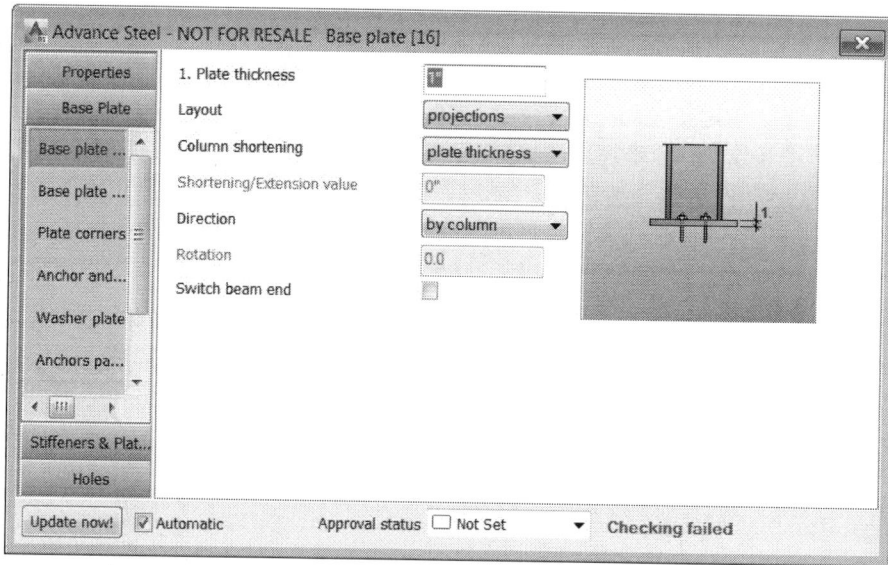

Figure 2 The Base plate layout tab of the Base Plate category

tab. Selecting **from center** option from this list allows you to specify the size of the base plate in terms of its projection from the center of the column. Selecting **by anchors** from this list allows you to specify the size of the base plate in terms of its projections from the anchors.

Column shortening
This list allows you to specify how the selected column will be shortened to accommodate the plate thickness. By default, the **plate thickness** option is selected. As a result, the column is automatically shortened by the value equal to the thickness of the plate. Selecting **none** from this list will not shorten the column. Selecting **value** from this list allows you to manually enter the column shortening value in the edit box below this list.

What I do
*When inserting the **Plate at beam** joint at the bottom of the column standing on top of a concrete footing, I make sure the **plate thickness** option is selected from the list above. This will ensure the base plate sits on top of the concrete footing and not inserted inside it. Also, the column is automatically shortened to be placed on top of the base plate.*

Shortening/Extension value
This edit box is only available if you select the **value** option from the above list. The required value can be entered in this edit box.

Direction
This list allows you to specify the direction of the base plate to be inserted. By default, the **by column** option is selected. As a result, the plate is inserted normal to the column. Selecting the **horizontal** option will insert the base plate horizontal, irrespective of the orientation of the column. This is useful when the column you have inserted in inclined. Selecting the **vertical** option inserts a vertical base plate. Figure 3 shows a horizontal base plate inserted for an inclined column and Figure 4 shows a base plate with the direction defined by the column.

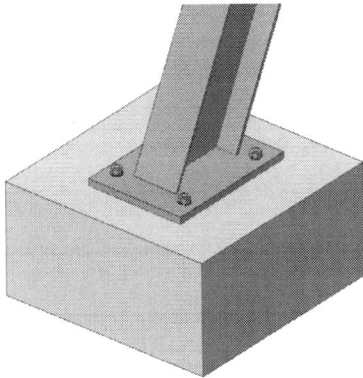

Figure 3 *Base plate with horizontal direction*

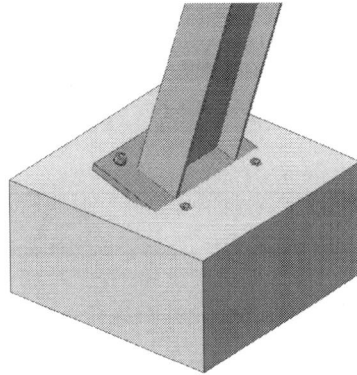

Figure 4 *Base plate with direction by column*

Rotation

This edit box is used to enter the rotation angle of the base plate and is only available if you select any option other than **projection** from the **Layout** list above.

Switch beam end

This tick box is used to switch the end of the column where the base plate will be inserted.

Base Plate Category > Base plate dimensions Tab

The **Base plate dimensions** tab, shown in Figure 5, is used to enter the dimensions of the base plate. The options in this tab will be active or inactive, depending on the option selected from the **Layout** list in the above discussed tab. The options available in this tab are discussed next.

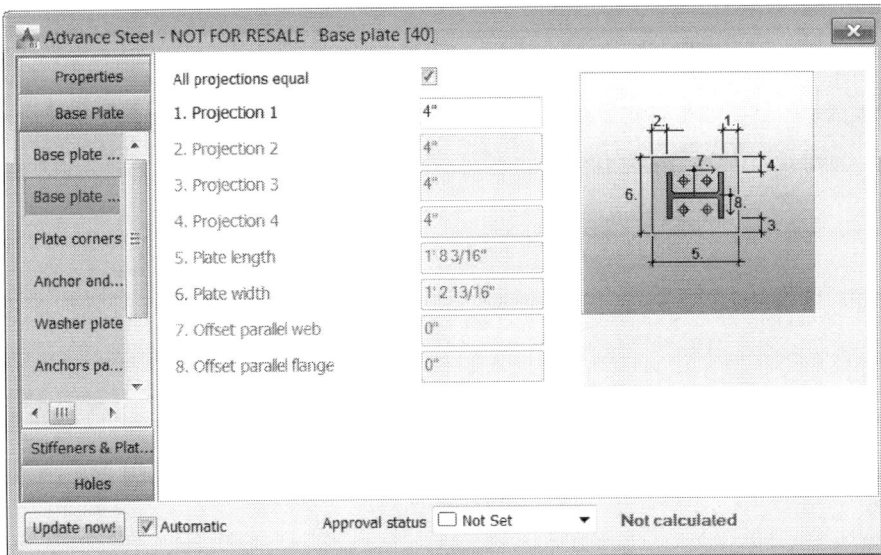

Figure 5 *The **Base plate dimensions** tab of the **Base Plate** category*

All projections equal

If this tick box is selected, the projections of the base plate on all sides of the column will be equal. As a result, you can specify only one of the projection values in the **Projection 1** edit box.

Projection 1/Projection 2/Projection 3/Projection 4

These edit boxes are used to specify the projections of the base plate. As mentioned above, if the **All projections equal** tick box is selected, then only the **Projection 1** edit box will be active. The value entered in this edit box will automatically be used for the remaining three projections as well.

*Tip: For asymmetric columns such as I-sections, you can enter a square plate by selecting the **from center** option from the **Layout** list in the **Base plate layout** tab. The size of the square plate can then be entered in the **Projection 1** edit box.*

Plate length/Plate width

These edit boxes are only available if you select the **total** option from the **Layout** list in the **Base plate layout** tab. In this case, you can enter the length and width of the base plate in these edit boxes.

Offset parallel web/Offset parallel flange

These edit boxes are only available if you select the **total** or **from center** option from the **Layout** list in the **Base plate layout** tab. In those cases, you can offset the base plate parallel to the web or flange of the column by entering the values in these edit boxes.

Base Plate Category > Plate corners Tab

The **Plate corners** tab, shown in Figure 6, is used to specify the type of plate corners. These options are discussed next.

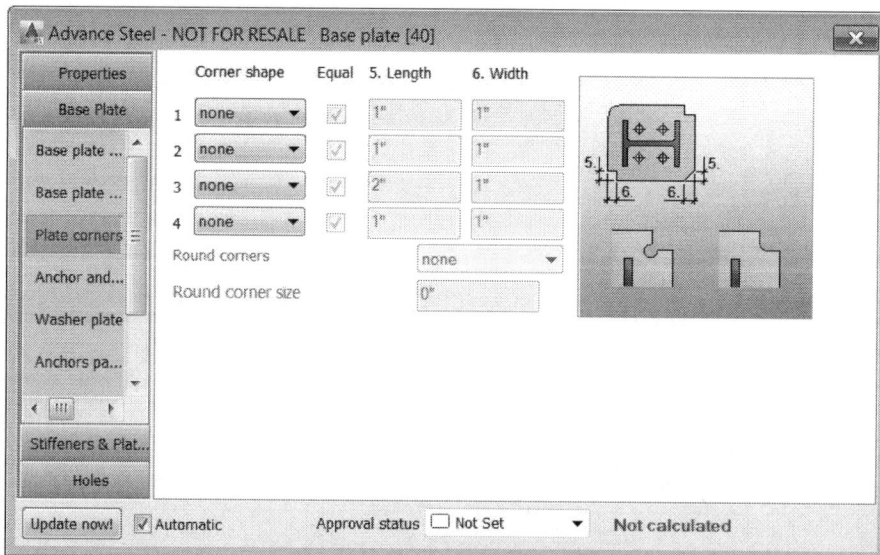

Figure 6 *The **Plate corners** tab of the **Base Plate** category*

Corner shape
These lists are used to specify the shape of the four corners of the base plate. By default, **none** is selected from this list. As a result, there is no corner treatment defined. Selecting the **fillet** option will insert a chamfer cut. Selecting the **cut** option will insert a square cut. Selecting the **convex** option will insert a fillet.

Equal
If these tick boxes are selected, the dimensions of fillet and cut will be equal along both directions.

Length
Enter the length of the fillet, cut, or convex corner treatment in these edit boxes.

Width
These edit boxes will only be selected if the **Equal** tick box is cleared. In that case, enter the width of the fillet or cut corner treatment in these edit boxes.

Base Plate Category > Anchor and holes Tab
The **Anchor and holes** tab, shown in Figure 7, is used to specify the parameters related to the anchors and holes to be inserted on the plate. These options are discussed next.

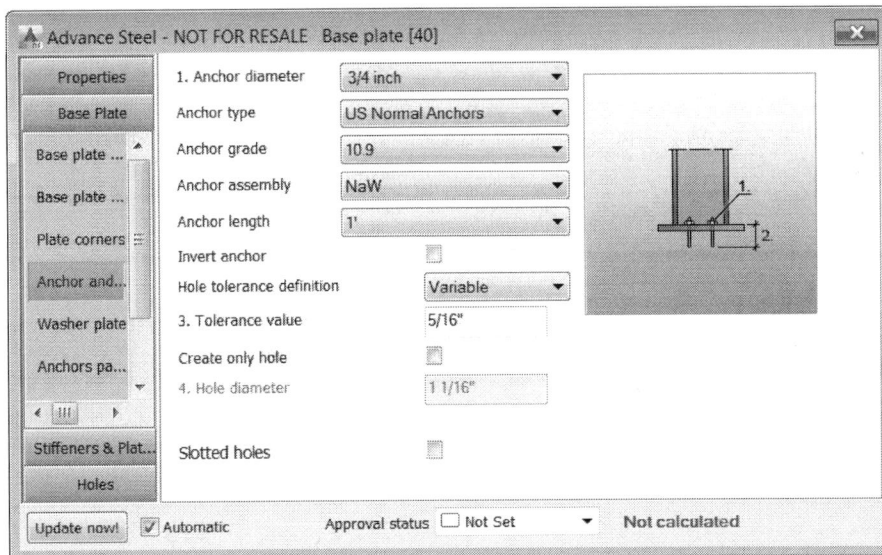

*Figure 7 The **Anchor and holes** tab of the **Base Plate** category*

Anchor diameter
This list is used to select the diameter of anchor bolt.

Anchor type
This list is used to select the type of anchor bolt.

Anchor assembly
This list is used to select the anchor assembly to be used for the joint.

Anchor length
This list is used to select the length of the anchor bolt.

Invert anchor
Selecting this tick box will invert the anchor bolts in the joint.

Hole tolerance definition
This list is used to specify the hole tolerance definition. By default, the **Variable** option is selected. As a result, you can enter the tolerance value in the edit box available below this list. Selecting the **Automatic** option from this list will automatically select the tolerance value for the anchor bolt.

Tolerance value
This edit box is only available if the **Variable** option is selected from the above list. In that case, you can enter the tolerance value in this edit box.

Create only hole
Selecting this tick box will only create holes on the base plate. No anchor bolts will be inserted in that case.

Hole diameter
While creating only holes, you can enter the hole diameter in this edit box.

Slotted holes
Select this tick box if you want to create slotted holes on the base plate.

Base Plate Category > Washer plate Tab
The **Washer plate** tab, shown in Figure 8, is used to specify the size of washer plates to be inserted between the anchor washers and the base plate. These options are discussed next.

Create washer plate
Select this tick box if you want to insert a washer plate between the anchor washer and the base plate. Figure 9 shows a base plate joint with washer plates created.

Thickness
This edit box is used to enter the thickness of the washer plates.

Size
This edit box is used to enter the length and width of the washer plates.

Hole tolerance
This list allows you to specify the tolerance of the holes on the washer plates. By default, the **same as base plate** option is selected. You can also manually enter the tolerance by selecting the **variable** option from this list.

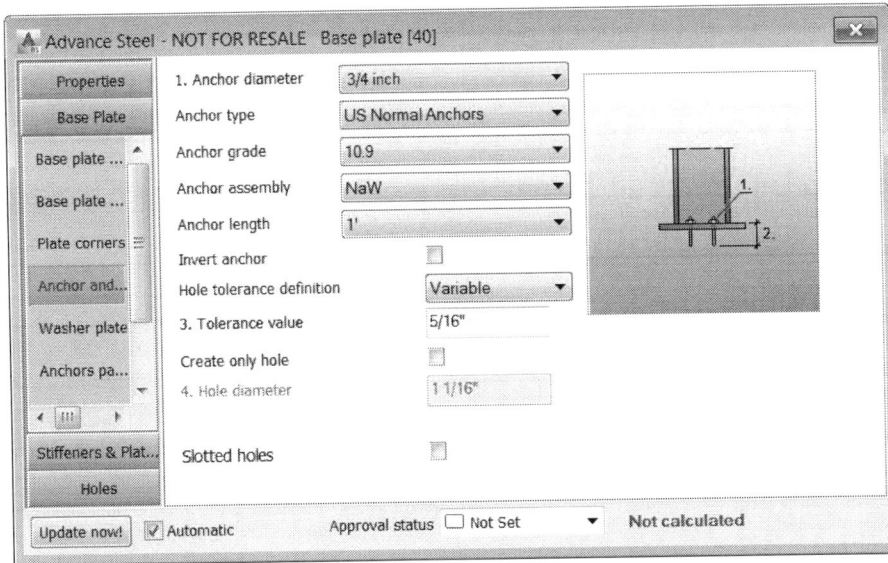

Figure 8 The **Washer plate** tab of the **Base Plate** category

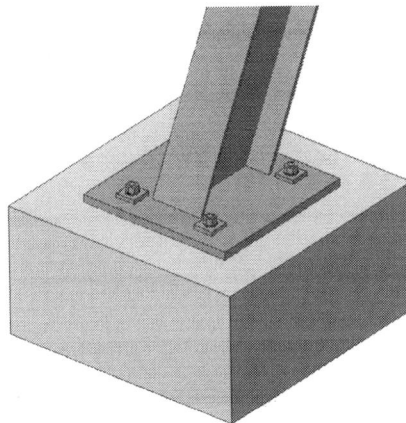

Figure 9 Base plate joint with washer plates

Tolerance value
This edit box is only available if the **variable** option is selected from the above list. In that case, you can enter the tolerance value in this edit box.

Hole diameter
If you selected the **Create only hole** option from the previous tab, you can enter the diameter of the holes on the washer plate in this edit box.

Create underneath
Select this tick box to insert the washer plates on the under side of the base plate.

Base Plate Category > Anchor parallel web Tab

The **Anchor parallel web** tab, shown in Figure 10, is used to specify the parameters related to the anchor bolts to be inserted parallel to the web of the column. These options are discussed next.

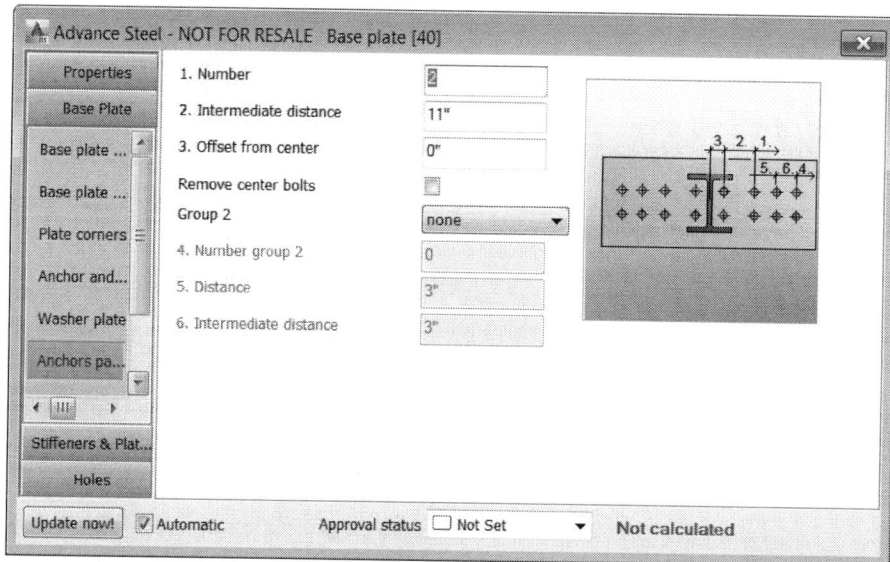

*Figure 10 The **Anchors parallel web** tab of the **Base Plate** category*

Number

Enter the number of anchor bolts to be inserted parallel to the column web in this edit box.

Intermediate distance

Enter the distance between the anchor bolts to be inserted parallel to the column web in this edit box.

Offset from center

If the anchor bolts to be inserted parallel to the column web need to be at certain offset from the center, you can enter the offset value in this edit box.

Remove center bolts

Select this tick box if you want the center bolts to be removed from the joint.

Group 2

While creating joints in Advance Steel, you can insert multiple groups of bolts. In this case, you can use this list to insert additional groups of bolts on the left side, right side, or both sides. The dimensions of the additional groups can be entered in the edit boxes available below this list. Figure 11 shows a base plate joint with two additional groups of bolts inserted on both sides parallel to the web.

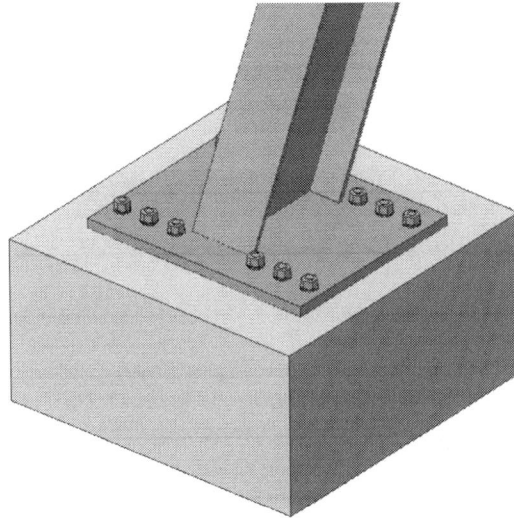

Figure 11 *The base plate joint with two additional bolt groups inserted on both sides parallel to the web*

Base Plate Category > Anchor parallel flange Tab

The **Anchor parallel flange** tab, shown in Figure 12, is used to specify the parameters related to the anchor bolts to be inserted parallel to the web of the column. These options are discussed next.

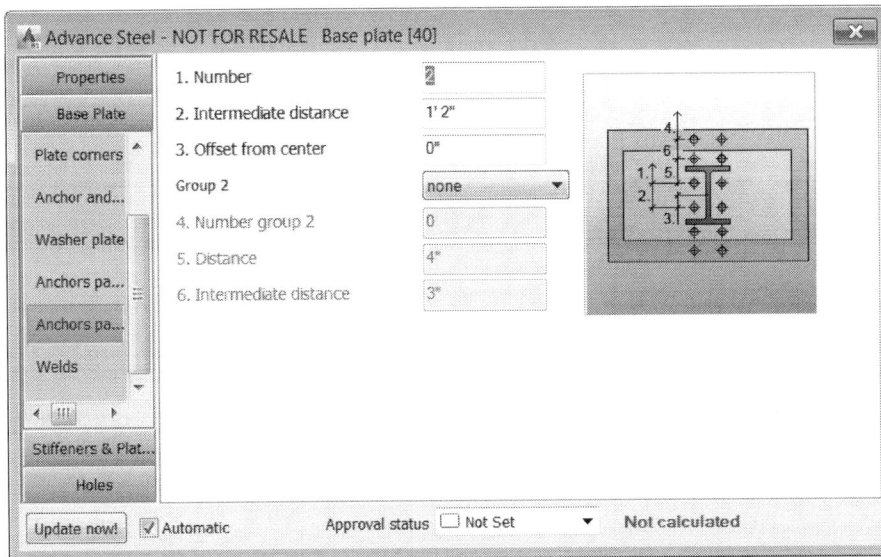

Figure 12 *The **Anchors parallel flange** tab of the **Base Plate** category*

Number

Enter the number of anchor bolts to be inserted parallel to the column flange in this edit box.

Intermediate distance

Enter the distance between the anchor bolts to be inserted parallel to the column flange in this edit box.

Offset from center

If the anchor bolts to be inserted parallel to the column flange need to be at certain offset from the center, you can enter the offset value in this edit box.

Remove center bolts

Select this tick box if you want the center bolts to be removed from the joint.

Group 2

You can use this list to insert additional groups of bolts on top, bottom, or both sides. The dimensions of the additional groups can be entered in the edit boxes available below this list. Figure 13 shows a base plate joint with two additional groups of bolts inserted on both sides parallel to the flange.

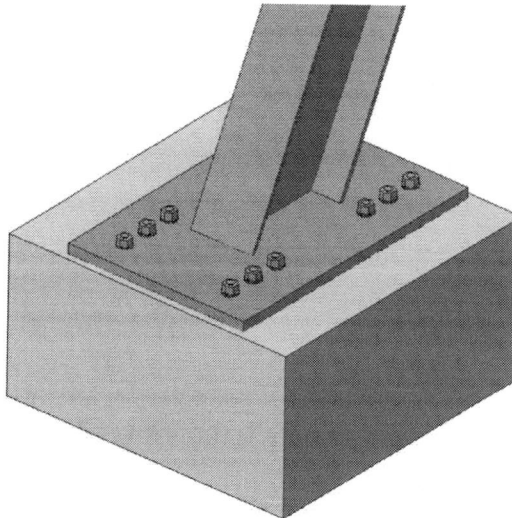

Figure 13   *The base plate joint with two additional bolt groups inserted on both sides parallel to the flange*

Base Plate Category > Weld Tab

The **Weld** tab, shown in Figure 14, is used to specify the parameters that will be used to weld the base plate to the column. You can enter the weld thicknesses at the web, flange top, and flange bottom in their respective edit boxes. The **Weld type** lists are used to select the types of welds at the three locations.

Stiffeners & Plates Category > Levelling plate Tab

The **Levelling plate** tab, shown in Figure 15, is used to specify the parameters related to the levelling plate. These options are discussed next.

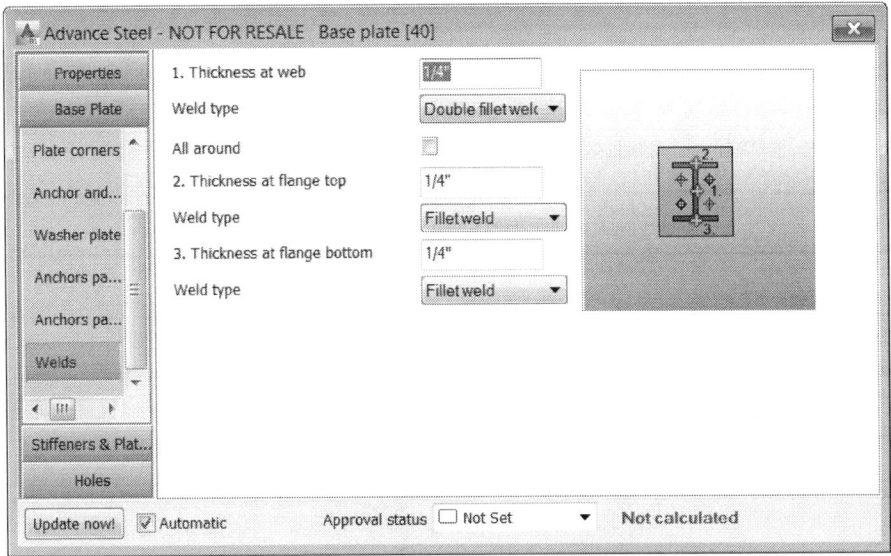

Figure 14 *The Weld tab of the Base Plate category*

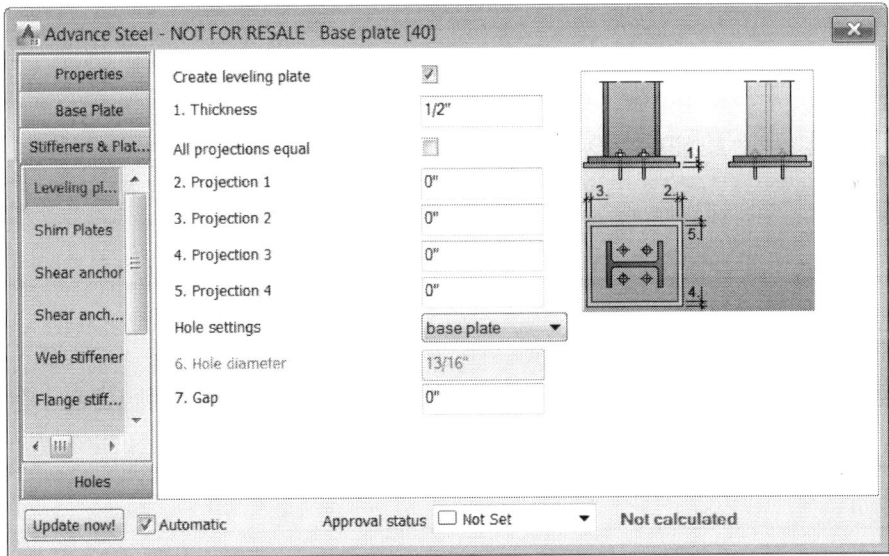

Figure 15 *The Levelling plate tab of the Stiffeners & Plates category*

Create levelling plate

By default, this tick box is not selected. As a result, there is no levelling plate created. However, if you want to insert a levelling plate, select this tick box.

Note: To insert the levelling plate between the concrete footing and the base plate, you need to select value from the Column shortening list of the Base Plate > Base plate layout tab. Then enter the sum of the base plate thickness and levelling plate thickness as the value in the Shortening/Extension value edit box.

Thickness
Enter the levelling plate thickness in this edit box.

All projections equal
Select this tick box if you want to insert the levelling plate with equal projections. Note that the levelling plate projections are measured from the end of the base plate.

Projection 1/Projection 2/Projection 3/Projection 4
These edit boxes are used to specify the projections of the levelling plate from the base plate. As mentioned above, if the **All projections equal** tick box is selected, then only the **Projection 1** edit box will be active. The value entered in this edit box will automatically be used for the remaining three projections as well. Figure 16 shows a joint with levelling plate inserted with equal projections from the base plate.

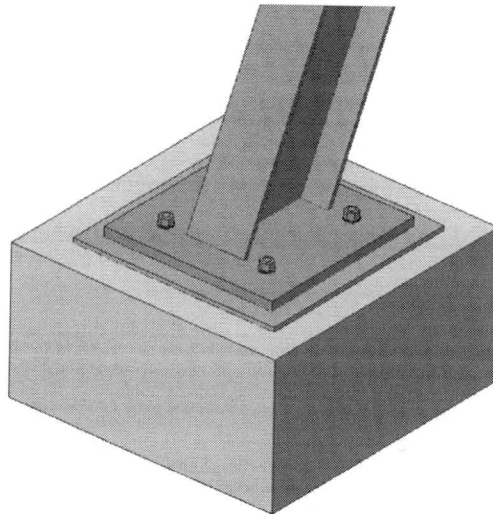

Figure 16 The base plate joint with the equal projection levelling plate

Hole settings
This list allows you to select the parameters related to the holes to be inserted on the levelling plate. By default, the **base plate** option is selected. As a result, the holes on the levelling plate are the same as those on the base plate. You can select the **variable** option from this list and manually enter the hole diameter in the **Hole diameter** edit box below this list.

Gap
You can insert a gap between the levelling plate and base plate by entering the gap value in this edit box.

Stiffeners & Plates Category > Shim plates Tab
The **Shim plates** tab, shown in Figure 17, is used to specify the parameters related to the shim plates. These options are discussed next.

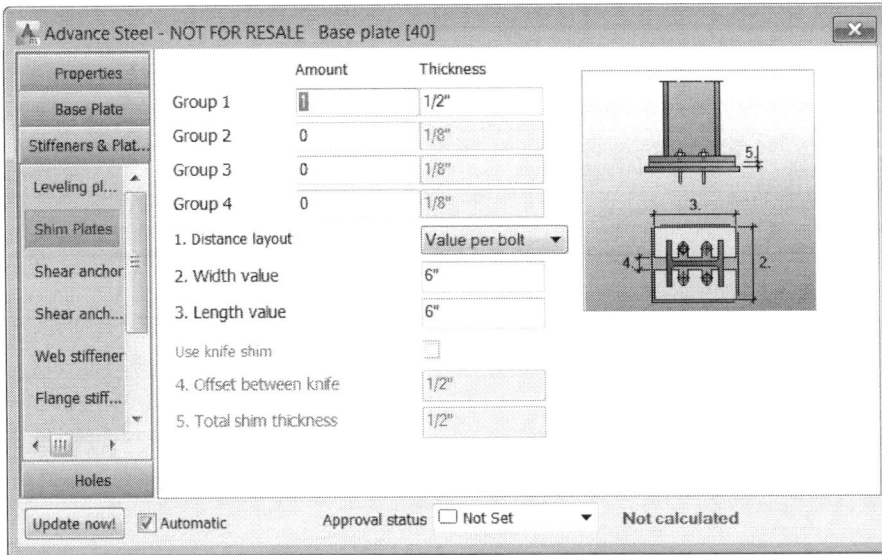

*Figure 17 The **Shim plates** tab of the **Stiffeners & Plates** category*

Group 1/Group 2/Group 3/Group 4
You can enter the amount and thicknesses of shim plate groups to be inserted in these edit boxes.

Distance layout
This list is used to specify how the distance layout of the shim plates will be defined. By default, the **Same as plate** option is selected. As a result, the shim plate is the same dimension as the base plate. You can also select the **Offset**, **Value**, or **Value per bolt** option and enter the values in the **Width value** and **Length value** edit boxes available below this list.

*Tip: While using the **Offset** option from the above list, you can enter a negative value to offset the shim plates so that they are bigger than the base plate.*

Use knife shim
Select this tick box to split the shim plates. The split gap is defined in the **Offset between knife** edit box below this list.

What I do
While creating various joints in Advance Steel, a number of times I need to isolate an item of the joint to preview it before closing the dialog box. To do this, I simply select the item I need to isolate and then orbit the model by holding down the SHIFT key and wheel mouse button. On doing this, the selected object is isolated and everything else is turned off temporarily. Once you stop orbiting, the visibility of everything will be restored. This is a workaround to isolate objects without terminating a tool/dialog box in Advance Steel.

Stiffeners & Plates Category > Shear anchor Tab

The **Shear anchor** tab, shown in Figure 18, is used to specify the parameters related to the shim plates. These options are discussed next.

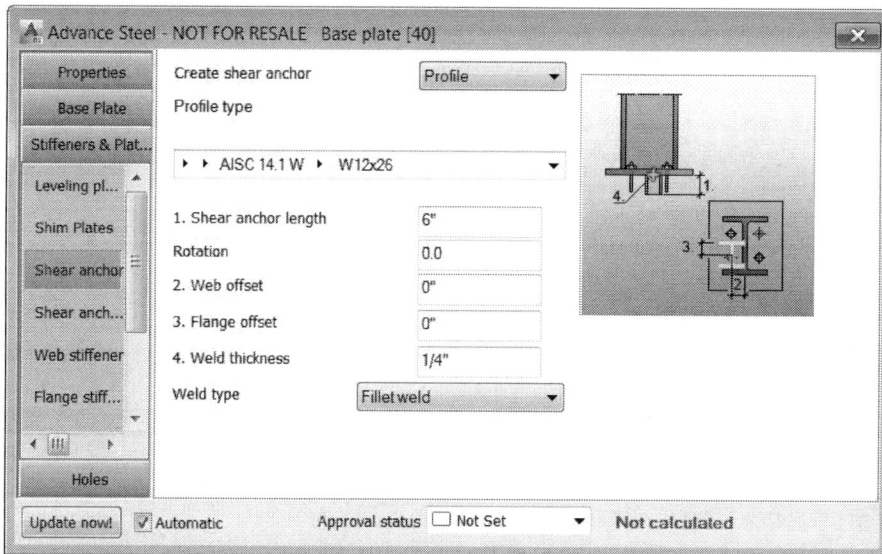

Figure 18 The **Shear anchor** *tab of the* **Stiffeners & Plates** *category*

Create shear anchor

This list is used to specify the type of shear anchor to be created. By default, the **None** option is selected from this list. As a result, the shear anchor is not created. You can select **Profile** from this list to insert a profile shear anchor. Alternatively, you can select **Plate** from this list to insert plates as shear anchors. For both these options, you can enter their values in the exit boxes available below this list. Figure 19 shows a profile being used for shear anchor and Figure 20 shows plates being used for shear anchors. In both these figures, the visibility of the concrete footing is turned off.

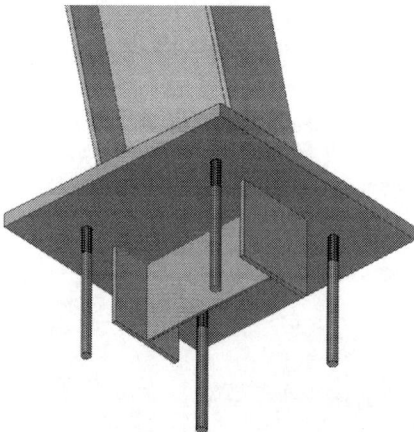

Figure 19 Profile being used for shear anchor

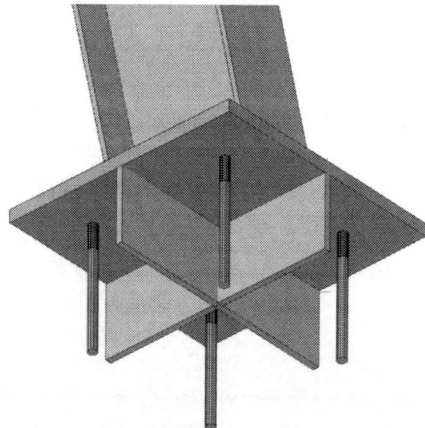

Figure 20 Plates being used for shear anchor

Stiffeners & Plates Category > Shear anchor dimensions Tab

While inserting plates as the shear anchors, you can enter the parameters related to those plates in the **Shear anchor dimensions** tab, shown in Figure 21. The options in this tab are similar to those discussed in earlier tabs.

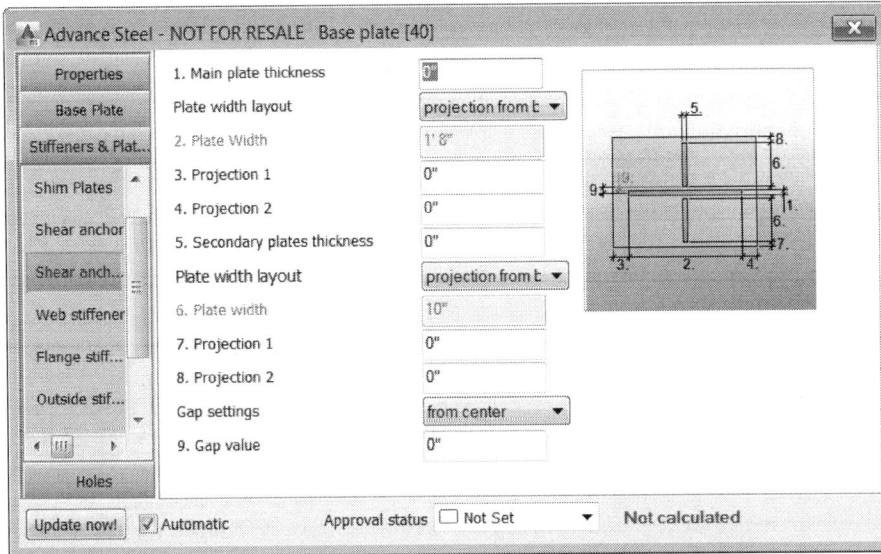

*Figure 21 The **Shear anchor dimensions** tab of the **Stiffeners & Plates** category*

Stiffeners & Plates Category > Web stiffener/Flange stiffener Tabs

The **Web stiffener** tab, shown in Figure 22, and **Flange stiffener** tab are used to insert web and flange stiffeners welded to the column.

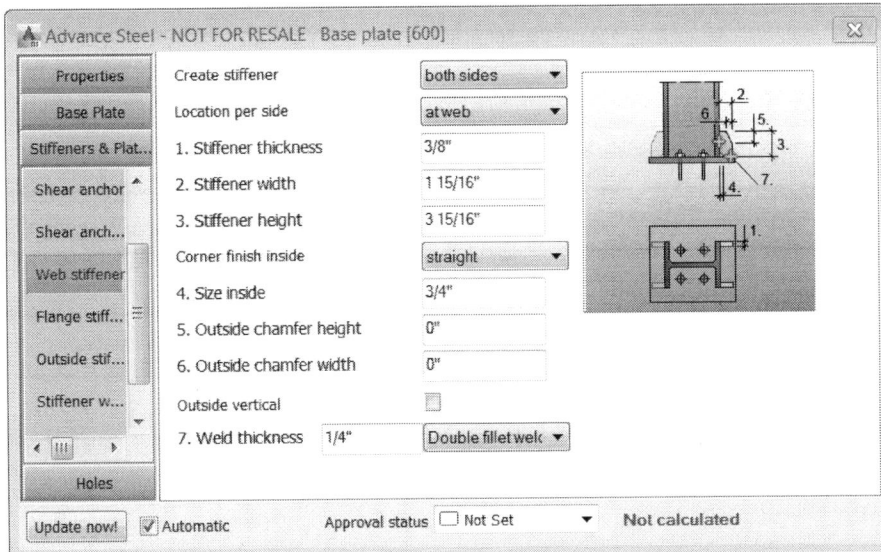

*Figure 22 The **Web stiffener** tab of the **Stiffeners & Plates** category*

The options available in both these tabs are similar. Some of these options are discussed next. The remaining options are similar to those discussed in the other tabs of this dialog box.

Create stiffener

This list is used to specify whether or not the web stiffener will be created. By default, **none** is selected from this list. As a result, the stiffener is not created. However, you can select **one side**, **other side**, or **both sides** to insert the web stiffeners. The parameters related to the stiffeners can be entered in the edit boxes available below this list.

Outside vertical

While creating the base plate joint on an inclined column, you can select this tick box to cut the stiffener such that the outside edge of it is vertical. Figure 23 shows a base plate joint with the flange stiffener created on both sides. In this figure, the **Outside vertical** tick box is not selected. Figure 24 shows the same joint with the **Outside vertical** tick box selected.

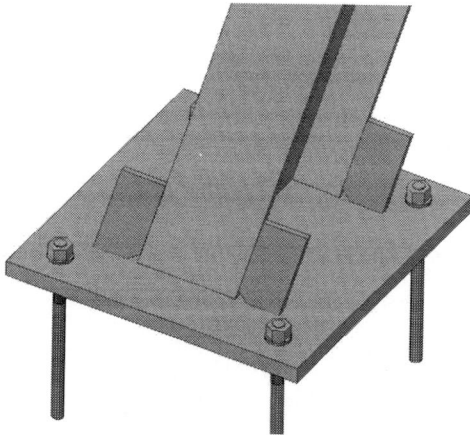

*Figure 23 Flange stiffeners created with **Outside vertical** tick box cleared*

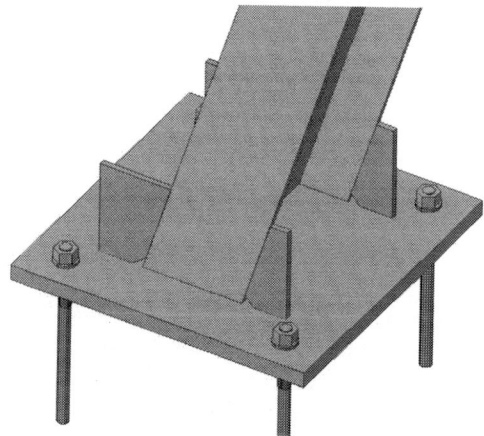

*Figure 24 Flange stiffeners created with **Outside vertical** tick box selected*

Stiffeners & Plates Category > Outside stiffener Tab

The **Outside** tab, shown in Figure 25, is used to insert an outside stiffeners welded to the column. To insert the outside stiffener, select the **Create stiffener** tick box. From the **Layout settings** list, select the option to define the dimensions of the stiffener from the plate or from the column. Figure 26 shows a base plate joint with the outside stiffener created.

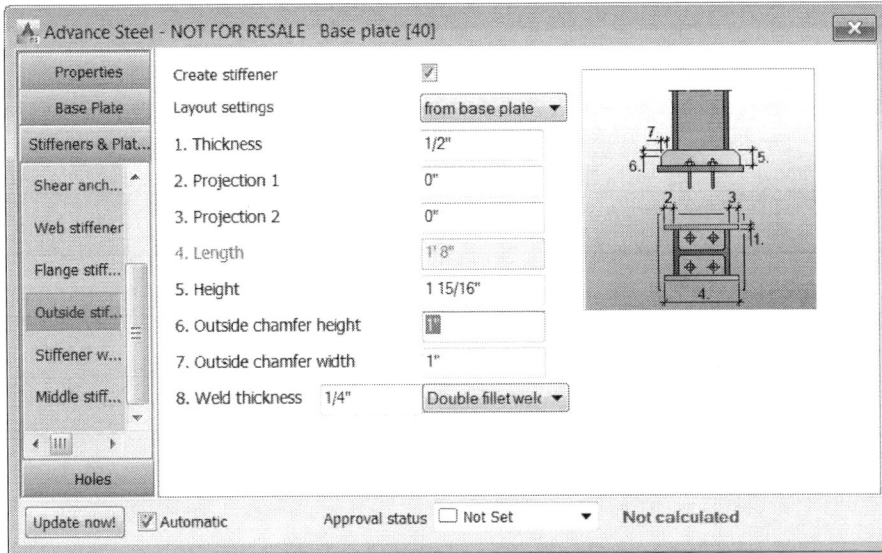

Figure 25 *The* **Outside stiffener** *tab of the* **Stiffeners & Plates** *category*

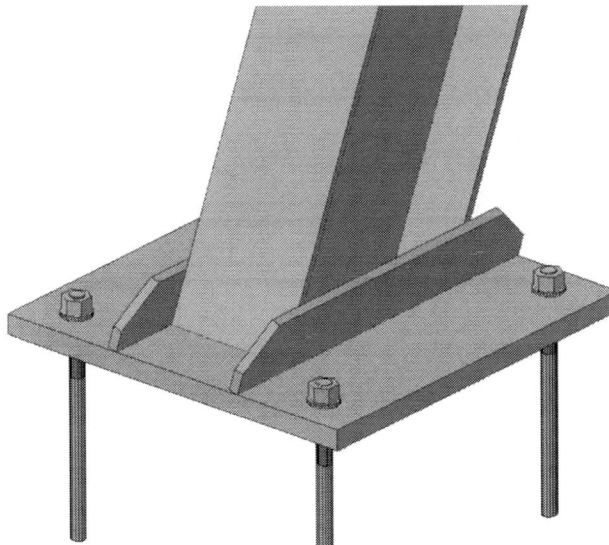

Figure 26 *The base plate joint with the outside stiffener created*

Stiffeners & Plates Category > Stiffener weld preparation Tab

The options in the **Stiffener weld preparation** tab, shown in Figure 27, are only available when you are creating the flange stiffeners. In this case, all the parameters related to weld preparations of the stiffeners can be configured in this tab.

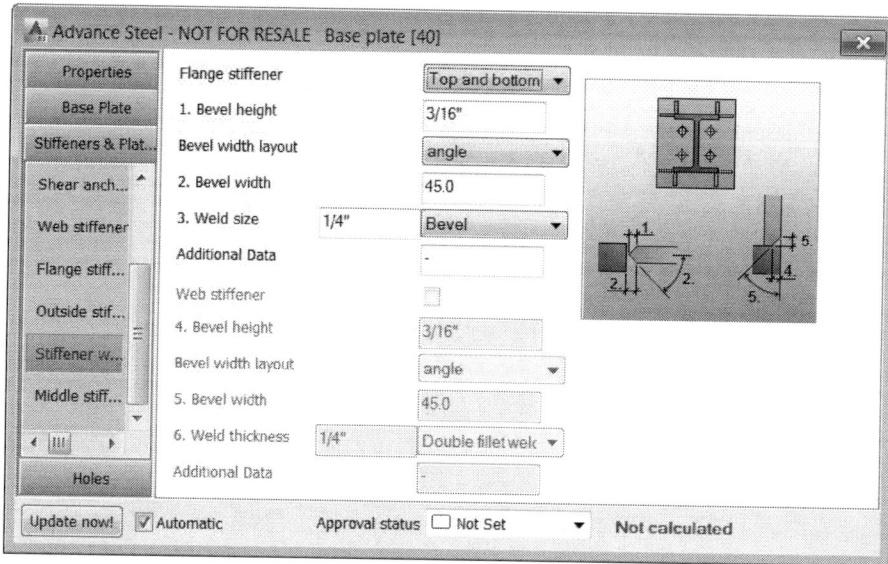

*Figure 27 The **Stiffener weld preparation** tab of the **Stiffeners & Plates** category*

Stiffeners & Plates Category > Middle stiffener Tab

The options in the **Middle stiffener** tab, shown in Figure 28, are used to insert middle stiffeners welded to the column. The options available in this tab are similar to the **Web stiffener** and **Flange stiffener** tabs.

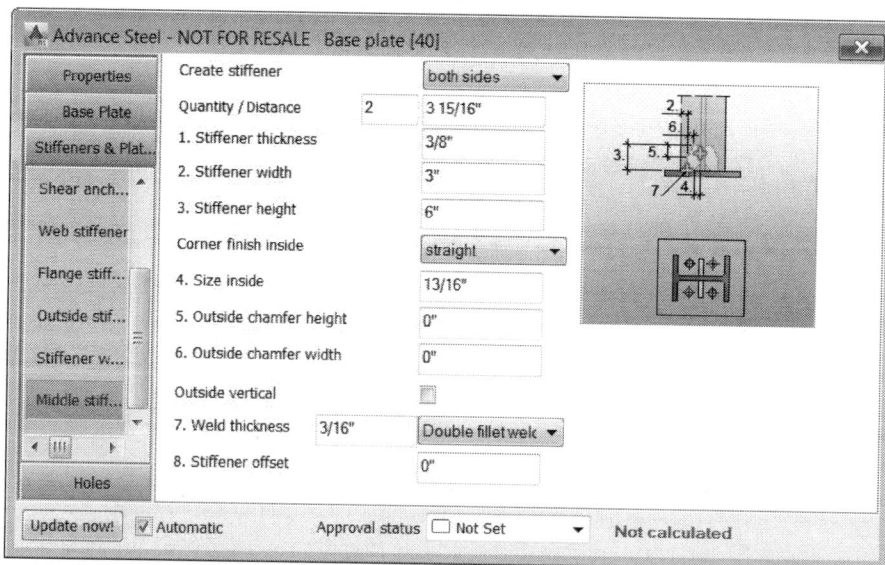

*Figure 28 The **Middle stiffener** tab of the **Stiffeners & Plates** category*

Figure 29 shows a base plate joint with two middle stiffeners created.

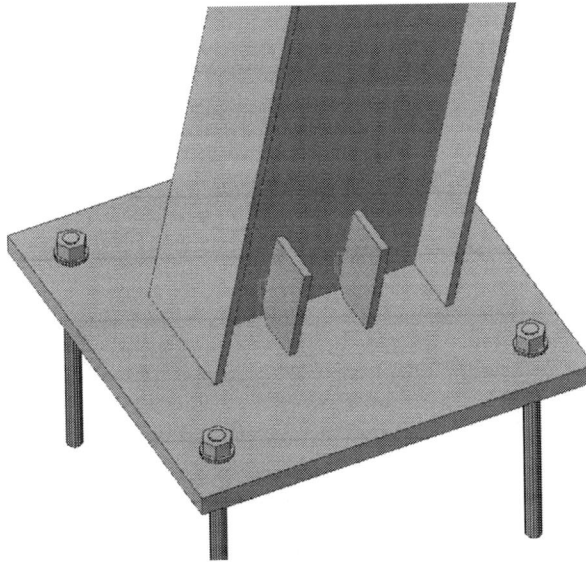

Figure 29 *The base plate joint with two middle stiffener created*

Holes Category > Galvanizing holes Tab

The **Galvanizing holes** tab, shown in Figure 30, is used to specify the parameters related to the galvanizing holes. To create the galvanizing holes, select the **Create galvanizing holes** tick box and then enter the parameters in the options available below this tick box. If you are creating outside stiffeners, you can also create holes in them by selecting the **Create in outside stiffener** tick box.

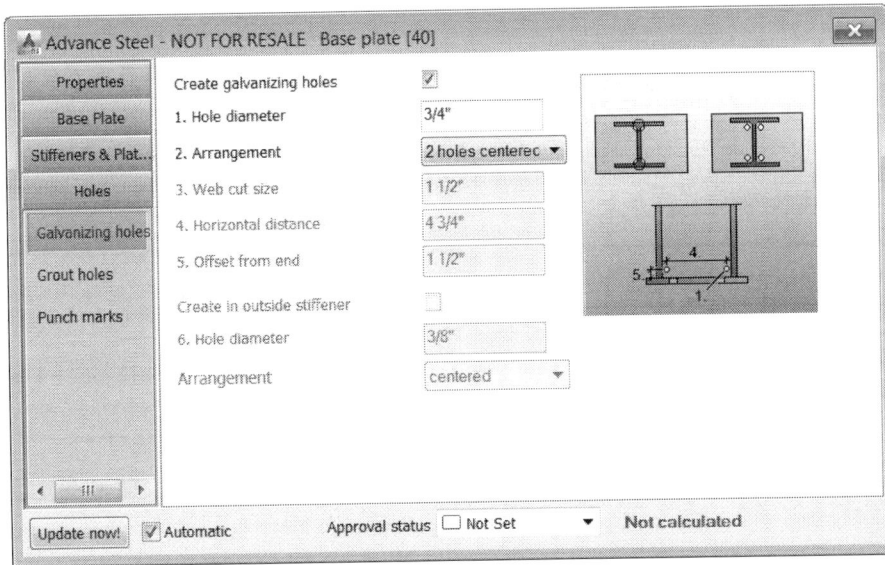

Figure 30 *The* ***Galvanizing holes*** *tab of the* ***Holes*** *category*

Holes Category > Grout holes Tab

The **Grout holes** tab, shown in Figure 31, is used to specify the parameters related to the grout holes on the base plate. To create the grout holes, select the **Create grout holes** tick box and then enter the parameters in the options available below this tick box.

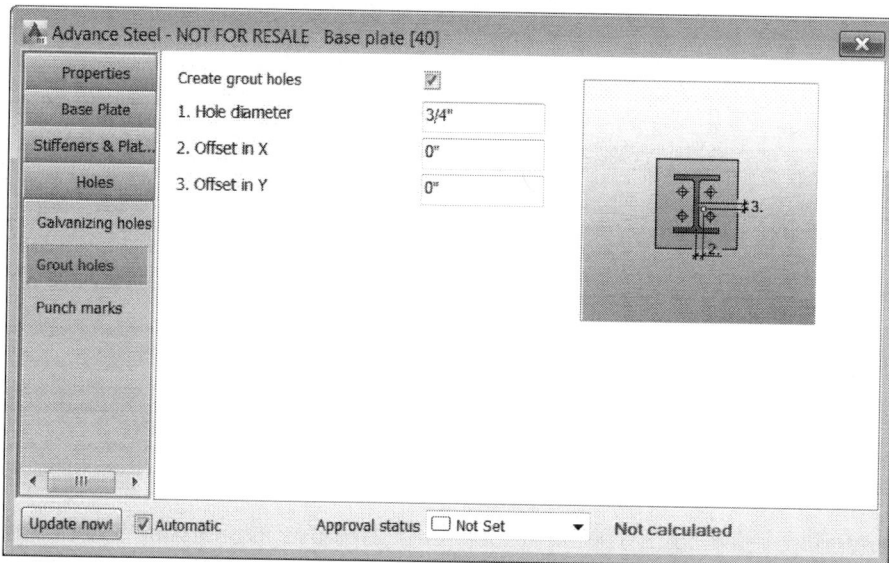

*Figure 31 The **Grout holes** tab of the **Holes** category*

Holes Category > Punch marks Tab

The **Punch marks** tab, shown in Figure 32, is used to specify the parameters related to the punch marks on the base plate.

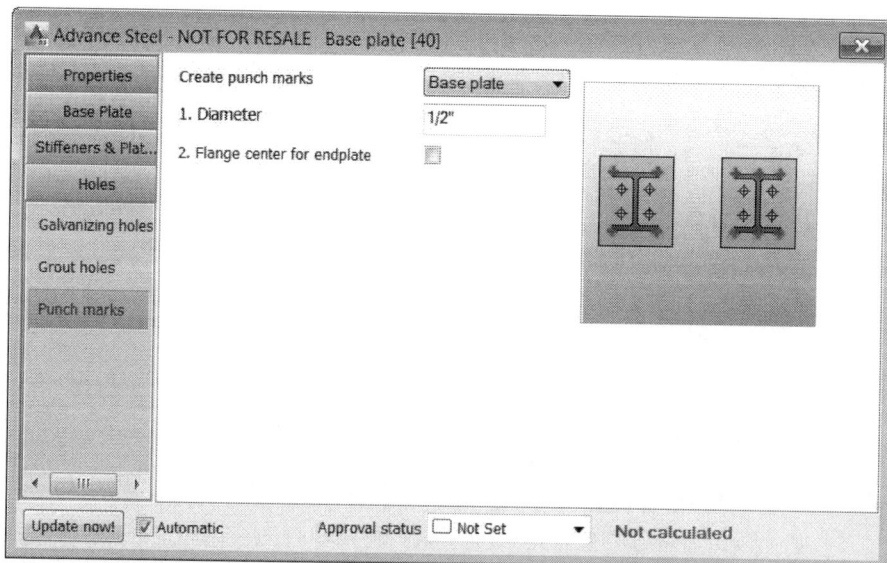

*Figure 32 The **Punch marks** tab of the **Holes** category*

To create the punch marks, select the **Base plate** from the **Create punch marks** list and then enter the parameters in the options available below this list.

Tube base plate Joint

Tube base plate

As the name suggests, this joint type only allows a tube section, such as round hollow section, rectangular hollow section, square hollow section, and so on to be used as the column. Figure 33 shows a tube plate joint of a circular hollow section. In this joint, the washer plates are also created. The dialog box options to create this type of joint are similar to those discussed in the **Base plate** joint.

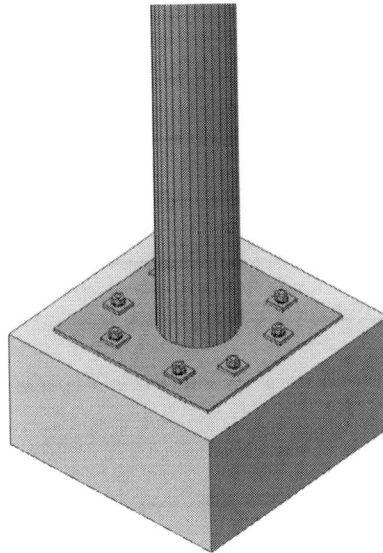

Figure 33 *The tube base plate joint with washer plates*

Corner base plate Joint

Corner base plate

This joint type is generally used for the corner columns and allows any section type to be used as the column. Figure 34 shows a corner base plate joint of an I-section. The dialog box options to create this type of joint is similar to those discussed in the **Base plate** joint.

Base plate cut Joint

Base plate cut

Figure 35 shows an example of the base plate cut joint of an I-section with the washer plate and levelling plate created. The dialog box options to create this type of joint is similar to those discussed in the **Base plate** joint.

The rest of the base plate joints are similar to those discussed above and so they are not discussed.

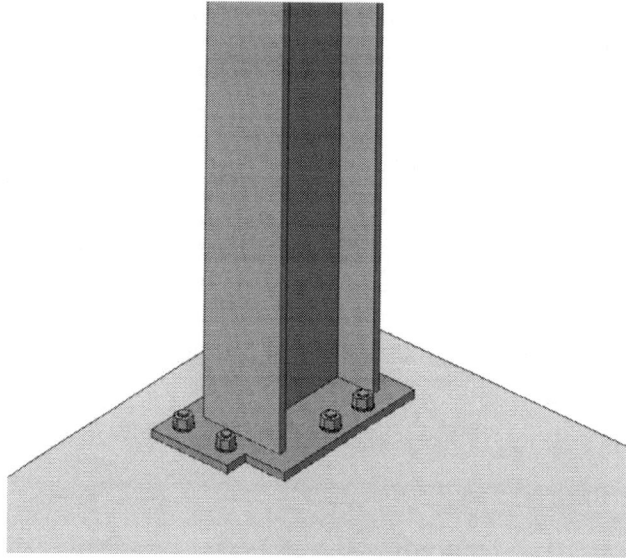

Figure 34 *The corner base plate joint*

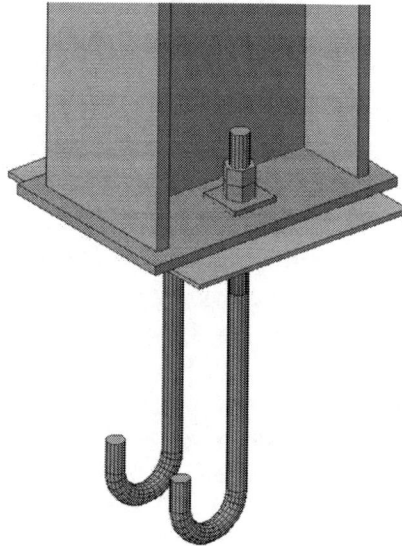

Figure 35 *The base plate cut joint with washer and levelling plates*

INSERTING THE COLUMN - BEAM JOINTS

This category lists various joints that can be added between a column and a beam. These joint types are discussed next.

Note: It is important to mention here that most of the joints in this category are similar. Therefore, only certain joint types are discussed in detail.

Knee of frame bolted, with haunch Joint

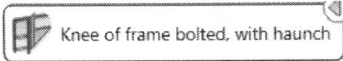

This type of joint is created between the flange of a column and a beam and a haunch is created from the beam or from the plates. When you invoke this tool, you will be prompted to select a column. Once you select the column and press the ENTER key, you will be prompted to select a beam. Upon selecting the beam, press the ENTER key; the **Advance Steel - Knee of frame bolted, with haunch [X]** dialog box will be displayed with the **Properties Category > Properties** tab active. Also, the model in the drawing window will be updated and will show the joint created with the default values. The tabs in the **Properties** category are similar to those discussed for the previous joints. The tabs of the remaining categories are discussed next.

General Category > Haunch - Rafter Tab

When you click on the **General** category in the left pane of the dialog box, the **Haunch - Rafter** tab is the first tab, as shown in Figure 36. The options available in this tab are discussed next.

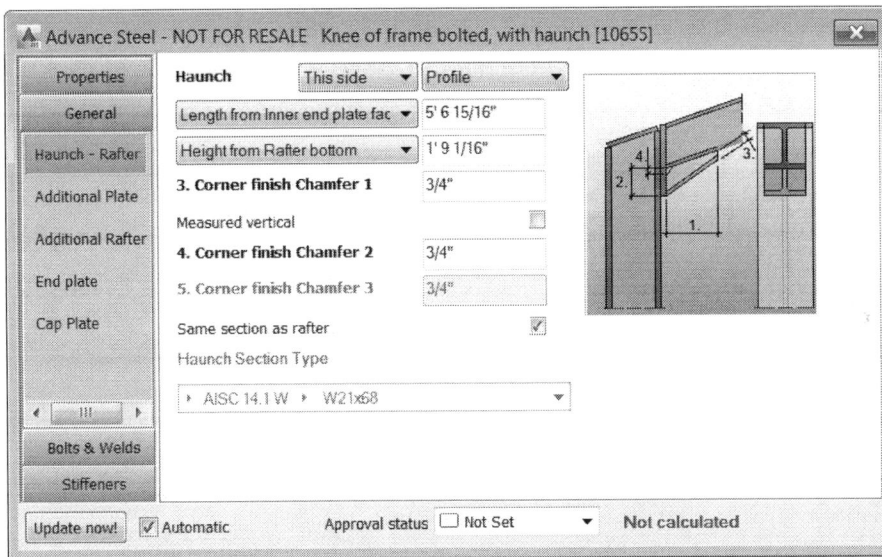

*Figure 36 The **Haunch - Rafter** tab of the **General** category*

Haunch

The first list in this area is used to define the side at which the haunch will be created. The column will be trimmed based on the side you select to insert the haunch. The second list is used to specify whether the haunch will be created based on the profile or the various plates welded together. Selecting **None** from the second list will ensure no haunch is created. Figure 37 shows a joint with the haunch and Figure 38 shows a joint without the haunch.

Length from Inner end plate face

This list is used to specify where will the length of the haunch be measured from. By default, it is measured from the inner end plate face. You can also measure it from the column flange,

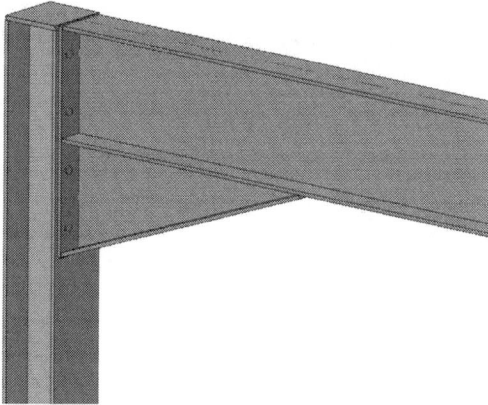

Figure 37 *The column and beam joint created with a haunch*

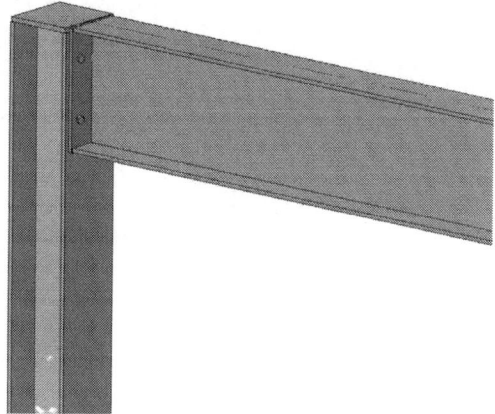

Figure 38 *The column and beam joint created without a haunch*

column axis, or in terms of slope from the end plate. The value is entered in the edit box available on the right of this list.

Height from Rafter bottom
This list is used to specify where will the height of the haunch be measured from. By default, it is measured from the bottom of the rafter. You can also measure it from rafter top, rafter axis, from web depth, or insert it as a single cut haunch. The length value can be entered in the edit box available on the right of this list.

Corner finish chamfer 1/Corner finish chamfer 2/Corner finish chamfer 3
These edit boxes are used to specify the corner finish chamfer values. Note that **Corner finish chamfer 3** edit box will only be available when you are inserting haunch based on plates and not profile.

Measured vertical
Select this tick box to measure the chamfer value at the end of the haunch vertically.

Same section as rafter
While inserting the haunch based on profile, select this tick box to insert the haunch with the same section as that of the rafter. If you clear this tick box, you can select a different profile for the haunch from the list below this tick box.

General Category > Additional Plate Tab
The **Additional Plate** tab, shown in Figure 39, is used to insert additional plates under the haunch and at the end of it. To insert these plates, select their respective tick boxes and then enter the plate parameters in the edit boxes available in this tab.

General Category > Additional Rafter Tab
The **Additional Rafter** tab, shown in Figure 40, is used to insert an additional rafter welded on the other side of the column. To insert this additional rafter, select the **Additional Rafter** tick box

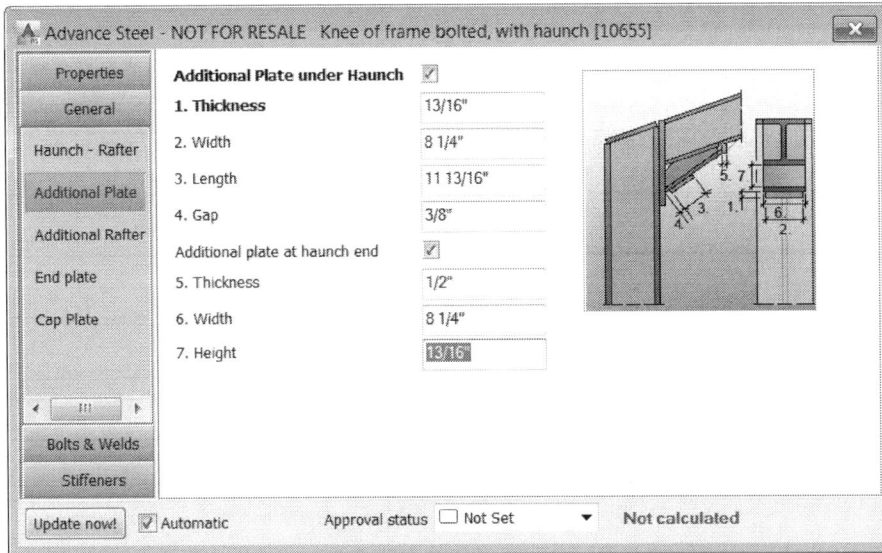

Figure 39 The *Additional Plate* tab of the *General* category

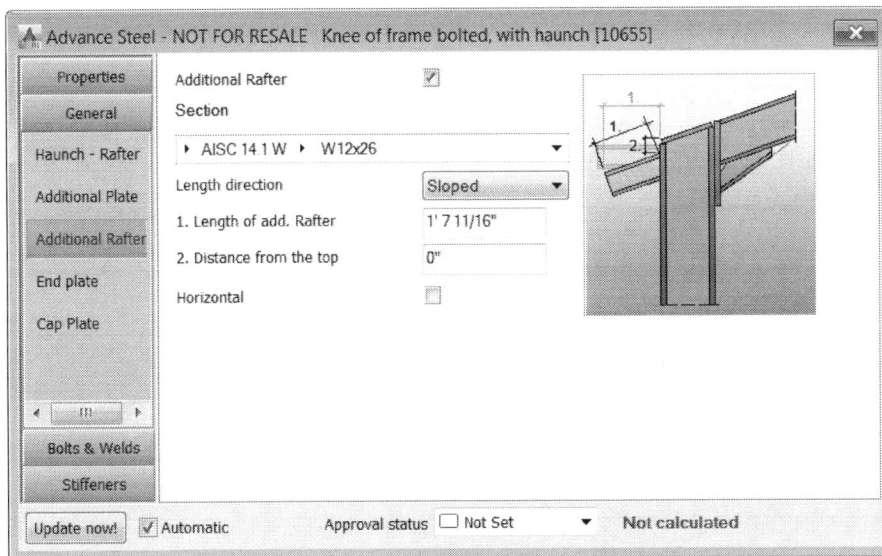

Figure 40 The *Additional Rafter* tab of the *General* category

and then enter the parameters in the edit boxes available in this tab. Figure 41 shows a joint with an additional rafter aligned with the original rafter and Figure 42 shows the additional rafter inserted in the horizontal orientation by selecting the **Horizontal** tick box.

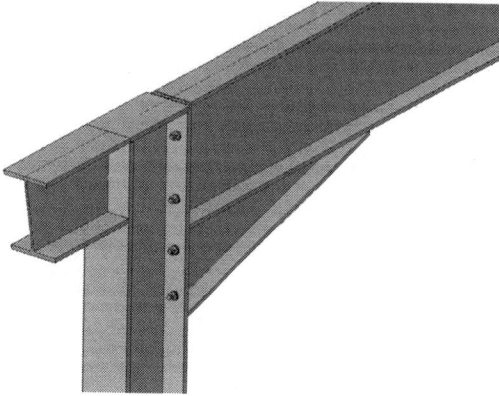

Figure 41 *Additional rafter aligned with the parent rafter*

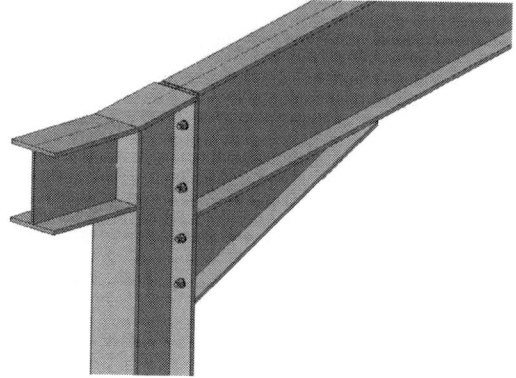

Figure 42 *Additional rafter inserted in the horizontal orientation*

General Category > End plate Tab

The **End plate** tab, shown in Figure 43, is used to specify the parameters related to the end plate. The length and width of the end plate can be defined in terms of the projections from the rafter and haunch or in terms of exact dimension values.

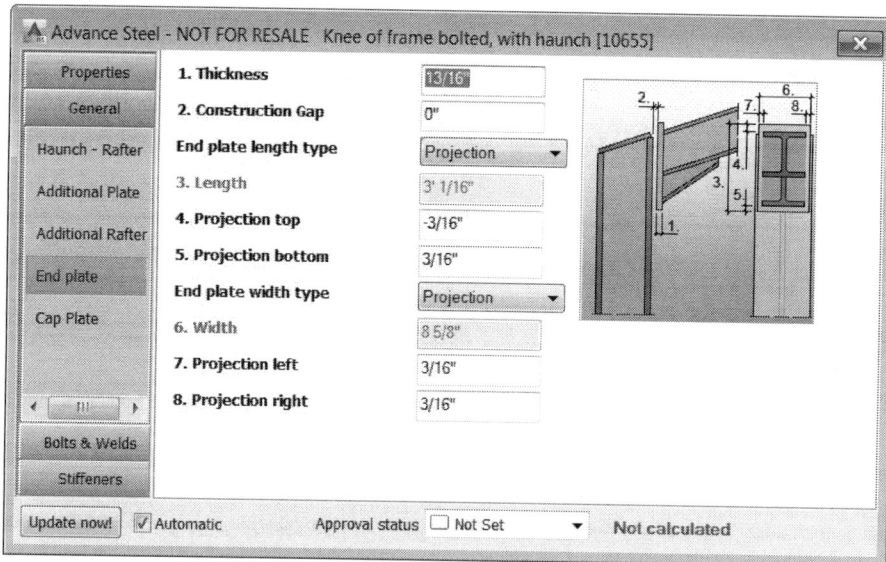

Figure 43 *The End plate tab of the General category*

General Category > Cap plate Tab

The **Cap plate** tab, shown in Figure 44, is used to specify the parameters related to the cap plate at the top of the column. By default, **Full** is selected from the **Cap plate type** list. As a result, the full cap plate is created. You can also select the option to insert half cap plate, cap plate based on

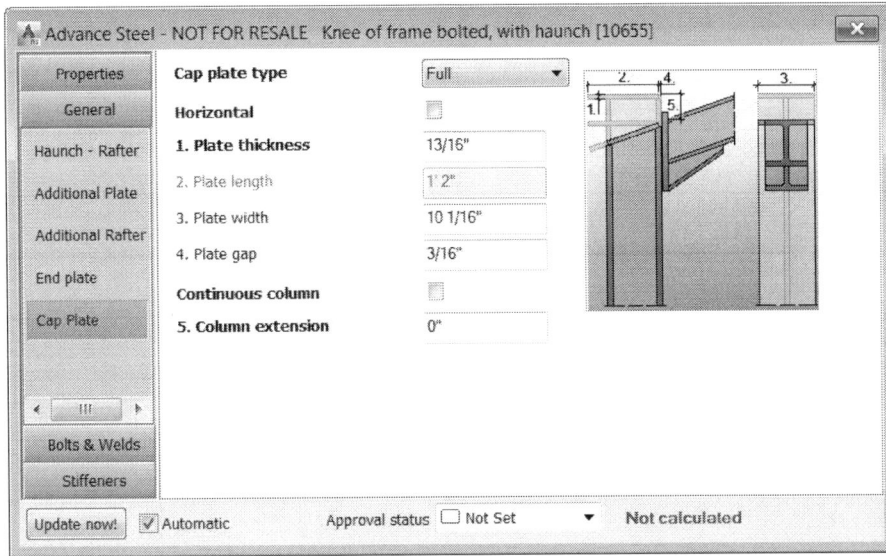

Figure 44 *The* **Cap plate** *tab of the* **General** *category*

dimension values, or no cap plate. Selecting the **Horizontal** tick box will cut the column horizontal, as shown in Figure 45. Clearing this tick box will ensure the column is cut in the same slope as the rafter, as shown in Figure 46.

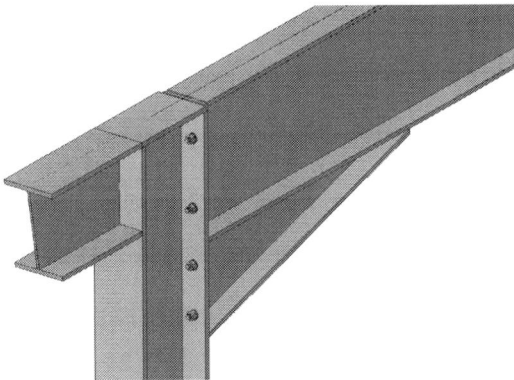

Figure 45 *Column cut aligned with the rafter slope*

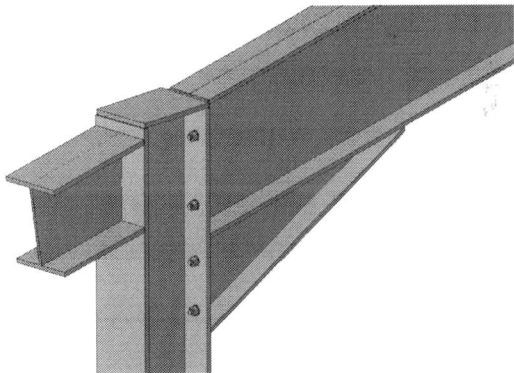

Figure 46 *Column cut horizontal*

Selecting the **Continuous column** tick box will disable the options to insert the cap plate and also ensures that the column is not trimmed.

Bolts & Welds Category > Bolts Tab

The **Bolts** tab, shown in Figure 47, is used to specify the parameters related to the bolts to be used for the joint. These options are discussed next.

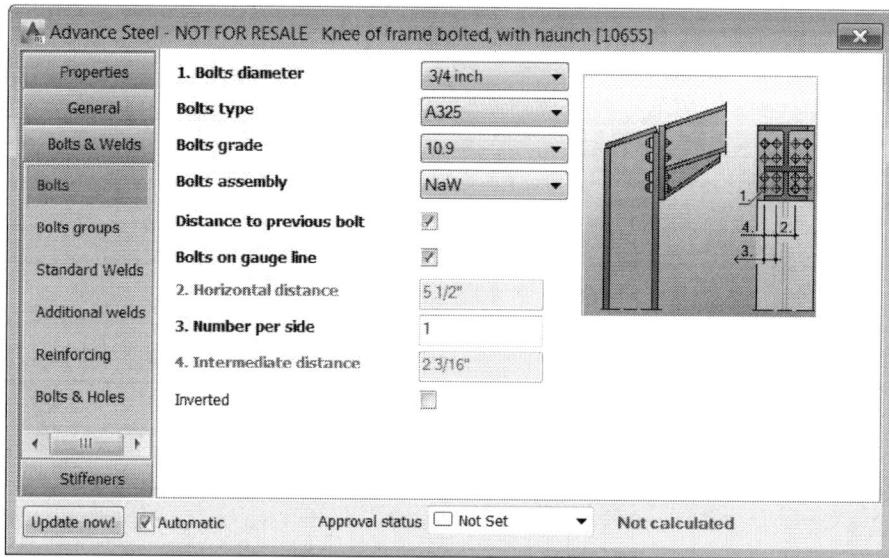

*Figure 47 The **Bolts** tab of the **Bolts & Welds** category*

Bolts diameter
Select the bolt diameter to be used from this list.

Bolts type
Select the bolt type to be used from this list.

Bolts grade
Select the bolt grade to be used from this list.

Bolts assembly
Select the bolt assembly to be used from this list.

Distance to previous bolt
Selecting this tick box will ensure that the dimension values entered on the **Bolts groups** tab are measured from the previous bolt. If you clear this tick box, all the values will have to be entered from the top of the end plate.

Bolts on gauge line
Select the this tick box to insert the bolt on the gauge line. If you clear this tick box, you will be able to insert the horizontal distances between the bolts in the edit box available below this tick box.

Horizontal distance
This tick box is only available when you not are inserting the bolts on the gauge line. In that case, enter the horizontal distance between the bolts in this edit box.

Number per side

Enter the number of bolt columns to be inserted on each side of the rafter.

Intermediate distance

This edit box is only available if you are inserting more than one column of bolts on each side of the rafter. In that case, enter the intermediate distance between the bolt columns in this edit box. Figure 48 shows two columns of bolts inserted on each side of the rafter. In this case, the bolts are not inserted on the gauge line.

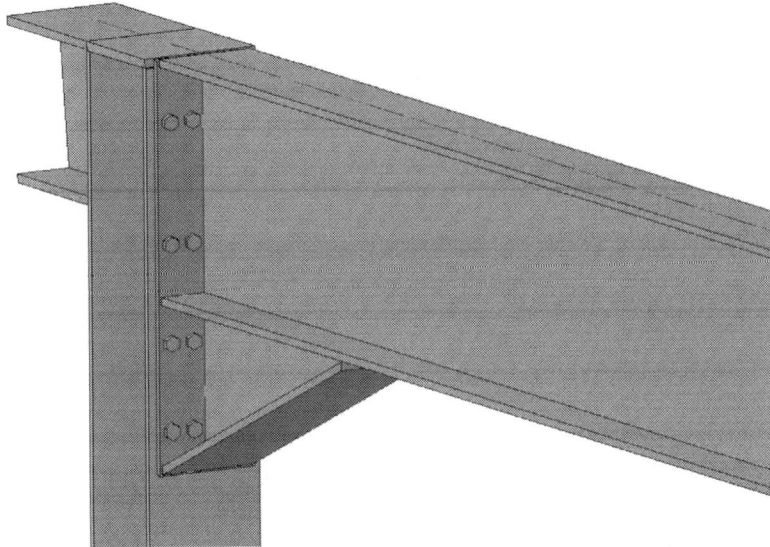

Figure 48 *Two columns of bolts inserted on each side of the rafter*

Inverted

Selecting this tick box will invert the direction of bolts.

Bolts & Welds Category > Bolts groups Tab

The **Bolts groups** tab, shown in Figure 49, is used to specify the number of groups and lines of bolts to be inserted. By default, there are two groups of bolts inserted with each group containing two lines of bolts. Their distances are specified in the **Start dist.** and **Interm. dist.** edit boxes. You can add additional groups of bolts and lines, if required by entering their values in the edit boxes available in this tab. Figure 50 shows a joint with group one containing three lines of bolts between the rafter flanges and group two containing two lines of bolts between the haunch flanges.

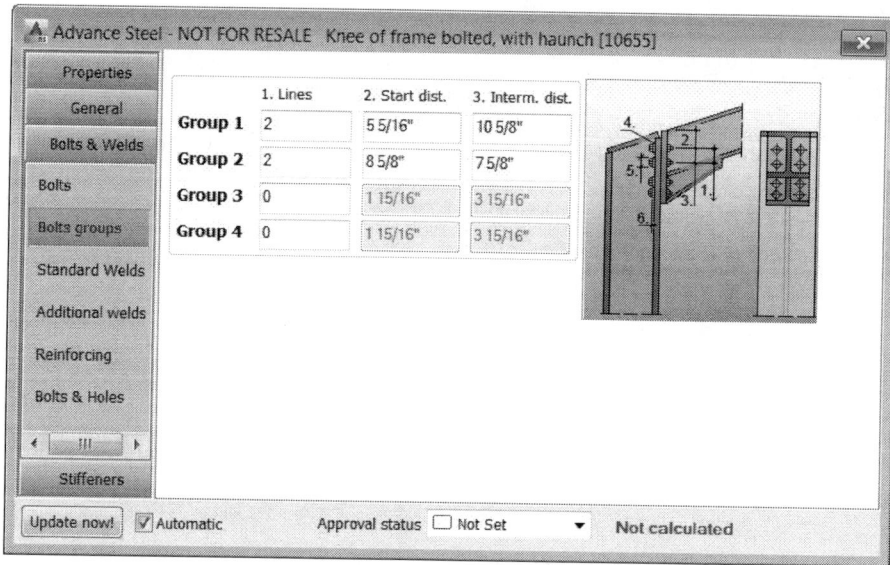

Figure 49   *The* **Bolts groups** *tab of the* **Bolts & Welds** *category*

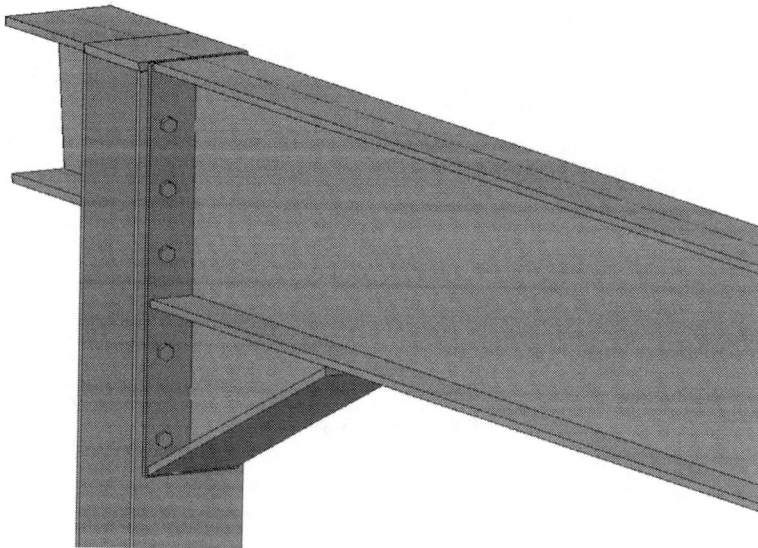

Figure 50   *Bolt groups inserted for the joint*

Bolts & Welds Category > Standard Welds/Additional welds Tabs

The **Standard Welds** tab, shown in Figure 51, and the **Additional welds** tab are used to specify the values of various welds to be created for the joint.

Bolts & Welds Category > Reinforcing Tab

The **Reinforcing** tab, shown in Figure 52, is used to insert reinforcing plates between the nut washer and the column. You can select the options to insert the reinforcing plates for all the bolt

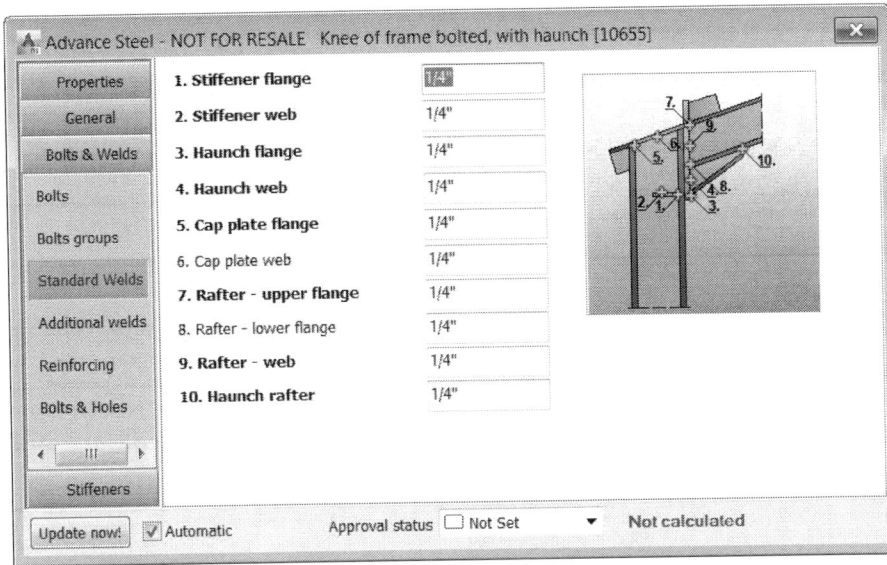

*Figure 51 The **Standard Welds** tab of the **Bolts & Welds** category*

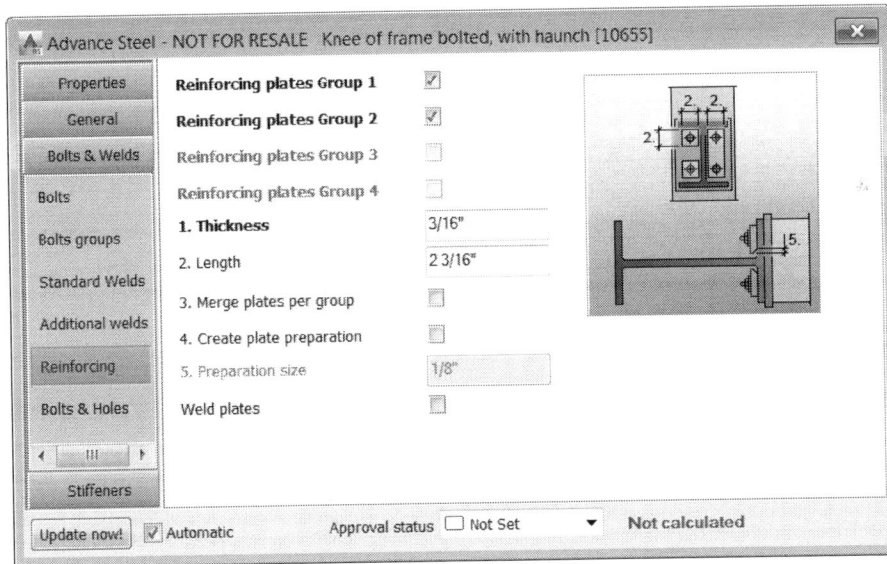

*Figure 52 The **Reinforcing** tab of the **Bolts & Welds** category*

groups and then enter their sizes in the edit boxes available in this tab. Selecting the **Merge plates per group** tick box will insert a single plate per group, as shown in Figure 53.

Bolts & Welds Category > Bolts & Holes Tab

The **Bolts & Holes** tab is used to insert slotted holes on the column, end plate, or both. By default, standard holes are inserted because **none** is selected from the **Slotted part** list. To insert slotted holes on column, end plate, or both, select the required option from this list.

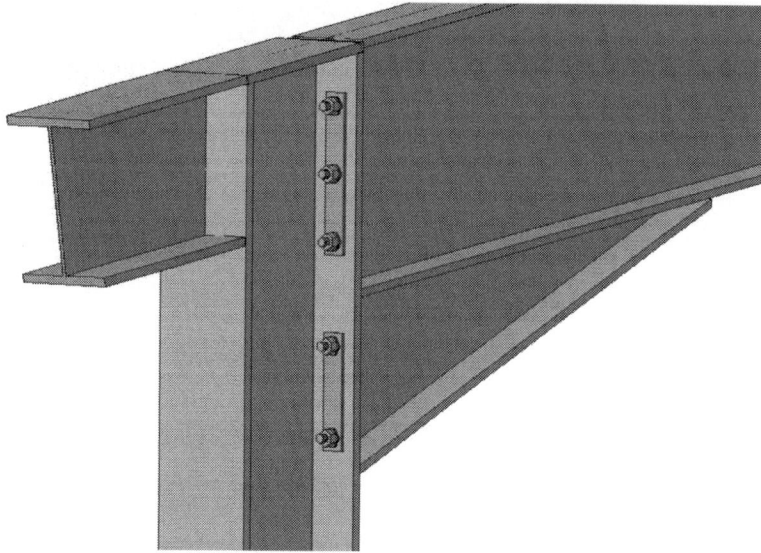

Figure 53 Reinforcing plates merged for the two groups of bolts

Stiffeners Category > Stiffeners 1 Tab

The **Stiffeners 1** tab, shown in Figure 54, is used specify the type of stiffeners to be inserted on the column and rafter.

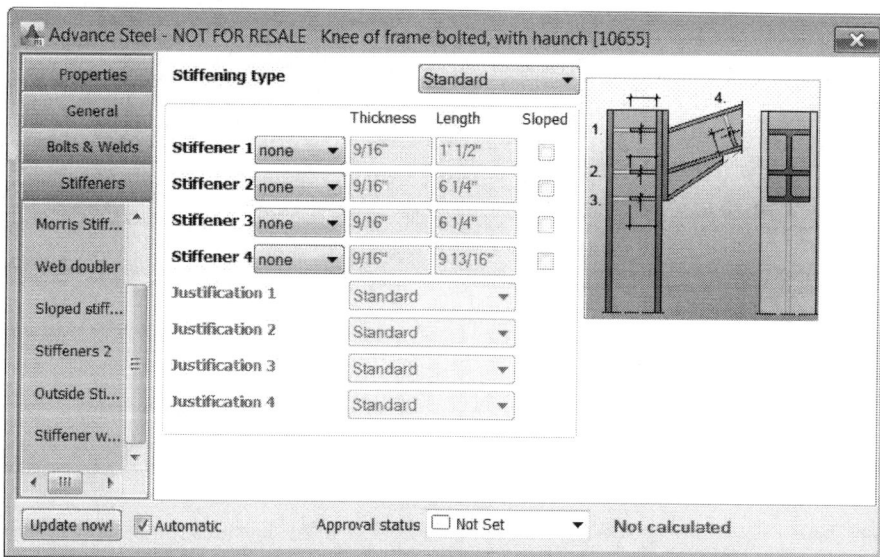

*Figure 54 The **Stiffener 1** tab of the **Stiffeners** category*

By default, **Standard** is selected from the **Stiffening type** list. Also, from the **Stiffener 1, 2, 3, 4** lists are set to **none**. As a result, no stiffener is inserted. You can select to insert half, full, or 3/4 stiffeners from these lists or select the option to insert the stiffener based on a value. Figure 55 shows full stiffeners 2, 3, and 4 inserted on the column and rafter. Selecting **Morris stiffener** from the **Stiffening type** list will insert the morris stiffener. You can enter the values of this stiffener in the **Morris Stiffeners** tab of this category. Figure 56 shows the morris stiffener inserted with the full stiffener 3 added from the **Stiffener 1** tab.

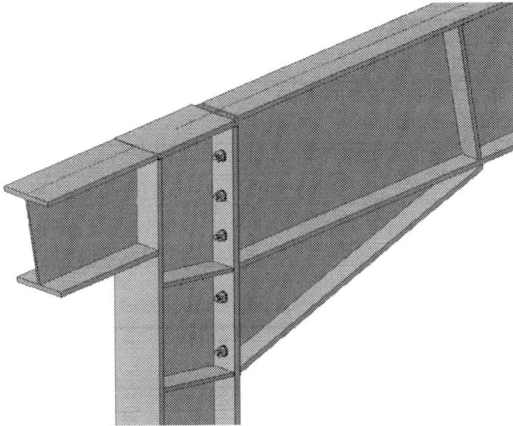

Figure 55 *Full standard 2, 3, and 4 stiffeners inserted*

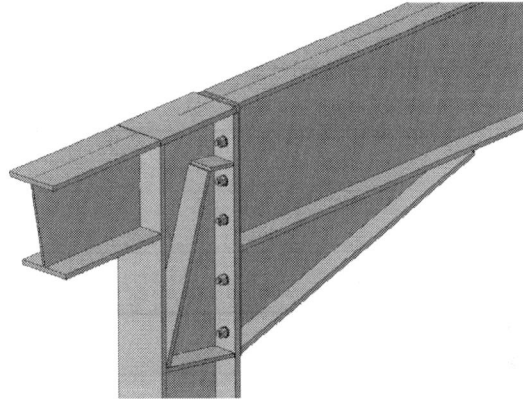

Figure 56 *Morris stiffener and standard stiffener 3 inserted*

Selecting **Web doubler** from the **Stiffening type** list will insert the web doubler stiffener on the column flange. You can enter the values of this stiffener in the **Web doubler** tab of this category. Selecting **Sloped stiffener** from the **Stiffening type** list will insert a sloped stiffener. You can enter the values of this stiffener in the **Sloped stiffeners** tab of this category. Figure 57 shows the sloped stiffener inserted with the full stiffener 3 added from the **Stiffener 1** tab.

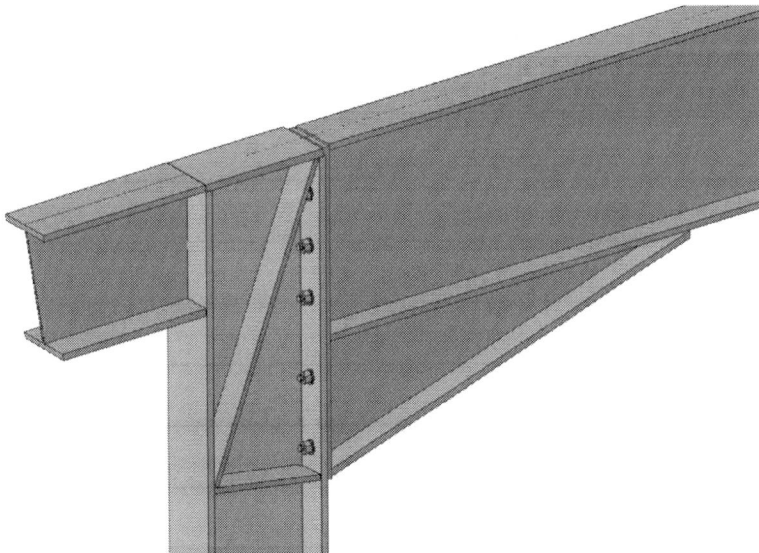

Figure 57 *Sloped stiffener and standard stiffener 3 inserted*

Knee of frame at web, with haunch Joint

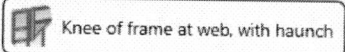

📊 Knee of frame at web, with haunch

This type of joint is created between the web of a column and a beam, and a haunch is created from the beam or from the plates. The options to create this type of joint are similar to those discussed in the previous joint. Figure 58 shows a knee of frame at web with a haunch joint.

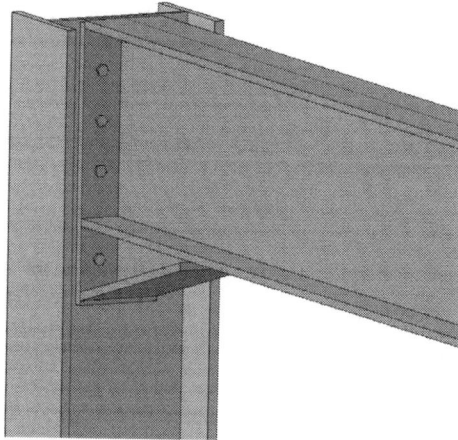

Figure 58 Knee of frame at web, with haunch joint

Gable wall end plate Joint

🔧 Gable wall end plate

This type of joint is specifically used for the gable frames and is created between the rafter and the gable post. Figure 59 shows this type of joint with a full centered stiffener.

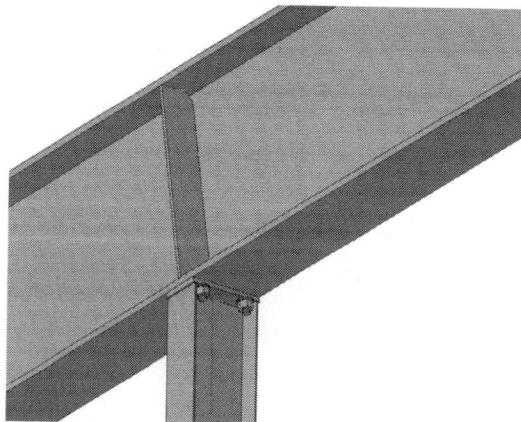

Figure 59 Gable wall end plate joint with full centered stiffener

The process of applying the remaining Column - Beam joints is similar to those discussed above. As a result, they are not discussed in detail.

ADDING JOINTS TO FAVORITES CATEGORY

The **Favorites** category is used to list the joints that you create on regular basis. This category can be invoked by click the first button in the **Joint Category** area of the **Connection vault** palette. To add a joint to the **Favorites** category, click on the star icon on the right of that joint name, as shown in Figure 60. When you do this, the star icon changes to Yellow and the joint is listed in the **Favorites** category.

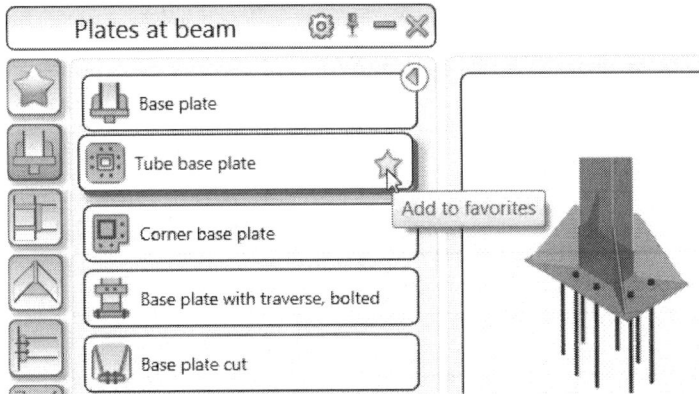

Figure 60 *Adding a joint to the* ***Favorites*** *category*

EDITING JOINTS

Advance Steel allows you to edit joints after you have inserted them. However, the process of editing the joints is different, depending on whether the **Connection boxes** (Imperial) or **Joint box** (Metric) layer that controls the visibility of the joint boxes around the joints is turned on or off. If this layer is turned on, the connection box around the joint is displayed, as shown in Figure 61. In that case, double-click on this box to edit the joint.

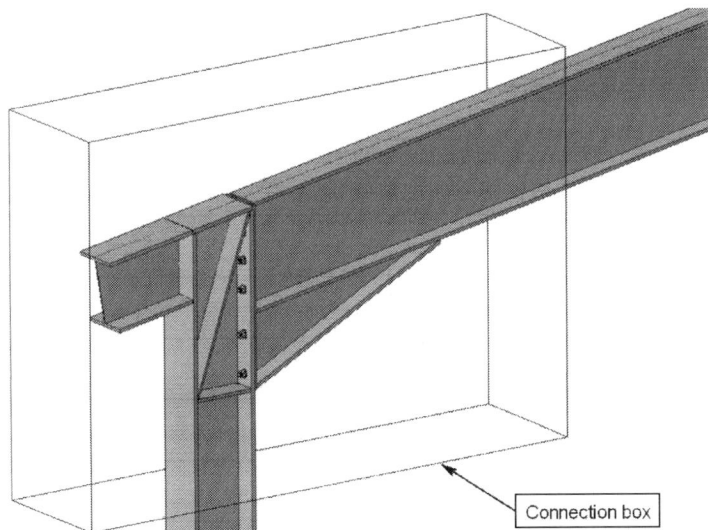

Figure 61 *Connection box displayed around the joint*

In a complex model, if the **Connection boxes** (Imperial) or **Joint box** (Metric) layer is turned off for reducing the objects on the screen, you can select any item of the joint such as the bolt, plate, or stiffener and then right-click to display the shortcut menu. From this menu, select **Advance Joint Properties**, as shown in Figure 62, to edit the joint.

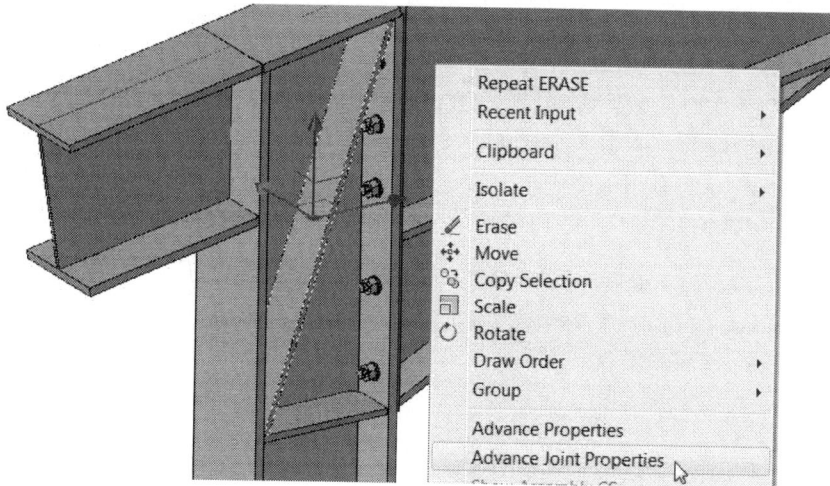

Figure 62 Editing the joint using the shortcut menu

COPYING JOINTS

In a complex model, you need to create similar joints at various places. In Advance Steel, instead of creating all those joints individually, you can create one joint and then copy it at the rest of the locations. This allows you to create the "master-slave" relationship. This relationship ensures that if the master joint changes, the slave joints are updated automatically. If required, you can also break the master-slave relationship between the joints. The joints are copied using the **Tools** tab of the **Advance Steel** tool palette. This tool palette provides you an option to copy the joint at a single location or multiple locations. The processes of creating both these joint copies are discussed next.

Note: *It is important to mention here that while copying the joints, you need to remember the selection sequence of the source joint. The same sequence will be used to select the objects while copying the joints.*

Process for Creating a Single Copy of the Joint

If you need to create only a single copy of the joint, you can use the **Create joint in a joint group** tool. The process for creating a single copy of the joint using this tool is discussed next.

1. From the **Advance Steel** tool palette > **Tools** tab, click the **Create joint in a joint group** tool; you will be prompted to select the connection part.

2. Select one of the items from the source connection, such as the bolt, end plate, stiffener, and so on and then press the ENTER key; you will be prompted to select objects.

Tip: If you press the **F2** key, the **AutoCAD Text Window** will be displayed that will list the type of object to be selected for the first and second selection.

3. Select the first object and then press ENTER; you will be prompted to select the second object.

4. Select the second object and press the ENTER key; the joint will be copied and you will again be prompted to select the first object.

Process for Creating Multiple Copies of the Joint

If you need to create only multiple copies of the joint, you can use the **Create joint in a joint group, multiple** tool. The process for creating multiple copies of the joint using this tool is discussed next.

1. From the **Advance Steel** tool palette > **Tools** tab, click the **Create joint in a joint group, multiple** tool; you will be prompted to select the source joint.

2. Select one of the items from the source connection, such as the bolt, end plate, stiffener, and so on and then press the ENTER key; you will be prompted to select beams corresponding to the source joints input (1/2).

3. Select all the first set of objects to be used for copying the joint and then press the ENTER key; you will be prompted to select beams corresponding to the source joints input (2/2).

4. Select all the second set of objects to be used for copying the joint and then press the ENTER key; the joints will be copied between all the selected objects and you will again be prompted to select the first object.

UPGRADING JOINT TO MASTER

Once you have created master-slave relationship between the joints, you cannot edit the slave joint as the options in the dialog box are grayed out. This is because the slave joints are automatically updated when you update the master joint. However, on a large project that spans over a few months, you may not remember which joint was the master that was used to create copies of the joint. In that case, Advance Steel allows you to upgrade any slave joint to master so you can edit that joint and also all the other joints in the group. The process of upgrading a joint to master is discussed next.

1. If the connection box of the slave joint is displayed, double-click on it to edit the joint. Else, select any item of the joint. Then right-click and select **Advance Joint Properties** from the shortcut menu; the dialog box of the joint will be displayed.

2. From the **Properties** category > **Properties** tab of the dialog box, select the **Upgrade to master** tick box.

Tip: If the **Upgrade to master** tick box is grayed out, that means either the selected joint is the master joint or there are no slave joints associated to that joint.

ADDING OR REMOVING JOINTS FROM THE JOINT GROUP

You can use the tools available on the **Advance Steel** tool palette > **Tools** tab to add a joint to the joint group or remove a joint from the joint group. The processes for both these activities are discussed next.

Process for Adding a Joint to the Joint Group

To add a joint to the joint group, you can use the **Add joint to the joint group** tool. The process of using this tool to add a joint to the joint group is discussed next.

1. From the **Advance Steel** tool palette > **Tools** tab, click the **Add joint to the joint group** tool; you will be prompted to select the joint.

2. Select one of the items from the joint that needs to be added to the group, such as the bolt, end plate, stiffener, and so on and then press the ENTER key; you will be prompted to select the master joint.

3. Select any item in the master joint and then press the ENTER key; the joint will be added to the master joint group.

Process for Removing a Joint from the Joint Group

If you want to independently modify a joint without affecting the other joints of that group, you need to first remove that joint from the group. This can be done using the **Remove joint from a joint group** tool. The process of doing this is discussed next.

1. From the **Advance Steel** tool palette > **Tools** tab, click the **Remove joint from a joint group** tool; you will be prompted to select connection part.

2. Select one of the items from the joint that needs to be removed from the group and then press the ENTER key; the joint will be removed.

Tip: To ensure a joint is removed from the group, edit the joint and check if the **Upgrade to master** tick box is grayed out or not. If it is grayed out, the joint is removed from the group.

Hands-on Tutorial (STRUC/ BIM)	*In this tutorial, you will complete the following tasks:* 1. *Open the file containing the frames.* 2. *Insert the base plate joints, as shown in Figures 63 and 64.* 3. *Insert the **Knee of frame bolted, with haunch** joint, as shown in Figure 65.* 4. *Insert the **Gable wall end plate** joint, as shown in Figure 66.* 5. *Copy all the joints at other places in the model.* 6. *Create additional joints in the Elevator Tower or Hospital Structure model.* 7. *Modify the base plate joints in the Elevator Tower or the Hospital Structure model.*

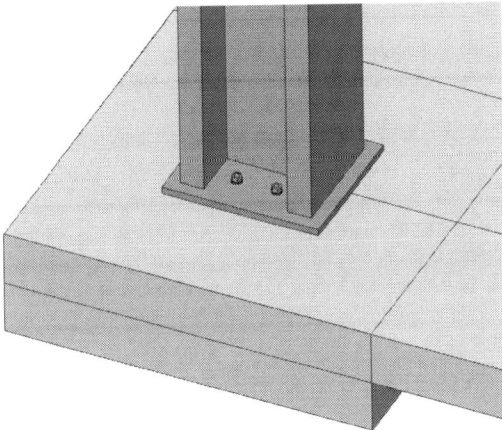

Figure 63 The base plate joint for the portal and gable frame columns

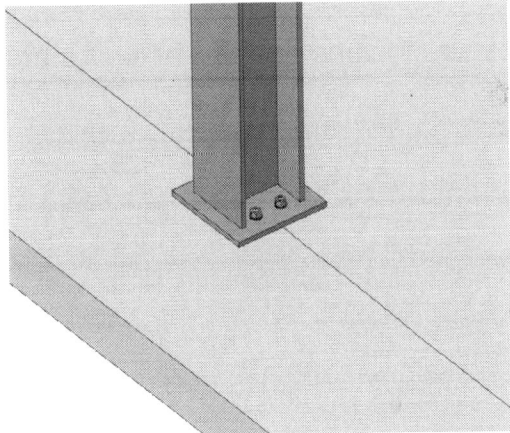

Figure 64 The base plate joint for the gable posts

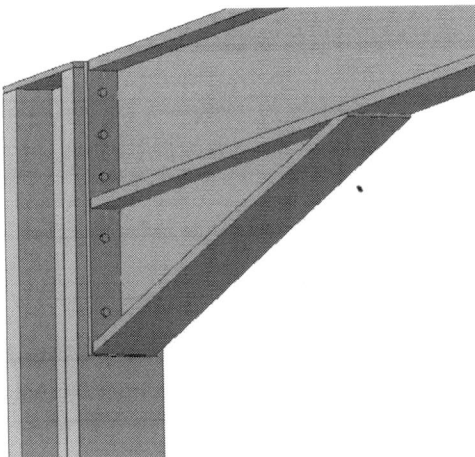

Figure 65 The knee of frame bolted, with haunch joint for the columns and rafters

Figure 66 The gable post with rafter joint

Base Plate Joint Parameters

The following are the parameters for the base plate joint for the frame columns. The dimensions not listed below need to accepted with their default values. **Note that if you are using the Metric units, do not enter the unit string (mm) while typing the dimension values.**

	Columns	Gable Posts
Plate Thickness	1" / 20 mm	3/4" / 16mm
Base Plate Projection 1	2" / 50mm	1" / 25mm
Anchor dia	3/4" / 16mm	3/4" / 16mm
Anchors Parallel Web No.	2	2
Intermediate distance	10" / 180mm	8" / 120mm
Anchors Parallel Flange No.	2	2
Intermediate distance	6" / 150mm	4" / 75mm

Knee of frame bolted, with haunch Joint Parameters

The following are the parameters for the knee of frame bolted, with haunch joint between the columns and rafters. The dimensions not listed below will be accepted with their default values.

End Plate Thickness	1" / 20 mm
Cap Plate Thickness	1" / 20 mm
Bolt Dia	1" / 20mm
Horizontal Distance	7" / 100mm
Bolt Group 1	3 Lines
Bolt Group 1 Start Dist.	4 7/8" / 65mm
Bolt Group 1 Interm. Dist.	6 5/8" / 150mm
Bolt Group 2	2 Lines
Bolt Group 1 Start Dist.	9 5/8" / 200mm
Bolt Group 1 Interm. Dist.	11 5/8" / 220mm

Gable wall end plate Joint Parameters

The following are the parameters for the gable wall end plate joint between the rafters and gable posts. The dimensions not listed below will be accepted with their default values.

Plate Thickness	1/2"/10 mm
Number of bolts in X	2
Intermediate distance X	5" / 75mm
Number of bolts in Y	2
Intermediate distance X	5" / 75mm
Stiffener type	Full
Stiffener arrangement	Centered

Section 1: Opening the File and Inserting the Plate at Beams Joints

In this section, you will open **C04-Imperial-Frame.dwg** or **C04-Metric-Frame.dwg** file and then insert the base plate joints to one of the columns and one of the gable posts.

1. From the **C04 > Struc-BIM** folder, open the **C04-Imperial-Frame.dwg** or **C04-Metric-Frame.dwg** file, based on the required units. The model for this tutorial is shown in Figure 67.

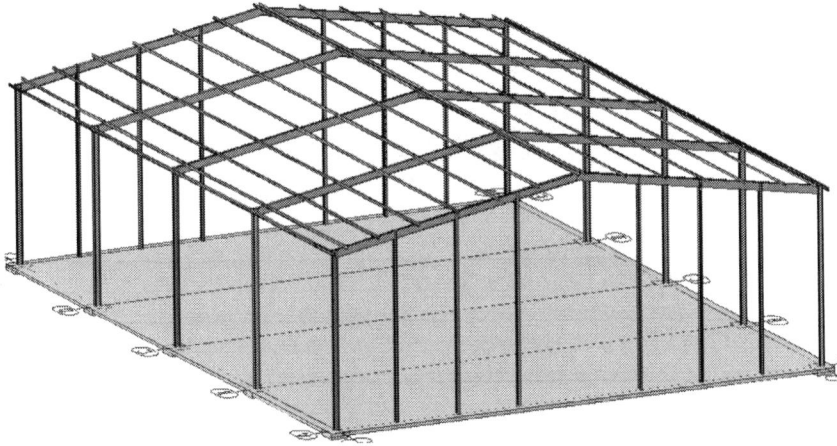

Figure 67 *The model for the tutorial*

Notice that this file is similar to the one you created in Chapter 3. You will now add the base plate joint to the column at the **A1** grid intersection point.

2.　Zoom close to the **A1** grid intersection point.

3.　From the **Home** ribbon tab > **Extended Modeling** ribbon panel, invoke the **Connection vault** tool; the **Connection vault** palette is displayed.

4.　From the **Joint Categories** area, click the **Plates at beam** button, which is the second button; various plates and beam joints are displayed in the **Joint Type** area.

Before you invoke the joint tool, you will add it to the **Favorites** category.

5.　Click the star icon on the right of the **Base plate** button; the joint is added to the **Favorites** category.

Next, you will invoke this tool to insert the joint.

6.　Click the **Base plate** button to invoke this joint tool; you are prompted to select the column.

Before you select the column, it is better to hide the **Connection vault** palette. If you do not hide this palette, it will interfere with the joint dialog box.

7.　From the top right of the **Connection vault** title bar that displays **Plates at beam**, click the **minus "-"** sign to hide the palette.

8.　Now, select the column at the **A1** grid intersection point.

9.　Press the ENTER key; the **Advance Steel - Base plate [X]** dialog box is displayed.

10. From the left pane of the dialog box, click the **Base Plate** category; the **Base plate layout** tab is activated.

11. In the **Plate thickness** edit box, enter **1"** as the value if you are using Imperial units or **20** as the value if you are using Metric units.

12. From the **Layout** list, make sure **projections** is selected.

13. Activate the **Base plate dimensions** tab.

14. Make sure the **All projections equal** tick box is selected.

15. Enter **2"** or **50** as the value in the **Projection 1** edit box.

16. Activate the **Anchor and holes** tab.

17. From the **Anchor diameter** list, select **3/4"** or **16.00 mm**.

18. Activate the **Anchor parallel web** tab.

19. In the **Number** edit box, enter **2**.

20. In the **Intermediate distance** edit box, enter **10"** or **180**.

21. Activate the **Anchor parallel flange** tab.

22. In the **Number** edit box, enter **2**.

23. In the **Intermediate distance** edit box, enter **6"** or **150**.

What I do

*The joints that I regularly use in multiple projects, I add them to the library using the **Library** tab of the joint dialog box. The process is similar to that discussed in earlier chapters.*

---

24. Save this joint in the library with the name **Col-Frame-C04** and then close the dialog box.

    Next, you will insert the base plate joint for the gable post on the right of the column at the **A1** grid intersection point.

25. From the top right of the **Connection vault** title bar, click the **plus "+"** sign to expand the palette.

26. Click the **Base plate** button to invoke this joint type tool; you are prompted to select the column.

    Before you select the column, you will hide the **Connection vault** palette.

27. Hide the **Connection Vault** palette by clicking the **minus "-"** sign on the top right of its title bar.

28. Select the gable post on the right of the column at the **A1** grid intersection point; you are again prompted to select objects.

29. Press the ENTER key; the **Advance Steel - Base plate [X]** dialog box is displayed.

30. From the left pane of the dialog box, click the **Base Plate** category; the **Base plate layout** tab is activated.

31. Enter **3/4"** or **16** as the value, depending on whether you are using the Imperial units or Metric units, in the **Plate thickness** edit box.

32. From the **Layout** list, make sure **projections** is selected.

33. Activate the **Base plate dimensions** tab.

34. Make sure the **All projections equal** tick box is selected.

35. Enter **1"** or **25** as the value in the **Projection 1** edit box.

36. Activate the **Anchor and holes** tab.

37. From the **Anchor diameter** list, select **3/4"** or **16.00 mm**.

38. Activate the **Anchor parallel web** tab.

39. In the **Number** edit box, enter **2**.

40. In the **Intermediate distance** edit box, enter **8"** or **120**.

41. Activate the **Anchor parallel flange** tab.

42. In the **Number** edit box, enter **2**.

43. In the **Intermediate distance** edit box, enter **4"** or **75**.

44. Save this joint in the library with the name **Col-Gable-C04** and then close the dialog box.

    The model, after inserting the two base plate joints, is shown in Figure 68.

Section 2: Inserting the Joint between the Column and the Rafter

In this section, you will insert the **Knee of frame bolted, with haunch** joint to the column at the **A1** grid intersection point and the rafter connecting to it.

1. From the top right of the **Connection vault** title bar, click the **plus "+"** sign to expand the palette.

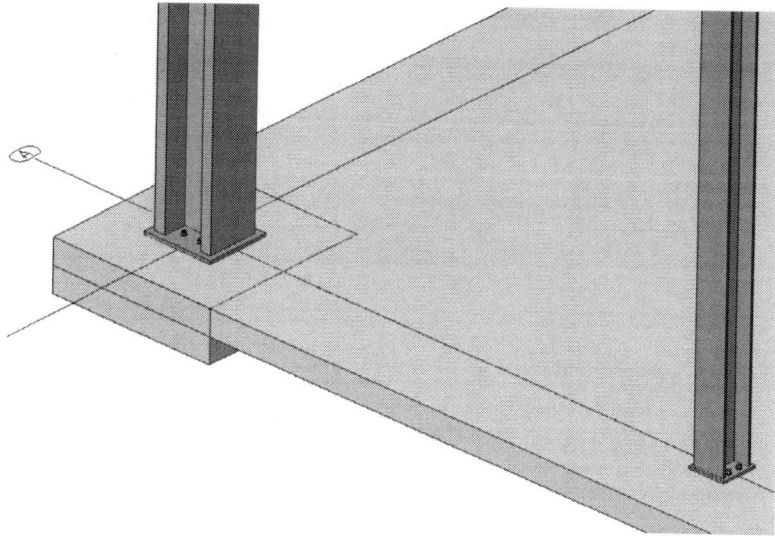

Figure 68 The model after inserting the two base plate joints

2. From the **Joint Categories** area, click the **Column - Beam** button, which is the third button; various column - beam joints are displayed in the **Joint Type** area.

    Before you invoke the joint tool, you will add it to the **Favorites** category.

3. Click the star icon on the right of the **Knee of frame bolted, with haunch** button; the joint is added to the **Favorites** category.

4. Click the **Knee of frame bolted, with haunch** button to invoke this joint type tool; you are prompted to identify the column.

    Before you select the column, you need to hide the **Connection vault** palette.

5. Hide the **Connection Vault** palette by clicking the **minus "-"** sign on the top right of its title bar.

6. Select the column at the **A1** grid intersection point; you are again prompted to select objects.

7. Press the ENTER key; you are prompted to identify the rafter.

8. Select the rafter connecting to the selected column and then press ENTER; the **Advance Steel - Knee of frame bolted, with haunch [X]** dialog box is displayed.

9. From the left pane of the dialog box, activate the **General** category > **End plate** tab.

10. In the **Thickness** edit box, enter **1"** if you are using Imperial units or **20** as the value if you are using Metric units.

11. Activate the **Cap plate** tab.

12. In the **Plate thickness** edit box, enter **1"** or **20** as the value.

13. From the left pane of the dialog box, activate the **Bolts & Welds** category; the **Bolts** tab is activated by default in this category.

14. From the **Bolts diameter** list, select **1 inch** or **20.00 mm**.

15. Clear the **Bolts on gauge line** tick box.

16. In the **Horizontal distance** edit box, enter **7"** or **100** as the value.

17. Activate the **Bolts groups** tab.

18. For **Group 1 > Lines**, enter **3** as the value.

19. For **Group 1 > Start dist.**, enter **4 7/8"** or **65** as the value.

20. For **Group 1 > Interm. dist.**, enter **6 5/8"** or **150** as the value.

21. For **Group 2 > Lines**, make sure the value is entered as **2**.

22. For **Group 2 > Start dist.**, enter **9 7/8"** or **200** as the value.

23. For **Group 2 > Interm. dist.**, enter **11 5/8"** or **220** as the value.

    Because this is a complex joint, it is better to add its settings to the library.

24. From the left pane of the dialog box, activate the **Properties** category > **Library** tab.

25. Save the values of the current joint into the library with the name **Col-Raf-C04**.

26. Close the dialog box.

    The model after adding this joint is shown in Figure 69.

Section 3: Inserting the Joint between the Rafter and the Gable Post

In this section, you will insert the **Gable wall end plate** joint between the previously selected rafter and the first gable post on the right of the column at the **A1** grid intersection point.

1. From the top right of the **Connection vault** title bar, click the **plus "+"** sign to expand the palette.

    The **Column - Beam** joint category is still active. You will use the **Gable wall end plate** joint from this category. However, before you invoke the joint tool, you will add it to the **Favorites** category.

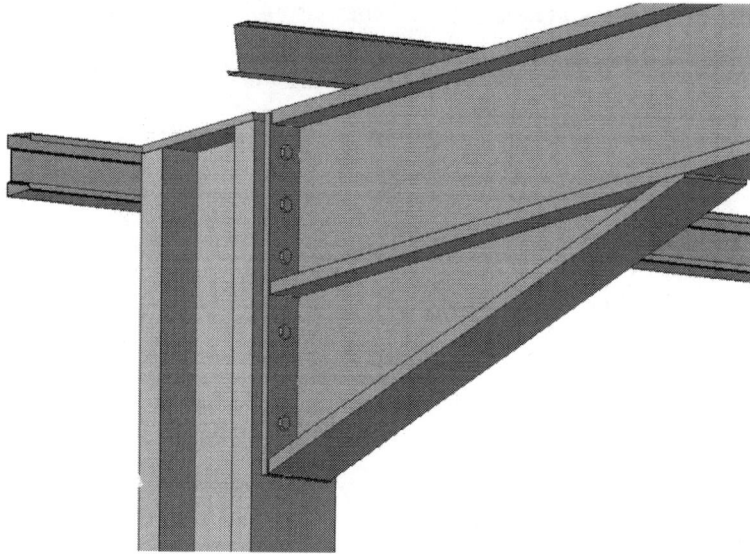

Figure 69 *The model after inserting the joint between the column and the rafter*

2. Click the star icon on the right of the **Gable wall end plate** button; the joint is added to the **Favorites** category.

3. Click the **Gable wall end plate** button to invoke this joint type tool; you are prompted to select the rafter.

    Before you select the rafter, you will close the **Connection vault** palette.

4. Close the **Connection Vault** palette by clicking the **X** sign on the top right of its title bar.

5. Select the rafter to which you applied the joint in the previous section; you are again prompted to select objects.

6. Press the ENTER key; you are prompted to select the beam.

7. Select the first gable post on the right of the column at the **A1** grid intersection point.

8. Press the ENTER key; the **Advance Steel - Gable wall end plate [X]** dialog box is displayed.

9. From the left pane of the dialog box, invoke the **Plate & bolts** category; the **Plate** tab is activated by default.

10. In the **Plate thickness** edit box, enter **1/2"** as the value if you are using the Imperial units or **10** as the value if you are using the Metric units.

11. Activate the **Bolt distances** tab.

12. Make sure the value of the **Number of bolts in x** is set to **2**.

13. In the **Intermediate distance x** edit box, enter **5"** or **75** as the value.

14. Make sure the value of the **Number of bolts in Y** is set to **2**.

15. Clear the **On gauge line** tick box.

16. In the **Intermediate distance y** edit box, enter **5"** or **75** as the value.

17. Activate the **Stiffeners** tab.

18. From the **Type** list, select **full**.

19. From the **Arrangement** list, make sure **centered** is selected.

    Next, you will add these joint settings to the library.

20. From the left pane of the dialog box, activate the **Properties** category > **Library** tab and add these joint settings to the library with the name **Gab-Raf-C04**.

21. Close the dialog box.

    The model after adding this joint is shown in Figure 70.

Figure 70 The model after inserting the joint between the rafter and the gable post

Section 4: Copying the Joints

In this section, you will create multiple copies of the joints you have added in the previous sections. This is done using the **Advance Steel** tool palette. Therefore, you need to invoke this tool palette, if it is not turned on by default.

1. If the **Advance Steel** tool palette is not turned on, turn it on from the **Home** ribbon tab > **Extended Modeling** ribbon panel > **Advance Steel Tool Palette** tool.

2. From the left of this palette, invoke the **Tools** tab.

   Because you need to create multiple copies of the joints, you will use the **Create joint in a joint group, multiple** tool.

3. From the tool palette, invoke the **Create joint in a joint group, multiple** tool; you are prompted to select the source joint.

   The first joint that you will copy is the first base plate joint you added to the column.

4. Select the base plate of the base plate joint you added to the column at the **A1** grid intersection point; you are again prompted to select objects.

   Tip: You can also select other items, such as a bolt, from the source joint to create copies of the joint.

5. Press the ENTER key; you are prompted to select beams corresponding to the source joints input (1/1).

6. Select the remaining nine columns of the gable and portal frames.

7. Press the ENTER key; the base plate joint is added to all the selected columns.

   Next, you will copy the base plate connection of the gable post.

8. From the tool palette, invoke the **Create joint in a joint group, multiple** tool; you are prompted to select the source joint.

9. Select the base plate of the base plate joint you added to the gable post; you are again prompted to select objects.

10. Press the ENTER key; you are prompted to select beams corresponding to the source joints input (1/1).

11. Select the remaining five gable posts on the front gable frame and the six gable posts on the rear gable frame.

12. Press the ENTER key; the base plate joint is added to all the selected gable posts.

    Next, you will copy the joint that was inserted between the column and the rafter.

13. From the tool palette, invoke the **Create joint in a joint group, multiple** tool; you are prompted to select the source joint.

14. Select the haunch of the joint; you are again prompted to select objects.

15. Press the ENTER key; you are prompted to select beams corresponding to the source joints input (1/2).

 Remember that you need to select the objects in the same sequence that was used to insert the joint. In this case, you need to select the columns first and then the rafters.

16. One by one, select the remaining four columns on the left side of the model and then select the five columns on the right side of the model; you are again prompted to select objects.

17. Press the ENTER key; you are prompted to select beams corresponding to the source joints input (2/2).

18. One by one, select the four rafters on the left side of the model and the five rafters on the right side of the model; you are again prompted to select objects.

19. Press the ENTER key; the **Knee of frame bolted, with haunch** joint is copied between all the columns and rafter.

 Finally, you will copy the joints between the rafter and gable post.

20. From the tool palette, invoke the **Create joint in a joint group, multiple** tool; you are prompted to select the source joint.

21. Select stiffener of the joint between the rafter and the gable post; you are again prompted to select objects.

22. Press the ENTER key; you are prompted to select beams corresponding to the source joints input (1/2).

23. One by one, select the two rafters of the gable frame at the front and then the two rafters of the gable frame at the rear; you are again prompted to select objects.

24. Press the ENTER key; you are prompted to select beams corresponding to the source joints input (2/2).

25. One by one, select the five gable posts of the gable frame at the front and then the six gable posts of the gable frame at the rear; you are again prompted to select objects.

26. Press the ENTER key; the **Gable wall end plate** joint is copied between all the gable posts and the rafters, as shown in Figure 71.

Section 5a (STRUCTURE ONLY): Inserting Joints in the Elevator Tower Model

In this section, you will open the model of the Elevator Tower and then insert one base plate joint. You will use the joint information saved in the library in the previous sections to insert this joint. After inserting one joint, you will then copy it at the remaining locations.

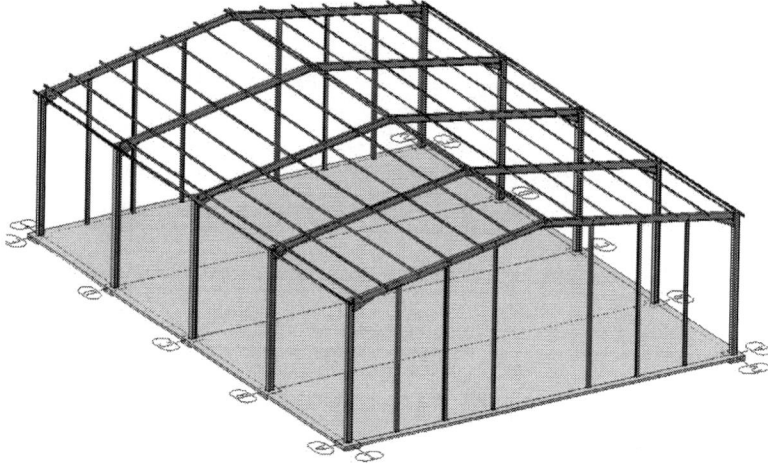

Figure 71 *The model after copying all the joints*

1. From the **C04 > Struc-BIM** folder, open the **C04-Imperial-Elev-Tower.dwg** or **C04-Metric-Elev-Tower.dwg** file, based on the preferred units; the model looks similar to the one shown in Figure 72.

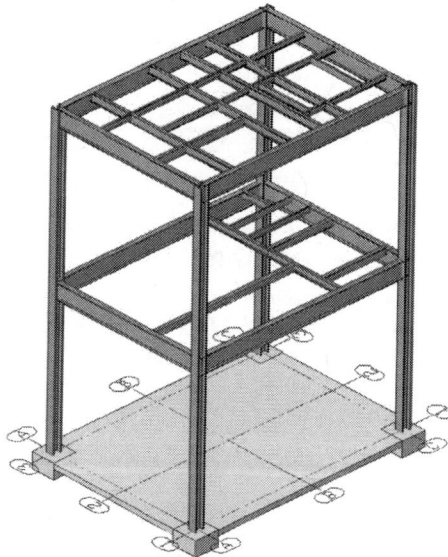

Figure 72 *The Elevator Tower model for inserting base plate joints*

You will now add the base plate joint to the column at the **A1** grid intersection point. You will use the settings saved in the library with the name **Col-Frame-C04** in the earlier sections to insert this joint.

2. Invoke the **Connection vault** palette using the **Home** ribbon tab > **Extended Modeling** ribbon panel > **Connection vault** tool.

   Because the **Base plate** joint was added to the **Favorites** category, you can invoke it directly from there.

3. From the **Joint Categories** area, click the **Favorites** button, which is the first button; various joints added to **Favorites** are displayed in the **Joint Type** area.

4. Click the **Base plate** button to invoke this joint type tool; you are prompted to select the column.

   Before you select the column, it is better to hide the **Connection vault** palette.

5. From the top right of the **Connection vault** title bar that displays **Favorites**, click the **minus "-"** sign to hide the palette.

6. Now, select the column at the **A1** grid intersection point; you are again prompted to select objects.

7. Press the ENTER key; the **Advance Steel - Base plate [X]** dialog box is displayed.

8. From the left pane of the dialog box, invoke the **Library** tab.

9. From the list of saved joints, click on **Col-Frame-C04**; the joint on the column is modified to match the saved values.

10. Close the dialog box.

    Next, you will copy this joint at the remaining columns.

11. If the **Advance Steel** tool palette is not turned on, invoke it from the **Home** ribbon tab > **Extended Modeling** ribbon panel > **Advance Steel Tool Palette** tool.

12. From the left of this palette, invoke the **Tools** tab.

    Because you need to create multiple copies of the joints, you will use the **Create joint in a joint group, multiple** tool.

13. From the tool palette, invoke the **Create joint in a joint group, multiple** tool; you are prompted to select the source joint.

14. Select the base plate of the base plate joint you added to the column at the **A1** grid intersection point; you are again prompted to select objects.

15. Press the ENTER key; you are prompted to select column.

16. Select the remaining three columns.

17. Press the ENTER key; the base plate joint is added to all the selected columns. The zoomed in view of the model, after adding all the base plate joints, is shown in Figure 73.

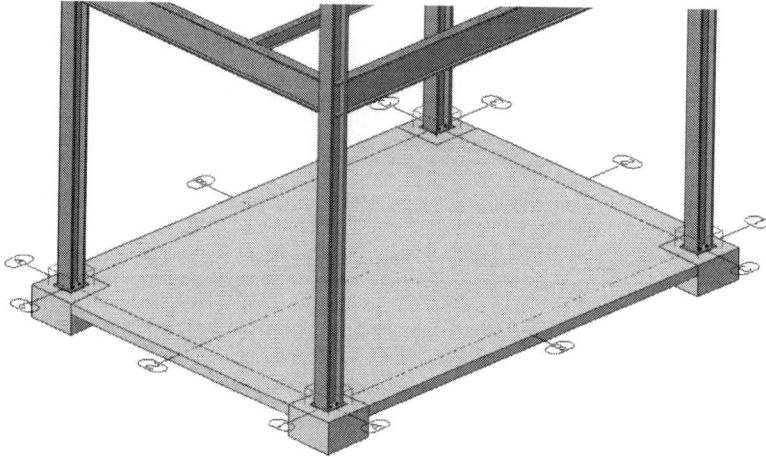

Figure 73 *The zoomed in view of the Elevator Tower model after inserting the base plate joints*

18. Save the file.

Section 5b: Editing the Joints

In this section, you will edit the base plate joint of the column at the **C1** grid intersection point. Because all the joints are a part of the same group, the remaining joints will be automatically modified when you modify one of the joints. However, before editing the joint, you need to upgrade it to master.

1. Zoom close to the **C1** grid intersection point.

2. Select the plate of the base plate joint.

3. Right-click in the blank area of the graphics window and select **Advance Joint Properties** from the shortcut menu; the **Advance Steel - Base plate [X]** dialog box is displayed with the **Properties** category > **Properties** tab active.

    Notice that the **Upgrade to master** tick box is active. This is because the selected joint is a slave joint created by copying the master joint. If you do not select this tick box, you will not be able to edit this joint.

4. Select the **Upgrade to master** tick box.

5. With the dialog box still displayed, zoom out so you can see all the four base plate joints.

6. From the left pane of the dialog box, select the **Base Plate > Base plate layout** tab.

7. In the **Plate thickness** edit box, enter **1 1/4"** or **30**.

8. Activate the **Base plate dimensions** tab.

9. In the **Projection 1** edit box, enter **6"** or **150**.

   Notice that the size of the base plate of all four joints update dynamically in the drawing window.

10. Activate the **Anchor parallel web** tab.

11. In the **Number** edit box, enter **4**.

12. In the **Intermediate distance** edit box, enter **6"** or **150**.

13. Select the **Remove center bolts** tick box.

14. Activate the **Anchor parallel flange** tab.

15. In the **Number** edit box, enter **4**.

16. In the **Intermediate distance** edit box, enter **7"** or **175**.

17. From the left pane of the dialog box, select the **Stiffeners & Plates > Web stiffener** tab.

18. From the **Create stiffener** list, select **both sides**; the web stiffeners are created on all the joints.

19. In the **Stiffener thickness** edit box, enter **1"** or **25**.

20. In the **Stiffener width** edit box, enter **6"** or **150**.

21. In the **Stiffener height** edit box, enter **1' 4"** or **400**.

22. From the **Corner finish inside** list, make sure **straight** is selected.

23. In the **Outside chamfer height** edit box, enter **10"** or **250**.

24. In the **Outside chamfer width** edit box, enter **4"** or **100**.

25. From the left pane of the dialog box, select the **Stiffeners & Plates > Middle stiffener** tab.

26. From the **Create stiffener** list, select **both sides**; the middle stiffeners are created on all the joints.

27. In the **Stiffener thickness** edit box, enter **1"** or **25**.

28. In the **Stiffener width** edit box, enter **8"** or **200**; the stiffener width is modified in all the joints.

29. In the **Stiffener height** edit box, enter **1' 4"** or **400**; the stiffener height is modified in all the joints.

30. From the **Corner finish inside** list, make sure **straight** is selected.

31. In the **Outside chamfer height** edit box, enter **10"** or **250**.

32. In the **Outside chamfer width** edit box, enter **5"** or **125**.

33. Close the dialog box. The model, after modifying the base plate joints, is shown in Figure 74.

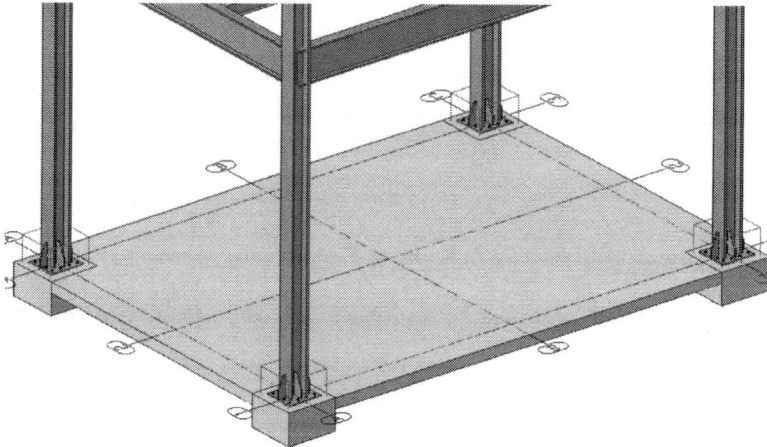

Figure 74 The zoomed in view of the Elevator Tower model after modifying the base plate joints

34. Save and close the file.

Section 6a (BIM ONLY): Inserting Joints in the Hospital Structure Model

In this section, you will open the Hospital structure model and then insert one base plate and one column-beam joint. You will use the joint information saved in the library in the earlier sections to insert this joint. After inserting one joint of each category, you will then copy them at the remaining locations.

1. From the **C04 > Struc-BIM** folder, open the **C04-Imperial-Hospital.dwg** or **C04-Metric-Hospital.dwg** file, based on the preferred units; the model looks similar to the one shown in Figure 75.

   You will now add the base plate joint to the column at the **A1** grid intersection point. You will use the settings saved in the library with the name **Col-Frame-C04** in the earlier sections to insert this joint.

2. Invoke the **Connection vault** palette using the **Home** ribbon tab > **Extended Modeling** ribbon panel > **Connection vault** tool.

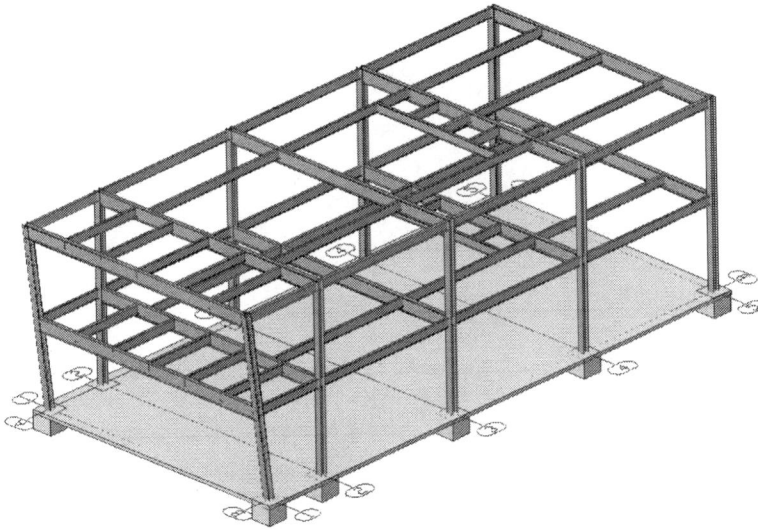

Figure 75 *The Elevator Tower model for inserting base plate joints*

Because the **Base plate** joint was added to the **Favorites** category, you can invoke it directly from there.

3. From the **Joint Categories** area, click the **Favorites** button, which is the first button; various joints added to **Favorites** are displayed in the **Joint Type** area.

4. Click the **Base plate** button to invoke this joint type tool; you are prompted to select the column.

   Before you select the column, it is better to hide the **Connection vault** palette.

5. From the top right of the **Connection vault** title bar that displays **Favorites**, click the **minus "-"** sign to hide the palette.

6. Now, zoom close to bottom of the column at the **A1** grid intersection point and then select it; you are again prompted to select objects.

7. Press the ENTER key; the **Advance Steel - Base plate [X]** dialog box is displayed.

8. From the left pane of the dialog box, invoke the **Library** tab.

9. From the list of saved joints, click on **Col-Frame-C04**; the joint on the column is modified to match the saved values.

   Notice that although the joint matches the values saved in the library, however, the base plate is placed normal to the inclined column, instead of being horizontal, see Figure 76. Therefore, you need to fix the orientation of this base plate.

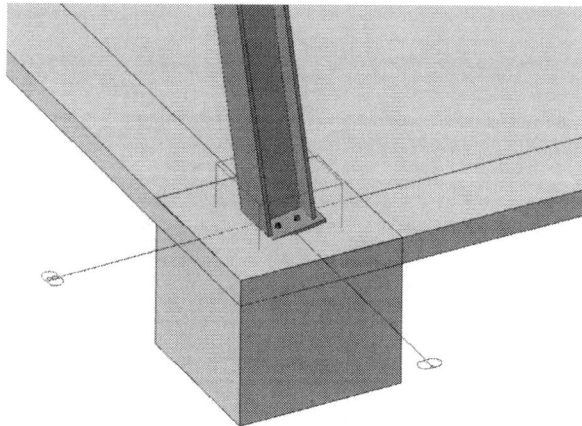

Figure 76   *The base plate inserted normal to the column*

10. From the left pane of the dialog box, click the **Base Plate** category; the **Base plate layout** tab is invoked.

11. From the **Direction** list in the right pane, select **horizontal**; the base plate is oriented correctly now.

12. Close the dialog box.

13. Similarly, add the base plate joint to the column at the **B2** grid intersection point. Use the settings saved in the library to insert this joint.

    Next, you will insert the column-beam joint between the column at the **B1** grid intersection point and the longer beam connected to it at level 2. You will use the joint settings saved in the library to insert this joint.

14. Zoom close to the top of the column at the **B1** grid intersection point where the longer beam intersects this column, refer to Figure 77.

15. Expand the **Connection vault**.

16. From the **Favorites** category, invoke the **Knee of frame bolted, with haunch** joint; you are prompted to identify the column.

17. Select the column at the **B1** grid intersection point and then press ENTER; you are prompted to identify the rafter.

18. Select the longer beam intersecting the selected column and then press ENTER; the joint is inserted with the default settings.

19. From the left pane of the dialog box, activate the **Library** tab.

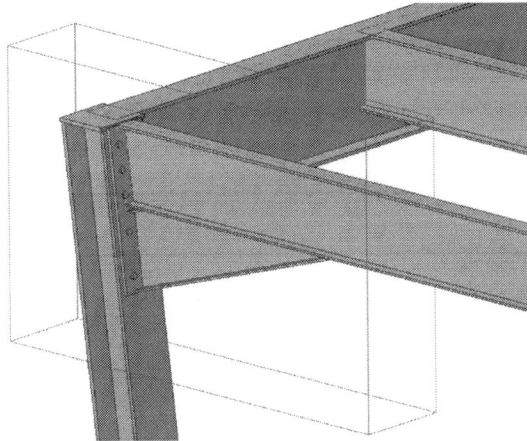

Figure 77 The joint inserted between the column and the beam

20. From the saved joints list, select **Col-Raf-C04**; the joint is modified to match the saved settings, as shown in Figure 77.

    Next, you will insert the same joint between the same column and the longer beam at level 1.

21. Insert the same joint between the same column and the longer beam intersecting this column at level 1. Use the settings of the **Col-Raf-C04** joint saved in the library.

    Notice that the joint trims the lower half of the column, as shown in Figure 78.

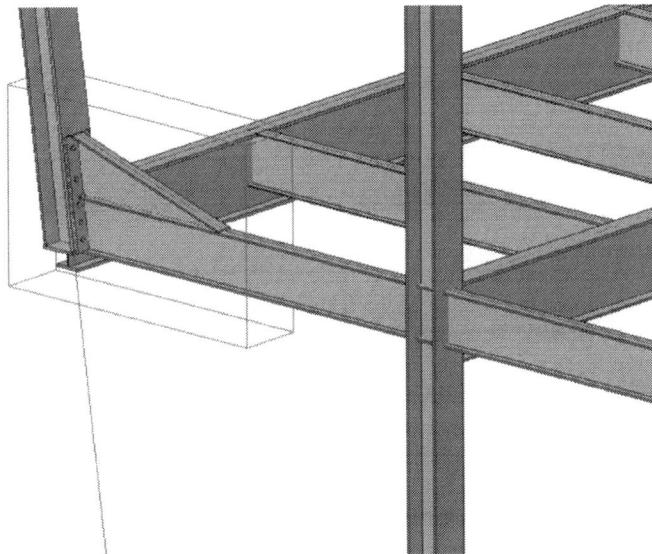

Figure 78 The lower half of the column trimmed by the joint

You will now fix this by editing the values in the dialog box. If you have closed the dialog box, you can double-click on the bounding box of the joint to redisplay it.

22. From the left pane of the dialog box, click on the **General** category; the **Haunch - Rafter** tab is activated.

23. From the **Haunch** list, select **Other side**; the upper half of the column is now trimmed.

24. Activate the **Cap plate** tab of the dialog box.

25. Select the **Continuous column** (Imperial) or **Column not trimmed** (Metric) tick box; the joint is updated and the column is restored, as shown in Figure 79.

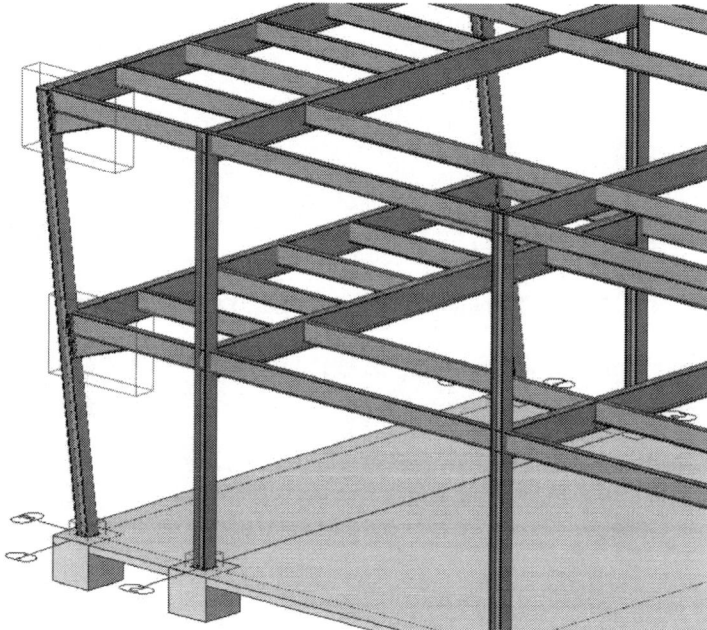

Figure 79 The model after updating the joint

26. Close the dialog box.

Next, you will copy the joints. However, because the two base plate joints and the two column-beam joints are different, you will copy them separately.

27. From the **Advance Steel** tool palette > **Tools** tab, invoke the **Create joint in a joint group** tool; you are prompted to select the source joint.

28. Select the base plate of the base plate joint you added to the column at the **B1** grid intersection point; you are again prompted to select objects.

29. Press the ENTER key; you are prompted to select column.

30. Select the column at the **A1** grid intersection point.

31. Press the ENTER key; the base plate joint is added to the selected column.

Next, you will use the same tool to copy the column-beam joint at level 2.

32. Invoke the **Create joint in a joint group** tool again; you are prompted to select the source joint.

33. Select the haunch of the column-beam joint you added between column at the **B1** grid intersection point and the beam at level 2; you are again prompted to select objects.

34. Press the ENTER key; you are prompted to identify the column.

35. Select the column at the **A1** grid intersection point and then press ENTER; you are prompted to identify the rafter.

36. Select the longer beam intersecting with the selected column and then press ENTER; the joint is copied.

37. Repeat the same process and copy the joint at level 1 to the other side.

Next, you will create multiple copies of the base plate joint added to the column at the **B2** grid intersection point.

38. From the **Advance Steel** tool palette > **Tools** tab, invoke the **Create joint in a joint group, multiple** tool; you are prompted to select the source joint.

39. Select the base plate of the base plate joint you added to the column at the **B2** grid intersection point; you are again prompted to select objects.

40. Press the ENTER key; you are prompted to select column.

41. Select the remaining columns.

42. Press the ENTER key; the base plate joint is added to the selected column. The model, after inserting and copying the joints, is shown in Figure 80.

43. Save the file.

*Tip: In some cases, when you copy a joint, the connection box of the copied joint may not be displayed. In that case, right-click on any item of the copied joint and select **Advance Joint Properties** from the shortcut menu to display the joint dialog box. Then close the dialog box and the connection box will be displayed.*

Section 6b: Editing the Joints

In this section, you will edit the base plate joint of the column at the **B3** grid intersection point. Because all the joints of the vertical columns are a part of the same group, the remaining joints will be automatically modified when you modify one of the joints. However, before editing the joint, you need to upgrade it to master.

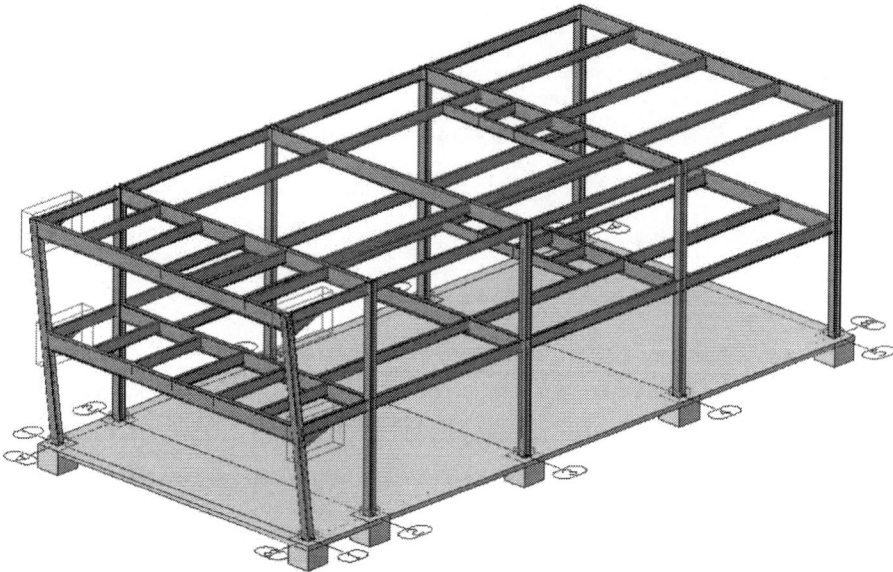

Figure 80 *The model after inserting and copying the joints*

1. Zoom close to the **B3** grid intersection point.

2. Select the plate of the base plate joint.

3. Right-click in the blank area of the graphics window and select **Advance Joint Properties** from the shortcut menu; the **Advance Steel - Base plate [X]** dialog box is displayed with the **Properties** category > **Properties** tab active.

   Notice that the **Upgrade to master** tick box is active. This is because the selected joint is a slave joint created by copying the master joint. If you do not select this tick box, you will not be able to edit this joint.

4. Select the **Upgrade to master** tick box.

5. With the dialog box still displayed, zoom out so you can see the base plate joints at the **B4** and **B5** grid intersection points as well.

6. From the left pane of the dialog box, select the **Base Plate > Base plate layout** tab.

7. In the **Plate thickness** edit box, enter **1 1/4"** or **30**.

8. Activate the **Base plate dimensions** tab.

9. In the **Projection 1** edit box, enter **6"** or **150**.

   Notice that the size of the base plate of the joints update dynamically in the drawing window.

10. Activate the **Anchor parallel web** tab.

11. In the **Number** edit box, enter **4**.

12. In the **Intermediate distance** edit box, enter **7"** or **175**.

13. Select the **Remove center bolts** tick box.

14. Activate the **Anchor parallel flange** tab.

15. In the **Number** edit box, enter **4**.

16. In the **Intermediate distance** edit box, enter **7"** or **175**.

17. From the left pane of the dialog box, select the **Stiffeners & Plates > Web stiffener** tab.

18. From the **Create stiffener** list, select **both sides**; the web stiffeners are created on all the joints.

19. In the **Stiffener thickness** edit box, enter **1"** or **25**.

20. In the **Stiffener width** edit box, enter **6"** or **150**.

21. In the **Stiffener height** edit box, enter **1' 4"** or **400**.

22. From the **Corner finish inside** list, make sure **straight** is selected.

23. In the **Outside chamfer height** edit box, enter **10"** or **250**.

24. In the **Outside chamfer width** edit box, enter **4"** or **100**.

25. From the left pane of the dialog box, select the **Stiffeners & Plates > Middle stiffener** tab.

26. From the **Create stiffener** list, select **both sides**; the middle stiffeners are created on all the joints.

27. In the **Stiffener thickness** edit box, enter **1"** or **25**.

28. In the **Stiffener width** edit box, enter **8"** or **200**; the stiffener width is modified in all the joints.

29. In the **Stiffener height** edit box, enter **1' 4"** or **400**; the stiffener height is modified in all the joints.

30. From the **Corner finish inside** list, make sure **straight** is selected.

31. In the **Outside chamfer height** edit box, enter **10"** or **250**.

32. In the **Outside chamfer width** edit box, enter **5"** or **125**.

33. Close the dialog box.

34. Similarly, using the same settings, modify the joint of the column at the **B1** grid intersection point. The zoomed in view of the model, after modifying the base plate joints, is shown in Figure 81.

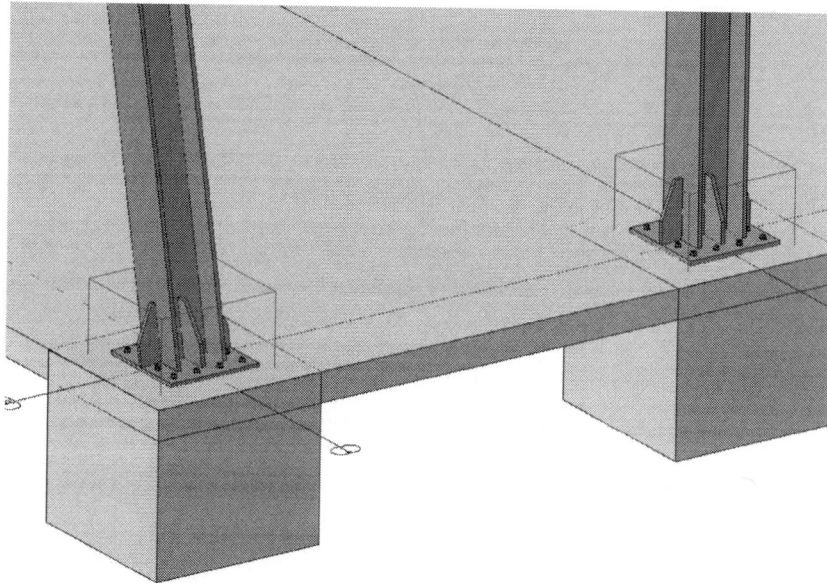

Figure 81 The model, after modifying the base plate joints

35. Save and close the file.

Skill Evaluation

Evaluate your skills to see how many questions you can answer correctly. The answers to these questions are given at the end of the book.

1. The **Connection vault** palette can be used to insert structural columns in the model. (True/False)

2. The commonly used joint settings cannot be saved in Advance Steel. (True/False)

3. There is only one type of column-beam connection available. (True/False)

4. While inserting the **Base plate** joint, the base plate can be aligned with the column or placed horizontal. (True/False)

5. Advance Steel allows you to copy a selected connection single or multiple times. (True/False)

6. Which tool palette is used to copy connections?

   (A) **Library**   (B) **Connection vault**
   (C) **Advance Steel**   (D) **Properties**

7. Which joint type only allows tubes to be used for the plates and beam joint types?

   (A) **Base plate**   (B) **Round plate**
   (C) **Base plate with haunch**   (D) **Tube base plate**

8. Which joint type is mainly created between the rafter and the gable post?

   (A) **Cross connection**   (B) **Gable wall end plate**
   (C) **Star plate**   (D) **Base plate**

9. Selecting the star icon on the right of the joint button adds that joint to which category?

   (A) **None**   (B) **Star**
   (C) **Favorites**   (D) **Super star**

10. The **Knee of frame bolted, with haunch** connection type is available in which category?

    (A) **Plates at beam**   (B) **End plate**
    (C) **Column-Beam**   (D) **Haunch**

Class Test Questions

Answer the following questions:

1. Explain briefly the process of copying a joint multiple times.

2. Explain the process of upgrading a joint to master.

3. What is the process of adding a joint to the group?

4. How do you add a joint to the **Favorites** category?

5. What is the advantage of adding a joint to the library?

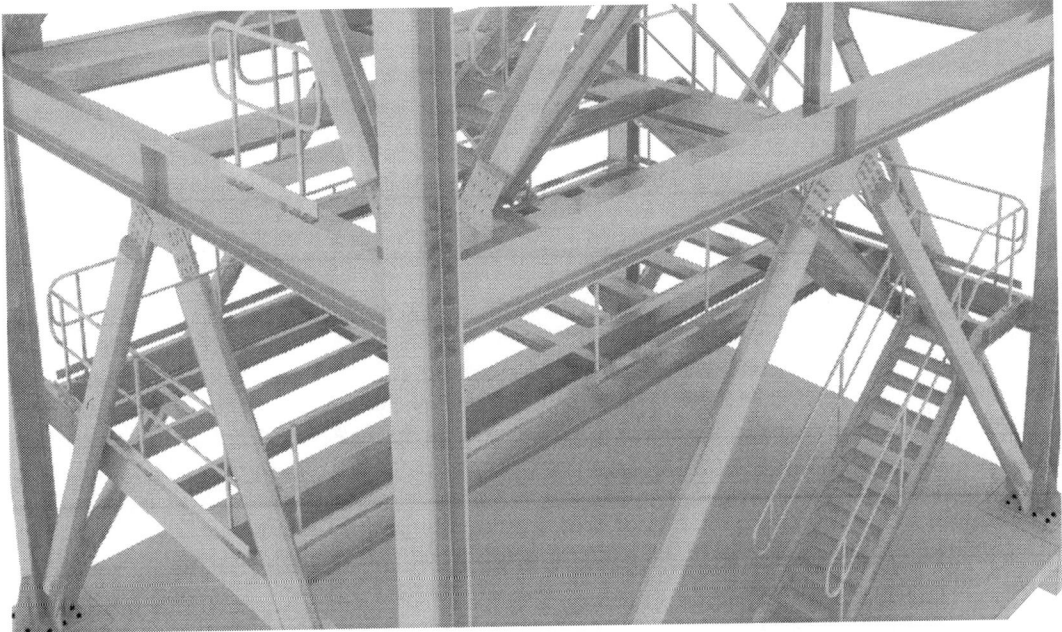

Chapter 5 – Inserting the Beam End to End, Platform Beam, and Purlin Joints

The objectives of this chapter are to:

√   *Explain what are **Beam End to End** joints*
√   *Explain the process of inserting the **Apex Haunch** and **Double Apex Haunch** joints*
√   *Explain the process of inserting the **Front plate splice** joint*
√   *Explain what are **Platform Beam** joints*
√   *Explain the process of inserting the **Clip Angle** and **Clip Angle - Skewed** joints*
√   *Explain the process of inserting the **Double Sided Clip Angle** joint*
√   *Explain the process of inserting the **Single and Double Sided End Plate** joints*
√   *Explain the process of inserting the **Shear Plate** joint*
√   *Explain the process of inserting various types of purlin joints*
√   *Explain the process of inserting various types of eaves beams joints*

INSERTING THE BEAM END TO END JOINTS

This category lists various joints that can be added between two sections to connect them end to end. Some of the joints in this category are discussed next. Remaining joints are similar, so are not discussed in detail.

Apex haunch Joint

This type of joint is used to connect two beams end to end by inserting the bolted end plates between them and also inserting haunches at the bottom. When you invoke the tool for this joint, you will be prompted to select the main beam and then the secondary beam. Once you select both the beams and press the ENTER key, the **Advance Steel - Apex bolted with haunch [X]** dialog box will be displayed with the **Properties** category > **Properties** tab active. For this joint type, most of the categories and tabs are similar to those discussed in the **Knee of frame bolted, with haunch** joint that was discussed in Chapter 4. Only the **General** category > **End plate** tab is different, which is discussed next.

General Category > End plate Tab

When you click on the **General** category in the left pane of the dialog box, the **End plate** tab is the first tab, as shown in Figure 1. The options available in this tab are discussed next.

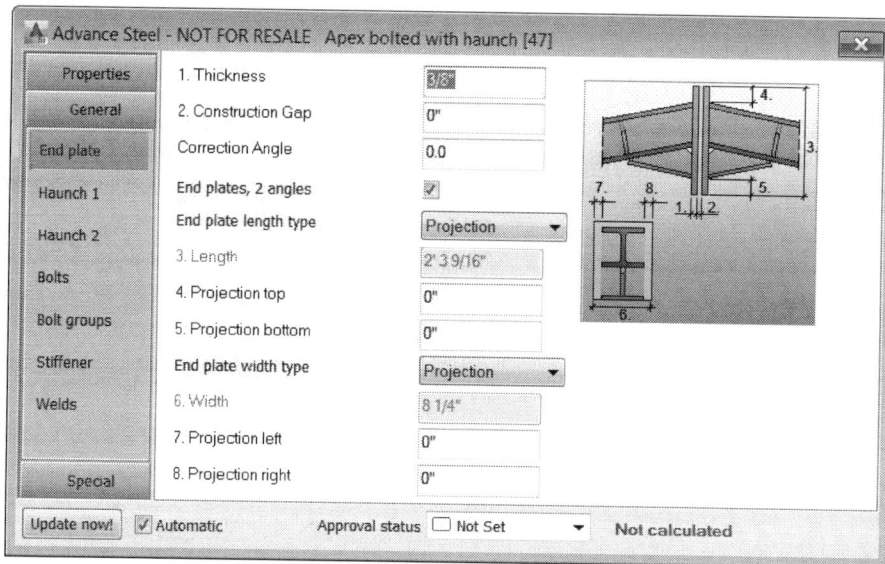

*Figure 1 The **End plate** tab of the **General** category*

Thickness
This edit box is used to enter the thickness of end plate that will be welded at the end of the two beams.

Construction Gap
This edit box is used to enter a gap between the two end plates.

Correction Angle
This edit box is used to enter an angle between the two end plates.

End plate, 2 angles
Select this tick box to insert the end plates with two angles.

End plate length type
This list is used to specify how the length of the end plates will be defined. By default, **Projection** is selected from this list. Therefore, you can specify the top and bottom projections in the edit boxes available below this list. Selecting **From top** from this list allows you to specify the projection of the plates from the top and its length. Selecting **From bottom** from this list allows you to specify the projection of the plates from the bottom and its length. Selecting **By bolts** from this list links the length of the plates to the bolt spacings. However, you can still enter the top and bottom projections for the plates.

End plate width type
This list is used to specify how the width of the end plates will be defined. By default, **Projection** is selected from this list. Therefore, you can specify the left and right projections in the edit boxes available below this list. Selecting **Exact value** from this list allows you to enter the width value of the plates in the edit box below this list. Selecting **By bolts** from this list links the width of the plates to the bolt spacings. However, you can still enter the left and right projections for the plates. Figure 2 shows the joint with no projections specified for the end plates and Figure 3 shows the same joint with projections defined for the plate on all four sides.

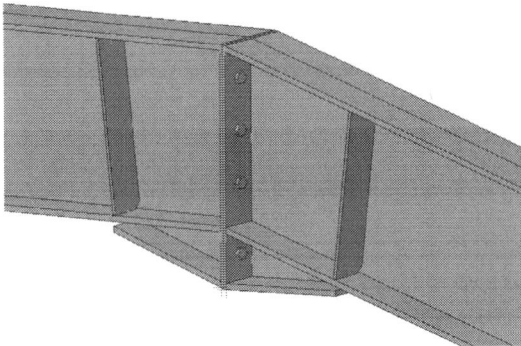

Figure 2 Joint with no projection defined for the end plates

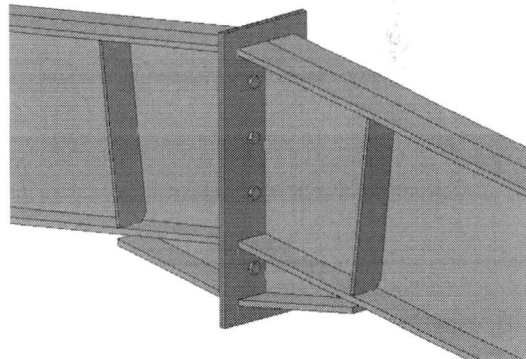

Figure 3 Joint with projection values defined on all four sides of the end plates

Double apex haunch Joint

[Double apex haunch]

This joint is similar to the one discussed above, with the only difference being that in this case, the haunches will be inserted above the beams as well. Figure 4 shows a double apex haunch joint.

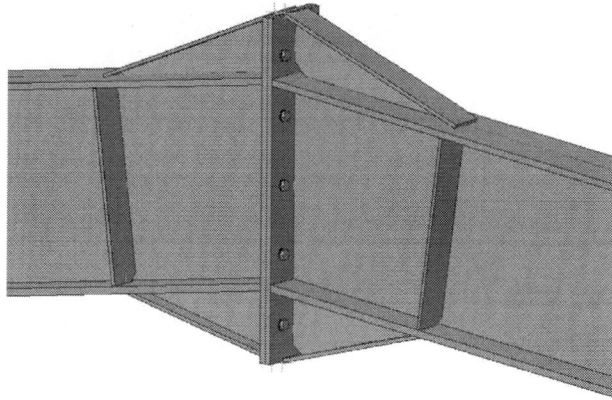

Figure 4 *Double apex haunch joint*

Tip: *While inserting the **Double apex haunch** joint, remember that in the **General** category > **Bolt groups** tab, **Group 1** defines the bolt distances for the bolts between the beams, **Group 2** defines the bolt distances for the bolts on the bottom haunches and **Group 3** defines the bolt distances for the bolts on the top haunches. The **Start dist.** value for the **Group 3** bolts is defined as a negative value because this value is defined from the last bolt on the bottom haunches.*

Front plate splice Joint

Front plate splice

This joint is similar to the ones discussed above, with the difference being that in this joint, haunches are not inserted. Only the end plates are inserted between the two beams. Most of the categories and tabs in this joint are similar to those discussed earlier. Only the **General** category > **Bolt groups** tab is different for this joint. This tab is discussed next.

General Category > Bolt groups Tab

The options available in this tab, shown in Figure 5, are used to define the options related to the bolt groups to be inserted for this joint type. These options are discussed next.

Number per side

This edit box is used to specify how many columns of bolts are to be inserted on each side of the beams. Figure 6 shows one column of bolts inserted on each side and Figure 7 shows two columns of bolts inserted on each side.

Gauge distance

This edit box is used to specify the horizontal distance between the holes on the two sides of the beams.

Intermediate distance

When inserting more than one column of bolts on each side, the intermediate spacing between the columns is specified in this edit box.

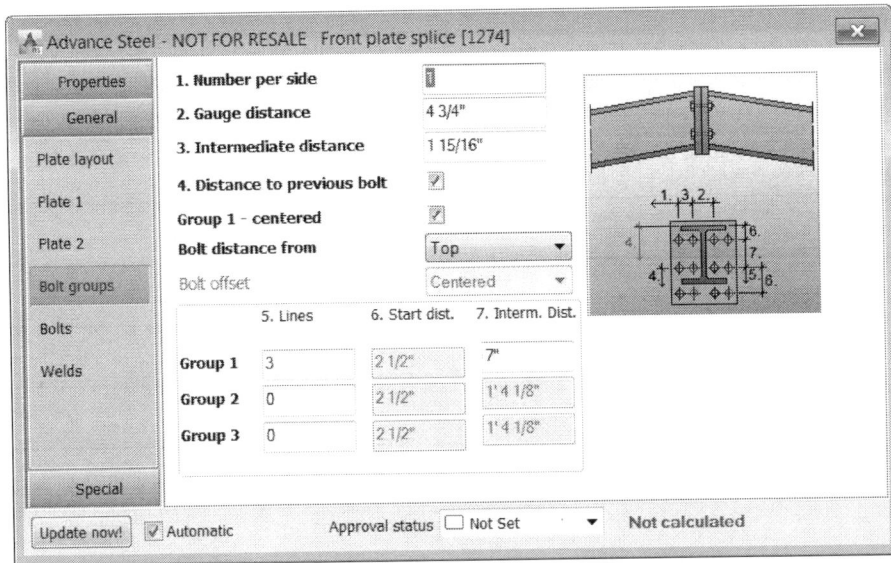

Figure 5 The **Bolt groups** tab of the **General** category

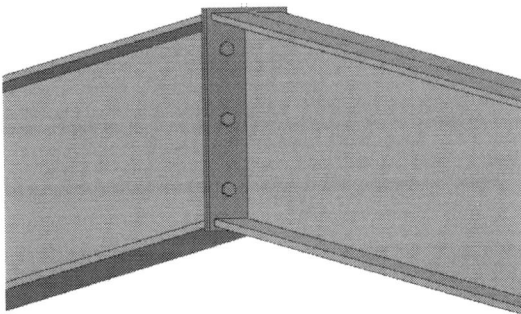

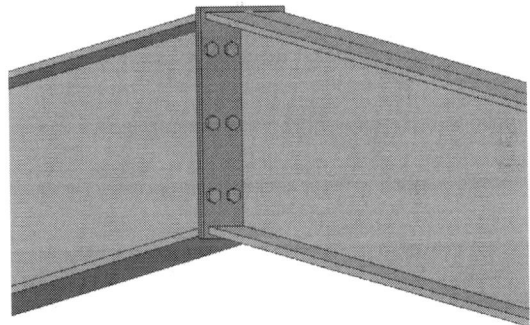

Figure 6 Joint with one column of bolts inserted on each side

Figure 7 Joint with two columns of bolts inserted on each side

Distance to previous bolts

While inserting multiple groups of bolts, if this tick box is selected, the dimension values defined for groups 2 and 3 are measured from the previous bolt. However, if this tick box is cleared, the dimension values will be measured from the top or bottom, depending on the option selected from the **Bolt distance from** list.

Group 1 centered

Select this tick box to place the group 1 bolts centrally aligned between the beams.

Bolt distance from
This list is used to specify whether the dimensions of the bolt groups are measured from the top of the beams or the bottom.

Bolt offset
This list is used to specify bolt offsets.

Group 1, Group 2, Group 3
The edit boxes in these areas are used to specify the number of bolts in the three groups, their start distances, and their intermediate distances.

INSERTING THE PLATFORM BEAMS JOINTS

This category lists various joints that can be added between two sections to connect them using an end plate. The end plate types that can be used for these joints include flat plates, equal or unequal angles, or even customized flat plates. Some of the joints in this category are discussed next. Remaining joints are similar, so are not discussed in detail.

Clip angle Joint

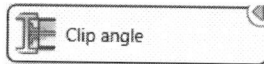

Clip angle

This type of joint is used to connect two beams or a beam and a column by inserting a clip angle between the two. You can use equal angles or unequal angles to insert. You can also decide to bolt the sections to the angles or weld them.

When you invoke the tool for this joint, you will be prompted to select the main beam and then the beam to attach. Note that the beam to attach will be trimmed using the main beam. Once you select both the beams and press the ENTER key, the **Advance Steel - Clip angle [X]** dialog box will be displayed with the **Properties** tab active. This tab is similar to those discussed for previous joints. The remaining tabs are discussed next.

Angle cleat Tab
The options available in the **Angle cleat** tab, shown in Figure 8, are used to define the options related to the angle cleat. These options are discussed next.

Number of angles
This list is used to specify how many angle cleats will be inserted in the joint. By default, **double** is selected from this list. As a result, the angle cleat will be inserted on both sides of the beam to be attached. You can also select the option to insert a single cleat on the left side or the right side of the beam to connect. Figure 9 shows the two beams to connect using this joint and Figure 10 shows the double angle cleats inserted. Note that in Figure 10, the **Display type** of the beam to connect is set to **Symbol**.

Angle profile size
Select the angle cleat type and size from the two flyouts available in this list.

Long leg side
If you are using unequal angles for the cleat profile, you can decide to place the longer side of the cleat on the main beam or the secondary beam using the options in this list.

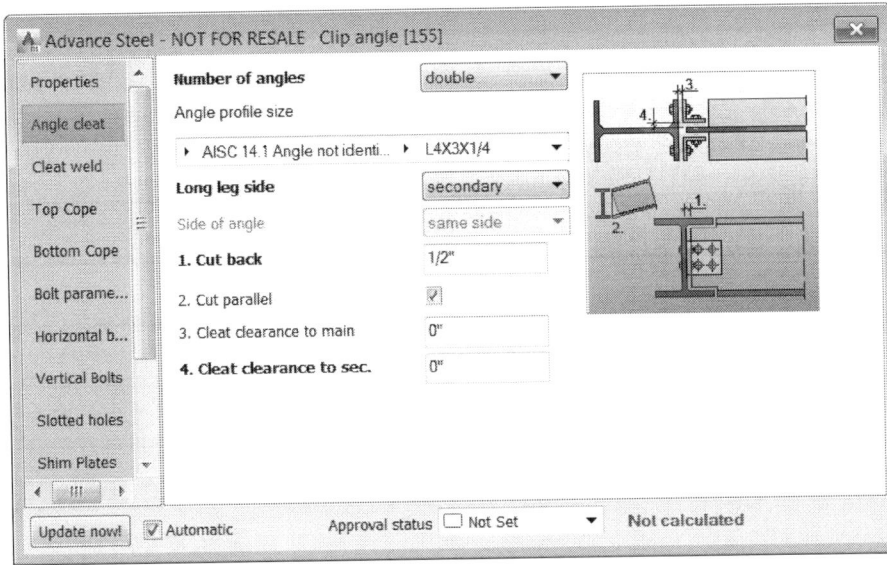

*Figure 8 The **Angle cleat** tab of the **Advance Steel - Clip angle [X]** dialog box*

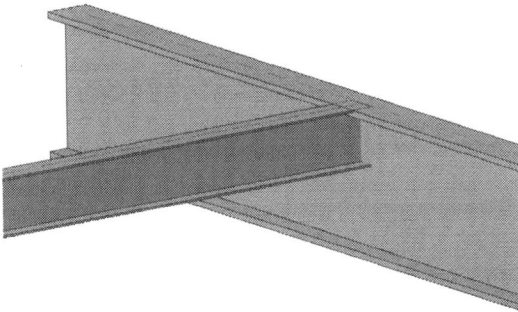

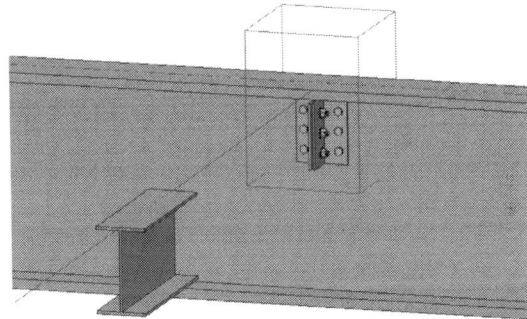

Figure 9 Two beams to apply the clip angle joint

Figure 10 Clip angle joint with the cleats inserted on both sides

Side of angle (for Imperial) or Side of web to fix cleat (for Metric)

This list is only available when you are inserting a single cleat and is used to specify which side of the secondary beam is the angle cleat to be attached. By default, the **same side** option is selected from this list. As a result, the angle cleat is attached to the secondary beam on the same side where it is attached to the main beam. You can also select the option to insert it on the opposite side.

Cut back

This edit box is used to specify the cut back distance of the secondary beam from the main beam.

Cut parallel

This tick box is used when secondary beam is at an angle to the main beam. In that case, you can select this tick box to cut the beam parallel to the main beam.

Cleat clearance to main / Cleat clearance to sec.

You can use these edit boxes to enter clearance between the cleat and the main beam and the secondary beam.

Cleat weld Tab (for Imperial) or Weld Tab (for Metric)

The name of this tab varies depending on the country standard you are using. The options available in the **Cleat weld** tab, shown in Figure 11, are used to specify whether the cleats will be bolted to the main and secondary beam or welded. By default, the **none** option is selected from the **Welded at** list in this tab. As a result, the cleats are bolted to both main beam and the secondary beam. However, you can select the option to weld the cleat to the main beam, secondary beam, or both beams from this list. If you select the option to weld the cleat, the parameters related to the welds can be defined in the edit boxes and lists available below the **Welded at** list.

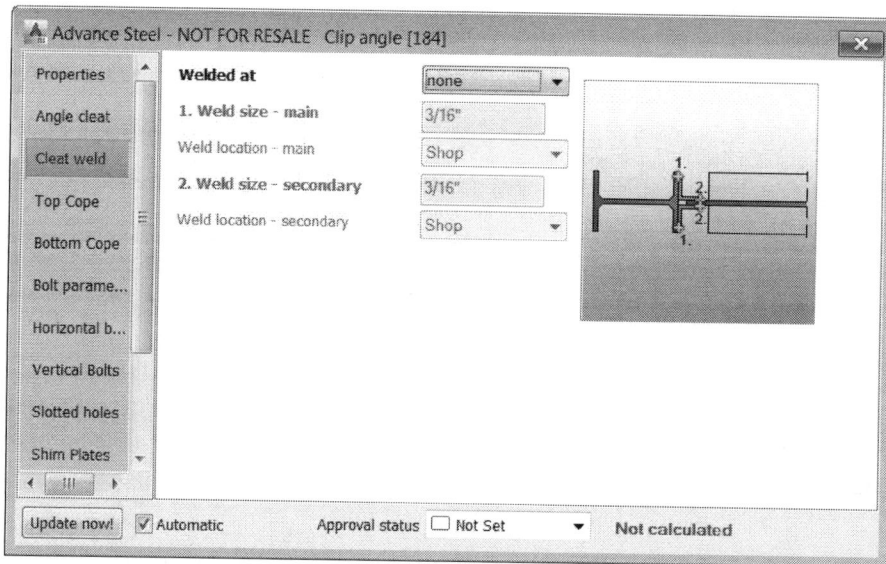

*Figure 11 The **Cleat weld** tab of the **Advance Steel - Clip angle [X]** dialog box*

Top cope Tab (for Imperial) or Top detail Tab (for Metric)

Coping is the process of cutting the secondary beam to match the shape of the main beam so that the two beams could be connected. The options available in the **Top cope** tab, shown in Figure 12, are used to specify the options related to the top cope. These options are discussed next.

Cope alignment

This list allows you to specify whether the top cope will be aligned or perpendicular to the main beam.

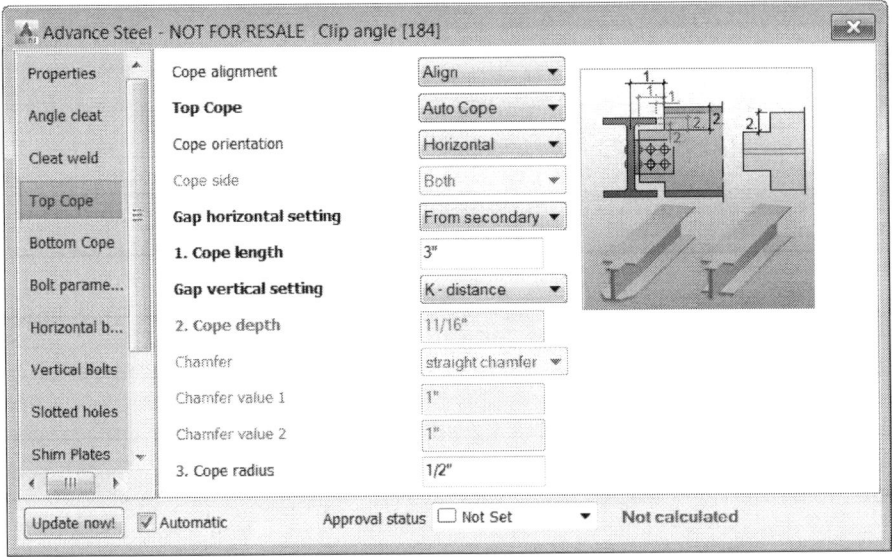

*Figure 12 The **Top cope** tab of the **Advance Steel - Clip angle [X]** dialog box*

Top Cope (for Imperial) or Top detail (for Metric)

This list is used to specify the type of top cope. By default, **Auto Cope** is selected from this list. You can select the option to not cope the beam, trim the beam, add a chamfer, or cut and flush the beam. Figure 13 shows a clip angle joint with the secondary beam defined **Auto Cope** option. Figure 14 shows the same beam with **None** option selected from the **Top Cope** list. In both these figures, the **Display type** of the main beam is set to **Symbol**.

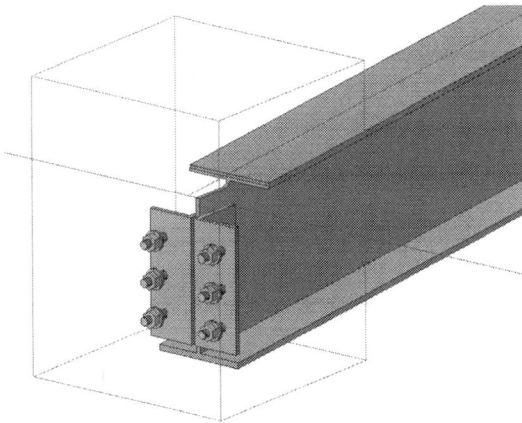

Figure 13 The secondary beam defined auto cope

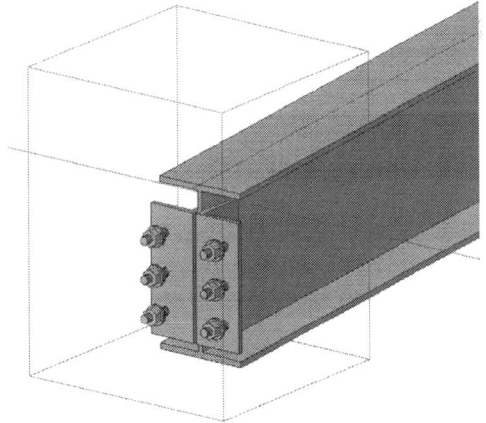

Figure 14 The secondary beam defined no cope

Cope orientation (for Imperial) or Notch orientation (for Metric)

This list is used to specify orientation of the cope. The cope orientation options available in this list are **Horizontal** or **Sloped**.

Cope Side (for Imperial) or Trim/chamfer side (for Metric)
This list is only available if you select the **Trim**, **Chamfer**, or **Cut and flush** option from the **Top Cope** list. In these cases, you can select to specify whether the cope needs to be created on both sides, left side or right side.

Gap horizontal setting (for Imperial) or Horizontal set out (for Metric)
This list is used to specify whether the cope length that you define is measured from the end of the secondary beam, the system axis of the secondary beam, or the flange of the main beam.

Cope length (for Imperial) or Detail length (for Metric)
This edit box is used to specify the length of the cope.

Gap vertical setting (for Imperial) or Vertical set out (for Metric)
This list is used to specify whether the vertical gap of the cope is to be measured from the outside of the main beam, inside of the main beam, outside of the secondary beam, or inside of the secondary beam. In all these cases, the vertical gap cab be defined using the **Cope depth** edit box available below this list. Alternatively, you can select the **k-distance** option from this list, which automatically defines the vertical cope gap.

Chamfer / Chamfer value 1 / Chamfer value 2
This options are used when you select **Chamfer** from the **Top Cope** list. In this case, you can specify the type of chamfer from the **Chamfer** list and enter its dimensions in the **Chamfer value 1** and **Chamfer value 2** edit boxes.

Cope radius (for Imperial) or Notch radius (for Metric)
This edit box is used to specify the cope radius.

Bottom Cope Tab (for Imperial) or Bottom detail Tab (for Metric)

The options in this tab are used when the secondary beam also requires the cope to be created at the bottom to connect to the main beam. These options are similar to those discussed in the **Top Cope** tab.

Bolt parameters Tab

The options in this tab are similar to those discussed in earlier joints.

Horizontal bolts Tab

The options in the **Horizontal bolts** tab, shown in Figure 15, are used specify the options related to the horizontal bolt layout. These options are discussed next.

Set out at main
This list is used to specify the horizontal bolts set out at the main beam. By default, **Angle back** is selected from this list. As a result, the bolt set out is defined from the angle back. Select **Bolt centers** from this list to define the set out from the bolt centers. You can also select **Total & angle gauge** or the **DASt (gap)** to specify the set out based on these parameters.

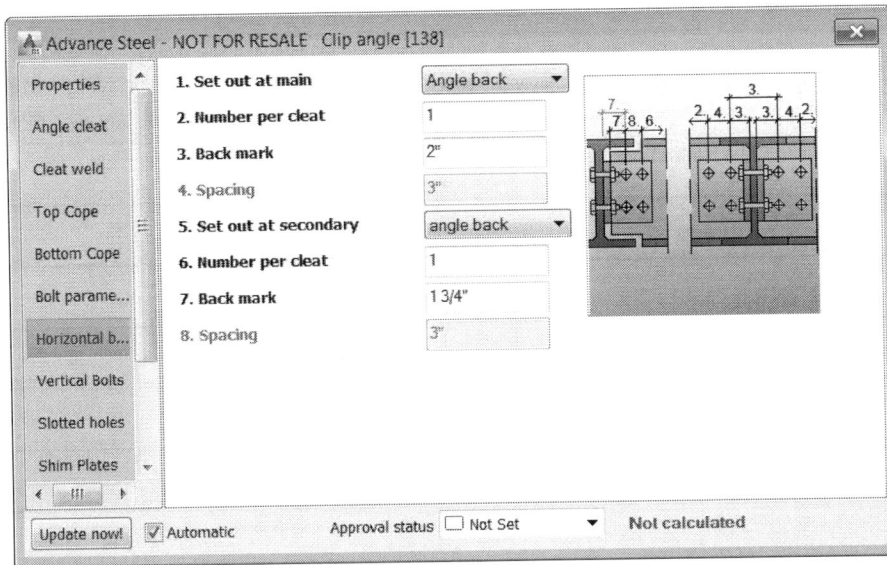

Figure 15 *The **Horizontal bolts** tab of the **Advance Steel - Clip angle [X]** dialog box*

Number per cleat
This edit box is used to specify the columns of bolts to be inserted per cleat on the main beam. Note that if you need more than one number of columns, you may need to use a bigger cleat size.

Back mark
This edit box is used to specify the back mark distance for the bolts on the main beam.

Spacing
While creating more than one bolts per cleat, you can enter the spacing between them in this edit box.

Set out at secondary
This list is used to specify the horizontal bolts set out at the secondary beam. By default, **angle back** is selected from this list. As a result, the set out is measured from the angle back. Selecting **beam end** from this list will measure the set out from the end of the secondary beam.

Number per cleat
This edit box is used to specify the columns of bolts to be inserted per cleat on the secondary beam.

Back mark
This edit box is used to specify the back mark distance for the bolts on the secondary beam.

Vertical Bolts Tab
The options in the **Vertical Bolts** tab, shown in Figure 16, are used specify the options related to the vertical bolt layout. These options are discussed next.

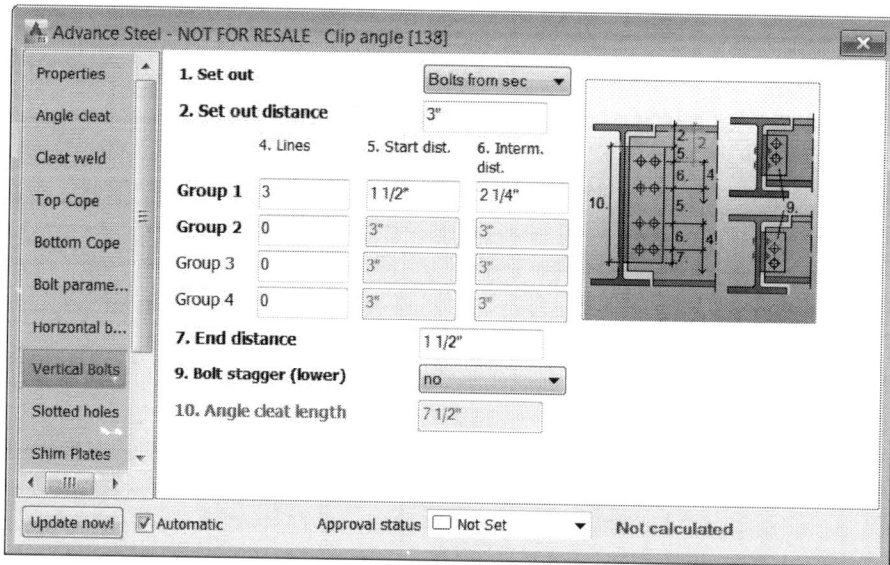

Figure 16 *The Vertical Bolts* tab of the *Advance Steel - Clip angle [X]* dialog box

Set out

This list is used to specify where will the vertical set out distance be measured from. By default, **Bolts from sec** is selected from this list. As a result, the distance you specify in the **Set out distance** edit box will be measured from the top of the secondary beam to the first bolt. If you select **Angle from sec** from this list, the distance will be measured from the top of the secondary beam to the top of the angle cleat. Selecting **Middle** from this list will place the bolts centrally aligned to the secondary beam.

Set out distance

This is the edit box where you specify the set out distance for the vertical bolts.

Group 1 / Group 2 / Group 3 / Group 4

These are the areas where you specify the number of bolts, start distance, and the intermediate distances of the four bolt groups.

End distance

This edit box is used to specify the distance between the center of the last bolt and the bottom of the angle cleat.

Bolt stagger (lower)

This list is used to specify if and how the bolts will be staggered. By default, **no** is selected from this list. As a result, the bolts are not staggered. To stagger the bolt on the main beam or the secondary beam, you can select their respective options from this list. Figure 17 shows the bolts staggered using the **main - 1** option and Figure 18 shows the bolts staggered using the **secondary - 1** option.

The options in the remaining tabs of the Advance Steel - Clip angle [X] dialog box are the same as those discussed in the earlier joints.

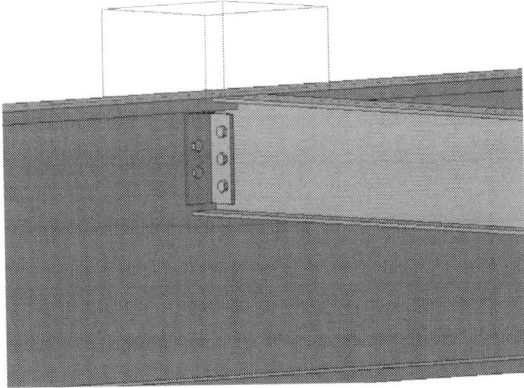

Figure 17 Bolts staggered on the main beam

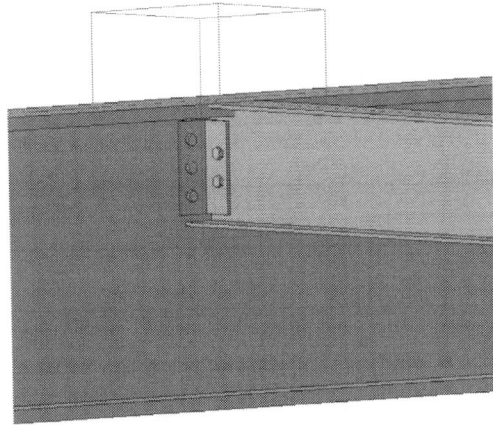

Figure 18 Bolts staggered on the secondary beam

Clip angle - Skewed Joint

Clip angle - Skewed

This joint is similar to the **Clip angle** joint and is used if the beams are at an angle to each other. The process of creating this type of joint is similar to that of the **Clip angle** joint. Figures 19 and 20 show a couple of examples of the **Clip angle - Skewed** joint.

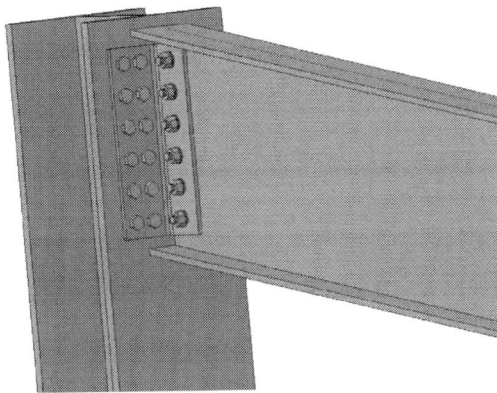

Figure 19 An example of the clip angle - skewed joint

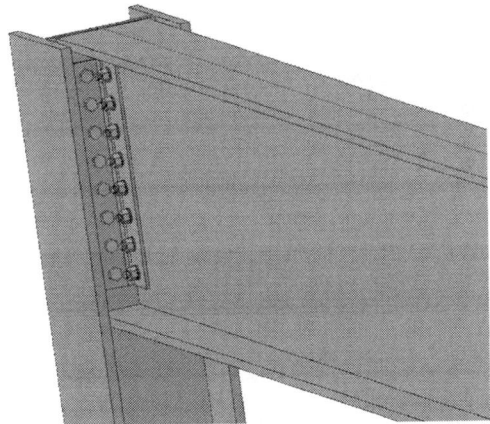

Figure 20 Another example of the clip angle - skewed joint

Double side clip angle Joint

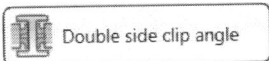

Double side clip angle

This joint is used to connect two beams that are intersecting with another beam or column. When you invoke this tool, you will be prompted to select the main beam. Next, you will be prompted to select the first beam to connect and then finally the second beam to connect. On making all the selections and pressing ENTER, the **Advance Steel - Double Side Clip Angle** dialog box will be displayed. The options in this dialog box are similar to those discussed in the earlier joints. Figure 21 shows an example of the double side clip angle joint. In this figure, the **Display type** of the main beam is set to **Symbol** for better understanding of the joint.

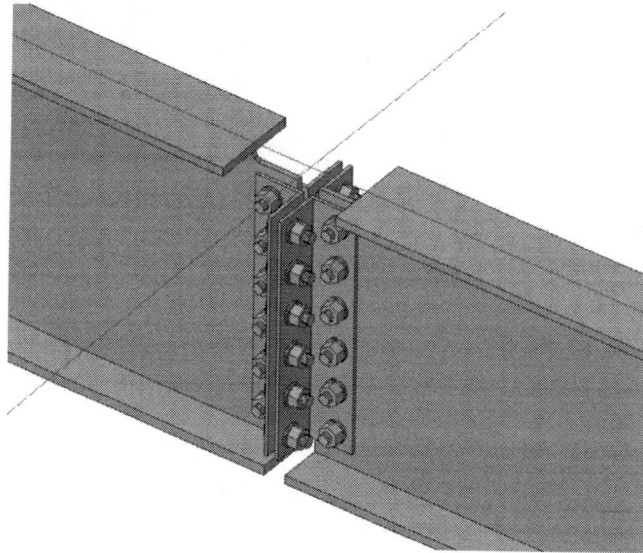

*Figure 21 Double side clip angle joint with the main beam **Display type** set to **Symbol***

Single side end plate Joint

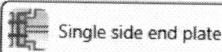

Single side end plate

This joint is used to connect a main beam or column to a secondary beam using an end plate. This end plate will be welded to the secondary beam and bolted to the main beam or column. The options in the dialog box to create this joint are similar to those discussed in the earlier joints. Figure 22 shows a single side end plate joint with six vertical bolts in group 1 and center bolt layout. Figure 23 shows a single side end plate joint with three bolts in group 1 and two bolts in group 2 and the bolt layout set by plate 1 to align the plate to the top.

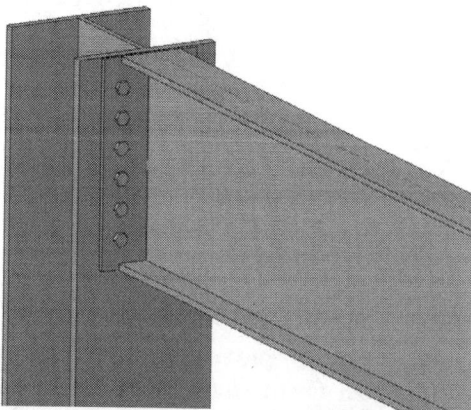

Figure 22 Single side end plate joint with six vertical bolts in group 1 and center bolt layout

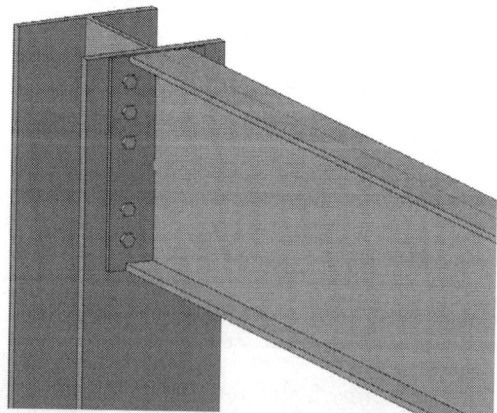

Figure 23 Single side end plate joint with three vertical bolts in group 1 and two bolts in group 2

Double side end plate with safety bolt Joint

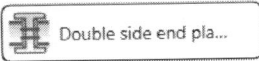

[icon] Double side end pla...

This joint is used to connect a main beam or column to two secondary beams using end plates. The end plates will be welded to the secondary beams and bolted to the web of the main beam or column. Because this connection is created on the web, one set of bolts are used to bolt both the secondary beams to the main beam or column. The options in the dialog box to create this joint are similar to those discussed in the earlier joints. Figure 24 shows the main beam and the two secondary beams to connect and Figure 25 shows the joint created between the three beams. In this joint, there is no safety bolt inserted. Note that in Figure 25, the **Display type** of the main beam is set to **Symbol**.

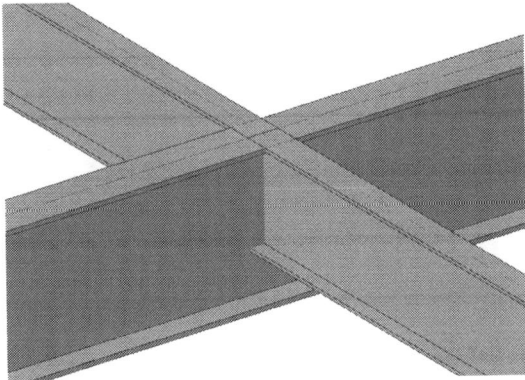

Figure 24 The main beam and the two secondary beams to connect

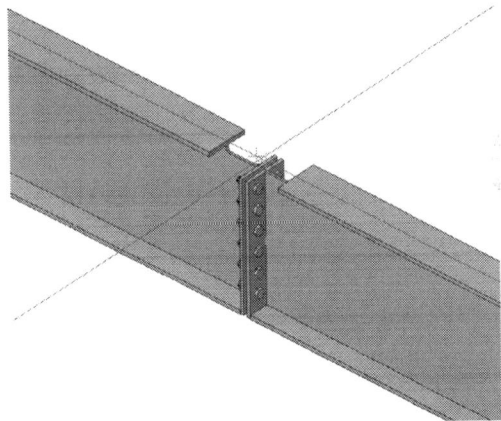

*Figure 25 The joint created between the beams with the **Display type** of main beam set to **Symbol***

Shear plate Joint

[icon] Shear plate

This is one of the most commonly used joints and is used to connect a main beam to a secondary beam using a plate. This plate will be welded to the main beam and bolted to the secondary beam. Most of the options to insert this type of joint are similar to those discussed in the previous joints. However, the options in the **Plate & bolts** category > **Plate layout** tab are different. This tab is discussed next.

Plate & bolts Category > Plate layout Tab

The **Plate layout** tab, shown in Figure 26, is used to define the shear plate to be used for the joint. The options in this tab are discussed next.

Thickness
Enter the plate thickness in this edit box.

Change side
This list is used to specify the side on which the plate will be inserted. You can select the option to insert the plate on the left side, right side, or both sides.

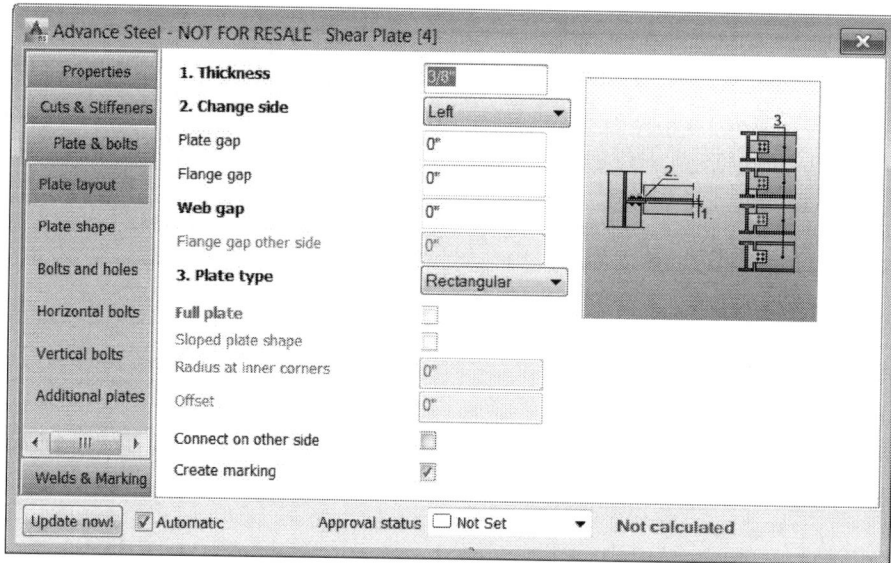

*Figure 26 The **Plate layout** tab of the **Plate & bolts** category*

Plate gap
This edit box is used to specify a gap between the plate and the secondary beam web.

Flange gap
This edit box is used to specify a gap between the plate and the main beam flange. This is used when you are inserting a plate with flange on top or both sides using the options in the **Plate type** list below.

Web gap
This edit box is used to specify a gap between the plate and the main beam web.

Flange gap other side
This edit box is used to specify a gap between the plate and the main beam flange. This is used when you are inserting a plate with flange on bottom or both sides using the options in the **Plate type** list below.

Plate type
This list is used to specify the type of the plate to be used for the joint. Be default, **Rectangular** is selected from the list. As a result, a rectangular plate is used by default. Selecting the **Flange top** or **Flange bottom** option will insert a rectangular plate with a flange on top or flange on bottom. You can also select the **Flange both** option to insert a plate with the flange on both sides. Figure 27 shows a shear plate joint with a rectangular plate, two numbers of horizontal bolts, and three vertical bolts in group 1. Figure 28 shows the same joint that uses a plate with flange on both sides, one number of horizontal bolt, and three vertical bolts in group 1.

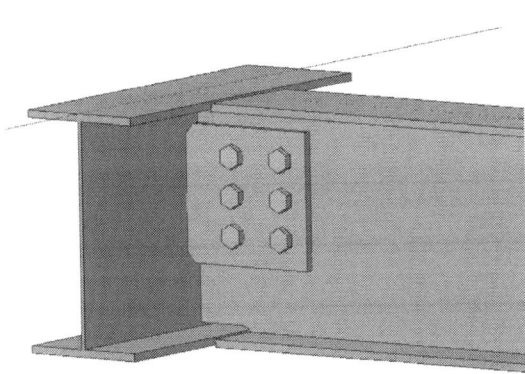

Figure 27 *Shear plate joint with a rectangular plate, two numbers of horizontal bolts, and three vertical bolts in group 1*

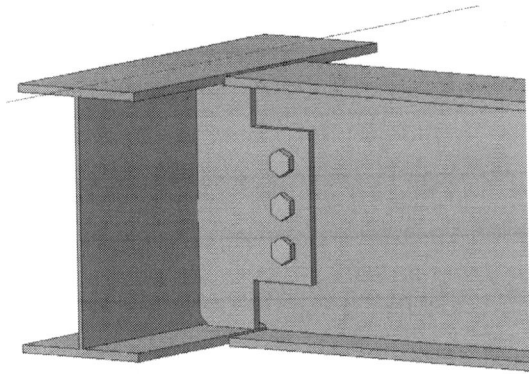

Figure 28 *Shear plate joint that uses a plate with flange on both sides, one number of horizontal bolt, and three vertical bolts in group 1*

Full plate
This tick box is only available if you are creating a flanged plate using the options in the **Plate type** list. In that case, selecting this tick box will convert a flanged plate into a full rectangular plate.

Sloped plate shape
This tick box is only available if you are creating a flanged plate using the options in the **Plate type** list. In that case, selecting this tick box will apply a slope to the top flange of the plate.

Radius at inner corners
When creating a plate with the flanges, you can specify the radius for the inner flange corners by entering a value in this edit box.

Offset
This edit box is used to specify the offset value for the plate flanges.

Connect on other side
Select this tick box to connect the plate on the other side.

Create marking
This tick box is selected to create the plate markings.

The options in the rest of the tabs of this dialog box are similar to those discussed in previous joints.

INSERTING THE PURLINS AND COLD ROLLED JOINTS

This category lists various joints that can be added to purlins and cold rolled sections. Some of the joints in this category are discussed next. Remaining joints are similar, so are not discussed in detail.

Purlin connection Joint

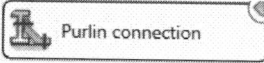

Purlin connection

This type of joint is used to connect a purlin to a rafter using a folded section called a shoe. The shoe is bolted to the purlin and can be bolted or welded to the rafter. When you invoke this tool, you will be prompted to select the main element, which is a rafter. Once you select the rafter and press ENTER, you will be prompted to select a purlin. One selecting the purlin and pressing ENTER, the **Advance Steel - Purlin connection elements [X]** dialog box will be displayed with the **Properties** tab active. The options in this tab are similar to those discussed in the earlier joints. The remaining tabs are discussed next.

Distances Tab

The **Distances** tab, shown in Figure 29, provides the options that are used to specify various distances in the joint. The options are discussed next.

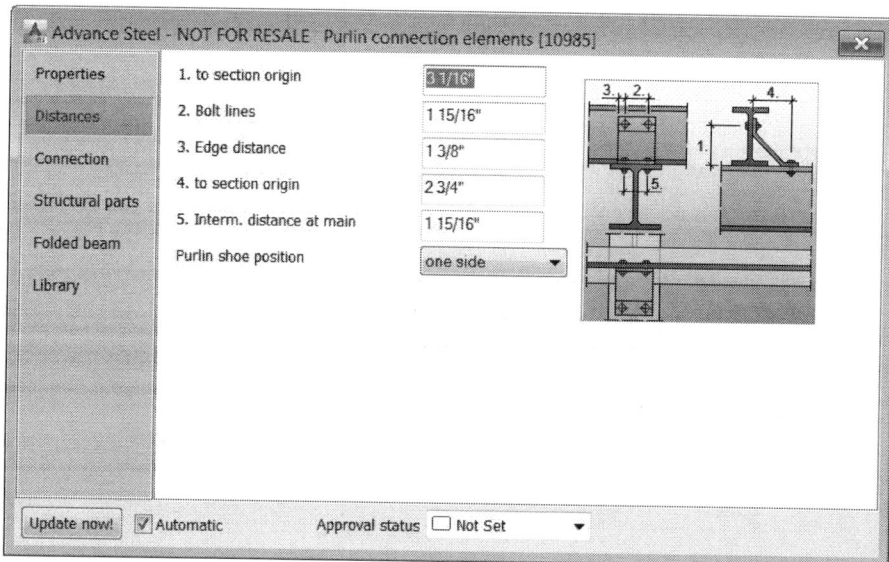

Figure 29 The Distances tab of the Advance Steel - Purlin connection element [X] dialog box

to section origin
This edit box is used to specify the location of the bolts on the purlin from the bottom face of the purlin.

Bolt lines
This edit box is used to specify the spacing between the bolts on the purlin.

Edge distance
This edit box is used to specify the distance between the bolt holes and the edges of the folded section.

to section origin
This edit box is used to specify the location of the bolts on the rafter from the face of the purlin.

Interm. distance at main
This edit box is used to specify the spacing between the bolts on the rafter. By default, this value is set the same as the value of the **Bolt lines** edit box.

Purlin shoe position
This list is used to specify the purlin side on which the shoe will be placed. You can place it on either side of the purlin or on both sides. Figure 30 shows the purlin shoe placed using the **other side** option from this list.

Connection Tab
Most of the options available in this tab are similar to those discussed in the earlier joints. The only option that is different is the **Shoe - beam connection type**. This list is used to specify how the shoe is connected to the rafter. By default, **Bolt** is selected from the list. Alternatively, you can select **Weld** from this list to weld the shoe to the rafter, as shown in Figure 31. All the weld parameters can be defined in the options available below this list.

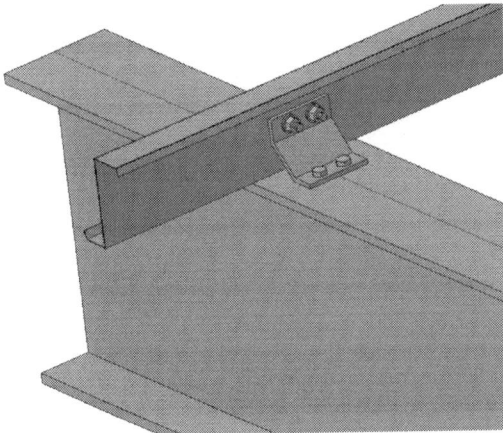

*Figure 30 A purlin joint with the shoe positioned using the **other side** option*

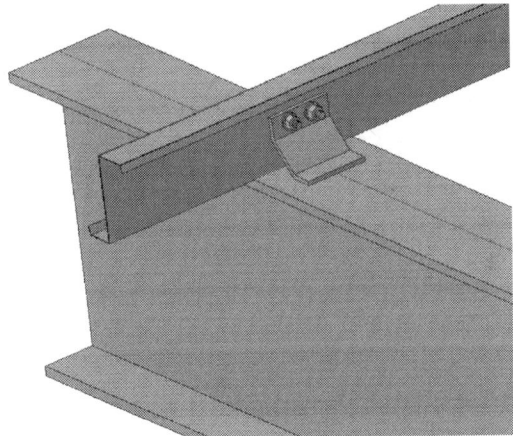

Figure 31 A purlin joint with the shoe bolted to the purlin and welded to the rafter

The options available in the remaining tabs of this joint are similar to those discussed in the earlier joints.

Purlin connection with plate Joint

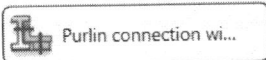

Purlin connection wi...

This type of joint is used to connect a purlin to a rafter using an angle shoe and a plate is an angle section. You can use equal angle or unequal angle as the shoe. The options to create this type of joint are similar to those discussed in the previous joint. Figure 32 shows a purlin connection with plate for a Z-section purlin and Figure 33 shows the same joint for an I-section purlin.

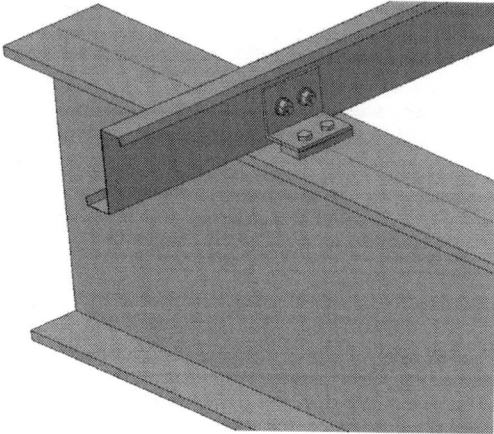

Figure 32 *A purlin connection with the plate for a Z-section purlin*

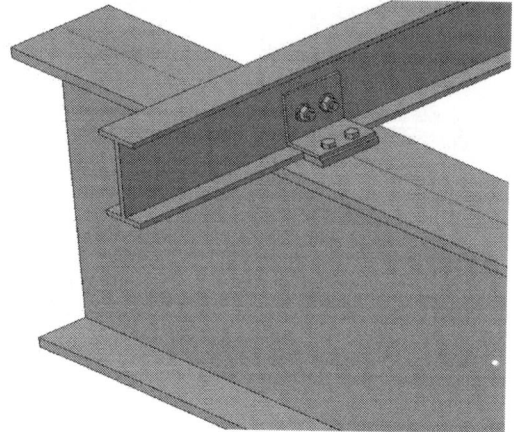

Figure 33 *A purlin connection with the plate for an I-section purlin*

Single purlin plate Joint

This type of joint is used to connect a single purlin to a rafter using a welded plate. This joint works is a little different from the rest. From the **Properties** tab > **Type** list of the **Advance Steel - Single purlin plate** dialog box, you can also select options to insert an angle, a custom cleat connection, a folded plate, or a Tee instead of inserting the plate. This joint also provides you with options to insert sleeves and stays between the purlin and the rafter. Figure 34 shows a single purlin plate joint with no rib stiffener. In this case, the **Edge Distance** value in the **Bolt dimensions** tab is set to **0** to ensure a single column of bolts. Figure 35 shows the same joint, but with the stays created on both sides.

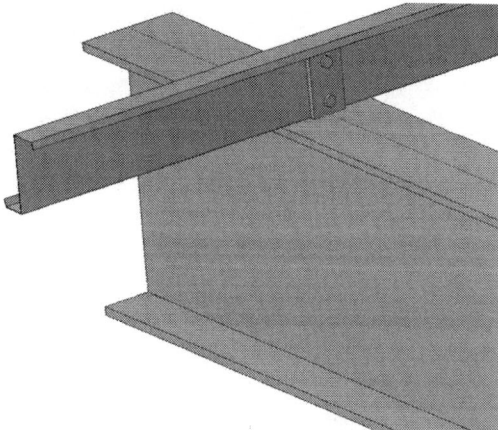

Figure 34 *Single purlin plate joint with no rib stiffener*

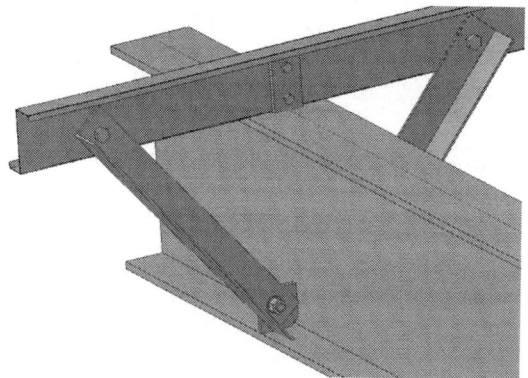

Figure 35 *Single purlin plate joint with stays on both sides*

Double purlin splice plate Joint

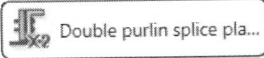

Double purlin splice pla...

This joint is used to connect two purlin to a single rafter. When you invoke this tool, you will be prompted to select the main element and then the two purlins. The options to create this joint are similar to the **Single purlin plate** joint. Figure 36 shows a double purlin plate joint with rib stiffener created on two overlapping purlin members. Figure 37 shows the same joint, but with the stays created on both sides.

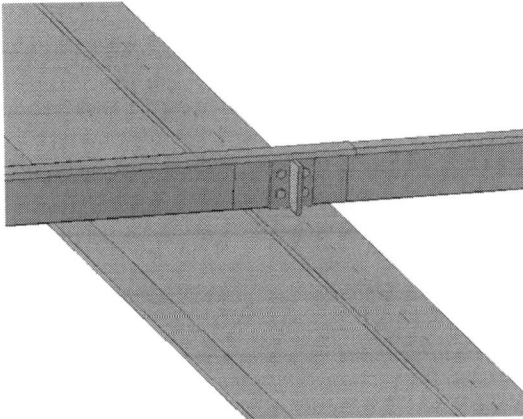

Figure 36 *Double purlin splice plate joint with rib stiffener on two overlapping purlin members*

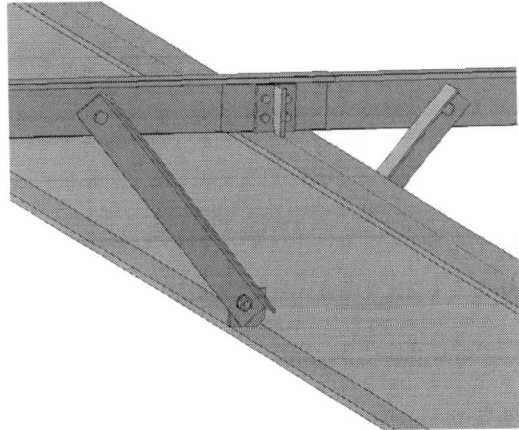

Figure 37 *Double purlin splice plate joint with rib stiffener and stays on two overlapping purlin members*

Single Eaves Beam Bracket from Plate with End Plate Joint

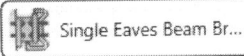

Single Eaves Beam Br...

This joint is specifically used to connect the eaves beam to the column with an eaves beam plate, a stool plate, and an end plate bolted to the column, as shown in Figures 38 and 39.

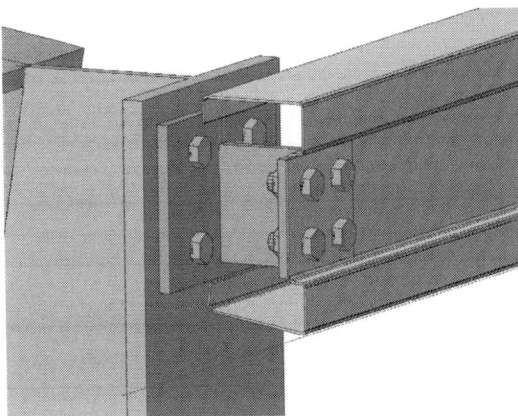

Figure 38 *Eaves beam connected to the column using an eaves beam plate, a stool plate, and an end plate*

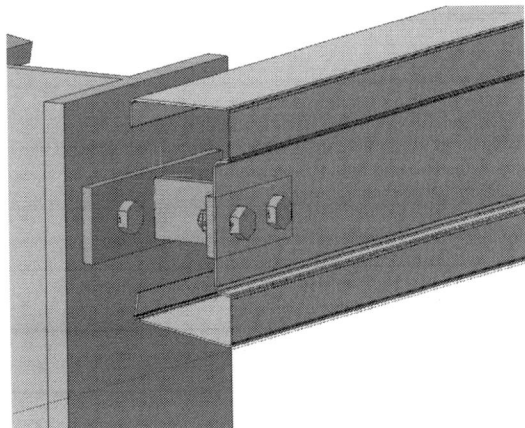

Figure 39 *Same connection with one set of vertical bolts*

Double Eaves Beam Bracket from Plate with End Plate Joint

 Double Eaves Beam...

This joint is used to connect two eaves beams to the column with an eaves beam plate, a stool plate, and an end plate bolted to the column, as shown in Figures 40 and 41. In these two figures, the column is made transparent for better visibility of the joint.

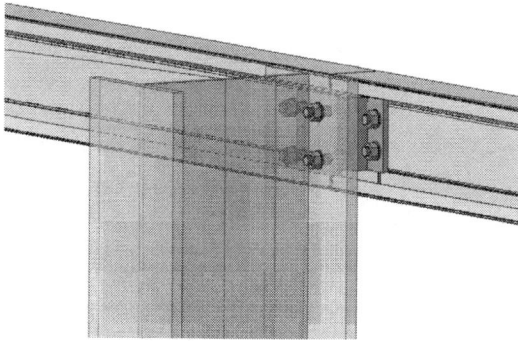

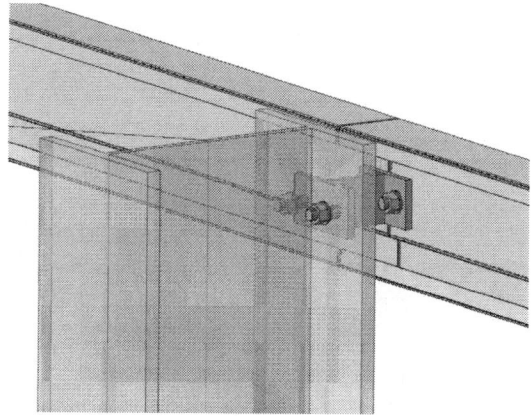

Figure 40 *Two eaves beam connected to the column using two sets of vertical bolts*

Figure 41 *Same connection with one set of vertical bolts*

Hands-on Tutorial (STRUC/BIM)

In this tutorial, you will complete the following tasks:
1. Open the file containing the frames.
2. Insert the **Apex Haunch** joint, as shown in Figures 42.
3. Insert the **Single purlin plate** joint to the end purlin sections, as shown in Figure 43.
4. Insert the **Double purlin splice plate connection** with stays, as shown in Figure 44. In this figure, the rafter display is set to symbol.
5. Insert the single and double eaves beam joints, as shown in Figure 45. In this figure, the eaves beams are made transparent to clearly display the joint.
6. Copy all the joints at other places in the model.

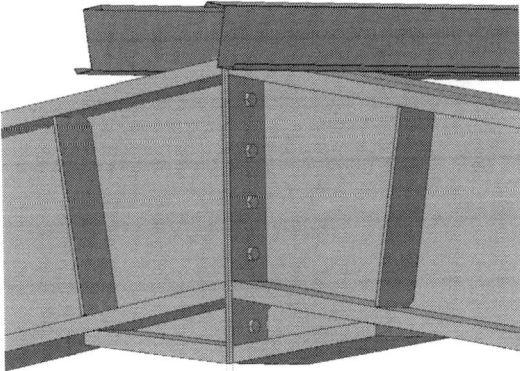

Figure 42 The apex haunch joint

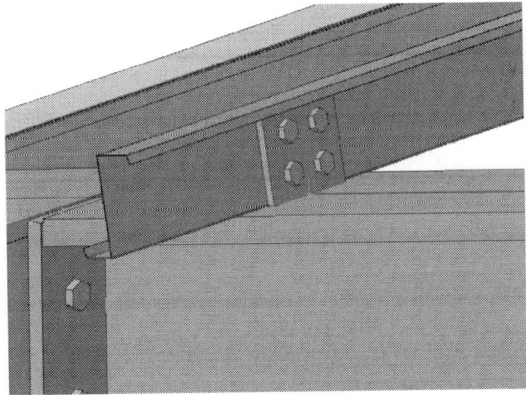

Figure 43 The single purlin plate joint

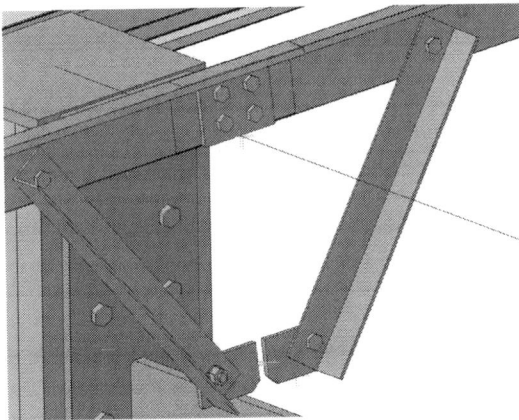

Figure 44 The double purlin splice plate joint with stays

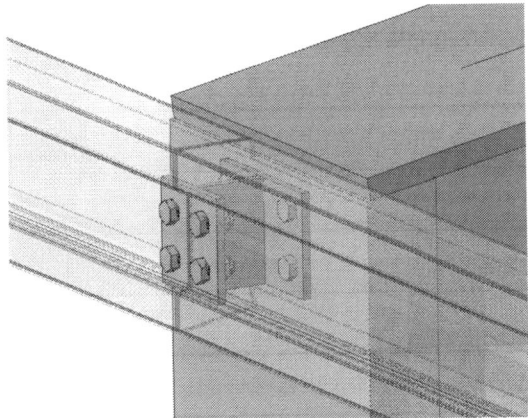

Figure 45 The double eaves beam joint

Apex Haunch Joint Parameters

The following are the parameters for the apex haunch joints. The dimensions not listed below will be accepted with their default values. **Note that if you are using the Metric units, do not enter the unit string (mm) while typing the dimension values.**

Plate Thickness	3/8" or 10mm		
Haunch Length	1' 5 11/16" or 450mm		
Haunch Height	6 5/16" or 160mm		
Horizontal Bolt Distance	6" or 100mm		
Bolt Group 1	4 Lines		
Bolt Group 1 Start Dist.	4" or 80mm	Interm. Dist.	5" or 100mm
Bolt Group 2	1 Line	Start Dist.	7" or 150mm

Single Purlin Plate Joint Parameters

The following are the parameters for the single purlin plate joint.

Plate Thickness	1/2" or 12 mm
Edge Distances	1 15/16" or 40 mm
Horizontal Bolt Layout	From Purlin
Edge Distance	1 1/2" or 40mm
Vertical Bolt Layout	From Purlin
Start Distance	1 1/2" or 40mm
No of Bolts	2
Intermediate Distance	2 1/2" or 60mm

Double Purlin Plate Joint with Stays Parameters

All the values are similar to the **Single purlin plate join** with default values of the stays.

Single Eaves Beam Joint Parameters

The following are the parameters for the single eaves beams.

Eaves Plate Thickness	3/8" or 10mm
Offset	0
Horizontal Bolt Layout	Total
Total Distance	2 1/4" or 60mm
Eaves Plate Layout	Total
End Plate Width	4 3/4" or 120mm
Additional Profile Projection	Overhang > 2' or 600mm
Eaves Plate Vertical Bolts	2
Start Distance	3" or 75mm
Intermediate Distance	2" or 50mm
Edge Distances	1 3/16" or 35mm
Stool Plate Thickness	3/8" or 10mm
Stool Height	Total from Top
Edge Distance 1	0
Total Height	4" or 100mm
End Plate Thickness	3/8" or 10mm

End Plate Length Type	**Projection**
Projections Top and Bottom	**1/2" or 12mm**
End Plate Horizontal Bolts	**Projection Layout**
Center Distance	**4" or 100mm**
Edge Distances 1 and 2	**1 3/16" or 35mm**
Offset	**0**
End Plate Vertical Bolts	**From Plate Top**
Group 1	**2 Lines**
Start Distance	**1 3/16" or 35mm**
Intermediate Distance	**2 1/2" or 55mm**

Double Eaves Beam Joint

All the values are similar to the **Single eaves beam joint**, except for the following values.

End Plate Horizontal Bolts	**Projection Layout**
Center Distance	**4" or 100mm**
Edge Distances 1 and 2	**1 3/16" or 35mm**
Offset	**0**
End Plate Vertical Bolts	**From Plate Top**
Group 1	**2 Lines**
Start Distance	**1 3/16" or 35mm**
Intermediate Distance	**2 1/4" or 55mm**

Section 1: Opening the File and Inserting the Apex Haunch Joint

In this section, you will open **C05-Imperial-Frame.dwg** or **C05-Metric-Frame.dwg** file and then insert one of the apex haunch joints.

1. From the **C05 > Struc-BIM** folder, open the **C05-Imperial-Frame.dwg** or **C05-Metric-Frame.dwg** file, based on the preferred units. The model for this tutorial is shown in Figure 46.

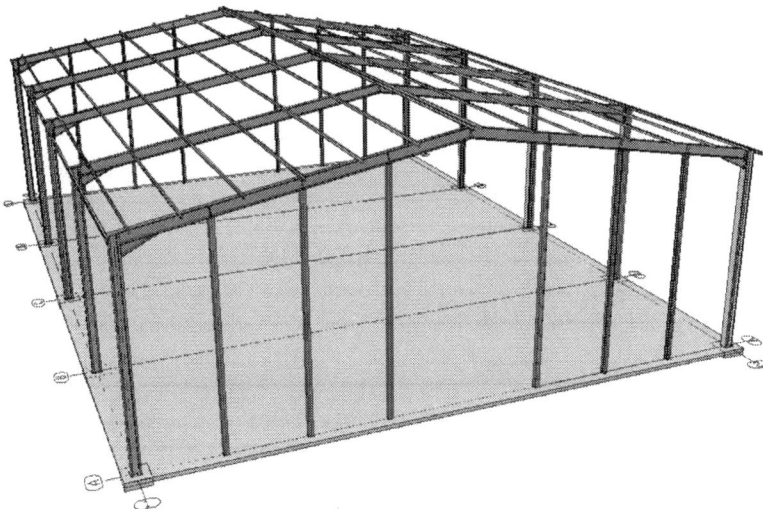

Figure 46 *The perspective view of the model for the tutorial*

Notice that this file is similar to the one in which you added joints in Chapter 4. You will now add the apex haunch joint to the rafters at the front.

2. Zoom to the apex of the front rafters, as shown in Figure 47.

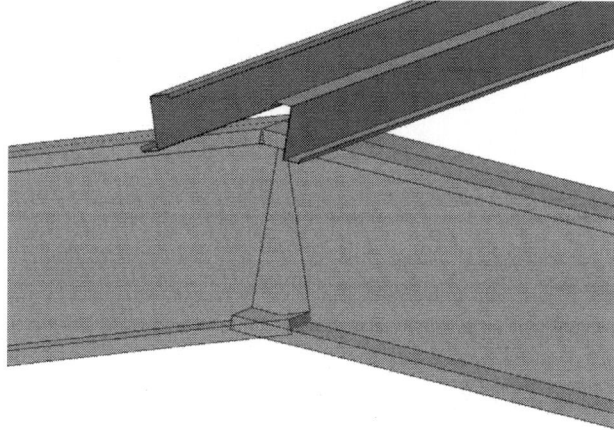

Figure 47 The apex of the rafters to add the joint

3. If the **Connection vault** palette is not displayed, turn it on from the **Home** ribbon tab > **Extended Modeling** ribbon panel > **Connection vault** tool.

4. From the **Joint Categories** area, click the **Beam end to end** button, which is the fourth button; various beam end to end joints are displayed in the **Joint Type** area.

   Before you invoke the joint tool, you will add it to the **Favorites** category.

5. Click the star icon on the right of the **Apex haunch** button; the joint is added to the **Favorites** category.

   Next, you will invoke this tool to insert the joint.

6. Click the **Apex haunch** button to invoke this joint tool; you are prompted to select the main beam.

   As mentioned in the previous chapter, it is better to hide the **Connection vault** palette before you start applying the joint.

7. From the top right of the **Connection vault** title bar that displays **Beam end to end**, click the **minus "-"** sign to hide the palette.

8. Now, select the left rafter and then press the ENTER key; you are prompted to select the secondary beam.

9. Select the right rafter and then press the ENTER key; the **Advance Steel - Apex bolted with haunch [X]** dialog box is displayed and the joint with the default values is created.

10. From the left pane of the dialog box, click the **General** category; the **End plate** tab is activated.

11. In the **Thickness** edit box, enter **3/8"** or **10**, depending on whether you are using Imperial or Metric units.

12. Activate the **Haunch 1** tab.

13. In the edit box on the right of the **Length from inner end plate face** list, enter **1' 5 11/16"** or **450**.

14. In the edit box on the right of the **Height from Rafter bottom** list, enter **1' 5 11/16"** or **160**.

15. Activate the **Bolts** tab.

    The default bolt diameter will be accepted for this joint.

16. From the **Bolts type** list, select **A325** (for Imperial) of **AS 1252** (for Metric).

17. In the **Horizontal distance** edit box, enter **6"** or **100**.

18. Activate the **Bolt groups** tab.

19. In the **Group 1 > Lines** edit box, enter **4**.

20. In the **Group 1 > Start dist.** edit box, enter **4"** or **80**.

21. In the **Group 1 > Interm. dist.** edit box, enter **5"** or **100**.

22. In the **Group 2 > Lines** edit box, enter **1**.

23. In the **Group 2 > Start dist.** edit box, enter **7"** or **150**.

    This defines all the required joint parameters. You will now add these joint settings to the library.

24. Activate the **Properties** category > **Library** tab and add the joint to the library. The model, after adding the joint, is shown in Figure 48.

Section 2: Inserting the Purlin Joints

In this section, you will first insert the **Single purlin plate** joint to the last to eave purlin on the left of the frame. Next, you will insert the **Double purlin splice plate connection** joint to the last to eave purlins on the left rafter of the first portal frame. This rafter connects to the column at the **B1** grid intersection point.

Note: You can also choose any other single and double purlins to insert the joints. At the end, you will copy the joints. Therefore, it will not matter which one you added as the master because you will be able to upgrade any joint to the master, if required.

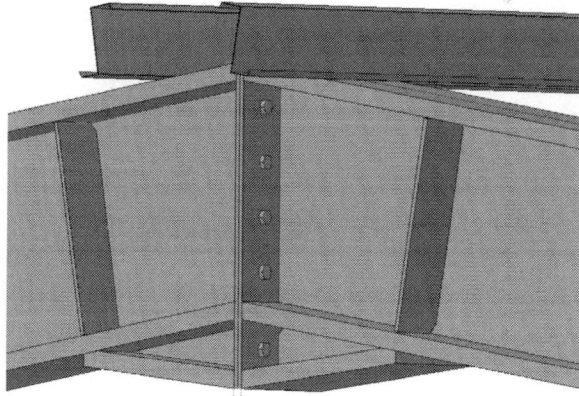

Figure 48 *The model after adding the apex haunch joint*

1. Zoom close to the top of the column at the **A1** grid intersection point so you can see the last to eave purlin and the rafter, as shown in Figure 49. In this figure, the rafter to which the joint is to be added is labeled as **1** and the purlin is labeled as **2**.

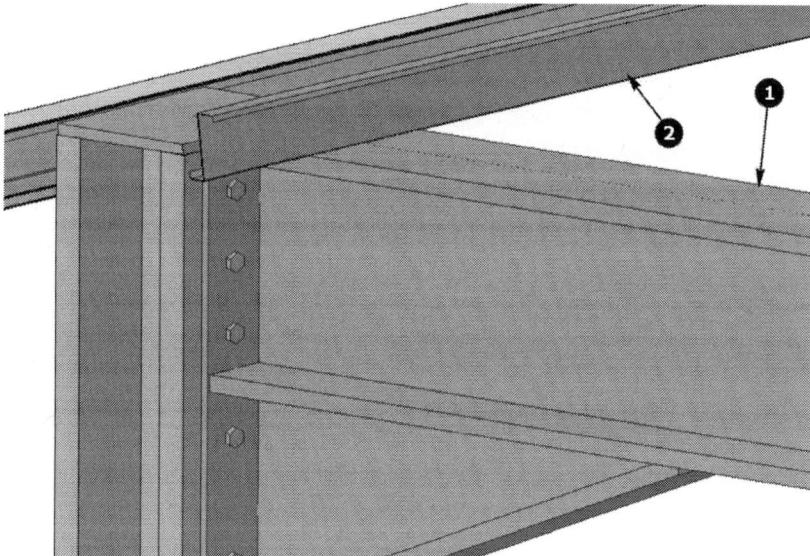

Figure 49 *The rafter and the last to eave purlin to insert the joint*

2. Expand the **Connection vault** and from the **Joint Categories** area, click the **Purlin & Cold rolled** button; various purlin and eaves beam joints are displayed in the **Joint Type** area.

3. Click the star icon on the right of the **Single purlin plate** button; the joint is added to the **Favorites** category.

4. Click the **Single purlin plate** button to invoke this joint tool; you are prompted to select the main element.

5. From the top right of the **Connection vault** title bar that displays **Purlin & Cold rolled**, click the **minus "-"** sign to hide the palette.

6. Now, select the rafter labeled as **1** in Figure 49 and then press the ENTER key; you are prompted to select the purlin.

7. Select the purlin labeled as **2** in Figure 49 and then press the ENTER key; the **Advance Steel - Single purlin plate [X]** dialog box is displayed and the joint with the default values is created.

8. Activate the **Plate cleat** tab.

9. In the **Plate thickness** edit box, enter **1/2"** or **12**.

10. In the **3. Edge distance** or **3. End distance** and **4. Edge distance** or **4. End distance** edit boxes, enter **1 15/16"** or **40**.

11. Activate the **Bolt dimensions** tab.

12. Make sure **From Purlin** is selected from the **Horizontal bolt layout** list.

13. In the **Edge distance** edit box, enter **1 1/2"** or **40**.

14. Make sure **From Purlin** is selected from the **Vertical bolt layout** list.

15. In the **Start distance** edit box, enter **1 1/2"** or **40**.

16. In the **Intermediate distance** edit box, enter **2 1/2"** or **60**.

17. Activate the **Bolts** tab.

18. From the **Purlin bolts** list, select **1/2 inch** or **12.00 mm**.

19. From the **Bolt Type** list, select **A325** (for Imperial) of **Laysaght Purlin bolts** (for Metric).

    Next, you need to turn off the rib stiffener so it is not created.

20. Activate the **Rib stiffener** tab.

21. Clear the **Create rib stiffener** tick box; the rib stiffener is no more created.

    This defines all the required joint parameters. You will now add these joint settings to the library.

22. Activate the **Library** tab and add the joint to the library. The model, after adding the joint, is shown in Figure 50.

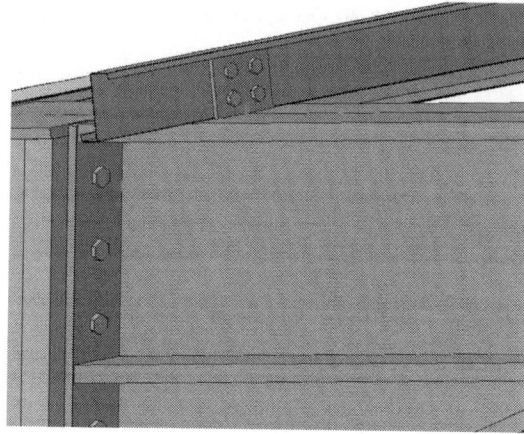

Figure 50 The model after adding the single purlin plate joint

Next, you will add the double purlin splice plate joint.

23. Zoom close to the top of the column at the **B1** grid intersection point so you can see the last to eave purlins and the rafter, as shown in Figure 51. In this figure, the rafter to which the joint is to be added is labeled as **1** and the two purlins are labeled as **2** and **3**.

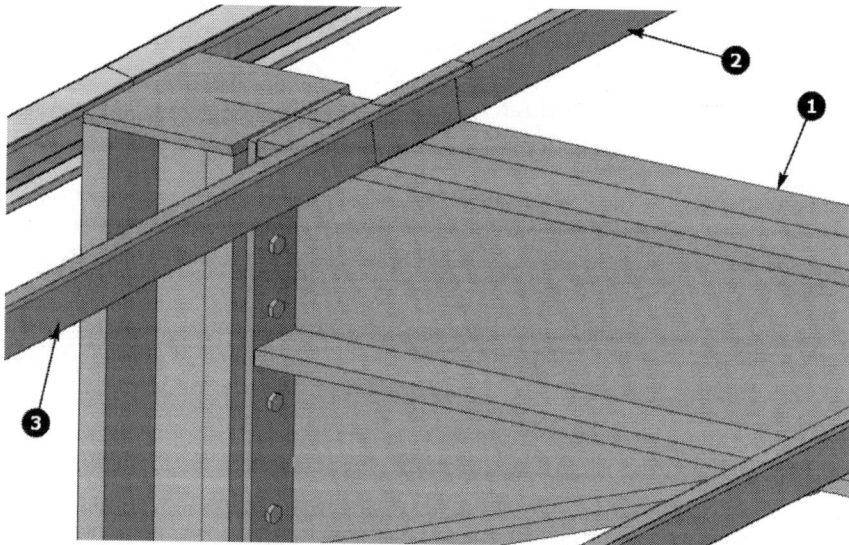

Figure 51 The rafter and the last to eave purlins to insert the joint

24. Expand the **Connection vault** and click the star icon on the right of the **Double purlin splice plate connection** button; the joint is added to the **Favorites** category.

25. Now, click the **Double purlin splice plate connection** button to invoke this joint tool; you are prompted to select the main element.

26. From the top right of the **Connection vault** title bar that displays **Purlin & Cold rolled**, click the **minus "-"** sign to hide the palette.

27. Now, select the rafter labeled as **1** in Figure 51 and then press the ENTER key; you are prompted to select the purlin.

28. Select the purlin labeled as **2** in Figure 51 and then press the ENTER key; you are again prompted to select the purlin.

29. Select the purlin labeled as **3** in Figure 51 and then press the ENTER key; the **Advance Steel - Double purlin splice plate connection [X]** dialog box is displayed and the joint with the default values is created.

30. Activate the **Plate cleat** tab.

31. In the **Plate thickness** edit box, enter **1/2"** or **12**.

32. In the **3. Edge distance** or **3. End distance** and **4. Edge distance** or **4. End distance** edit boxes, enter **1 15/16"** or **40**.

33. Activate the **Bolt dimensions** tab.

34. Make sure **From Purlin** is selected from the **Horizontal bolt layout** list.

35. In the **Edge distance** edit box, enter **1 1/2"** or **40**.

36. Make sure **From Purlin** is selected from the **Vertical bolt layout** list.

37. In the **Start distance** edit box, enter **1 1/2"** or **40**.

38. In the **Intermediate distance** edit box, enter **2 1/2"** or **60**.

39. Activate the **Bolts** tab.

40. From the **Purlin bolts** list, select **1/2 inch** or **12.00 mm**.

41. From the **Bolt Type** list, select **A325** (for Imperial) of **Laysaght Purlin bolts** (for Metric).

    Next, you need to turn off the rib stiffener so it is not created.

42. Activate the **Rib stiffener** tab.

43. Clear the **Create rib stiffener** tick box; the rib stiffener is no more created.

    Next, you need to add stays to the joint using the **Stays** tab. You will accept the default values of stay dimensions.

44. Activate the **Stays** tab.

45. From the **Create Stays** list, select **Both**.

46. If required, select the **Mirror** tick box to ensure the stays look similar to the one shown in Figure 52.

47. Activate the **Library** tab and add the joint to the library. The model, after adding the joint, is shown in Figure 52.

Figure 52 The model after adding the double purlin plate with stays joint

Section 3: Inserting the Eaves Beam Joints

In this section, you will first insert the single eaves beam joint between the column at the **A1** grid intersection point and eaves beam near the top of this column. Next, you will insert the double eaves beam joint between the column at the **B2** grid intersection point and the two eaves beams near the top of this column.

1. Zoom close to the top of the column at the **A1** grid intersection point so you can see the eave beam and the column, as shown in Figure 53. In this figure, the column to which the joint is to be added is labeled as **1** and the eave beam is labeled as **2**.

2. Expand the **Connection vault** and add the **Single Eaves Beam Bracket from Plate with End Plate** joint to the **Favorites** category.

3. Now, click the **Single Eaves Beam Bracket from Plate with End Plate** button; you are prompted to select the main element.

4. Hide the **Connection vault** palette.

5. Select the column labeled as **1** in Figure 53 and press ENTER; you are prompted to select the eaves beam.

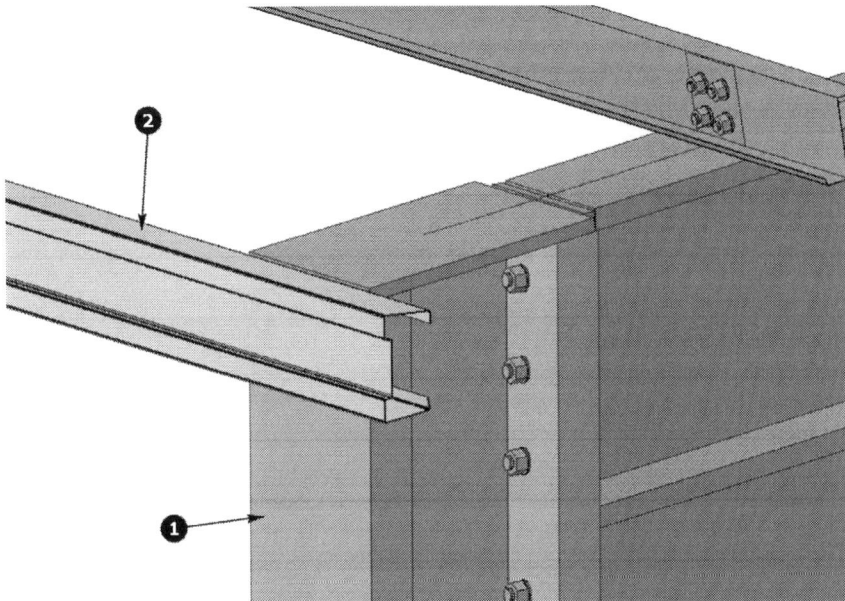

Figure 53 The column and the eaves beam to insert the joint

6. Select the eaves beam labeled as **2** in Figure 53 and press ENTER; the **Advance Steel - Single eaves beam - Bracket from plate with end plate [X]** dialog box is displayed and the joint with default values will be created.

7. Activate the **Eaves plate horizontal** tab.

8. In the **Plate thickness** edit box, enter **3/8"** or **10**.

9. Make sure **Total** is selected from the **Horizontal bolt layout** list.

10. In the **Total distance** edit box, enter **2 1/4"** or **60**.

11. Make sure **Total** is selected from the **Eaves plate layout** list.

12. In the **End plate width** edit box, enter **4 3/4"** or **120**.

13. Activate the **Additional profile** tab.

14. In the **Overhang** edit box, enter **2'** or **600**.

15. Activate the **Eaves plate vertical** tab.

16. In the **Number of bolts** edit box, enter **2**.

17. In the **Start distance** edit box, enter **3"** or **75**.

18. In the **Intermediate distance** edit box, enter **2"** or **50**.

19. In the two **Edge distance** edit boxes, enter **1"** or **25**.

20. Activate the **Stool plate** tab.

21. In the **Plate thickness** edit box, enter **3/8"** or **10**.

22. Make sure in the **Stool height** list, **Total from top** is selected.

23. Make sure the **Edge distance 1** value is set to **0**.

24. In the **Total height** edit box, enter **4"** or **100**.

25. Activate the **End plate** tab.

26. In the **Plate thickness** edit box, enter **3/8"** or **10**.

27. Make sure in the **End Plate Length Type** list, **Projection** is selected.

28. In the **Projection top** and **Projection bottom** edit boxes, enter **1/2"** or **12**.

29. Activate the **Horizontal bolts layout end plate** tab.

30. Make sure in the **Horizontal bolt layout** list, **Projection** is selected.

31. Clear the **On gauge line** tick box, if selected.

32. In the **Center distance** edit box, enter **4"** or **100**.

33. In the **Edge distance 1** and **Edge distance 2** edit boxes, enter **1 3/16"** or **35**.

34. Make sure the value of the **Offset** edit box is set to **0**.

35. Activate the **Vertical bolts layout end plate** tab.

36. Make sure in the **Main Bolt Layout** list, **From plate top** is selected.

37. In **Group 1 > Lines** edit box, enter **2**.

38. In **Group 1 > Start dist.**, enter **1 3/16"** or **35**.

39. In **Group 1 > Interm. dist.**, enter **2 1/2"** or **55**.

This defines all the required joint parameters. You will now add these joint settings to the library.

40. Activate the **Library** tab and add the joint to the library. The model, after adding the joint, is shown in Figure 54. In this figure, the eaves beam is made transparent for clearer visibility of the joint.

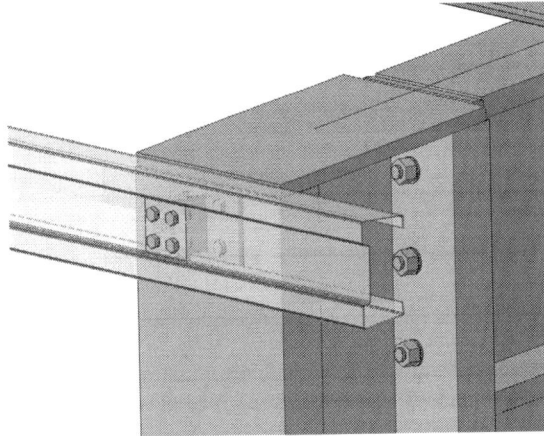

Figure 54 *The model after adding the single eaves beam joint*

Next, you will add the double eaves beam joint to the column at the **B1** grid intersection point and the two eaves beams near the top of this column.

41. Zoom close to the top of the column at the **B1** grid intersection point, as shown in Figure 55. In this figure, the column is labeled as **1** and the two eaves beams to be connected to the column are labeled as **2** and **3**.

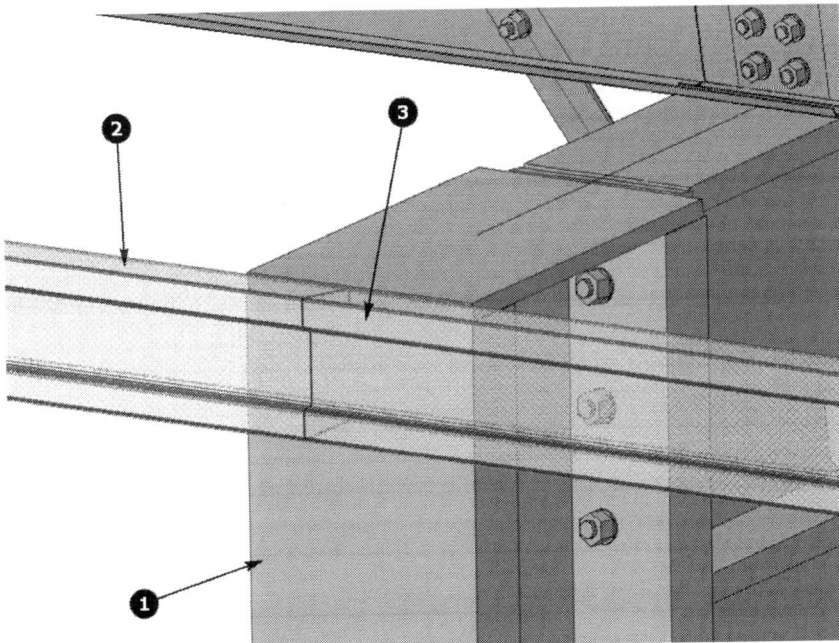

Figure 55 *The column and the eaves beams to insert the joint*

42. Expand the **Connection vault** and add the **Double Eaves Beam Bracket from Plate with End Plate** joint to the **Favorites** category.

43. Now, click the **Double Eaves Beam Bracket from Plate with End Plate** button; you are prompted to select the main element.

44. Hide the **Connection vault** palette.

45. Select the column labeled as **1** in Figure 55 and press ENTER; you are prompted to select the eaves beam.

46. Select the eaves beam labeled as **2** in Figure 55 and press ENTER; you are again prompted to select the eaves beam.

47. Select the eaves beam labeled as **3** in Figure 55 and press ENTER; the **Advance Steel - Double eaves beam - Bracket from plate with end plate [X]** dialog box is displayed and the joint with default values will be created.

48. Activate the **Eaves plate horizontal** tab.

49. In the **Plate thickness** edit box, enter **3/8"** or **10**.

50. Make sure **Total** is selected from the **Horizontal bolt layout** list.

51. In the **Total distance** edit box, enter **2 1/4"** or **60**.

52. Make sure **Total** is selected from the **Eaves plate layout** list.

53. In the **End plate width** edit box, enter **4 3/4"** or **120**.

54. Activate the **Eaves plate vertical** tab.

55. In the **Number of bolts** edit box, enter **2**.

56. In the **Start distance** edit box, enter **3"** or **75**.

57. In the **Intermediate distance** edit box, enter **2"** or **50**.

58. In the two **Edge distance** edit boxes, enter **1"** or **25**.

59. Activate the **Stool plate** tab.

60. In the **Plate thickness** edit box, enter **3/8"** or **10**.

59. From the **Stool height** list, select **Total from top**.

60. In the **Edge distance 1** edit box, enter **0**.

61. In the **Total height** edit box, enter **4"** or **100**.

62. Activate the **End plate** tab.

63. In the **Plate thickness** edit box, enter **3/8"** or **10**.

64. Make sure in the **End Plate Length Type** list, **Projection** is selected.

65. In the **Projection top** and **Projection bottom** edit boxes, enter **1/2"** or **12**.

66. Activate the **Horizontal bolts layout end plate** tab.

67. Make sure in the **Horizontal bolt layout** list, **Projection** is selected.

68. Clear the **On gauge line** tick box, if selected.

69. In the **Center distance** edit box, enter **4"** or **100**.

70. In the **Edge distance 1** and **Edge distance 2** edit boxes, enter **1 3/16"** or **35**.

71. Make sure the value of the **Offset** edit box is set to **0**.

72. Activate the **Vertical bolts layout end plate** tab.

73. Make sure in the **Main Bolt Layout** list, **From plate top** is selected.

74. In **Group 1 > Lines** edit box, enter **2**.

75. In **Group 1 > Start dist.**, enter **1 3/16"** or **35**.

76. In **Group 1 > Interm. dist.**, enter **2 1/2"** or **55**.

77. Activate the **Library** tab and add the joint to the library. The model, after adding the joint, is shown in Figure 56. In this figure, the eaves beam is made transparent for clearer visibility of the joint.

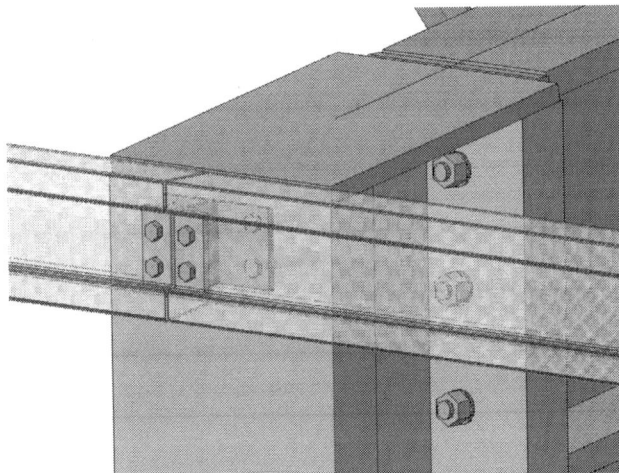

Figure 56 The model after adding the double eaves beam joint

Section 4: Creating the Joints on the Top Purlins

The top purlins need to be applied joints separately so you can copy them without any error. In this section, you will apply one single purlin joint and one double purlin joint to the top purlins and the rafters.

1. Zoom close to the apex of the front gable frame, as shown in Figure 57.

2. From the **Connection vault**, invoke the **Single purlin plate** tool; you are prompted to select the main element.

3. Select the rafter labeled as **1** in Figure 57 and then press the ENTER key; you are prompted to select the purlin.

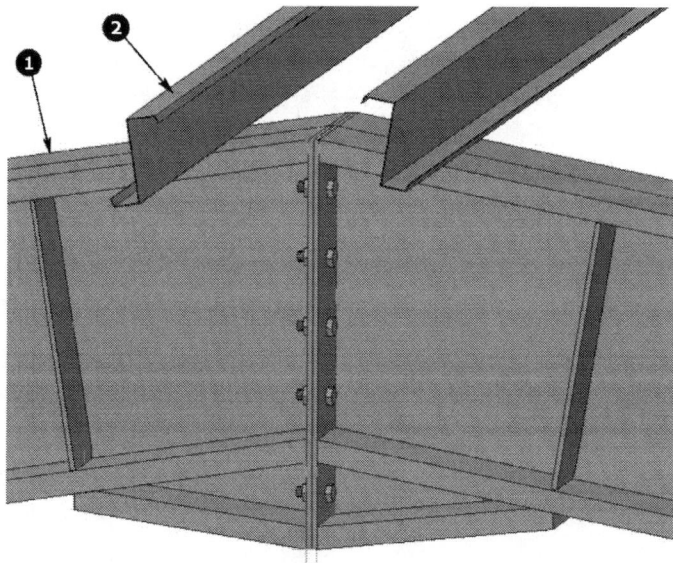

Figure 57 The rafter and the purlin to be selected

4. Select the purlin labeled as **2** in Figure 57 and then press the ENTER key; the **Advance Steel - Single purlin plate [X]** dialog box is displayed and the joint with the default values is created.

5. Activate the **Library** tab and select the joint that you saved in section 2; the joint is modified the match the library settings.

6. Close the dialog box.

Tip: If you forgot to add the joint to the library, you can at any point of time, edit the joint and then add it to the library.

Next, you will add the double purlin joint to the top purlin and then modify it to remove the stays.

7. Zoom close to the apex of the first portal frame, as shown in Figure 58.

8. Invoke the **Double purlin splice plate connection** tool; you are prompted to select the main element.

9. Select the rafter labeled as **1** in Figure 58 and then press the ENTER key; you are prompted to select the purlin.

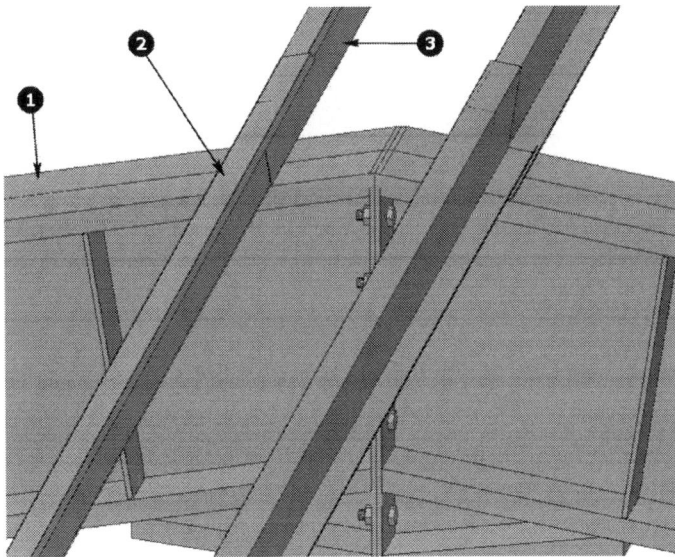

Figure 58 The rafter and the purlins to be selected

10. Select the purlin labeled as **2** in Figure 58 and then press the ENTER key; you are again prompted to select the purlin.

11. Select the purlin labeled as **3** in Figure 58 and then press the ENTER key; the **Advance Steel - Double purlin splice plate connection [X]** dialog box is displayed and the joint with the default values is created.

12. Activate the **Library** tab and select the joint that you saved in section 2; the joint is modified the match the library settings.

    Notice that the stays of this joint are intersecting with the bolts of the apex haunch joint. Therefore, you need to edit this joint to remove the stays.

13. Activate the **Stays** tab.

14. From the **Create Stay** list, select **None**.

15. Close the dialog box. The model, after inserting this joint, is shown in Figure 59.

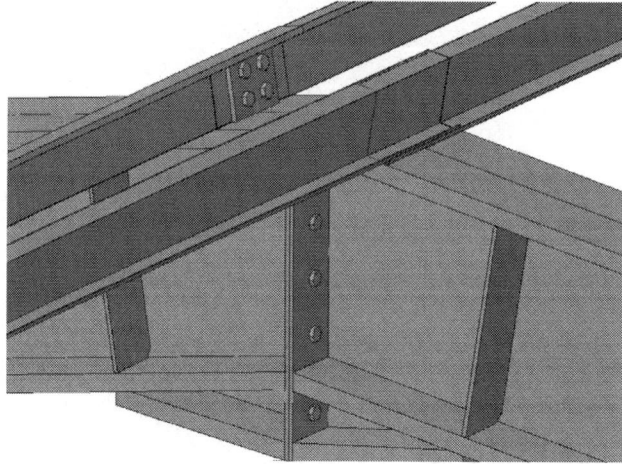

Figure 59 The model after inserting the double purlin joint at the top purlins

Section 5: Copying Joints

In this section, you will use the **Create joint in a joint group, multiple** tool to create multiple copies of all the joints you have created in the previous sections.

1. From the **Advance Steel** tool palette > **Tools** tab, invoke the **Create joint in a joint group, multiple** tool; you are prompted to select the source joint.

   The first joint you need to copy is the apex haunch joint. As mentioned in the previous chapter, you can select any element of the joint while selecting the master joint. In most cases, it is easier to select the plate or the bolts from the joints.

2. Select one of the plates from the apex haunch joints and press ENTER; you are prompted to select beams corresponding to the source joints input (1/2).

3. Select the left rafters of all three portal frames and the gable frame at the rear and then press ENTER; you are prompted to select beams corresponding to the source joints input (2/2).

4. Select the right rafters of all the three portal frames and the gable frame at the rear and then press ENTER; the joint is copied at the remaining four frames as well.

   Tip: As mentioned in the previous chapter, you can edit any of the copied joints and upgrade them to master to modify the original master joint and the other associated joints.

   The next joint you will copy is the single purlin joint. Note that you will not select the top purlins. The joints to the top purlins will be copied separately.

5. Invoke the **Create joint in a joint group, multiple** tool again; you are prompted to select the source joint.

6. Select the plate of the single purlin joint and press ENTER; you are prompted to select beams corresponding to the source joints input (1/2).

7. Select all the rafters of the front and rear gable frames and then press ENTER; you are prompted to select beams corresponding to the source joints input (2/2).

    Next, you need to select all the end purlins. However, are mentioned earlier, you will not select the top purlins as there are separate joints added to those purlins. They will be copied separately.

8. Select all the end purlins on the front and rear gable frames, **except the top purlins**. There will be a total of 23 purlins that need to be selected.

9. Once you have selected all the end purlins, press ENTER; all the single purlin joints are created. If you get a warning about the joints being different, click **No** in the dialog box.

10. Similarly, copy the remaining joints. If you get a warning about the joints being different, click **No** in the dialog box. The completed model, after copying all the joints, is shown in Figure 60.

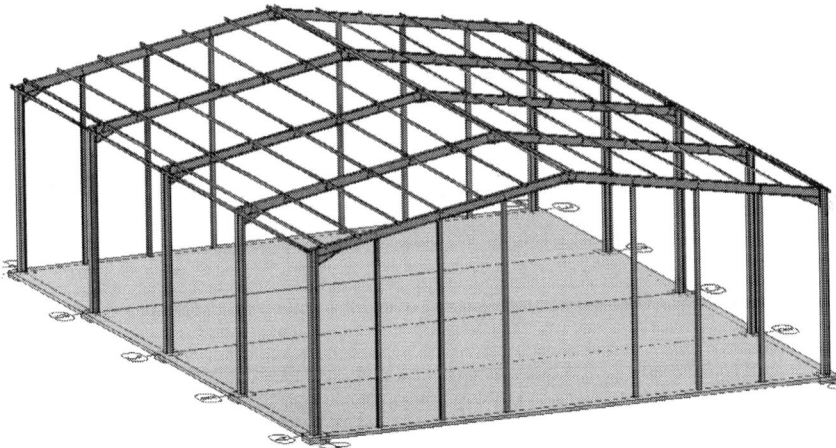

Figure 60 *The completed model, after copying all the joints*

Section 6a (STRUCTURE ONLY): Inserting and Copying the End Plate Joints in the Elevator Tower Model

In this section, you will open the model of the Elevator Tower and then insert a single side end plate joint between the flange side of the column at the **C1** grid intersection point and the perimeter beam. Next, you will insert another single side end plate joint between the web side of the same column and the other perimeter beam connecting to that column. Finally, you will copy these joints at all the required locations on levels 1 and 2.

1. From the **C05 > Struc-BIM** folder, open the **C05-Imperial-Elev-Tower.dwg** or **C05-Metric-Elev-Tower.dwg** file, based on the preferred units; the model looks similar to the one shown in Figure 61.

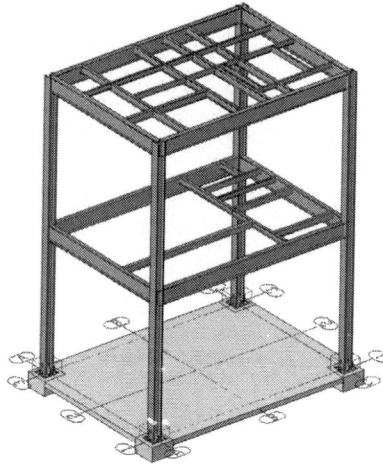

Figure 61 The Elevation Tower model to insert joints

2. Zoom to the area shown in Figure 62.

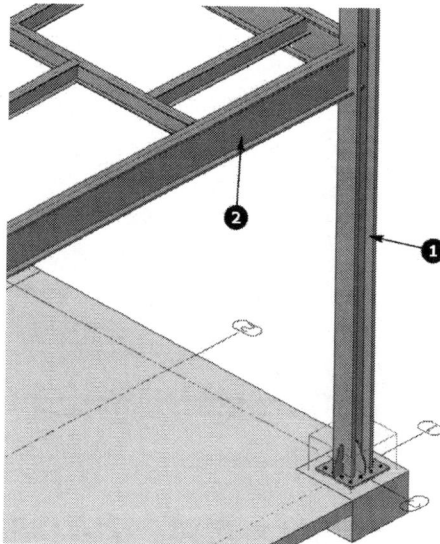

Figure 62 The zoomed in view to insert the joint

3. From the **Joint Categories** area of the **Connection vault** palette, click the **Platform beams** button, which is the fifth button; various platform beam joints are displayed in the **Joint Type** area.

> **Note**: *In the joints you will create next, the names of various dialog box options in this joint vary depending on the country standard you selected while installing the software. In the following steps, the names in the **English US** and **English Australia** installations are used. These may be different from your country installation.*

4. Click the **Single side end plate** button to invoke this joint tool; you are prompted to select the main beam.

5. Select the column labeled as **1** in Figure 62.

6. Press the ENTER key; you are prompted to select the secondary beam.

7. Select the beam labeled as **2** in Figure 62.

8. Press the ENTER key; the **Advance Steel - Single Side End Plate [X]** dialog box is displayed and the joint with the default values is created.

9. From the left pane of the dialog box, select the **Plate & bolts > Plate - Alignment** tab (for Imperial) or the **Plate & bolts > End plate** tab (for Metric).

10. In the **Plate thickness** edit box, enter **3/4"** or **20** as the value.

11. Activate the **Bolts and holes** tab.

12. From the **Diameter** list, select **3/4 inch** or **20.00 mm**.

13. From the **Bolt Type** list, select **A325** for Imperial units or **AS 1252** from the Metric units.

14. Activate the **Horizontal bolts** tab.

15. From the **Plate layout** or **Plate set out** list, make sure **Projection** is selected.

16. Clear the **Bolts on gauge lines** tick box.

17. In the **Center distance** edit box, enter **5 1/2"** or **120**.

18. In the **Edge distance / Proj. 1** edit box, enter **1 1/4"** or **20**.

19. In the **Edge distance / Proj. 2** edit box, enter **1 1/4"** or **20**.

20. Activate the **Vertical bolts** tab.

21. From the **Layout** or **Set out** list, select **Center** or **Middle**.

22. From the **Plate height by** or **Plate length by** list, select **Total from middle**.

23. In the **Plate height** or **Plate length** edit box, enter **2' 1/2"** or **528**.

24. In the **Group 1 > Lines** edit box, enter **5**.

    The **Group 1 > Start distance** edit box is grayed out because of the center plate layout.

25. In the **Group 1 > Intermed. Dist** edit box, enter **4 1/4"** or **95**.

This completes all the settings for this joint. Next, you will add this joint to the library.

26. Activate the **Properties > Library** tab of the dialog box and add the joint to the library with the name **C05-Flange-Tower**.

27. Close the dialog box. The model, after inserting this joint, is shown in Figure 63.

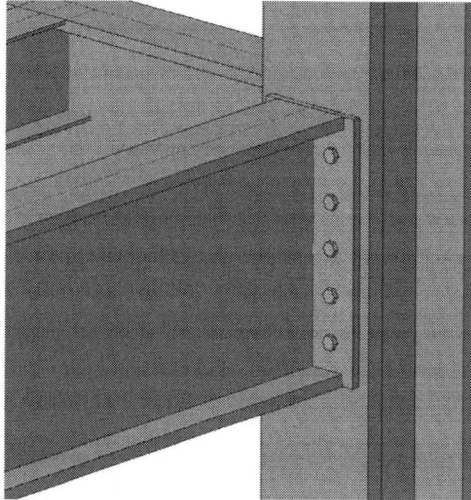

Figure 63 *The model, after inserting the end plate joint*

Next, you will insert the single side end plate joint between the same column and the beam connecting to the web of it. But because the settings of this joint is different from the previous joint, you will not copy it. Instead, you will insert a new joint.

28. Orbit the model to a view similar to the one shown in Figure 64.

29. From the **Connection vault** palette, invoke the **Single side end plate** joint tool again; you are prompted to select the main beam.

30. Select the column labeled as **1** in Figure 64 and then press ENTER; you are prompted to select the secondary beam.

31. Select the beam labeled as **2** in Figure 64 and then press ENTER; the **Advance Steel - Single Side End Plate [X]** dialog box is displayed and the joint with the default values is created.

32. From the left pane of the dialog box, select the **Plate & bolts > Plate - Alignment** tab or **Plate bolts > End plate** tab.

33. In the **Plate thickness** edit box, enter **3/4"** or **20** as the value.

34. Activate the **Bolts and holes** tab.

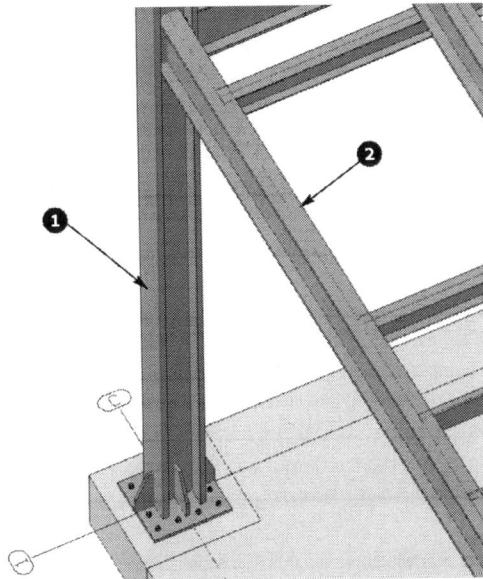

Figure 64 *The zoomed in view to insert the joint*

35. From the **Diameter** list, select **3/4 inch** or **20.00 mm**.

36. From the **Bolt Type** list, select **A325** for Imperial units or **AS 1252** from the Metric units.

37. Activate the **Horizontal bolts** tab.

38. From the **Plate layout** or **Plate set out** list, select **Total**.

39. In the **Width** edit box, enter **8"** or **210**.

40. In the **Center distance** edit box, enter **4"** or **100**.

41. Activate the **Vertical bolts** tab.

42. From the **Layout** or **Set out** list, select **Secondary** or **Bolts from sec**.

43. From the **Plate height by** or **Plate length by** list, select **Bolt distance from top**.

    In this joint, you will extend the plate above the beam in order to connect the bracing joint plate in the later chapters.

44. In the **Layout distance** or **Set out distance** edit box, enter **-9"** or **-230**.

45. In the **Group 1 > Lines** or **Rows** edit box, enter **2**.

46. In the **Group 1 > Start distance** edit box, enter **3"** or **75**.

47. In the **Group 1 > Intermed. Dist** edit box, enter **6"** or **150**.

48. In the **Group 2 > Lines** or **Rows** edit box, enter **5**.

49. In the **Group 2 > Start distance** edit box, enter **6"** or **155**.

50. In the **Group 2 > Intermed. Dist** edit box, enter **4 1/2"** or **95**.

51. In the **Group 3 > Lines** or **Rows** edit box, enter **1**.

52. In the **Group 3 > Start distance** edit box, enter **6"** or **135**.

53. In the **End distance** edit box, enter **2"** or **50**.

> **Note:** *The reason you have extended the plate of this joint above the beam is because in later chapters, you will connect the bracing joint plate to this extended plate.*

54. Activate the **Properties > Library** tab of the dialog box and add the joint to the library with the name **C05-Web-Tower**.

55. Close the dialog box. The model, after inserting this joint, is shown in Figure 65.

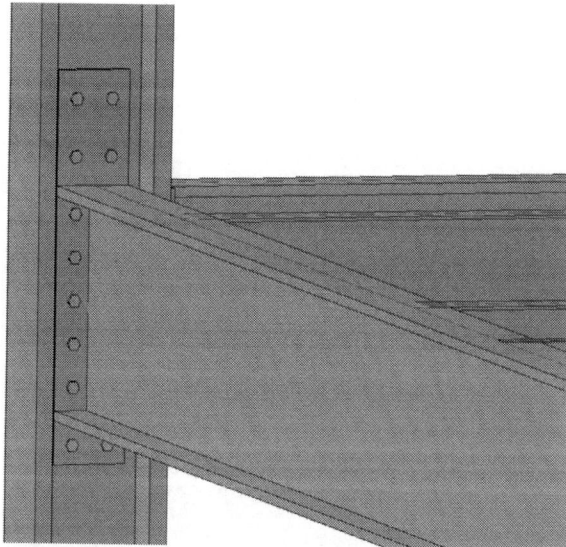

Figure 65 The model, after inserting the end plate joint

Next, you will copy the first end plate joint to the same column and the beam connecting to its web side on level 2. You will separate the copied joint from the parent joint and edit it to meet your requirements.

56. From the **Advance Steel Tool Palette > Tools** tab, invoke the **Create joint in a joint group** tool; you are prompted to select the connection part.

57. Select the plate of the first end plate joint you inserted and then press ENTER; you are prompted to select the main beam.

58. Select the column labeled as **1** in Figure 66 and then press ENTER; you are prompted to select the secondary beam.

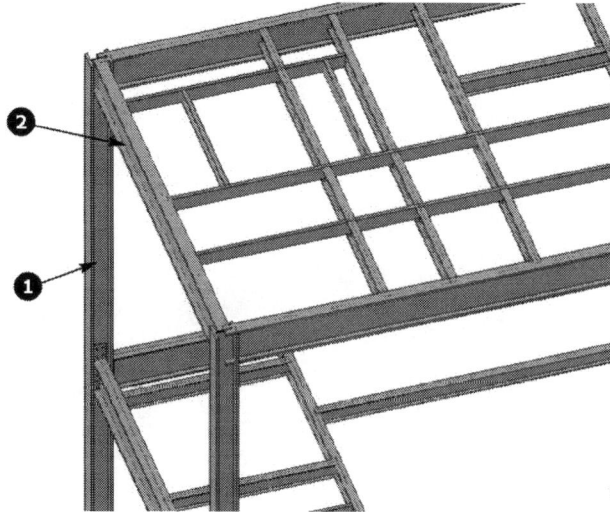

Figure 66 The zoomed in view to copy the joint

59. Select the beam labeled as **2** in Figure 66 and then press ENTER; you are again prompted to select the main beam.

60. Press ENTER; the **Warning** dialog box is displayed informing you that the joint is different from the master joint.

61. Choose the **Yes** button in the dialog box to separate the joint from the master. This will ensure that this joint is an independent joint.

62. Select the bolt of the copied joint and then right-click in the blank area of the drawing window to display the shortcut menu.

63. Select **Advance Joint Properties** from the shortcut menu; the **Advance Steel - Single Side End Plate [X]** dialog box is displayed.

64. Activate the **Plate & bolts > Horizontal bolts** tab.

65. From the **Plate layout** or **Plate set out** list, select **Total**.

66. In the **Width** edit box, enter **8"** or **210**.

67. In the **Center distance** edit box, enter **4"** or **100**.

68. Close the dialog box. The model, after copying and editing the joint, is shown in Figure 67.

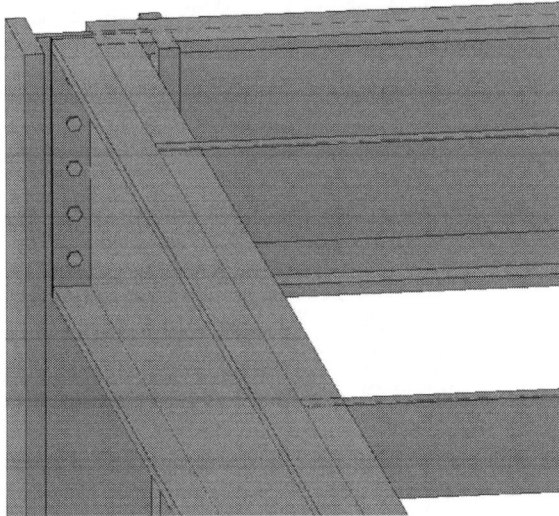

Figure 67 *The model, after copying and editing the joint*

Next, you will create multiple copies of these joints at the remaining column flange and perimeter beam intersections and column web and perimeter beam intersections.

69. Using the **Create joint in a joint group, multiple** tool, copy the first end plate joint at all the column flange and perimeter beam intersections labeled as **1** in Figure 68.

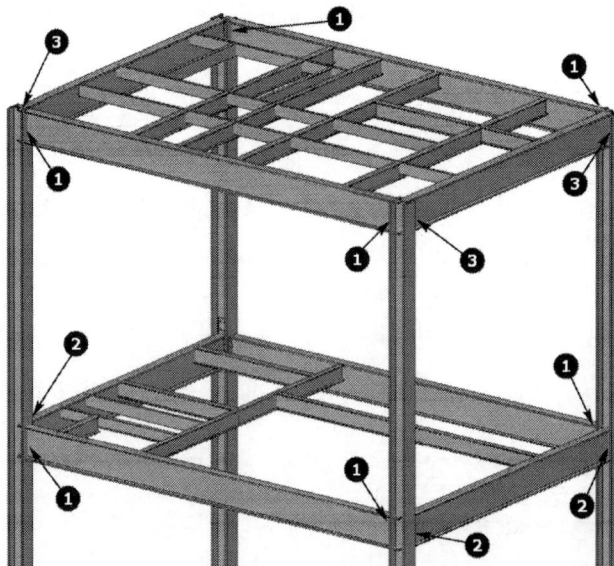

Figure 68 *The locations where the joints need to be copied*

70. Using the same tool, copy the second joint at all the column web and perimeter beams intersections labeled as **2** in Figure 68.

71. Similarly, copy the third joint from level 2 at the intersections labeled as **3** in Figure 68. The model, after copying all the joints, is shown in Figure 69.

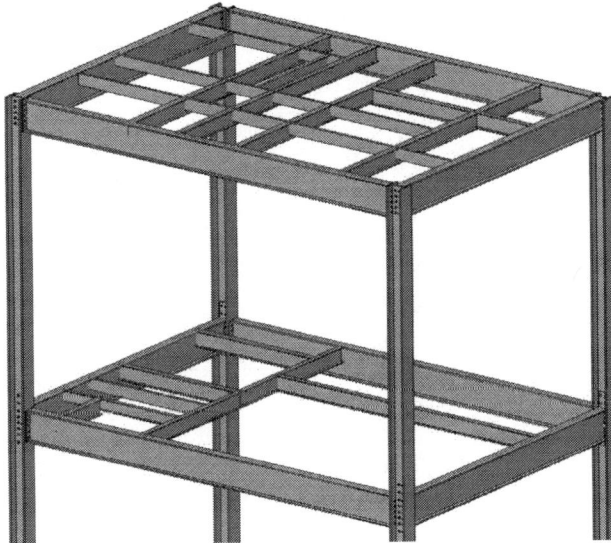

Figure 69 *The model, after copying all the joints*

Section 6b (STRUCTURE ONLY): Inserting and copying the Shear Plate Joint

In this section, you will insert a shear plate joint between the perimeter beam and the filter I-section at level 1. You will then apply the same joint between the perimeter beam and the channel at the same level.

1. Zoom to the area shown in Figure 70.

*Note: The dialog box and the names of options in this joint also vary, depending on the country standard selected while installing the software. In the following steps, the names in the **English US** and **English Australia** installations are used.*

2. In the **Connection vault** palette > **Platform beams** joints, scroll down and add the **Shear plate** joint to **Favorites**.

3. Now, invoke the **Shear plate** joint tool; you are prompted to select the main beam.

4. Select the beam labeled as **1** in Figure 70 and then press ENTER; you are prompted to select the secondary beam.

5. Select the beam labeled as **2** in Figure 70 and then press ENTER; the **Advance Steel - Shear Plate / Fin Plate [X]** dialog box is displayed.

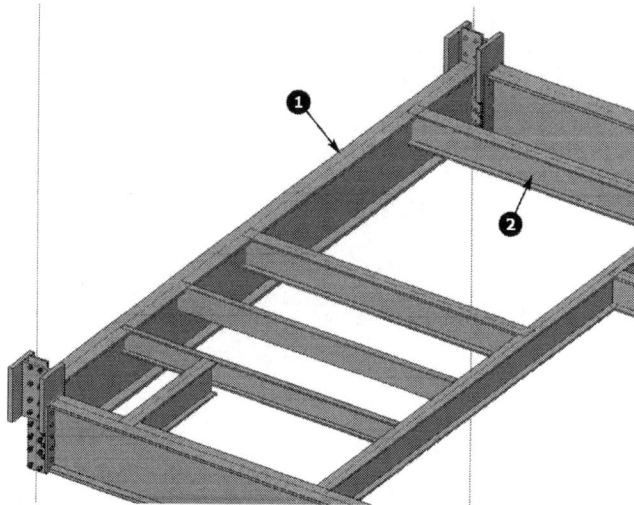

Figure 70 The zoomed in view to insert the joint

6. From the left pane of the dialog box, select the **Cuts & Stiffeners > Beam cut** tab.

   By default, **From face** is selected from the **Cut back layout** or **Cut back set out** list. This ensures the secondary beam is cut back from the face of the main beam.

7. In the **Cut back** edit box, enter **3/4"** or **15**.

   This ensures the beam is cut back by this value from the face of the main beam.

8. Activate the **Top Cope** or **Top notch** tab.

9. From the **Gap vertical setting** or **Vertical set out** list, select **Inside main**.

   This ensures the depth of the top cope is measured from the inside of the main beam.

10. In the **Cope depth** or **Detail depth** edit box, enter **1/2"** or **12**.

11. Activate the **Plate & bolts > Plate layout** tab.

12. In the **Thickness** edit box, enter **3/8"** or **10**.

13. From the **Plate type** list, make sure **Rectangular** is selected.

14. Activate the **Bolts and holes** tab.

15. From the **Diameter** list, make sure **3/4 inch** or **20.00 mm** is selected.

16. From the **Bolt Type** list, select **A325** (for Imperial) or **AS 1252** (for Metric).

17. Activate the **Horizontal bolts** tab.

18. Make sure from the **Layout** or **Set out** list, **plate** is selected.

19. In the **Edge distance** edit box, enter **4"** or **100**.

20. In the **Edge distance end** edit box, enter **1 1/2"** or **40**.

21. Activate the **Vertical bolts** tab.

22. From the **Layout** or **Set out** list, select **beam top** or **Bolts from sec**.

23. In the **Layout distance** edit box, enter **4"** or **75**.

24. In the **Group 1 > Lines** edit box, enter **2**.

25. In the **Group 1 > Start dist.** edit box, enter **2"** or **35**.

26. In the **Group 1 > Interm. dist.** edit box, enter **3"** or **75**.

    There is no bolt in group 2, so you will leave that as 0.

27. In the **End distance** edit box, enter **2"** or **35**.

    These are all the parameters that are required for this joint. You will now add this joint to the library.

28. Using the **Properties > Library** tab, add this joint to the library with the name **C05-I-sections-Tower**. The zoomed in view of the model, after inserting this joint, is shown in Figure 71.

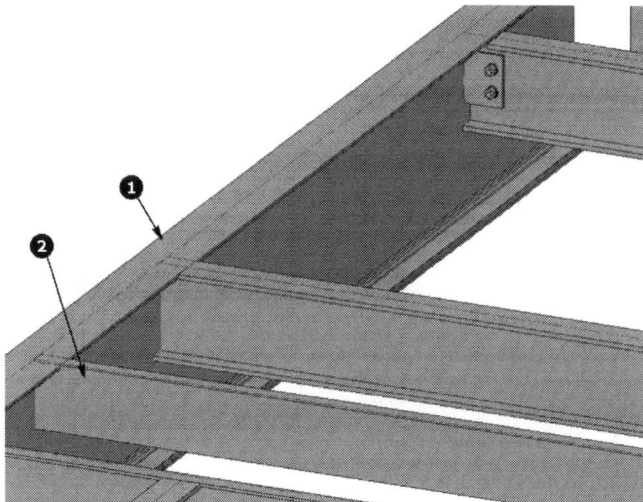

Figure 71 The zoomed in view after inserting the joint

Next, you will copy this joint between the beam labeled as **1** in Figure 71 and the channel labeled as **2**. You will then separate this joint from the master and edit it to create a new joint.

30. Using the **Create joint in a joint group** tool, copy the shear plate joint between the beam labeled as **1** and channel labeled as **2** in Figure 71.

31. If the **Warning** dialog box is displayed, click **Yes** to separate the joint from the master. If the **Warning** dialog box is not displayed, use the **Remove joint from a joint group** tool from the **Advance Steel Tool Palette > Tools** tab to remove the joint from the group.

32. Edit the copied joint to display the **Advance Steel - Shear Plate [X]** dialog box.

33. Activate the **Plate and bolts > Vertical bolts** tab.

34. In the **Layout distance** edit box, enter **3"** or **70**.

35. In the **Group 1 > Start dist.** edit box, enter **1 1/2"** or **40**.

36. In the **End distance** edit box, enter **1 1/2"** or **40**.

This completes the editing of this joint. You will now add this to the library.

37. Add this joint to the library with the name **C05-Channel-Tower**.

38. Close the dialog box. The model, after editing the joint, is shown in Figure 72.

Figure 72 The zoomed in view after copying and editing the joint

Next, you will create a new joint between the second channel and the perimeter beam. You cannot copy the joint because in the new joint, the plate needs to be placed on the other side of the channel. However, you will use the joint saved in the library to insert the new joint.

39. Invoke the **Shear plate** joint tool again.

40. Select the main beam labeled as **1** and secondary beam labeled as **2** in Figure 72.

41. Select the **C05-Channel-Tower** joint saved in the library for the dimensions.

42. From the **Plate & bolts > Plate layout** tab, select **Right** from the **Change side** list; the plate is moved to the correct side.

43. Close the dialog box.

   Next, you will copy these joints to the remaining intersections.

44. Using the **Create joint in a joint group, multiple** tool, copy the first shear plate joint at all the remaining perimeter beam and I-section intersections.

45. One by one, copy the all the joints created in the earlier steps to the remaining intersections. If the **Warning** dialog box is displayed, click **No** from the dialog box. The model, after copying all the joints, is shown in Figure 73.

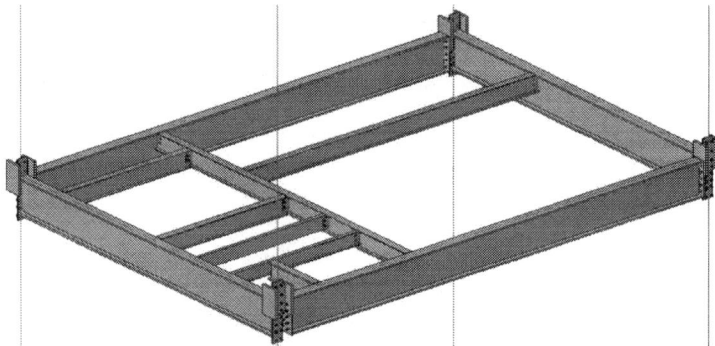

Figure 73   *The model, after copying all the joints*

Next, you will copy the shear plate joints from level 1 to level 2.

46. Using the **Create joint in a joint group, multiple** tool, copy the shear plate joints from level 1 to level 2. The model, after copying all the joints, is shown in Figure 74.

> *Tip*: While copying the shear plate joints added to the channel, if the plate is copied on the wrong side of the channel, you can use the **Advance Steel Tool Palette > Tools Tab > Remove joint from a joint group** tool to remove the joint from the group. Next, edit the joint and using the **Plate & bolts > Plate layout** tab, select the other side from the **Channel side** list.

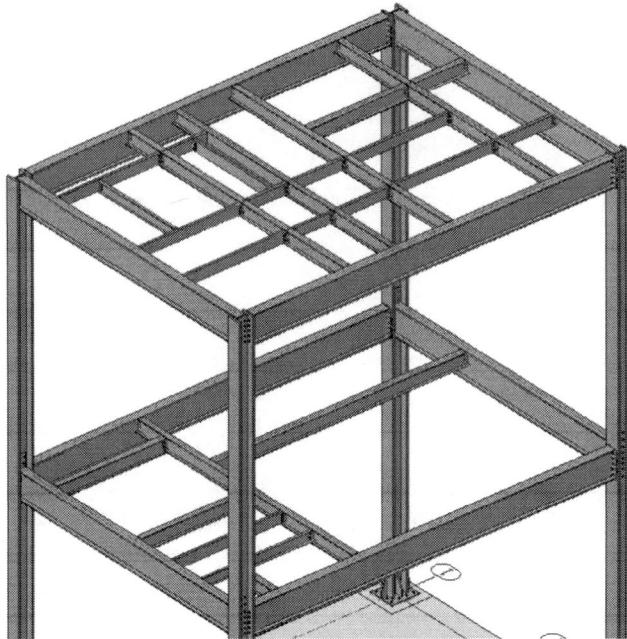

Figure 74 *The zoomed in view after copying and editing the joint*

Section 7a (BIM ONLY): Inserting and Copying the End Plate Joints in the Elevator Tower Model

In this section, you will open the model of the Hospital Structure and then insert a single side end plate joint between the flange side of the column at the **B2** grid intersection point and the perimeter beam on its left. Next, you will insert another single side end plate joint between the web side of the same column and the filter beam connecting to that column. Finally, you will copy these joints at all the required locations on levels 1 and 2.

1.   From the **C05 > Struc-BIM** folder, open the **C05-Imperial-Hospital.dwg** or **C05-Metric-Hospital.dwg** file, based on the preferred units; the model looks similar to the one shown in Figure 75.

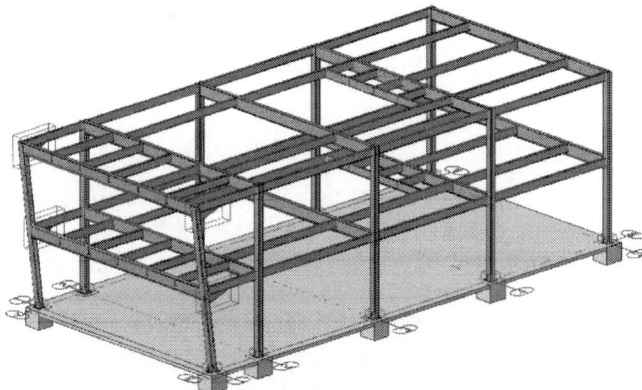

Figure 75 *The Hospital Structure model*

2. Zoom to the area shown in Figure 76.

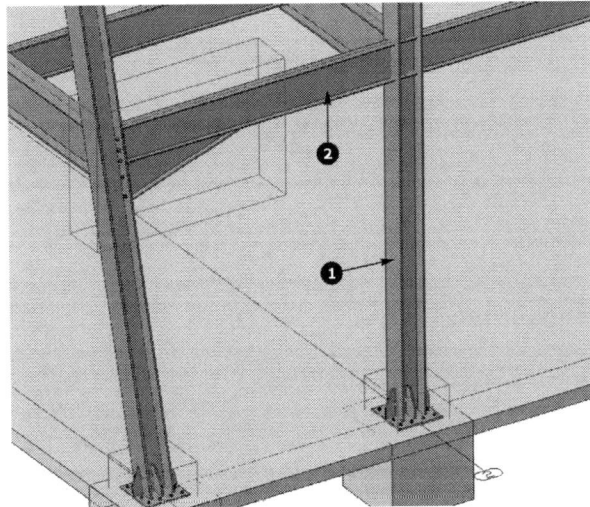

Figure 76 *The Hospital Structure model*

3. From the **Joint Categories** area of the **Connection vault** palette, click the **Platform beams** button, which is the fifth button; various platform beam joints are displayed in the **Joint Type** area.

> **Note**: *In the joints you will create next, the names of various dialog box options in this joint vary depending on the country standard you selected while installing the software. In the following steps, the names in the **English US** and **English Australia** installations are used. These may be different from your country installation.*

4. Click the star icon on the right of the **Single side end plate** button; the joint is added to the **Favorites** category.

5. Click the **Single side end plate** button to invoke this joint tool; you are prompted to select the main beam.

6. Now, select the column labeled as **1** in Figure 76.

7. Press the ENTER key; you are prompted to select the secondary beam.

8. Select the beam labeled as **2** in Figure 76.

9. Press the ENTER key; the **Advance Steel - Single Side End Plate [X]** dialog box is displayed and the joint with the default values is created.

10. From the left pane of the dialog box, select the **Plate & bolts > Plate - Alignment** or **End plate** tab.

11. In the **Plate thickness** edit box, enter **3/4"** or **20** as the value.

12. Activate the **Bolts and holes** tab.

13. From the **Diameter** list, select **3/4 inch** or **20.00 mm**.

14. From the **Bolt Type** list, select **A325** for Imperial units or **AS 1252** from the Metric units.

15. Activate the **Horizontal bolts** tab.

16. From the **Plate layout** or **Plate set out** list, make sure **Projection** is selected.

17. Clear the **Bolts on gauge line** tick box.

18. In the **Center distance** edit box, enter **5"** or **125**.

19. In the **Edge distance / Proj. 1** edit box, enter **1 1/4"** or **10**.

20. In the **Edge distance / Proj. 2** edit box, enter **1 1/4"** or **10**.

21. Activate the **Vertical bolts** tab.

22. From the **Layout** or **Set out** list, select **Center** or **Middle**.

23. From the **Plate height by** or **Plate length by** list, select **Total from middle**.

24. In the **Plate height** or **Plate length** edit box, enter **1' 10"** or **550**.

25. In the **Group 1 > Lines** or **Rows** edit box, enter **5**.

    The **Group 1 > Start distance** edit box is grayed out because of the center plate layout.

26. In the **Group 1 > Intermed. Dist** edit box, enter **3 3/4"** or **95**.

    This completes all the settings for this joint. Next, you will add this joint to the library.

27. Activate the **Properties > Library** tab of the dialog box and add the joint to the library with the name **C05-Hospital-Flange**.

28. Close the dialog box. The model, after inserting this joint, is shown in Figure 77.

    Next, you will copy the recently created end plate joint to the same column and the beam connecting to its web side on level 1. You will separate the copied joint from the parent joint and edit it to meet your requirements.

29. From the **Advance Steel Tool Palette > Tools** tab, invoke the **Create joint in a joint group** tool; you are prompted to select the connection part.

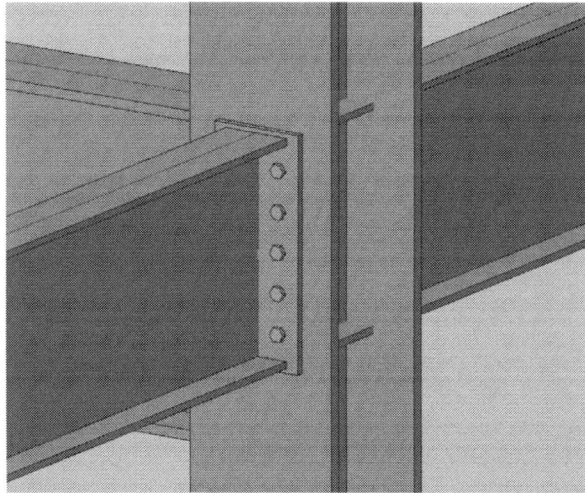

Figure 77 *The model, after inserting the end plate joint*

30. Select the plate of the first end plate joint you inserted and then press ENTER; you are prompted to select the main beam.

31. Select the column labeled as **1** in Figure 78 and then press ENTER; you are prompted to select the secondary beam.

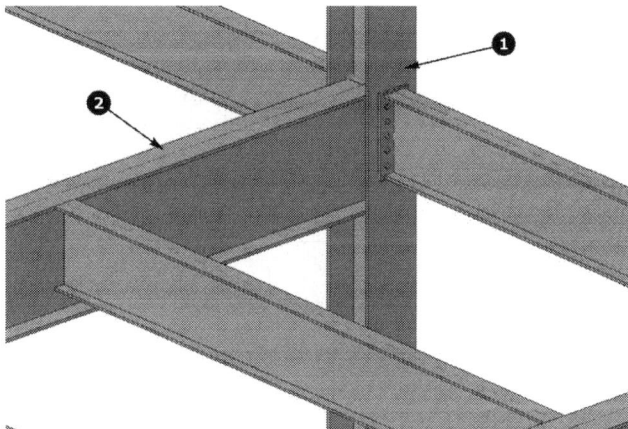

Figure 78 *Selecting the column and the beam to copy the joint*

32. Select the beam labeled as **2** in Figure 78 and then press ENTER; you are again prompted to select the main beam.

33. Press ENTER; the **Warning** dialog box is displayed informing you that the joint is different from the master joint.

34. Click **Yes** in the dialog box to separate the joint from the master. This will ensure that this joint is an independent joint.

35. Select one of the bolts of the copied joint and then right-click in the blank area of the drawing window to display the shortcut menu.

36. Select **Advance Joint Properties** from the shortcut menu; the **Advance Steel - Single Side End Plate [X]** dialog box is displayed.

37. Activate the **Plate & bolts > Horizontal bolts** tab.

38. From the **Plate layout** or **Plate set out** list, select **Total**.

39. In the **Width** edit box, enter **10 1/2"** or **250**.

40. In the **Center distance** edit box, enter **6"** or **120**.

41. Activate the **Vertical bolts** tab.

42. In the **Plate height** or **Plate length** edit box, enter **2'6"** or **790**.

43. In the **Group 1 > Lines** or **Rows** edit box, enter **6**.

44. In the **Group 1 > Intermed. Dist** edit box, enter **4 3/4"** or **120**.

This completes all the parameters for this joint. Next, you will save this joint in the library.

45. Save this joint in the library with the name **C05-Hospital-Web**.

46. Close the dialog box. The model, after copying and editing the joint, is shown in Figure 79. In this figure, the display type of some of the beams is changed to symbol for better visibility of the joint.

Figure 79 *The model, after copying and editing the joint*

Next, you will create multiple copies of these joints at the remaining column flange and perimeter beam intersections and column web and perimeter beam intersections.

47. Using the **Create joint in a joint group, multiple** tool, copy the first end plate joint to the four columns labeled as **1** and the three beams labeled as **2** in Figure 80.

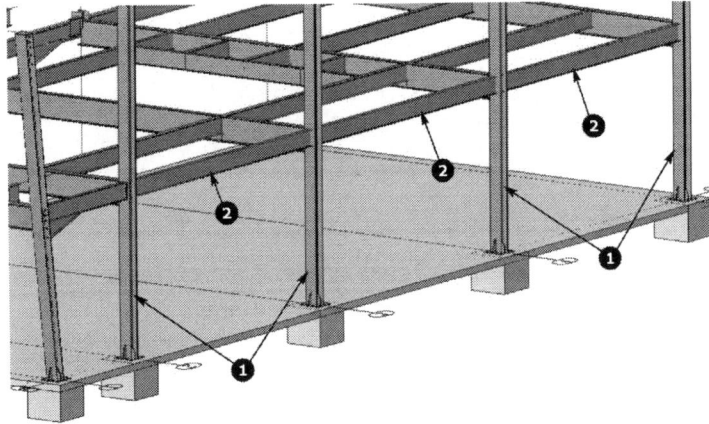

Figure 80 Selecting the columns and beams to copy the joint

48. Similarly, copy this joint to the columns **A2**, **A3**, **A4**, and **A5** and the perimeter beams intersecting the flanges of those columns. The zoomed in view of the model, after copying these joints, is shown in Figure 81. In this figure, the display type of some of the sections is changed to symbol.

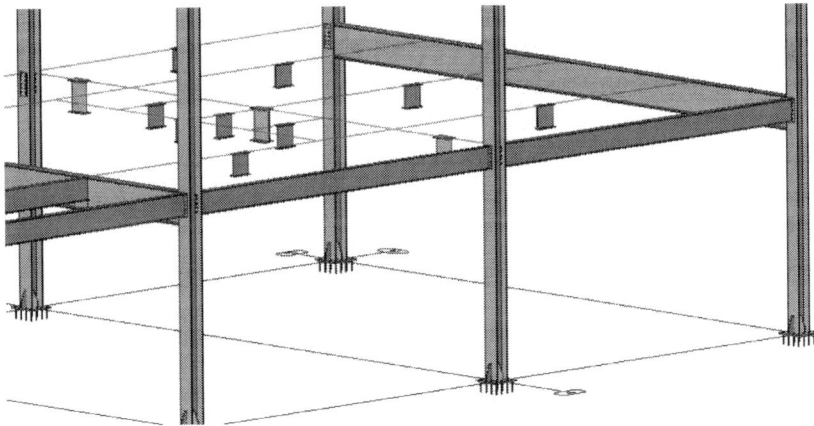

Figure 81 The zoomed in view of the model after copying the joints

49. Using the **Create joint in a joint group, multiple** tool, copy the second end plate joint between the columns labeled as **1** and the beams labeled as **2** in Figure 82.

50. Similarly, copy this joint to the columns **A2**, **A3**, **A4**, and **A5** and the filter beams intersecting the webs of those columns. The zoomed in view of the model, after copying these joints, is shown in Figure 83. In this figure, the display type of some of the sections is changed to symbol.

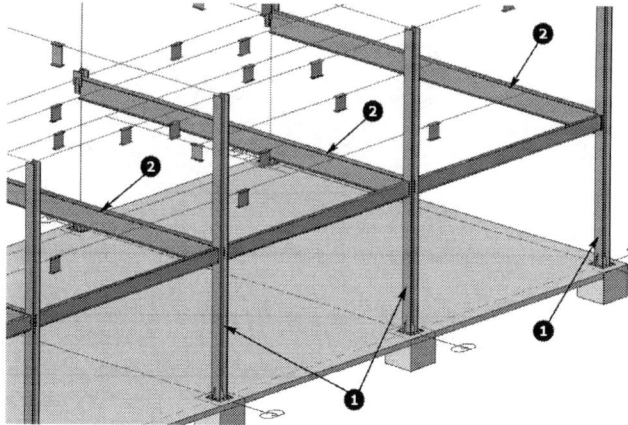

Figure 82   *Selecting the columns and beams to copy the joint*

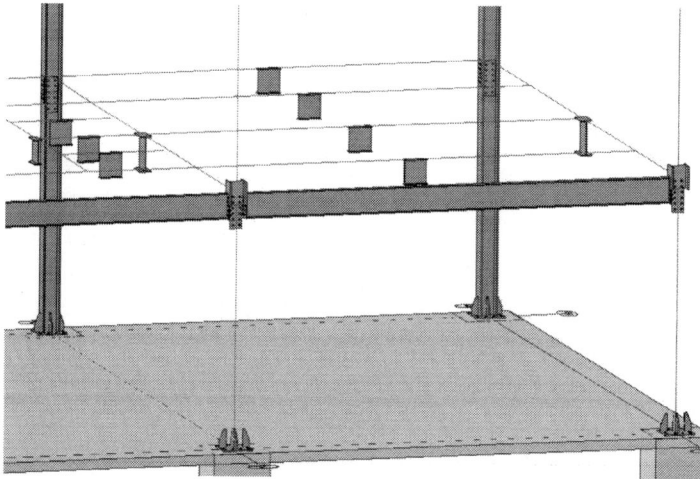

Figure 83   *Zoomed in view of the rear of the model after copying the joints*

Section 7b (BIM ONLY): Inserting and copying the Shear Plate Joint

In this section, you will insert a shear plate joint between the perimeter beam and the filter I-section at level 1. You will then apply the same joint between the perimeter beam and the channel at the same level.

1. Zoom to the front of the model, as shown in Figure 84.

2. From the **Connection vault** palette, scroll down in the **Platform beams** joints and add the **Shear plate** joint to **Favorites**.

> *Note: The dialog box and the names of options in this joint also vary, depending on the country standard selected while installing the software. In the following steps, the names in the **English US** and **English Australia** installations are used.*

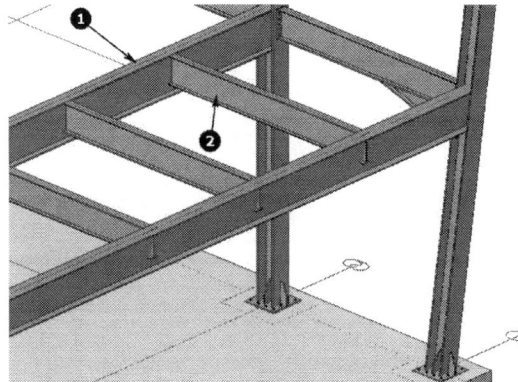

Figure 84 *Selecting the beams to insert the shear plate joint*

3. Now, invoke the **Shear plate** joint tool; you are prompted to select the main beam.

4. Select the beam labeled as **1** in Figure 84 and then press ENTER; you are prompted to select the secondary beam.

5. Select the beam labeled as **2** in Figure 84 and then press ENTER; the **Advance Steel - Shear Plate [X]** dialog box is displayed and the joint with the default values is created.

6. From the left pane of the dialog box, select the **Cuts & Stiffeners > Beam cut** tab.

    By default, **from face** is selected from the **Cut back layout** or **Cut back set out** list. This ensures the secondary beam is cut back from the face of the main beam.

7. In the **Cut back** edit box, enter **3/4"** or **20**.

    This ensures the beam is cut back by this value from the face of the main beam.

8. Activate the **Top Cope** or **Top notch** tab.

9. From the **Gap vertical setting** or **Vertical set out** list, select **Inside main**.

    This ensures the depth of the top cope is measured from the inside of the main beam.

10. In the **Cope depth** or **Detail depth** edit box, enter **1/2"** or **12**.

11. Activate the **Plate & bolts > Plate layout** tab.

12. In the **Thickness** edit box, enter **3/8"** or **10**.

13. From the **Plate type** list, select **Flange top**.

14. Activate the **Plate shape** tab.

15. From the **Corner finish** list, select **convex**.

16. In the **Dimension corner finish** edit box, enter **1"** or **25**.

17. Activate the **Bolts and holes** tab.

18. From the **Diameter** list, make sure **3/4 inch** or **20.00 mm** is selected.

19. From the **Bolt Type** list, select **A325** (for Imperial) or **AS 1252** (for Metric).

20. Activate the **Horizontal bolts** tab.

21. Make sure from the **Layout** or **Set out** list, **plate** is selected.

22. In the **Edge distance** edit box, enter **6"** or **150**.

23. In the **Edge distance end** edit box, enter **1 1/2"** or **40**.

24. Activate the **Vertical bolts** tab.

25. From the **Layout** or **Set out** list, select **beam top** or **Bolts from sec**.

26. In the **Layout distance** edit box, enter **5"** or **120**.

27. In the **Group 1 > Lines** edit box, enter **3**.

28. In the **Group 1 > Start dist.** edit box, enter **2"** or **50**.

29. In the **Group 1 > Interm. dist.** edit box, enter **3 1/2"** or **90**.

    There is no bolt in group 2, so you will leave that as 0.

30. In the **End distance** edit box, enter **2"** or **50**.

    These are all the parameters that are required for this joint. You will now add this joint to the library.

31. Using the **Properties > Library** tab, add this joint to the library with the name **C05-Hospital-Front**.

32. Create multiple copies of this joint between the beam labeled as **1** in Figure 84 and the remaining four beams on the left of the beam labeled as **2** in Figure 84. The zoomed in view of the model, after inserting and copying the joints, is shown in Figure 85.

    Next, you will copy one of these joints to the other end of the smaller beam that connects to the skewed perimeter beam. Because the perimeter beam is skewed, you will separate the copied joint from the master and then edit the parameters to create the required joint.

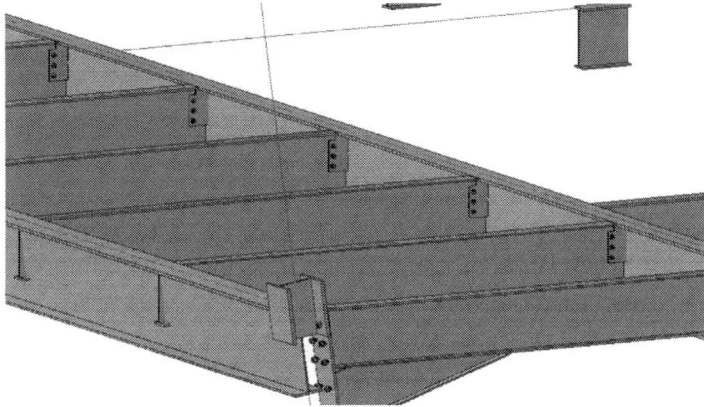

Figure 85 *The model, after inserting and copying the shear plate joints*

33. Using the **Create joint in a joint group** tool, copy the shear plate joint labeled as **1** to the beams labeled as **2** and **3** in Figure 86. When you press ENTER to finish the copy, the **Warning** dialog box is displayed. Click **Yes** in the dialog box to separate the copied joint from the parent joint.

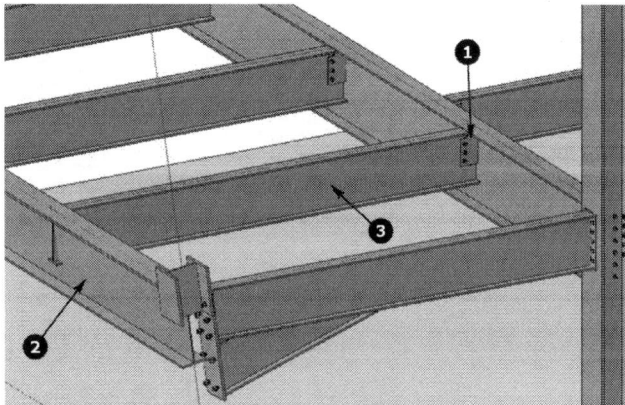

Figure 86 *Selecting the beams to copy the shear plate joint*

You will now edit the copied joint to increase the depth of the top cope and then save it in the library with the name **C05-Hospital-Front-Skewed**.

34. Edit the copied joint and change the **Cuts & Stiffeners > Top Cope > Cope depth** value to **1"** or **25**.

35. Select the **Straight Web Cut** tick box, if it is not already selected.

36. Save this joint in the library with the name **C05-Hospital-Front-Skewed**.

37. Create multiple copies of this joint between the beam labeled as **2** in Figure 86 and the remaining smaller beams on the left of the beam labeled as **3** in Figure 86. The zoomed in view of the model, after copying these joints, is shown in Figure 87.

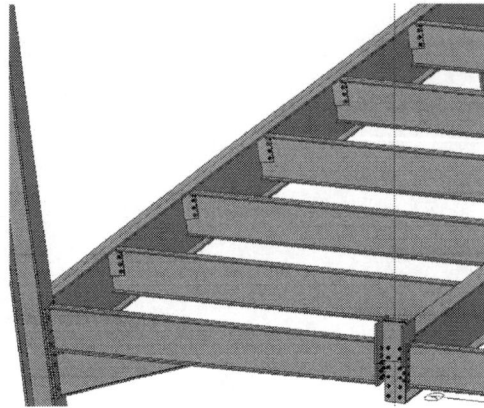

Figure 87 The model, after copying and editing the shear plate joints

Next, you will insert a new shear plate joint between the beams labeled as **1** in Figure 86 and the beam labeled as **2** in the same figure. You will use the settings saved in the library and then modify the joint to define the joint you need.

38. From the **Connection vault**, invoke the **Shear plate** joint tool; you are prompted to select the main beam.

39. Select the beam labeled as **1** in Figure 88 and then press ENTER; you are prompted to select the secondary beam.

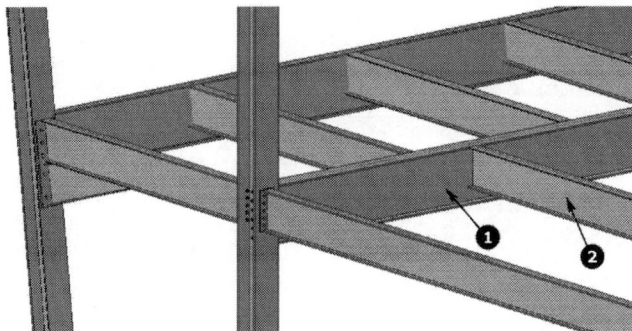

Figure 88 Selecting the beams to insert the shear plate joint

40. Select the beam labeled as **2** in Figure 88 and then press ENTER; the **Advance Steel - Shear Plate [X]** dialog box is displayed and the joint with the default values is created.

41. Activate the **Plate & bolts > Plate layout** tab.

42. From the **Change side** list, select **Right**.

43. Activate the **Horizontal bolts** tab.

Because this is a long heavy beam connecting to the another heavy beam, you will add another line of bolts to strengthen the joint and reduce the sideways deflection of the beam.

44. In the **Number of bolt** edit box, enter **2**.

45. In the **Edge distance end** edit box, enter **1 1/2"** or **40**.

46. Activate the **Vertical bolts** tab.

47. In the **Group 1 > Lines** edit box, enter **4**.

48. In the **Group 1 > Interm. dist.** edit box, enter **2 3/4"** or **70**.

   These are all the parameters that are required for this joint. You will now add this joint to the library.

49. Add this joint to the library with the name **C05-Hospital-Middle**.

50. Copy this joint at all the similar intersections. If a warning message is displayed informing you that the copied joint is different, click **No** in the dialog box. The zoomed in view of the model, after copying all these joints, is shown in Figure 89.

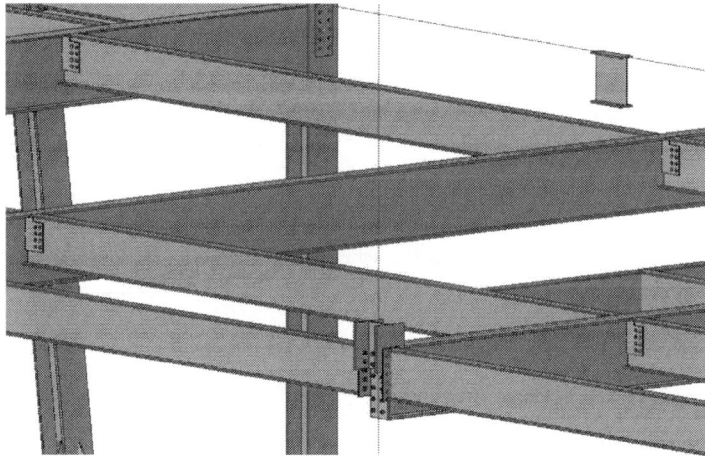

Figure 89 The model, after copying the shear plate joints

51. Save the file.

Section 7c (BIM ONLY): Inserting and copying the Clip Angle Joint

In this section, you will insert a clip angle joint and then create multiple copies of this joint.

1. Zoom to the area shown in Figure 90.

2. From the top of **Platform beams** joints, add the **Clip angle** joint to **Favorites**.

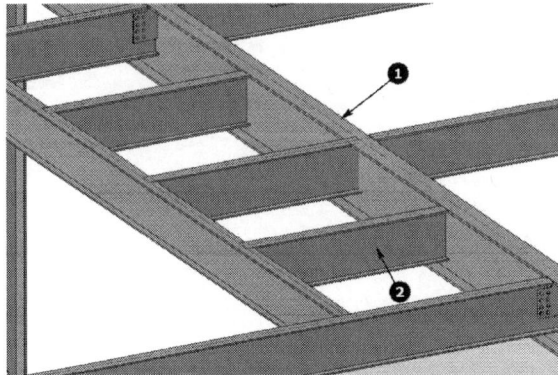

Figure 90 *Selecting the beams to insert the clip angle joint*

3. Now, invoke the **Clip angle** joint tool; you are prompted to select the main beam.

> **Note:** *The dialog box and the names of options in the **Clip angle** joint also vary, depending on the country standard selected while installing the software. In the following steps, the names in the **English US** and **English Australia** installations are used.*

4. Select the beam labeled as **1** in Figure 90 and then press ENTER; you are prompted to select the beam to be attached.

5. Select the beam labeled as **2** in Figure 90 and then press ENTER; the **Advance Steel - Clip angle [X]** or **Advance Steel - Angle cleats [X]** dialog box is displayed and the clip angle joint with the default values is created.

> **Note:** *While creating the clip angle joint, you can select the profile and size of the angle. However, its height is controlled by the number of vertical bolts. Therefore, you will now select the profile and size of the angle to be used.*

6. Activate the **Angle cleat** tab.

7. From the **Angle profile size** list, select **AISC 14.1 Angle identical > L4X4X1/4** (for Imperial) or **Australian Equal Angle > 100x100x6 EA** (for Metric).

8. Activate the **Top cope** or **Top detail** tab.

9. From the **Gap horizontal setting** or **Horizontal set out** list, select **From main flange**.

10. In the **Cope length** or **Detail length** edit box, enter **1/2"** or **12**.

11. From the **Gap vertical setting** or **Vertical set out** list, select **Inside main**.

12. In the **Cope depth** or **Detail depth** edit box, enter **3/4"** or **20**.

13. Activate the **Bolts parameters** tab.

14. From the **Diameter** list, make sure **3/4 inch** or **20.00 mm** is selected.

15. From the **Bolt Type** list, select **A325** (for Imperial) or **AS 1252** (for Metric).

16. Activate the **Vertical bolts** tab.

17. From the **Set out** list, select **Bolts from sec**.

18. In the **set out distance** edit box, enter **4"** or **100**.

19. In the **Group 1 > Lines** edit box, enter **4**.

20. In the **Group 1 > Start dist.** edit box, enter **1 1/2"** or **40**.

21. In the **Group 1 > Interm. dist.** edit box, enter **3"** or **75**.

    There are no more bolt groups, so you will leave them as 0.

22. In the **End distance** edit box, enter **1 1/2"** or **40**.

    These are all the parameters that are required for this joint. You will now add this joint to the library.

23. Add this joint to the library with the name **C05-Hospital-Angle**.

24. Copy this joint at the intersections of the beam labeled as **1** in Figure 90 and the two other beams on left of the beam labeled as **2** in Figure 90. The mode, after copying the joints, is shown in Figure 91.

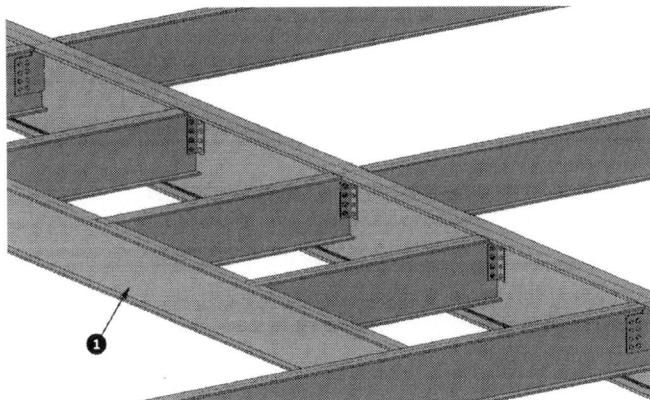

Figure 91   *The model, after inserting and copying the clip angle joint*

25. Copy the clip angle joint on the other ends of the smaller beams and the beam labeled as **1** in Figure 91. If a warning message is displayed informing you that the copied joints are different, click **No** in the dialog box.

Section 7d (BIM ONLY): Inserting and copying the End Plate Joint

The last joint for level 1 between the angled columns at the **A1** and **B1** grid intersection points and the beam at the front. In this section, you will insert a single side end plate joint between these sections. To define the parameters of this joint, you will use settings saved in the library with the name **C05-Hospital-Web**.

1. Invoke the **Single side end plate** joint tool; you are prompted to select the main beam.

2. Select the column labeled **1** in Figure 92 and press ENTER; you are prompted to select the secondary beam.

Figure 92 Selecting the column and the beam to insert the end plate joint

3. Select the beam labeled as **2** in Figure 92 and press ENTER; the **Advance Steel - Single Side end Plate** dialog box is displayed and the joint with the default values is created.

4. From the **Properties > Library** tab, select **C05-Hospital-Web**; the joint is modified to match the parameters saved in the library.

5. Close the dialog box. The model, after inserting this joint, is shown in Figure 93. In this figure, the display type of the beam is set to symbol for better visibility of the joint.

6. Copy this joint on the other side of the beam labeled as **2** in Figure 92 and the column at the **B1** grid intersection point.

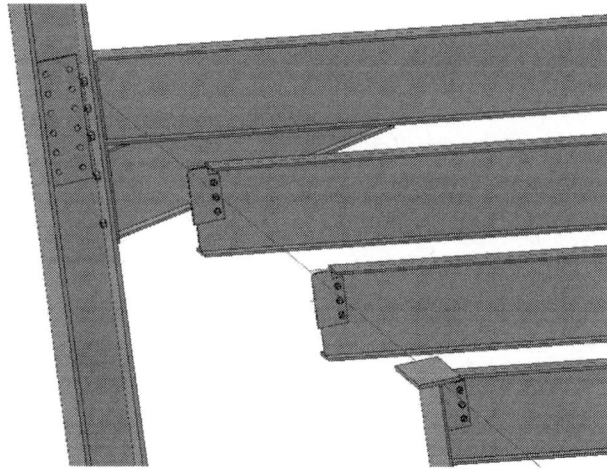

Figure 93   *The model after inserting the single side end plate joint*

This completes all the joints at level 1. The model, after inserting and copying all the joints at level 1, is shown in Figure 94.

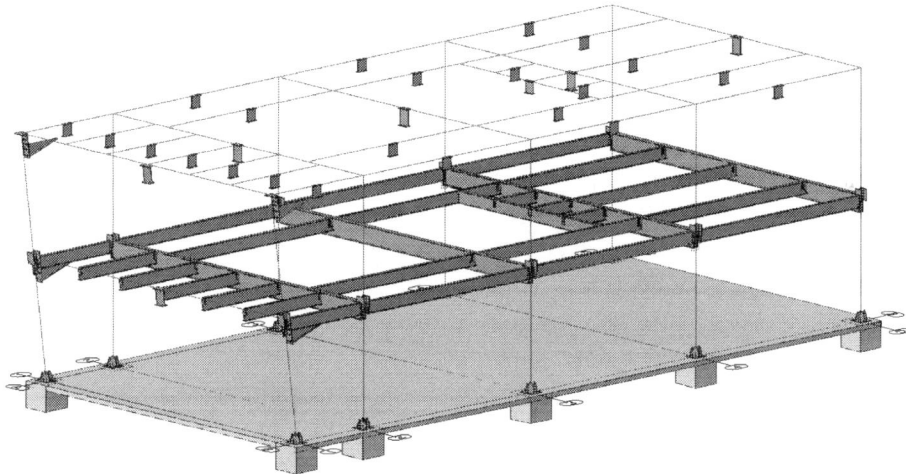

Figure 94   *The model after inserting and copying all the joints at level 1*

Section 7e (BIM ONLY): Copying all the Joints to Level 2

In this section, you will select the joints you created at level 1 and copy them to level 2.

1. Using the **Create joint in a joint group, multiple** tool, one by one select the end plate, shear plate, and the web angle joints created at level 1 and copy them to the beams at level 2.

TAKE A NOTE *Note: While copying the joints, if the **Warning** dialog box is displayed informing you that the copied joint is different from the master joint, click **Yes** in the dialog box to separate them. Also, while copying the shear plate joint, if the plate is inserted on the wrong side of the beam, you can use the **Remove joint from a joint group** tool to separate a joint from the master. Then, edit the joint to fix the location of the shear plate.*

2. Save the file. The model, after copying all the joints to level 2, is shown in Figure 95.

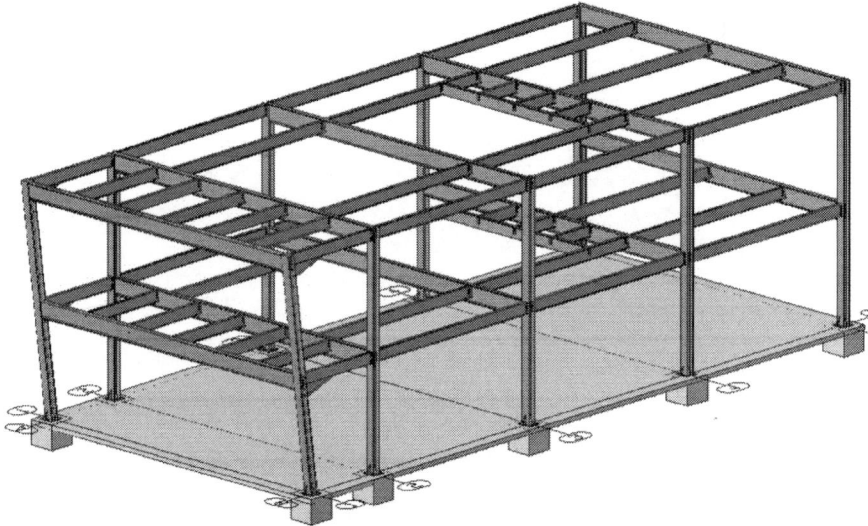

Figure 95 *The completed hospital structure model*

3. Close the file.

Skill Evaluation

Evaluate your skills to see how many questions you can answer correctly. The answers to these questions are given at the end of the book.

1. The **Apex Haunch** joint is generally added between a beam and a column. (True/False)

2. While inserting the shear plate joint, you can only add a rectangular plate. (True/False)

3. There are special joints available in the **Connection vault** to apply to the eaves beams. (True/False)

4. The **Front plate splice** joint is not a joint type available in Advance Steel. (True/False)

5. The joints created for purlins can also include stays. (True/False)

6. Which category is the purlin joints available in?

    (A) **Apex Joints**   (B) **Platform Beams**
    (C) **Miscellaneous**   (D) **Purlin & Cold rolled**

7. Which joint type allows you to insert a shear plate welded to the main beam and bolted to the secondary beam?

    (A) **Base plate**   (B) **Round plate**
    (C) **Shear plate**   (D) **Tube base plate**

8. Which joint type is used to connect to sections using an angle bolted to both the sections?

    (A) **Clip angle**   (B) **Purlin & Cold rolled**
    (C) **Apex Joints**   (D) **Base plate**

9. Which joint types are used to connect purlins to the rafters with a plate and stays?

    (A) **Single purlin plate**   (B) **Clip angle**
    (C) **Base plate**   (D) **Double purlin splice plate**

10. Shear plate joint is available in which category?

    (A) **Platform beams**   (B) **Beam end to end**
    (C) **Base plate**   (D) **Column - Beam**

Class Test Questions

Answer the following questions:

1. Explain briefly the process of inserting a purlin joint with stays.

2. Explain the process of inserting an eaves beam joint.

3. What is the process of inserting a clip angle joint?

4. How do you change the shear plate to include a top flange?

5. Which tab is used to specify the thickness of the end plate in the **Single side end plate joint**?

Chapter 6 – Advanced Structural Elements – II

The objectives of this chapter are to:

√ *Explain the process of inserting and bracings*
√ *Explain the process of inserting and editing joists*
√ *Explain the process of inserting and editing stairs*
√ *Explain the process of inserting and editing hand railings*

ADVANCED STRUCTURAL ELEMENTS

As mentioned in Chapter 3, the advanced structural elements behave as a single unit and can be edited using a single dialog box. Some of these tools were discussed in Chapter 3. The rest of them are discussed in this chapter.

INSERTING BRACINGS

Home Ribbon Tab > Extended Modeling Ribbon Panel > Bracing
Extended Modeling Ribbon Tab > Structural Elements Ribbon Panel > Bracing

Bracing is used to reinforce structural models so that their capacity to withstand gravity, seismic, and wind loads can be increased. Most of the structural models you create will require vertical or horizontal bracing. In Advance Steel, the bracing can be directly created using the **Bracing** tool. Note that the bracing is created in the current UCS plane. As a result, you will first need to align the XY plane of the UCS where the bracing is to be inserted.

When you invoke this tool, you will be prompted to select the two points, which are the two opposite corners of the bracing members. On specifying these two points, the **Advance Steel - Structural element - Bracing** dialog box will be displayed with the **Properties** tab active. The options in this tab and the **Library** tab are similar to those discussed in the earlier chapters. The remaining tabs of this dialog box are discussed next.

Type & Section Tab

The options in the **Type & Section** tab, shown in Figure 1, are used to specify the bracing type and sections to be used. These options are discussed next.

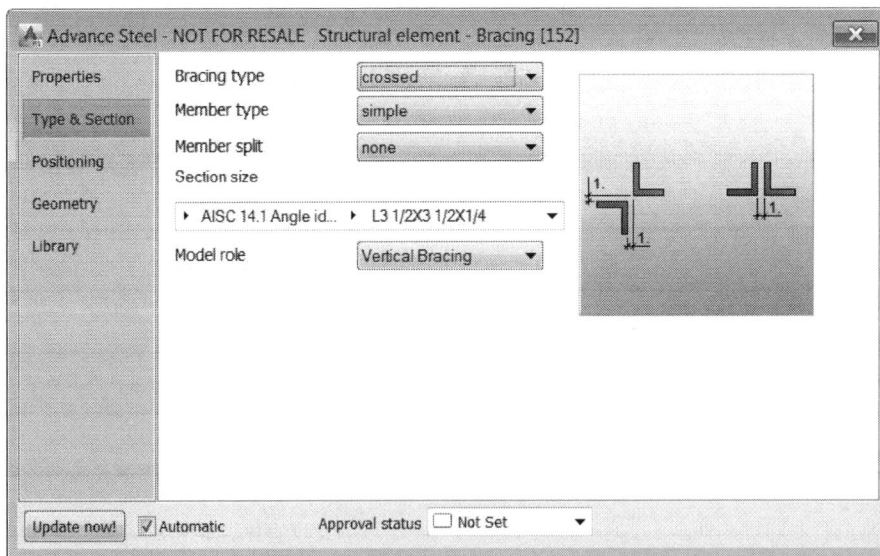

*Figure 1 The **Type & Section** tab of the **Advance Steel - Structural element - Bracing [X]** dialog box*

Bracing type

In Advance Steel, you can insert three types of bracings. They are crossed, single, and inserted. This list is used to select one of those types of bracing to be inserted. Figure 2 shows a cross bracing and Figure 3 shows an inserted bracing. The single bracing will simply insert a structural section between the two points you specify.

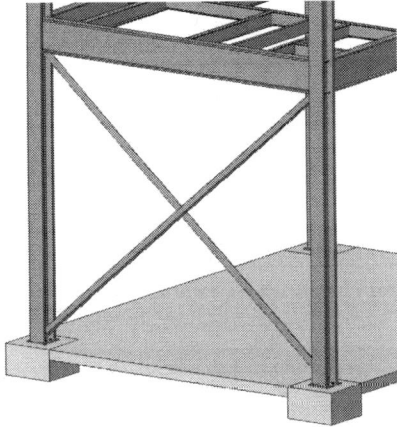

Figure 2 The cross bracing

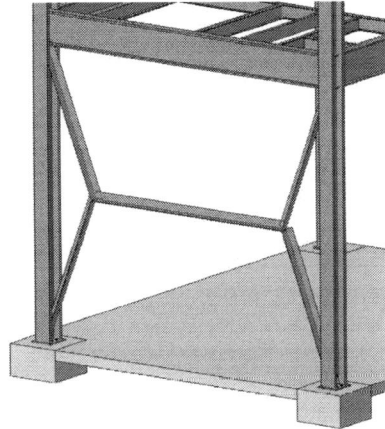

Figure 3 The inserted bracing

Member type

This list is used to specify the member type for the bracing. By default, **simple** is selected from this list. As a result, single member or members are inserted. Selecting **double mirrored** from this list will insert double mirrored sections, as shown in Figure 4. Selecting **double flipped** from this list will insert flipped double sections, as shown in Figure 5.

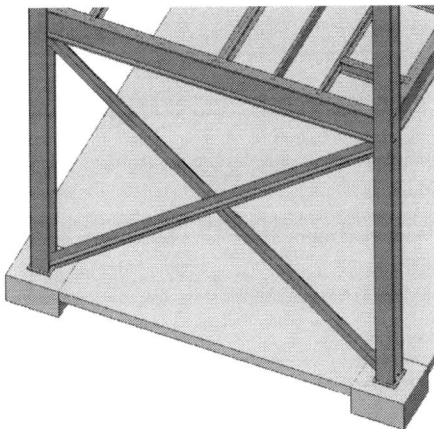

Figure 4 Double mirrored bracing sections

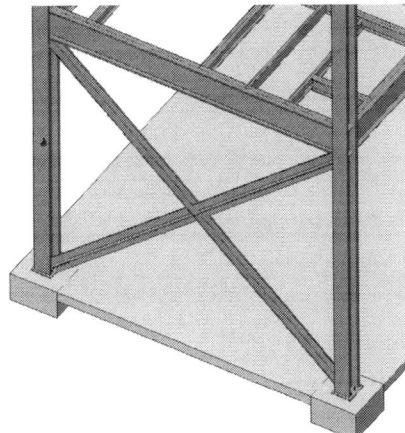

Figure 5 Double flipped bracing sections

Member none

While inserting crossed members, you can use this list to specify whether the bracing members will be split or not. Depending on the bracing joint that will be created, you can select the option to split one or both cross bracing sections from this list.

Section size

This list is used to select the section type and size.

Model role

This list is used to specify the model role for the bracing sections. By default, **Vertical Bracing** is selected from this list. If you are creating horizontal bracing, you can select **Horizontal Bracing** from this list.

Position Tab

The options in the **Position** tab, shown in Figure 6, are used to specify the positioning of the bracing sections. These options are discussed next.

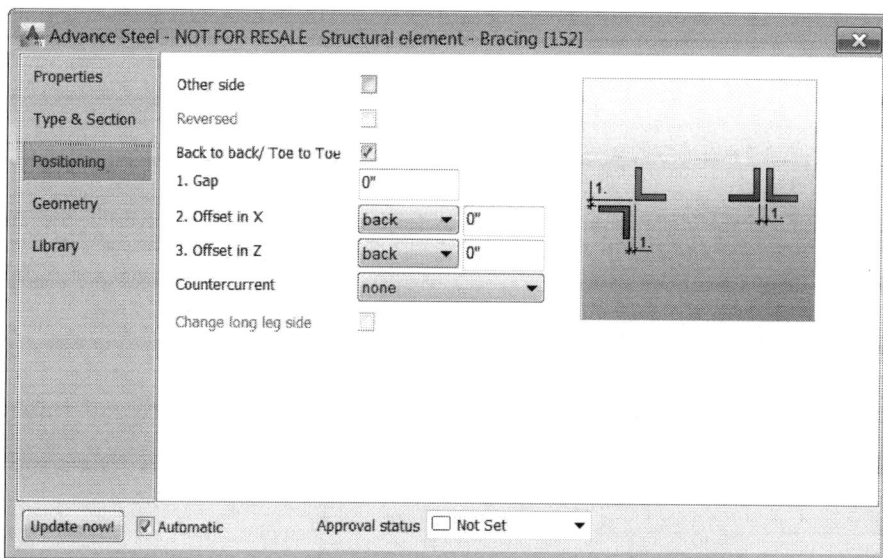

Figure 6 The Position tab of the Advance Steel - Structural element - Bracing [X] dialog box

Other side

This tick box is only available if you select **double mirrored** or **double flipped** from the **Member type** list. In these cases, selecting this tick box will place the mirrored or flipped section on the other side of the original bracing section.

Reversed

This tick box is only available while creating a single type bracing. In that case, selecting this tick box will insert the section between the other two corners instead of the two selected corners.

Back to back / Toe to Toe

This tick box is only available while creating crossed bracing type. In that case, selecting this tick box will ensure that the cross bracing sections are touching back to back. Figure 7 shows bracing channels sections inserted with this tick box selected and Figure 8 shows the same sections inserted with this tick box cleared.

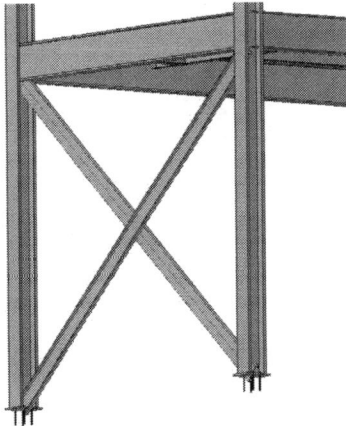

Figure 7 *The channel bracing sections with* **Back to back / Toe to Toe** *tick box selected*

Figure 8 *The channel bracing sections with* **Back to back / Toe to Toe** *tick box cleared*

Gap

This edit box is also available only while creating crossed bracing type and is used to specify a gap between the cross bracing members.

Offset in X / Offset in Z

These lists are used to specify the alignment of the system axes of the bracing sections with the placement points selected to insert the bracings. You can select to align the system axes to the **back**, **front**, **middle**, **center**, or **gauge line** of the bracing sections. You can also use the edit boxes on the right of these lists to enter an offset value.

Countercurrent

This list is used to specify whether the first or second bracing member will be placed in countercurrent position.

Change long leg side

If you are using unequal angle sections for bracing members, you can change the long leg side direction by selecting this tick box.

Geometry Tab

The options in the **Geometry** tab, shown in Figure 9, are used to specify the layout and offsets of the bracing sections. These options are discussed next.

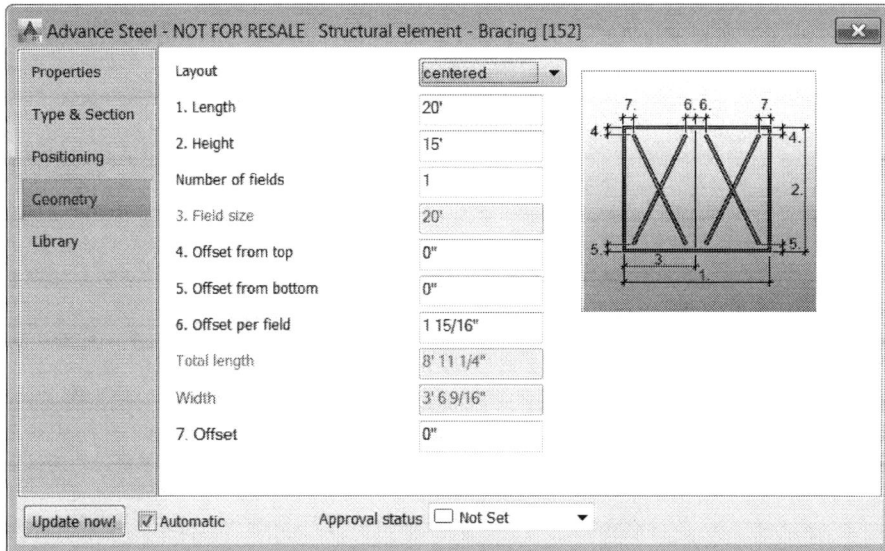

Figure 9 *The* *Geometry* *tab of the* *Advance Steel - Structural element - Bracing* *[X]* *dialog box*

Layout

This list allows you to define the layout of the bracing. You can select the bracing layout to be centered between the placement points or from the start or end points.

Length

This edit box is used to define the length of the bracing system. By default, the length is automatically calculated by the two corner points you specify to create the bracings.

Height

This edit box is used to define the height of the bracing system. By default, the height is also automatically calculated by the two corner points you specify to create the bracings.

Number of fields

This edit box is used to specify how many bracing systems you require between the selected corner points. Figure 10 shows a model with the value of this edit box set to **1** and Figure 11 shows the same model with the value of this field set to **2**.

Field size

While creating more than one field, you can set the size of the fields in this edit box.

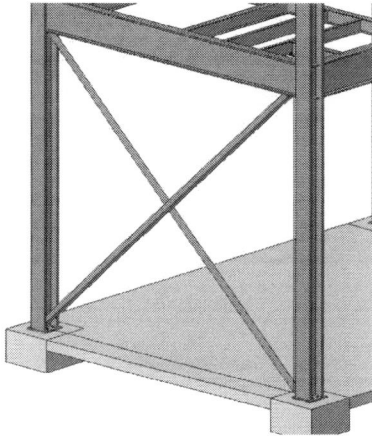

Figure 10 *The model with the **Number of fields** set to **1***

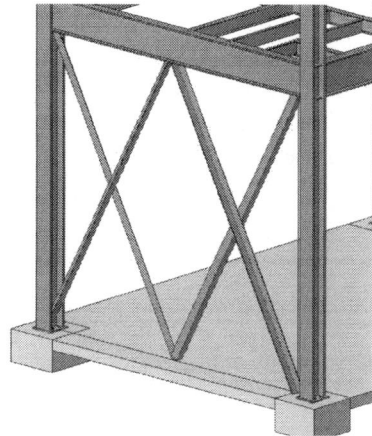

Figure 11 *The model with the **Number of fields** set to **2***

Offset from top
This edit box is used to specify the offset of the bracing sections from the top corner point you selected while creating the bracing. Figure 12 shows the bracing sections with offset from top.

Offset from bottom
This edit box is used to specify the offset of the bracing sections from the bottom corner point you selected while creating the bracing. Figure 13 shows the bracing sections with offset from bottom.

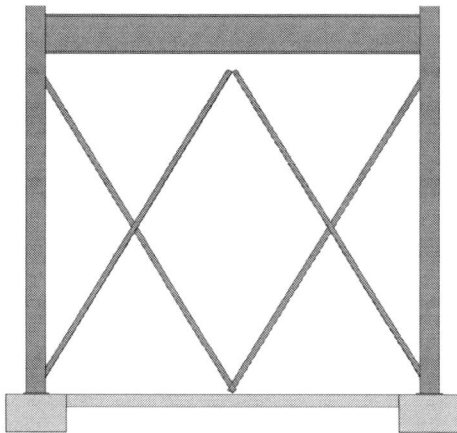

Figure 12 *Bracing sections with offset from top*

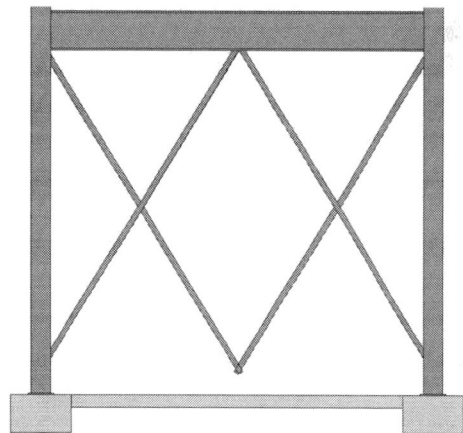

Figure 13 *Bracing sections with offset from bottom*

Offset per field
This edit box is used to specify the offset between multiple bracing fields. Figure 14 shows two bracing fields with the offset per field defined between them.

Offset

This edit box is used to specify the horizontal offset of the bracing sections from the placement points. Figure 15 shows the bracing sections with this offset defined.

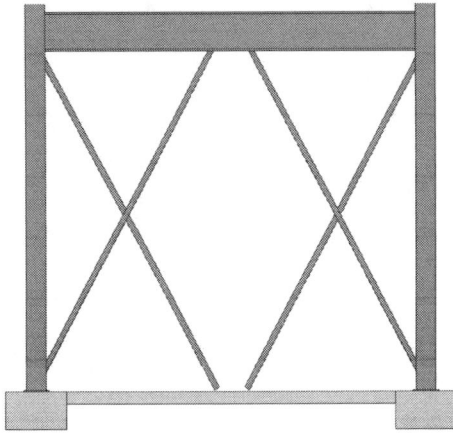

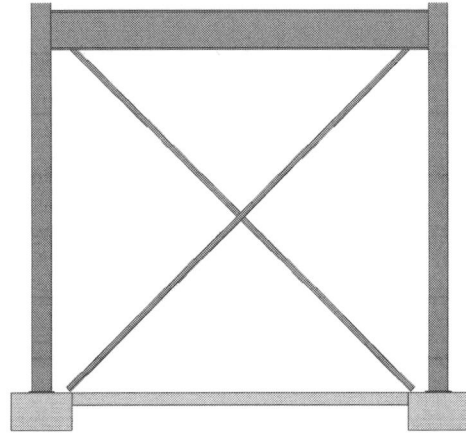

Figure 14 *Bracing sections with offset per field*   ***Figure 15*** *Bracing system with horizontal offset*

INSERTING JOISTS

Home Ribbon Tab > Extended Modeling Ribbon Panel > Joist
Extended Modeling Ribbon Tab > Structural Elements Ribbon Panel > Joist

---

Joists are generally used in structural models to support floor or ceilings and to transfer the loads to the structural frame. In Advance Steel, you can use the **Joist** tool to create floor or ceiling joist. When you invoke this tool, you will be prompted to specify the start point and the end point of the joist. After specifying these two points, you will be prompted to specify the point to define the height and orientation of the joist. This point needs to be specified going downward from the start and end points in order to orient the joist correctly. Once you specify the height and orientation point, the **Advance Steel - Joist [X]** dialog box will be displayed with the **Properties** tab active. This tab, the **Weld thickness** tab, and the **Library** tab of the dialog box are similar to those discussed in earlier chapters. The remaining tabs of the dialog box are discussed next.

Geometry Tab

The options in the **Geometry** tab, shown in Figure 16, are used to specify the geometry of the joist. These options are discussed next.

Joist name

This text box is used to specify the name of the joist. The name you specify here is automatically used when you save the joist settings to the library using the **Library** tab.

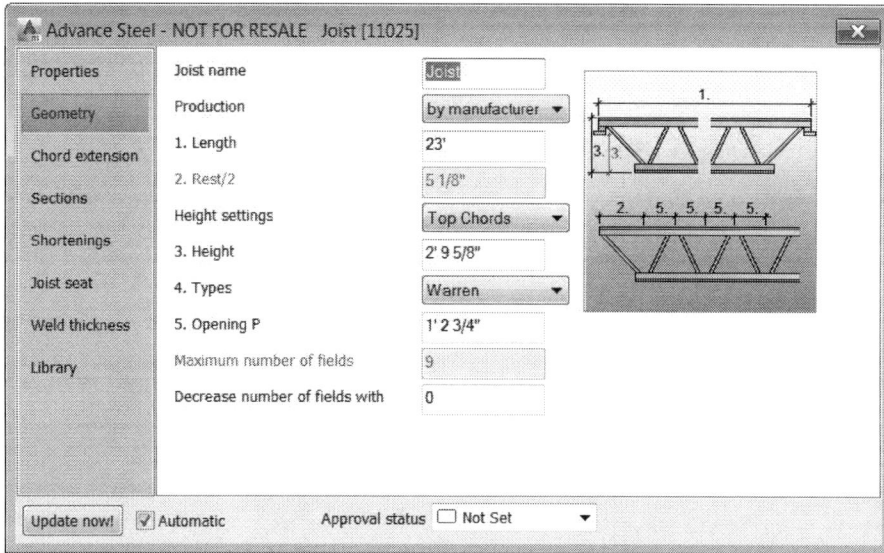

*Figure 16 The **Geometry** tab of the **Advance Steel - Joist [X]** dialog box*

Production

This list is used to define how the joist will be produced. You can select the **by manufacturer** or **by workshop** from this list. The option selected from this list will define how the drawings and the Bill of Material (BOM) of the joist will be generated.

Length

By default, the length is automatically defined by the start and end points that were used to define the joist. However, you can use this edit box to change the length of the joist.

Height settings

This list is used to specify how the joist will sit on the two points that were used to define the joist. If you select the **Top Chords** option from this list, the top of the top chords will be aligned with the start and end points that were used to create the joist, as shown in Figure 17. If you select **Support plates** from this list, the joist support plates will be aligned with the start and end points that were used to define the joist, as shown in Figure 18. This is generally the method used to insert the joists in a structural model.

Height

By default, the height is automatically defined by the point that was used to define the height and orientation of the joist. However, you can use this edit box to change the height of the joist.

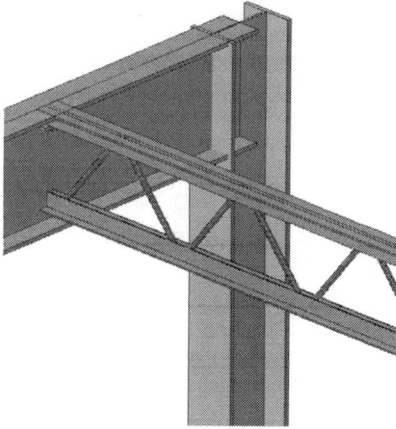

Figure 17 *The top chords aligned with the start point of the joist*

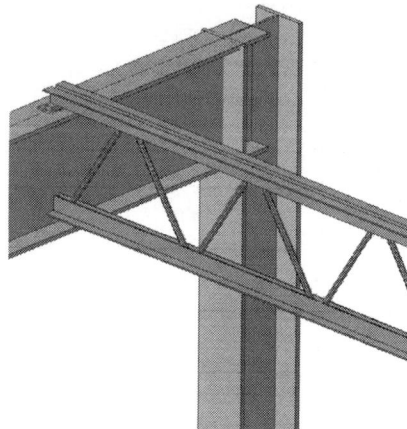

Figure 18 *The support plates aligned with the start point of the joist*

Types

This list is used to select the type of joist to be created. Selecting **Warren** from this list will create Warren type joist, as shown in Figure 19. Selecting **Pratt** from this list will create Pratt type joist, as shown in Figure 20.

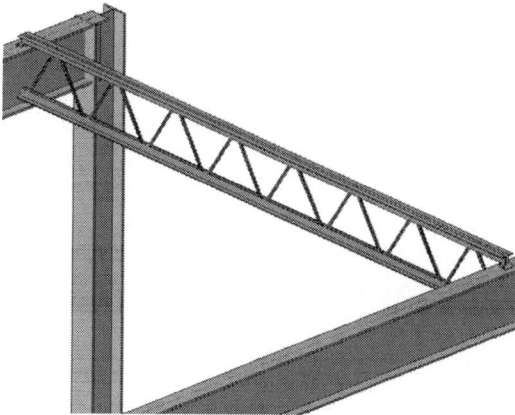

Figure 19 *Warren type joist*

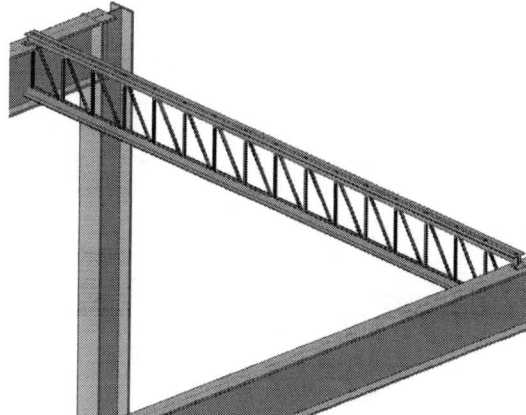

Figure 20 *Pratt type joist*

Selecting **Modified Warren** from this list will create a modified Warren type joist, as shown in Figure 21.

Opening P

This edit box is used to define the spacing between the joist fields.

Decrease number of fields with

This edit box is used to specify the number of fields to be decreased at the two joist ends.

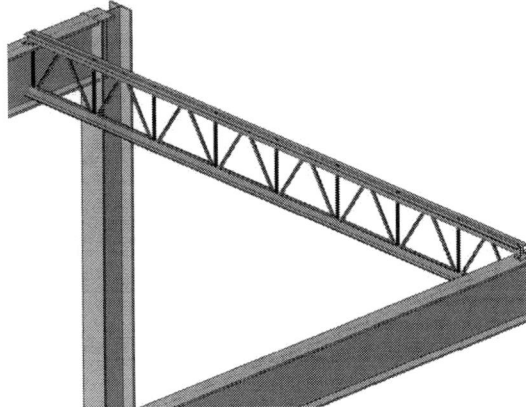

Figure 21 *Modified Warren type truss*

Figure 22 shows a joist with one field decreased and Figure 23 shows the same joist with two fields decreased.

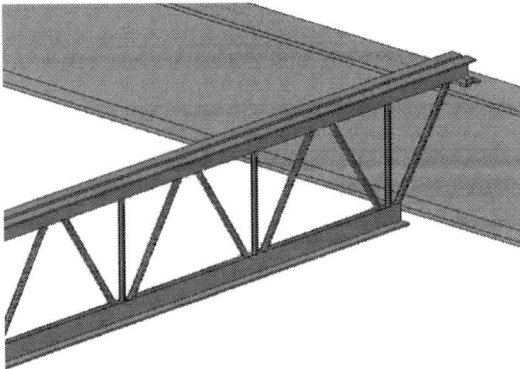

Figure 22 *Joist with one field decreased*

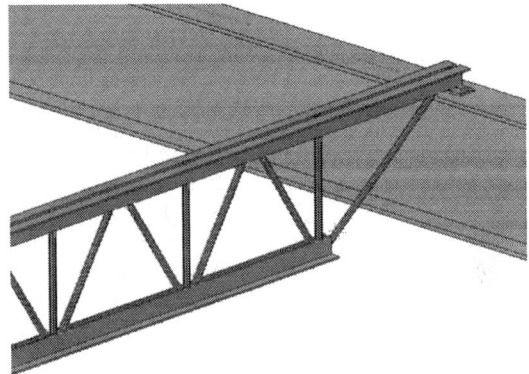

Figure 23 *Joist with two fields decreased*

Chord extension Tab

The options in the **Chord extension** tab, shown in Figure 24, are used to specify the information about extending the joist chords. These options are discussed next.

Main at start / Main at end

These edit boxes are used to specify the main chord extensions at the start and end of the joist.

Secondary at start / Secondary at end

These lists are used to specify the option of how the secondary chords will be extended at the start and end. You can select the **value** option from this list and define the extension values in the

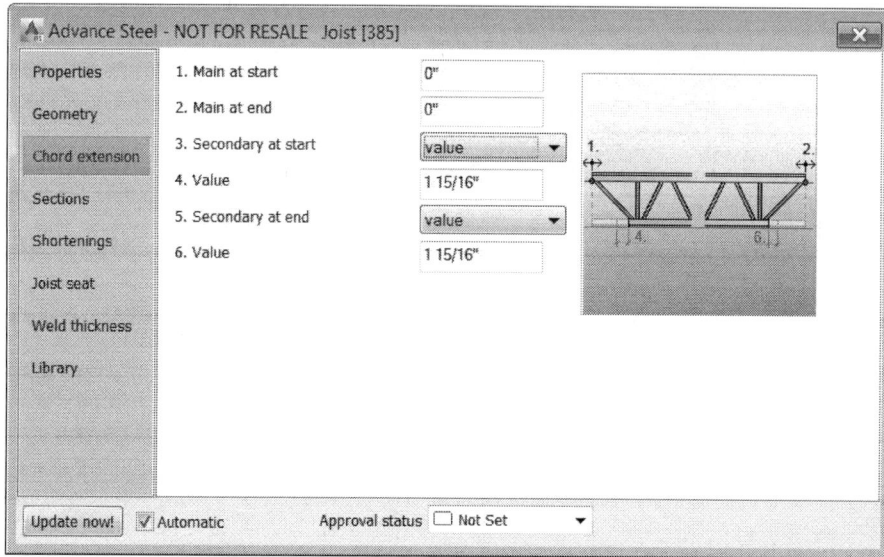

Figure 24 *The **Chord extension** tab of the **Advance Steel - Joist [X]** dialog box*

Value edit boxes available below these lists. Alternatively, you can select the **as main** option from this list to extend the bottom chords to the main chord level. If you do not want to extend the secondary chords, you can select **none** from these lists.

Sections Tab

The **Sections** tab, shown in Figure 25, is used to define the sections of the top chord, bottom chord, and vertical bars. If you select rectangular sections for the diagonal or vertical bars, you can rotate them by selecting the **Rotate** tick box.

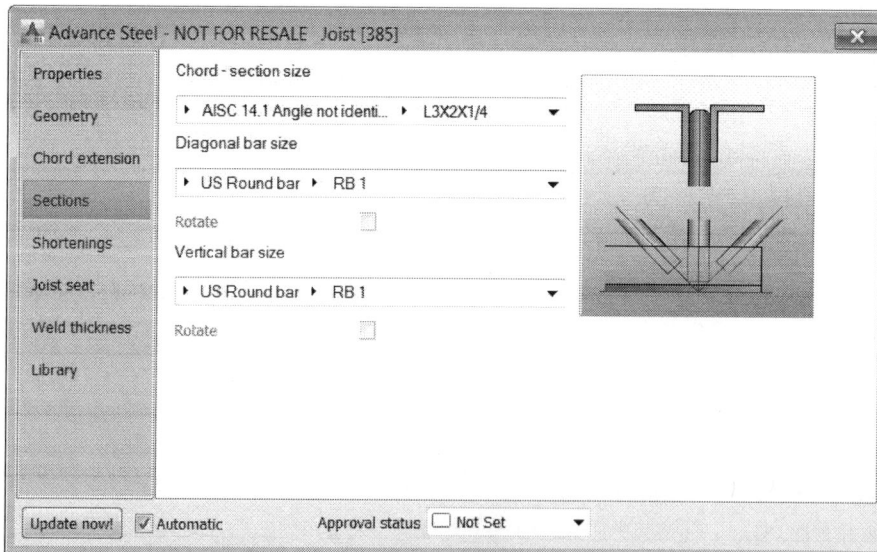

Figure 25 *The **Sections** tab of the **Advance Steel - Joist [X]** dialog box*

Shortenings Tab

The **Shortenings** tab, shown in Figure 26, is used to specify the diagonal and vertical bar shortenings at the upper and lower ends.

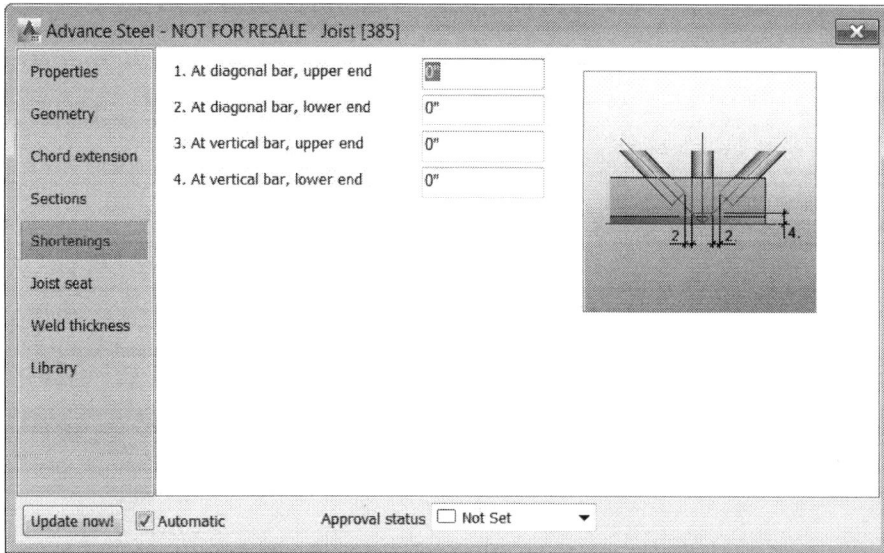

Figure 26 *The **Shortenings** tab of the **Advance Steel - Joist [X]** dialog box*

Joist seat Tab

The **Joist seat** tab, shown in Figure 27, is used to specify the parameters related to the joist seat. These are discussed next.

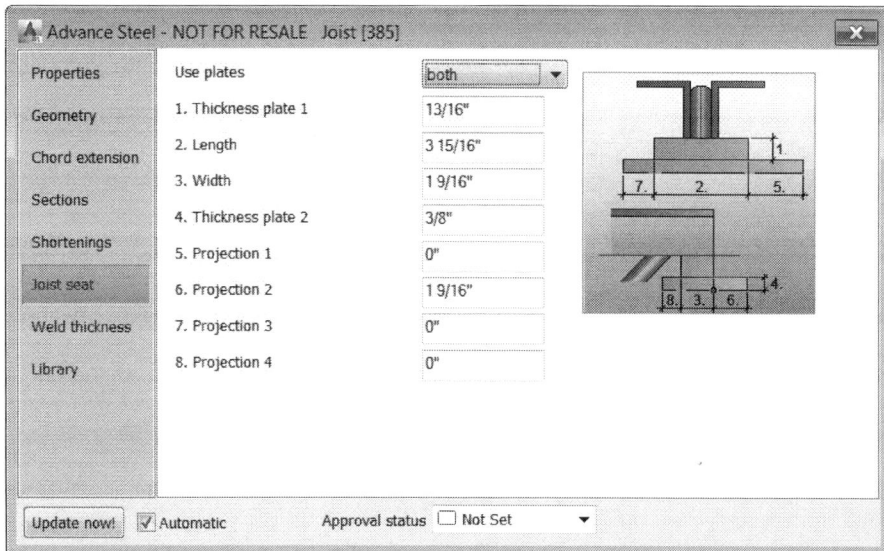

Figure 27 *The **Joist seat** tab of the **Advance Steel - Joist [X]** dialog box*

Use plates

This list is used to specify whether or not the plates will be used at the two ends of the joist. By default, **both** option is selected. As a result, the plates are created at both ends. You can also select the options to insert the plates at either end or to not insert the plates at all. Figure 28 shows one end of a joist with the plates inserted and Figure 29 shows the same end with plates not inserted.

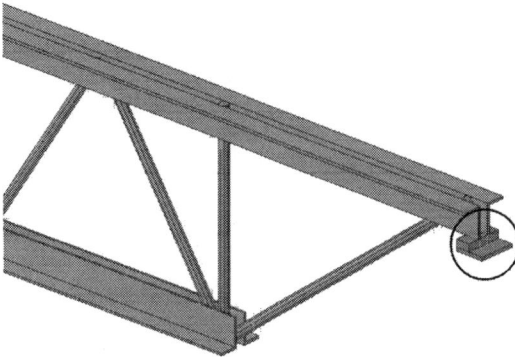

Figure 28 Joist with plates inserted   *Figure 29* Joist with plates not inserted

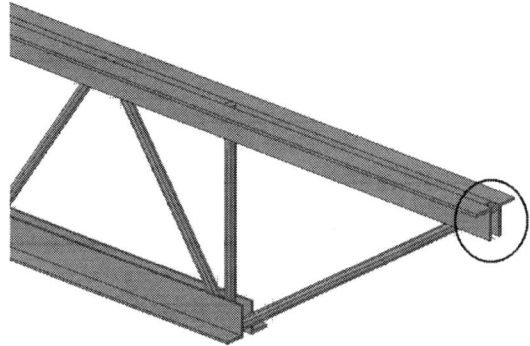

Thickness plate 1

This edit box is used to specify the thickness of plate 1.

Length

This edit box is used to specify the length of plate 1.

Width

This edit box is used to specify the width of plate 1.

Thickness plate 2

This edit box is used to specify the thickness of plate 2.

Projection 1 / Projection 2 / Projection 3 / Projection 4

These edit boxes are used to specify the projection of plate 2 from plate 1.

INSERTING STRAIGHT STAIRS

> **Home Ribbon Tab > Extended Modeling Ribbon Panel > Straight stair**
> **Extended Modeling Ribbon Tab > Structural Elements Ribbon Panel > Straight stair**

Advance Steel allows you to create straight stairs of various types. You can change the stringers and steps and also create the landing at top and bottom, if required. You can also use various tread types from the available library of treads. Figure 30 shows a stair with one of the tread types from the library and with no top landing and Figure 31 shows the stairs with a different tread type and with the landing.

Figure 30 *A stair type with no top cover*

Figure 31 *A stair type with top cover*

It is important to note that the **Stairs** tool will insert straight stairs in the current UCS plane. In Advance Steel, the straight stairs can be created based on the start and end points, length and angle, or height and angle. As a result, when you invoke this tool, the following prompt sequence will be displayed:

Start and end point=0 (default), Length and angle=1, Height and angle=2

In this prompt, you need to specify which one of these options do you want to use to create the stairs. By default, the start and end point option is selected. Therefore, when you press ENTER at the first prompt, you will be prompted to specify the start and end point of the stairs. Once you specify these two points, the following prompt sequence will be displayed:

Align stair : Left=0, Middle=1 (default), Right=2

In this prompt, you need to specify whether the stairs need to be aligned to the left of the two specified points, to the right, or use those points in the middle of the stairs. By default, the stairs are aligned to the middle. When you press ENTER at this prompt, the stairs with the default parameters will be created and the **Advance Steel - Stairs [X]** dialog box will be displayed with the **Properties** category > **Properties** tab active. This tab and the **Library** tab are similar to those discussed in earlier chapters. The remaining tabs are discussed next.

Distances + Stringer Category > Distances + stringer Tab

The options available in this tab, shown in Figure 32, are used to specify how the stairs will be defined. These options are discussed next.

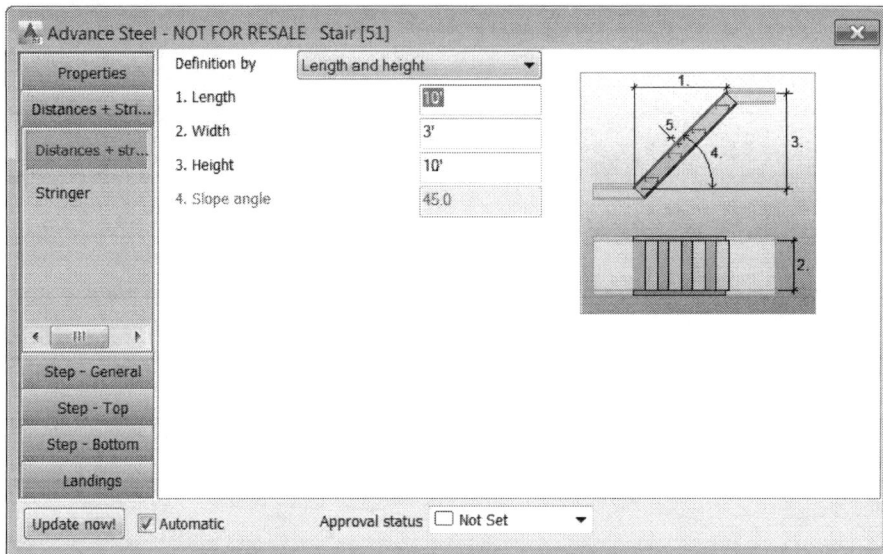

*Figure 32 The **Distances + stringer** tab of the **Distances + Stringer** category*

Definition by

This list is used to specify how the stairs will be created. By default, **Length and height** is selected from this list as this was the default option while creating the stairs. The length and height is automatically calculated from the two points you defined. However, you can override that using the **Length** and **Height** edit boxes in this tab. Alternatively, you can select **Length and angle** or **Height and angle** from this list and specify those values to create the stairs. The **Width** edit box can be used to specify the width of the stairs. Remember that some tread types selected from the **Step - General > Tread type** tab have predefined width. If you select those tread types, the **Width** edit box will be grayed out.

Distances + Stringer Category > Stringer Tab

The options available in this tab, shown in Figure 33, are used to specify the stringer parameters. These options are discussed next.

Stringer offset type

This list is used to specify how the stringer offset from the treads will be defined. By default, **Slope** is selected from this list. You can also select **Vertical** or **Horizontal** option from this list to define the stringer offset.

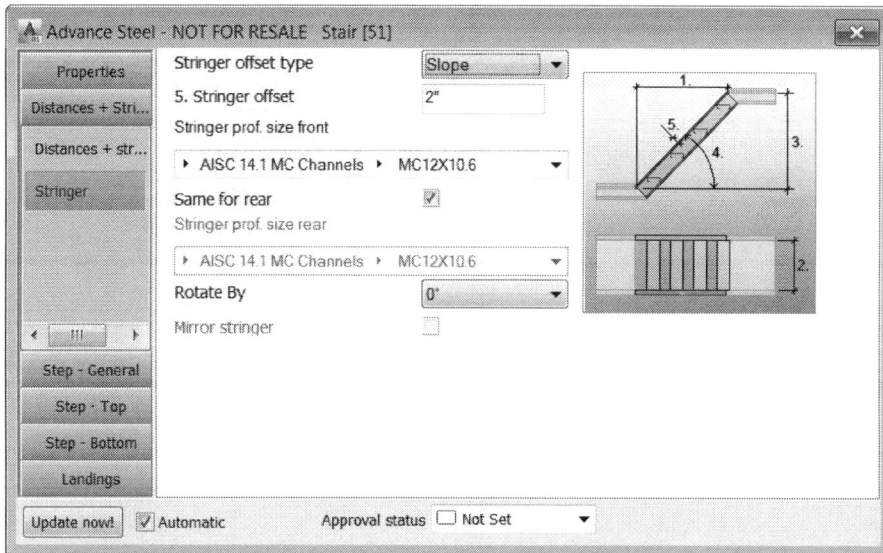

*Figure 33 The **Stringer** tab of the **Distances + Stringer** category*

Stringer offset
This is the edit box where you specify the stringer offset value to suit the grating or the plate.

Stringer pro. size front
The flyouts in this list are used to specify the type and size of the front stringer.

Same for rear
If this tick box is selected, the stringer type and size selected from the front will also be used for the rear stringer.

Rotated By
This list is used to rotate the stringer by the predefined angle values available in this list.

Mirror stringer
Select this tick box if you want to mirror the stringer.

Step - General Category > Step size Tab
The options available in this tab, shown in Figure 34, are used to specify the step size parameters. By default, the step size is defined by the slope relationship formula **2R+G**, where R is "riser" and G is "going". If you do not want to create the stairs based on this formula, you can clear the **Size from formula (2R+G)** tick box and then manually enter the values in this tab.

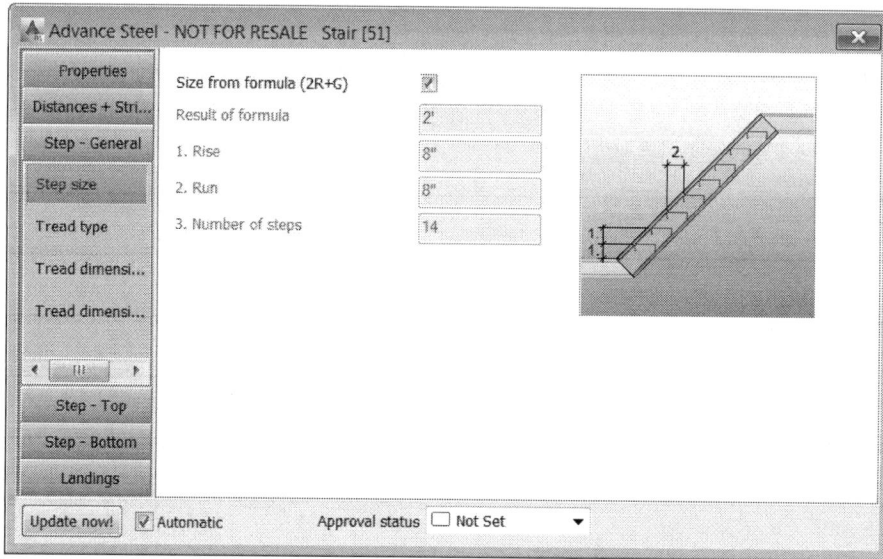

Figure 34 The Step size tab of the Step - General category

Step - General Category > Tread type Tab

The options available in this tab, shown in Figure 35, are used to specify the tread type parameters. These options are discussed next.

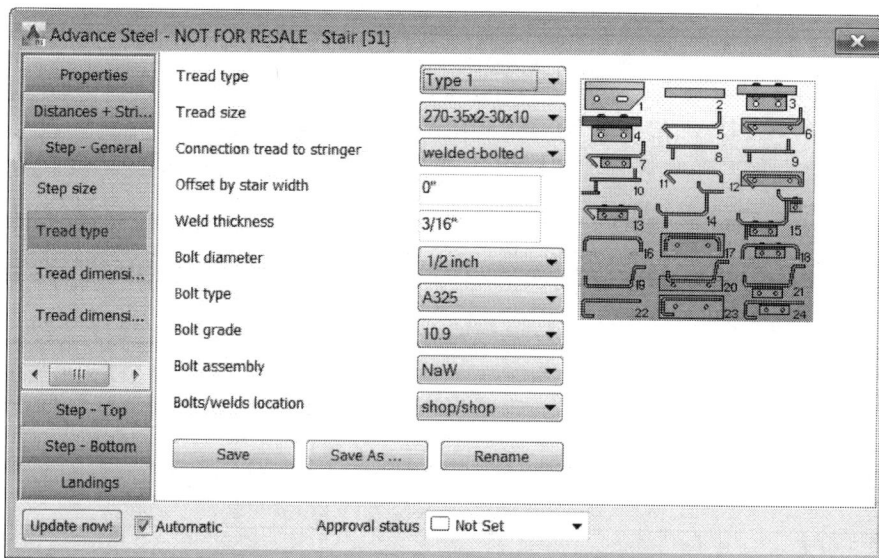

Figure 35 The Tread type tab of the Step - General category

Tread type

This list is used to select one of the available tread types. There are twenty-four tread types available by default for you to select from. The images of all those tread types are available in the preview window on the right of this list.

Tread size

This list is used to select the tread size. The options in this list will change, depending on the tread type selected from the **Tread type** list.

Connection tread to stringer

This list is used to specify whether the tread will be bolted to the stringer or welded. Note that this list may be grayed out for some of the tread types. Figure 36 shows a zoomed in view of the **Type 1 > 270-35x2-30x10** tread with **welded-bolted** connection. Figure 37 shows **Type 4 > Defaults 1** tread with **bolted-bolted** connection.

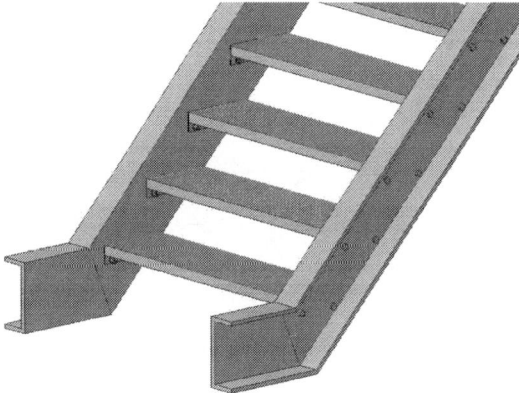

Figure 36 *Type 1 > 270-35x2-30x10 tread type with welded-bolted connection*

Figure 37 *Type 4 > Defaults 1 tread type with bolted-bolted connection*

Offset by stair width

This edit box is used to specify the offset value between the stringer and the treads. The width of the treads will be shortened on each side by this value.

Weld thickness

While creating a connection that involves welding, you can enter the weld thickness in this edit box.

Bolt diameter

While creating a connection that involves bolts, you can select the bolt diameter from this list.

Bolt type

This list is used to select the bolt type for the connections that involves bolts.

Bolt grade

This list is used to select the bolt grade for the connections that involves bolts.

Bolt assembly

This list is used to select the bolt assembly for the connections that involves bolts.

Bolt/welds location

This list is used to specify where will the bolted and welded be made. For example, selecting **site/shop** from this list will ensure the bolted connections are made on site, but the welding is done in the workshop.

Save

This button is used to override the current tread size in the **Tread size** list with the modified values in the **Tread type** tab. Once you specify the required parameters related to the tread type, tread size, and connections, you can click this button to override the default tread size parameters with the new ones.

Save As

When you click this button, the **Save as new dimension name** dialog box will be displayed. The name you specify in that dialog box will be saved as a new tread size and will be available in the **Tread size** list.

What I do

*I generally use the **Save As** button to save the preferred tread information with the initials of my company as a prefix. This way I can select my preferred tread information by simply selecting it from the **Tread size** list.*

Rename

This button is used to rename the current **Tread size** to a different name.

Step - General Category > Tread dimensions 1 Tab

The options available in this tab, shown in Figure 38, are used to specify the tread grating information. These options are discussed next.

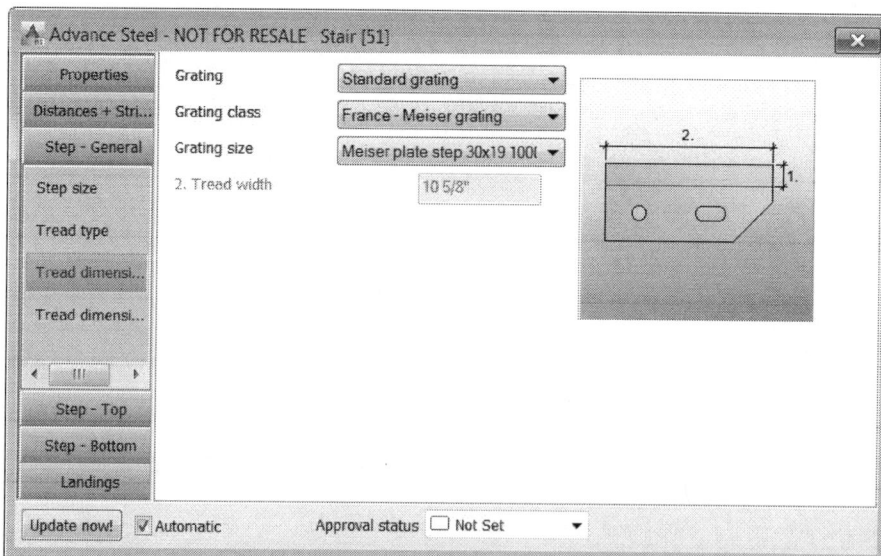

*Figure 38 The **Tread dimensions 1** tab of the **Step - General** category*

Grating

This list allows you to specify whether you want to use standard grating or variable grating.

*Tip: Selecting **Variable grating** from the **Grating** list will enable the **Width** edit box in the **Distances + stringer** tab, which will allow you to manually enter the width of the stairs.*

Grating class

Depending on whether you are creating a standard grating or a variable grating, you can select the grating class from this list.

Note: The preview window on the right side of this tab shows the side view of the grating.

Grating size

Depending on the grating class selected, you can select the size from this list.

Tread width

If you are creating a variable grating, you can enter its width in this edit box.

Step - General Category > Tread dimensions 2 Tab

The options available in this tab, shown in Figure 39, are used to specify the tread side, bolt location, and corner size. These options are discussed next.

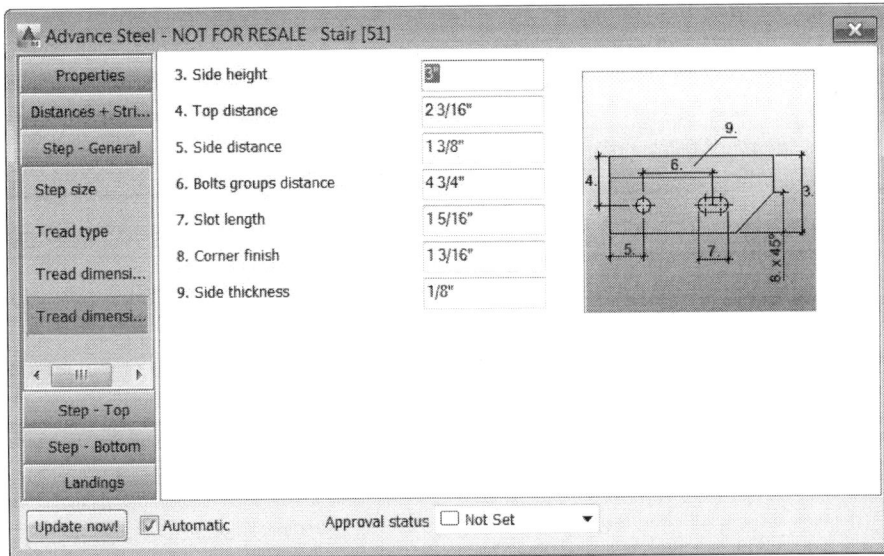

*Figure 39 The **Tread dimensions 2** tab of the **Step - General** category*

Side height

This edit box is used to specify the side height of the tread. Figure 40 shows a tread with side height of 2 1/2" and Figure 41 shows the same tread with the side height of 3 1/2".

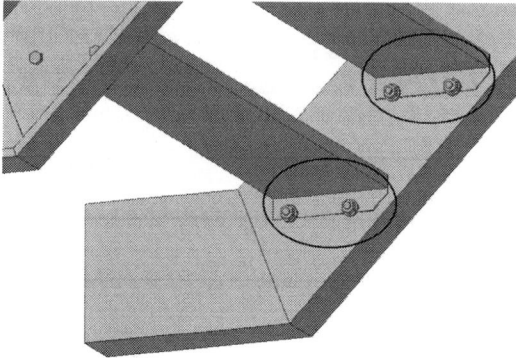

Figure 40 A tread with the side height of 2 1/2"　　*Figure 41* A tread with the side height of 3 1/2"

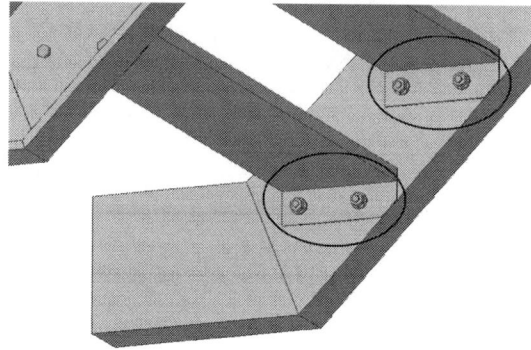

Top distance
This edit box is used to specify the vertical distance of the bolt location from the top of the tread.

Side distance
This edit box is used to specify the horizontal distance of the bolt location from the side of the tread.

Bolts groups distance
This edit box is used to specify the spacing between the bolt groups.

Slot length
This edit box is used to specify the overall length of the slot.

Corner finish
This edit box is used to specify the length of the corner chamfer of the tread side.

Side thickness
This edit box is used to specify the thickness of the side plate of the tread.

Step - Top / Step - Bottom Categories
The tabs and the options available in these categories are used to specify the top tread and the bottom tread. By default, the **Same as other steps** tick boxes are selected in both these tabs. As a result, all the options in these tabs are grayed out. However, if you want to create a different top and bottom treads, you can clear the **Same as other steps** tick box. On doing so, the options in various tabs of these categories will become available. These options are the same as those discussed in the **Step - General** category discussed above.

Landings Category > Top landing prof. Tab

The options in the **Top landing prof.** tab, shown in Figure 42, are used to specify whether or not the top landing will be created. If you are creating top landing, you can also select the profile of the top landing from this tab. These options are discussed next.

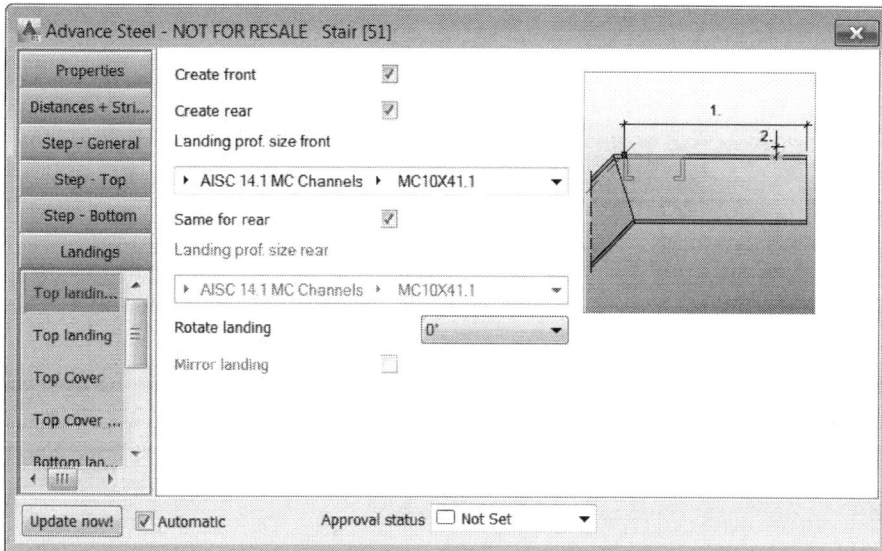

Figure 42 The Top landing prof. tab of the Landings category

Create front / Create rear

Select these tick boxes to create the front and rear landings at the top. Figure 43 shows a stairs without the front and rear landing at the top and Figure 44 shows the same stairs with the front and rear landings at the top.

Figure 43 Stairs without the front and rear landings at the top

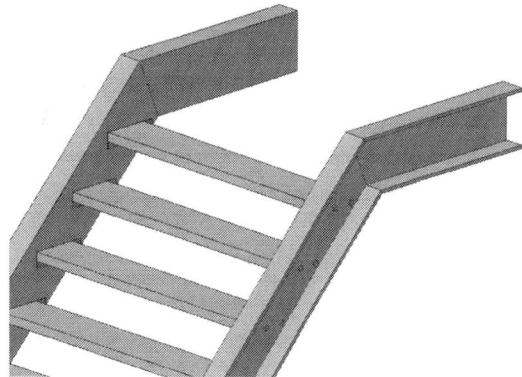

Figure 44 Stairs with the front and rear landings at the top

Landing prof. size front

The flyouts in this list are used to select the type and size of the landing profiles. Generally, the front and rear landing profiles are the same as that of the stringer.

Same for rear

If this tick box is selected, the type and size of the front landing will be used for the rear landing as well.

Rotate landing

This list provides predefined angles by which you can rotate the landings.

Mirror landing

If you want to mirror the landing, you can select this tick box.

Landings Category > Top landing Tab

The options in the **Top landing** tab, shown in Figure 45, are used to specify lengths and offset of the top landings. These options are discussed next.

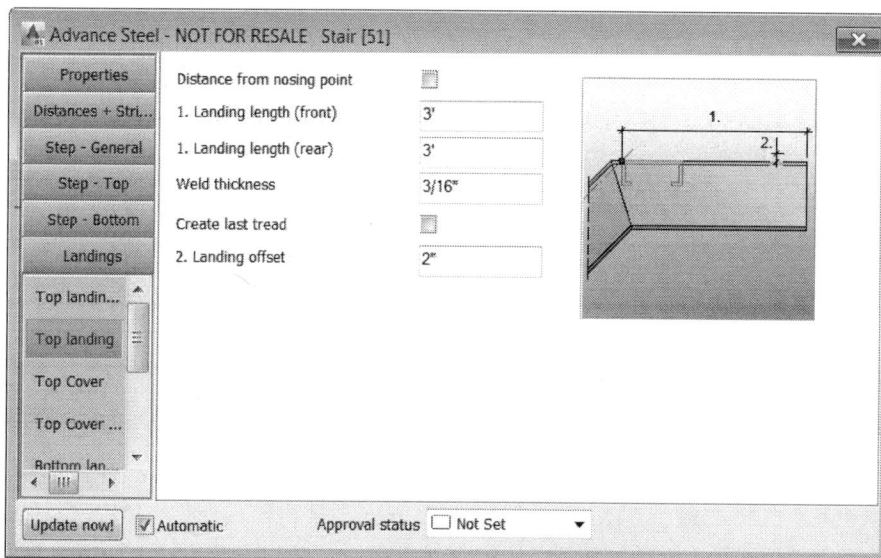

*Figure 45 The **Top landing** tab of the **Landings** category*

Distance from nosing point

Select this tick box if you want to measure the length of the landings from the nosing point.

Landing length (front) / Landing length (rear)

These edit boxes are used to specify the front and rear landing lengths at the top.

Weld thickness
The weld thickness of the welding between the top landings and the stringers can be specified in this edit box.

Create last tread
Select this tick box if you want to create the last tread. Figure 46 shows the top of the stairs without the last tread and Figure 47 shows the same stairs with the top tread.

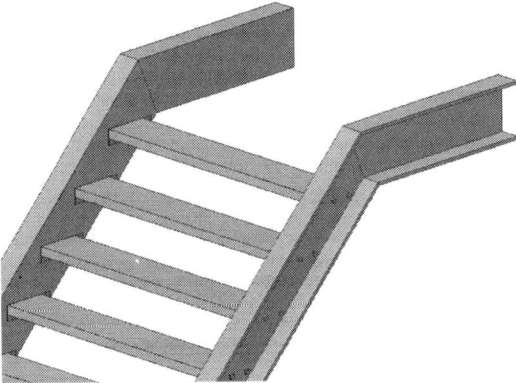

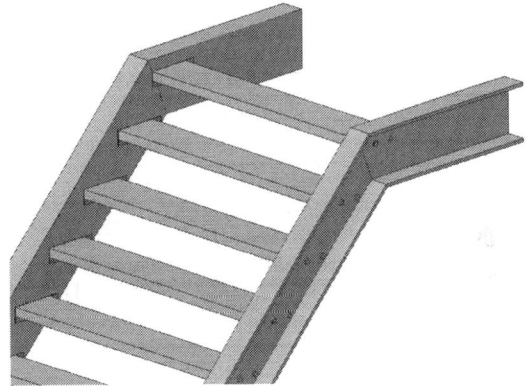

Figure 46 Stairs without the top tread   *Figure 47* Stairs with the top tread

Landing offset
This edit box is used to specify the landing offset from the stringer.

Landings Category > Top Cover Tab
The options in the **Top Cover** tab, shown in Figure 48, are used to specify whether or not the top cover will be created. If you decide to create the top cover, various options related to the top cover can also be defined using the options in this tab. These options are discussed next.

Cover made from
This list is used to specify how the top cover will be made. By default, **None** is selected from this list. As a result, the top cover is not created. Selecting **Plate** from this list will create the top cover using a plate. Selecting **Grate** from this list will create the top cover using grating. Selecting **Angle no plate** from this list will insert angle sections welded to the top landings, but no plate will be inserted. Figure 49 shows stairs the top cover created using a plate and Figure 50 shows stairs with the top cover created using grating.

Cover thickness
This edit box is used to specify the cover thickness while using a plate or grating.

Grating class / Grating size
If you are using grating as the top cover, you can select the grating class and size from these lists.

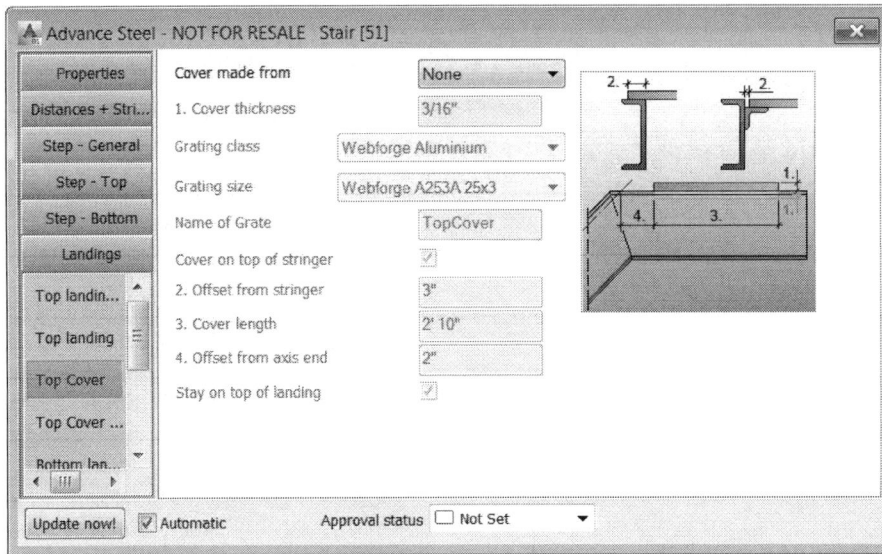

*Figure 48 The **Top Cover** tab of the **Landings** category*

Figure 49 Top cover created using a plate

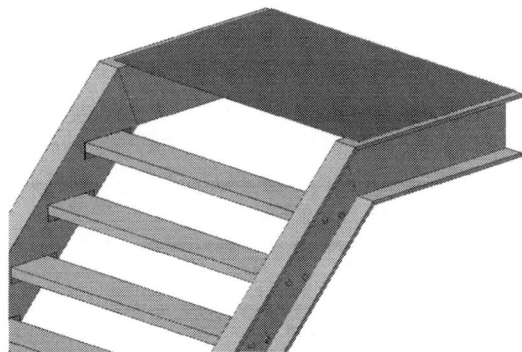

Figure 50 Top cover created using grating

Name of grating

While using grating as the top cover, you can enter its name in this edit box.

Cover on top of stringer

Selecting this tick box will ensure the cover is placed on top of the stringer, as shown in Figure 51. If this tick box is cleared, angle sections will be inserted welded to the stringers and the cover will be placed on top of those angle sections, as shown in Figure 52.

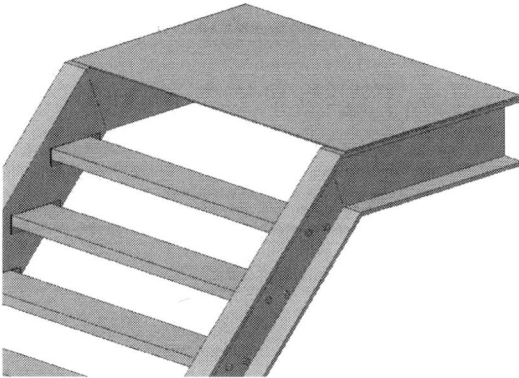

Figure 51 *Top cover placed on top of stringer*

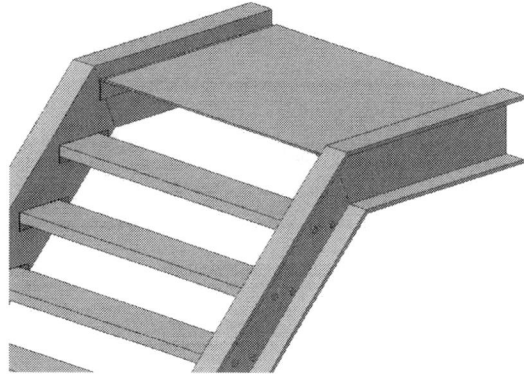

Figure 52 *Top cover placed on top of angles*

Offset from stringer

This edit box is used to specify the offset of the cover from the outside of the stringer if the cover is placed on top of the stringer. If you are placing the cover on an angle, this edit box defines the offset of the cover from the inside of the stringer.

Cover length

This edit box is used to define the length of the cover.

Offset from axis end

This edit box is used to define the offset of the cover from the end of the top landing axis.

Stay on top of landing

This tick box is generally used when you have selected the **Cover on top of stringer** tick box. In this case, selecting this tick box ensure the top stringer is aligned to the top point you selected while inserting the stair. As a result, the top cover will sit above the top point.

Landings Category > Top Cover - Angle Tab

While inserting the angle as the top cover or while placing the cover on an angle by clearing the **Cover on top of stringer** tick box, you can use the options in this tab to select the angle type and size, the weld thickness, and the angle offset value, as shown in Figure 53.

Landings Category > Bottom landing / Cover Tabs

The options available in all the **Bottom landing / Covers** tabs are similar to those discussed above, with the only difference that those tabs are used to define the landings and covers at the bottom of the stairs. Because these options are the same, they are not discussed again.

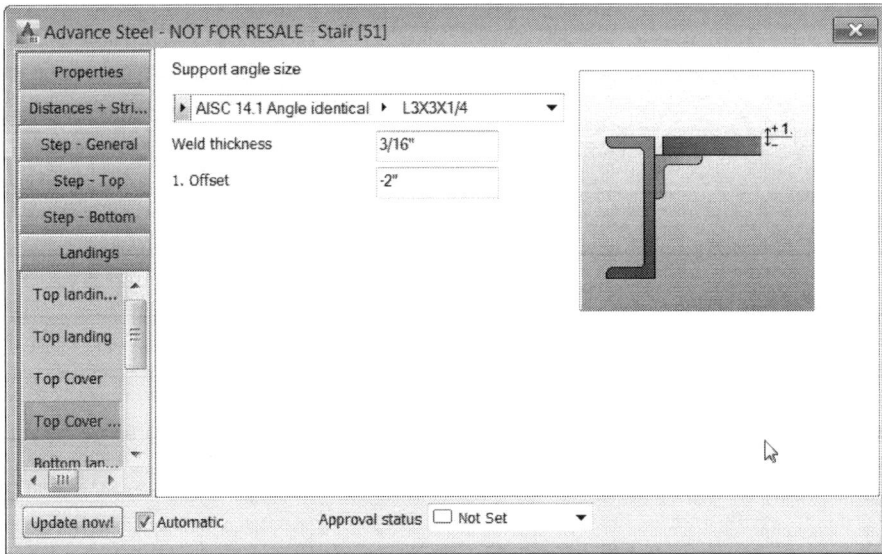

Figure 53 *The* **Top Cover - Angle** *tab of the* **Landings** *category*

INSERTING HAND-RAILINGS

Home Ribbon Tab > Extended Modeling Ribbon Panel > Hand-railing
Extended Modeling Ribbon Tab > Structural Elements Ribbon Panel > Hand-railing

In Advance Steel, you can use the **Hand-railing** tool to create various types of handrails. You can change the railing ends to meet your requirements and even define the post connection as welded or bolted. Figure 54 shows handrails with looped ends and post bolted and Figure 55 shows handrails with loop return ends and posts welded. Note that in both these figures, the handrails on the two sides of the stairs are created separately.

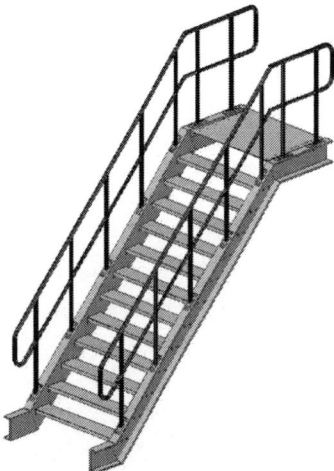

Figure 54 *Handrails with looped ends*

Figure 55 *Handrails with loop return ends*

When you invoke the **Hand-railing** tool, you will be prompted select base beams for railing. In this prompt, you need to select all the beams to which the posts of the handrails need to be connected. For example, in Figures 54 and 55, the stringer and the top landing is selected to create handrails. Once you have selected all the beams, press ENTER. Next, you will be prompted to start and end point of the railing. In this prompt, you can select the endpoints of the beams selected earlier. Once you have specified these two points, the following prompt sequence is displayed:

Do you want to select a nosing point relative to the start point? [Yes/No] <N>:

In this prompt sequence, you can either type **Y** to select a nosing point or press ENTER to accept the default value of not specifying the nosing point. On doing so, the **Advance Steel - Railing [X]** dialog box is displayed with the **Properties** category > **Properties** tab active. This tab and the **Library** tab are similar to those discussed in earlier chapters. The remaining tabs are discussed next.

Posts Category > Post Tab

The options available in this tab, shown in Figure 56, are used to specify the parameters related to the posts. These options are discussed next.

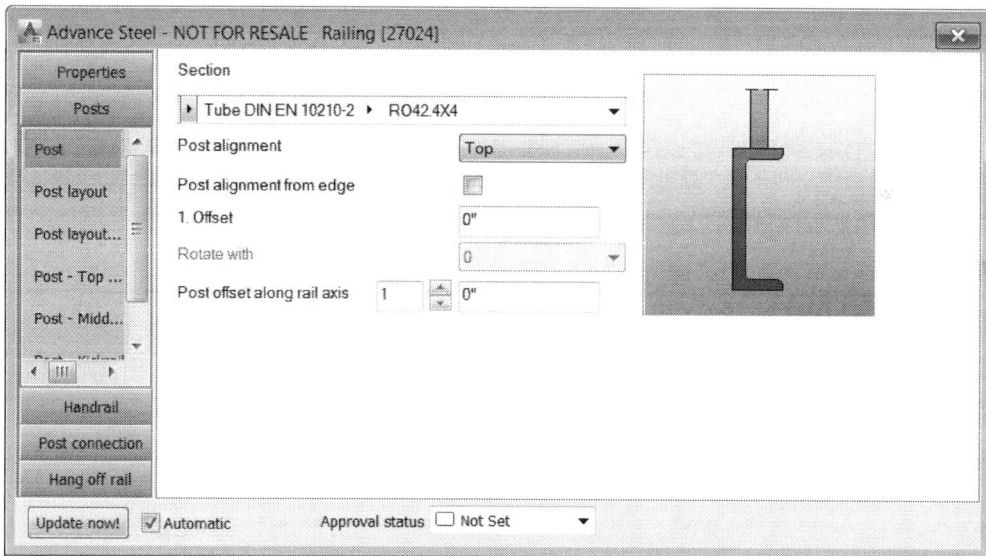

*Figure 56 The **Post** tab of the **Posts** category*

Section

The flyouts in this list are used to select the section type and size for the post.

Post alignment

This list is used to specify how the post will be aligned to the beam. By default, **Top** is selected. However, you can also select **Left** or **Right** to align the posts to the side of the beam.

Post alignment from edge

This tick box is only available when you are creating the post aligned at the top. In this case, you can select this tick box to define the location of the post as an offset value from the edge of the beam.

Offset

This edit box is only available when the post alignment is set to **Top** and is used to offset the posts from the center of the top face of the beam. Figure 57 shows the zoomed in view of the posts aligned at the top of the beam and Figure 58 shows the posts aligned with an offset from the center.

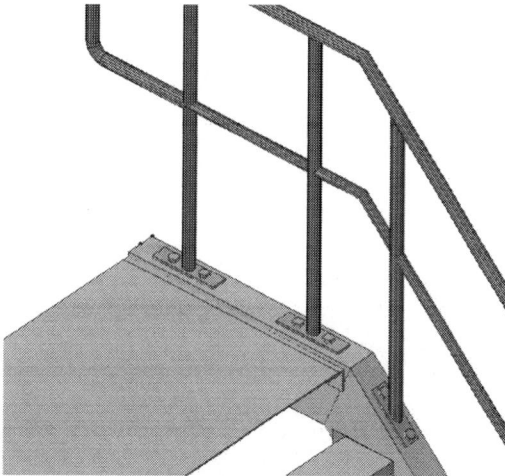

Figure 57 *Posts aligned at the top*

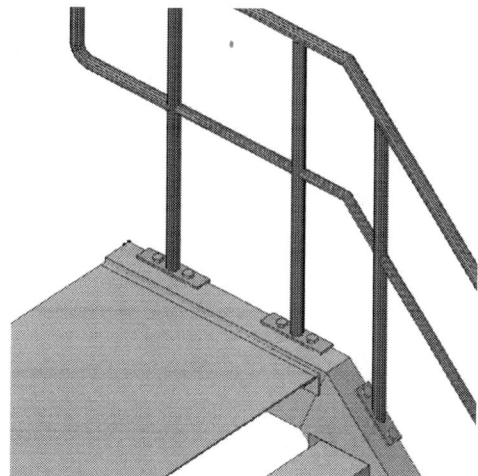

Figure 58 *Posts aligned at an offset from the edge*

Rotate with

This edit box is only available when you select a rectangular or an angular section for the post. In that case, you can use this list to rotate the post section.

Post offset along rail axis

This option allows you to select the railing posts and then offset them along the rail axis. You can use the spinner to go to the post number that needs to be offset and then enter the offset value in the edit box.

Posts Category > Post layout Tab

The options available in this tab, shown in Figure 59, are used to specify the post layout. These options are discussed next.

Points at position of Posts

If this tick box is selected, posts will be created at the start and end points that were defined while creating the handrails.

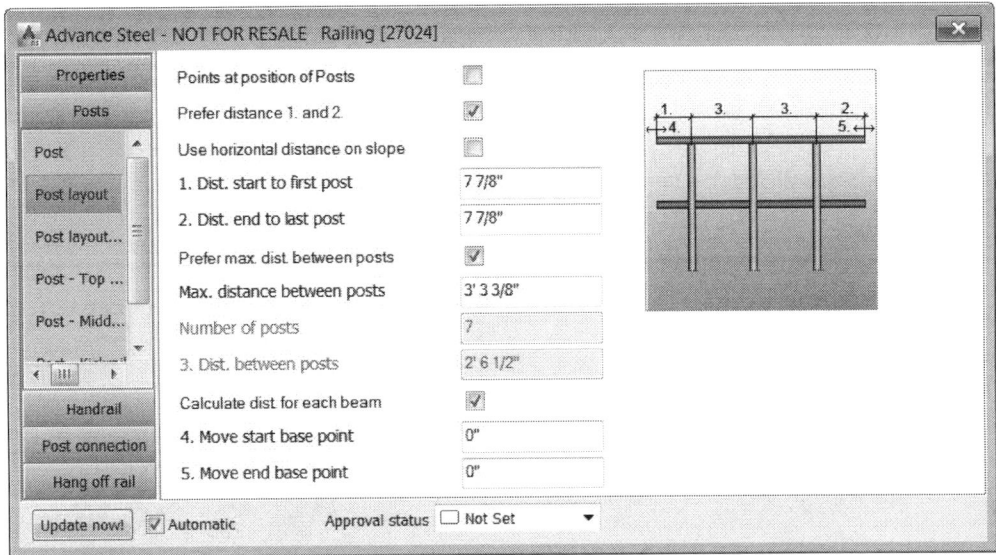

Figure 59 *The **Post layout** tab of the **Posts** category*

Prefer distance 1. and 2.
This tick box is selected to manually define the start distance to the first post and the end distance to the last post. These distances are defined in the **Dist. start to first post** and **Dist. end to last post** edit boxes.

Use horizontal distance on slope
If this tick box is selected, the horizontal distance between the posts will be used to define the distance between the posts on the slope as well.

Prefer max. distance between posts
This tick box is selected to define the number of posts in terms of the maximum distance between the posts. The preferred value is defined in the **Max. distance between posts** edit box available below this tick box.

Number of posts
This edit box is only available if the **Prefer max. distance between posts** tick box is cleared. In that case, you can enter the total number of posts in this edit box. Note that the total number of posts defined in this edit box will be defined on every beam selected to create railings. For example, in Figure 60, the number of posts are defined as 5. As a result, there are five posts created on the sloped beam and five posts created on the straight beam.

Distance between post
This edit box is used to define the individual spacing between the posts.

Calculate dist. for each beam

If this tick box is selected, the number of posts are calculated individually for each beam. If you clear this tick box, the total number of posts defined in the **Number of posts** edit box will be considered between the start and end points of the railing. For example, in Figure 61, the total number of posts are defined as 5. On clearing this tick box, that value is considered as a total value between the start and end points of the railing.

Figure 60 *Five posts created on the sloped as well as the horizontal beam*

Figure 61 *Total five posts created between the start and end points of the railing*

Move start base point

This edit box is used to define the distance by which you want to move the start base point of the railing.

What I do

*I use the **Move start base point** and **Move end base point** edit boxes to move back the start and endpoint of the railings. As a result, I do not have to use the object snap tracking to move back while defining the start and end points of railing.*

Move end base point

This edit box is used to define the distance by which you want to move the end base point.

Posts Category > Post - Top / Middle handrail and Kickrail Tabs

The options available in the **Post - Top handrail** tab, shown in Figure 62, **Post - Middle handrail** and **Post - Kickrail** tabs are used to specify how the top handrail will be connected to the post. These options are available in the **Connection type** lists available in these tabs. You can select the connection type from these lists and then enter the related values in the other options that appear in these tabs.

Handrail Category > Top handrail Tab

The options available in this tab, shown in Figure 63, are used to specify the section and distances

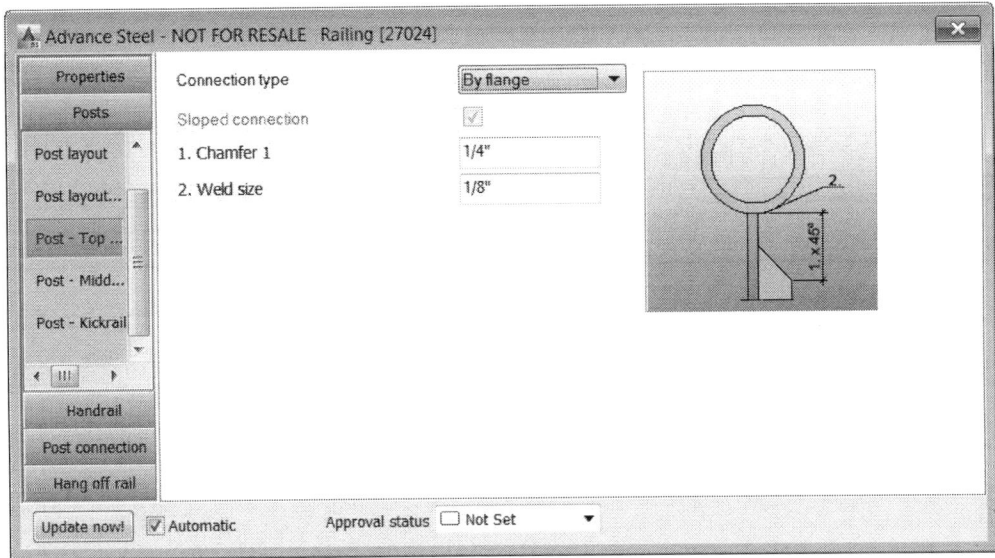

*Figure 62 The **Post - Top handrail** tab of the **Posts** category*

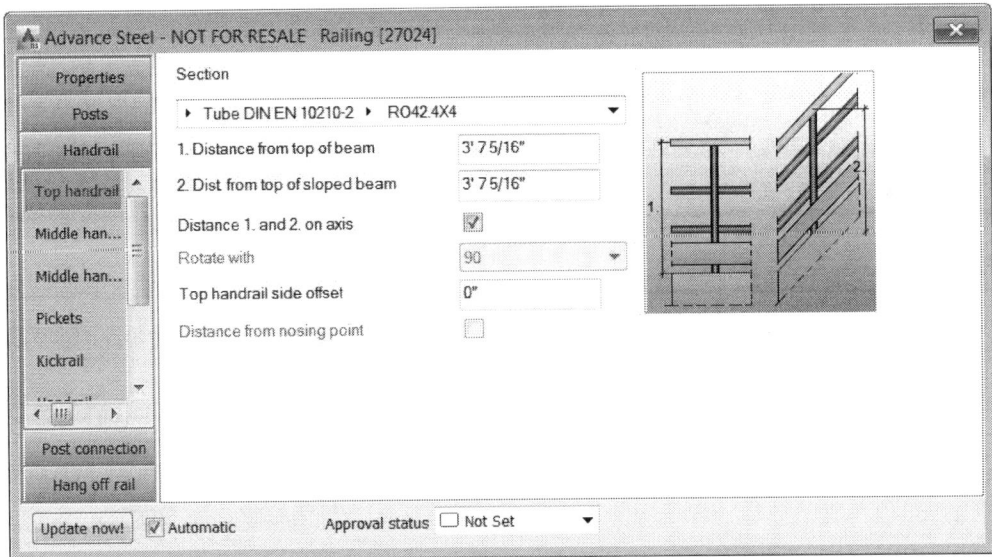

*Figure 63 The **Top handrail** tab of the **Handrail** category*

of the top handrail. You can also specify the side offset for the top handrail, if required, in this tab.

Handrail Category > Middle handrail horizontal Tab

The **Middle handrail horizontal** tab is used to specify the section and distances of the middle handrail, as shown in Figure 64. You can also create more than one middle handrail. In that case, you can specify the distance between the middle handrails also using the options in this tab.

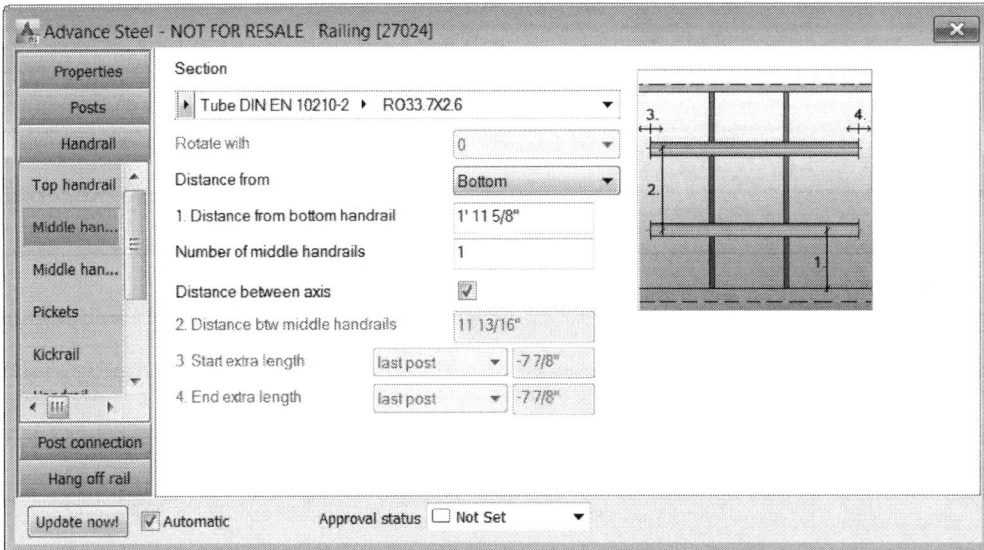

Figure 64 The **Middle handrail horizontal** *tab of the* **Handrail** *category*

Handrail Category > Middle handrail sloped Tab

The **Middle handrail sloped** tab, shown in Figure 65, is used to specify the middle handrail distances on the sloped beams. By default, the distance from the bottom handrail is measured in the vertical direction. However, you can select the **Measure dist. perp. to slope** tick box to measure this distance perpendicular to the slope.

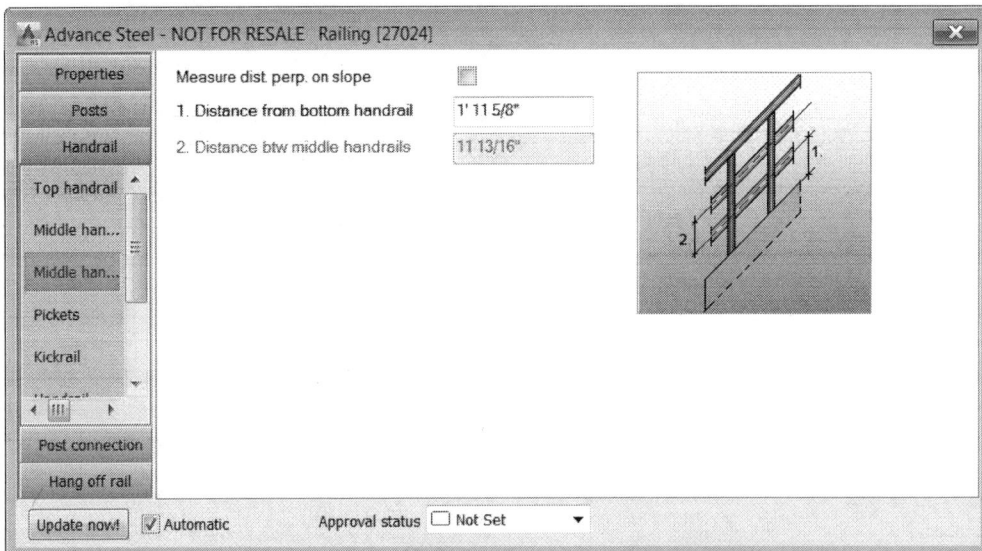

Figure 65 The **Middle handrail sloped** *tab of the* **Handrail** *category*

Handrail Category > Pickets Tab

The **Pickets** tab is used to specify the section and spacing between the pickets, as shown in Figure 66.

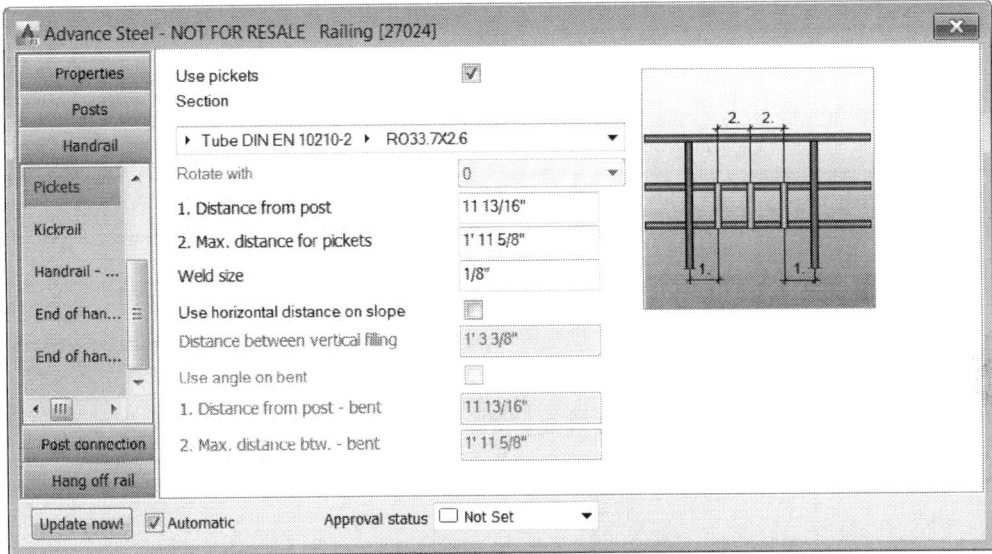

*Figure 66 The **Pickets** tab of the **Handrail** category*

By default, the pickets are not created. To insert pickets, select the **Use pickets** tick box and then specify their section and distances. Figure 67 shows a handrail without pickets and Figure 68 shows the same handrail with pickets.

Figure 67 Handrail without the pickets

Figure 68 Handrail with the pickets

Handrail Category > Kickrail Tab

The **Kickrail** tab, shown in Figure 69, is used to specify the section and spacing between of the kickrail.

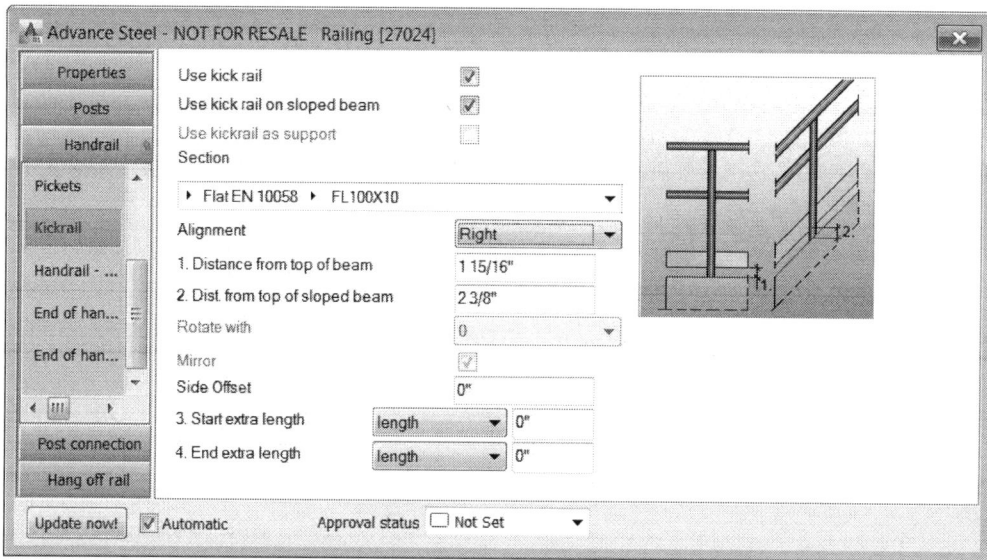

*Figure 69 The **Kickrail** tab of the **Handrail** category*

To create the kickrail, select the **Use kick rail** tick box and then select its section and define the distances. You can use the **Alignment** list to specify whether the kickrail will be inserted on the right side or the left side of the posts.

Handrail Category > Handrail - Handrail Tab

The options in the **Handrail - Handrail** tab are used to specify the connection between various handrail segments. By default, **Miter cut** is selected from the **Connection type** list. As a result, the handrail segments are miter cut and then welded together. Selecting **Knee** from the **Connection type** list allows you to insert a knee section between the straight handrail segments.

Handrail Category > End of handrail / End of handrails (end) Tabs

The options in the **End of handrail** tab, shown in Figure 70, and the **End of handrail (end)** tab are used to specify the end connection type of the handrails at the start and end of the handrail sections. You can select the required end connection type from the **Connection type** list and specify the related parameters using the options available in this tab. Selecting the **Same joint for both ends** tick box will ensure the same connection type is applied to both ends of the handrail. Figure 71 shows the bent end connection type at both ends of the handrails and Figure 72 shows the return connection type at both ends.

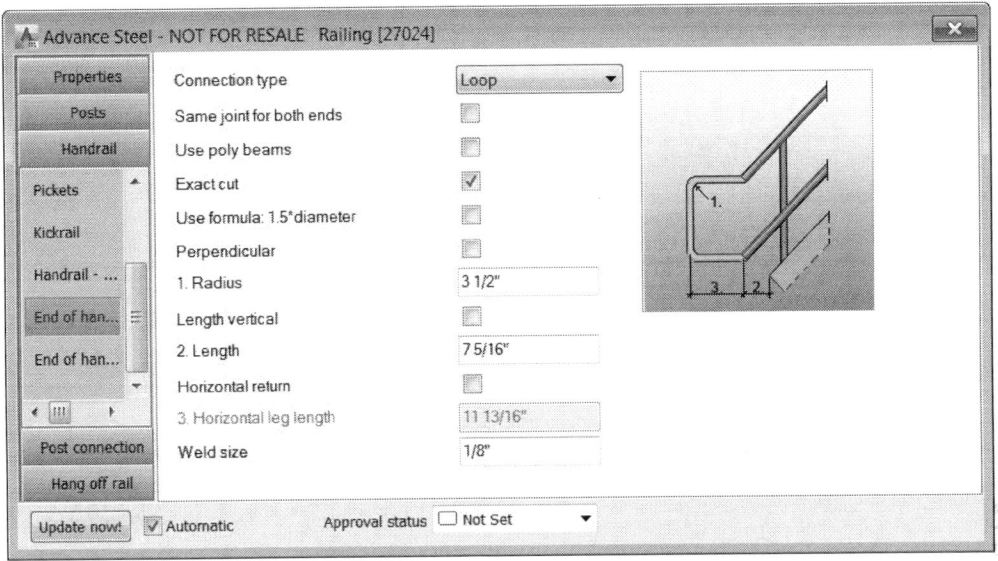

Figure 70 *The End of handrail tab of the Handrail category*

Figure 71 *Bent end connection type at the two ends*   *Figure 72* *Return connection type at the two ends*

Post connection Category

The tabs available in this category are used to specify how the post will be connected to the beam and all the related parameters. These options are parameters are similar to those discussed in earlier chapters while inserting joints.

Hang off rail Category

The tabs available in this category are used create a grab rail and specify its section, connection and position. These parameters are similar to those discussed in the earlier tabs of the **Handrailing** tool.

Hands-on Tutorial (STRUC/ BIM)	*In this tutorial, you will complete the following tasks:* 1. *Open the file containing the frames.* 2. *Insert vertical bracing and then create multiple copies of it, as shown in Figures 73.* 3. *Open the file containing trusses.* 4. *Insert joist and then create multiple copies of it, as shown in Figure 74.* 5. *Open the elevator tower file.* 6. *Insert stairs and handrails, as shown in Figure 75.*

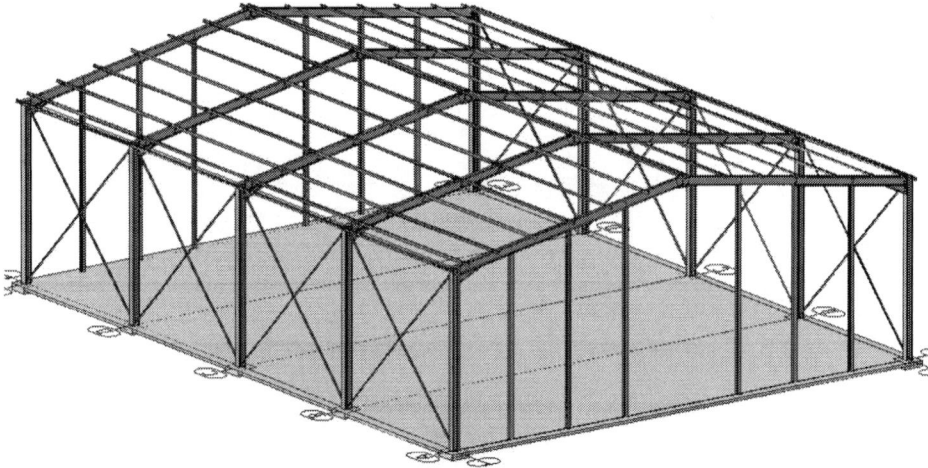

Figure 73 Frame model to insert bracing

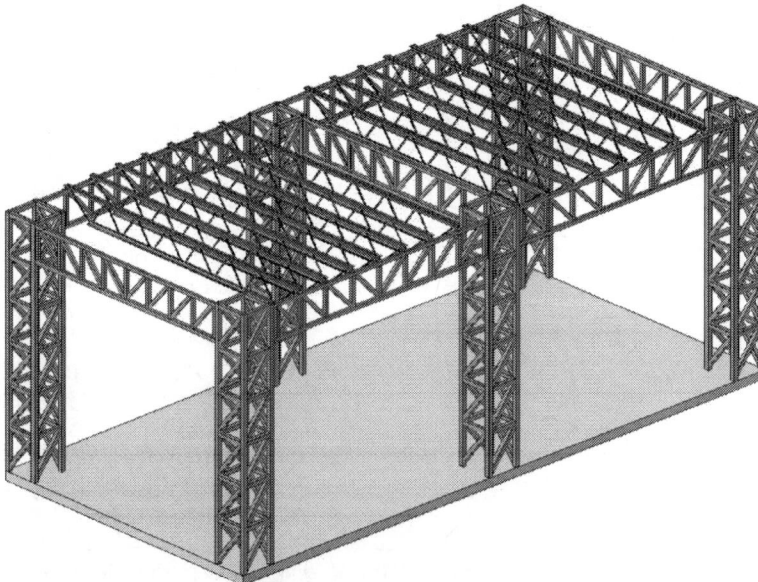

Figure 74 Model to insert joists

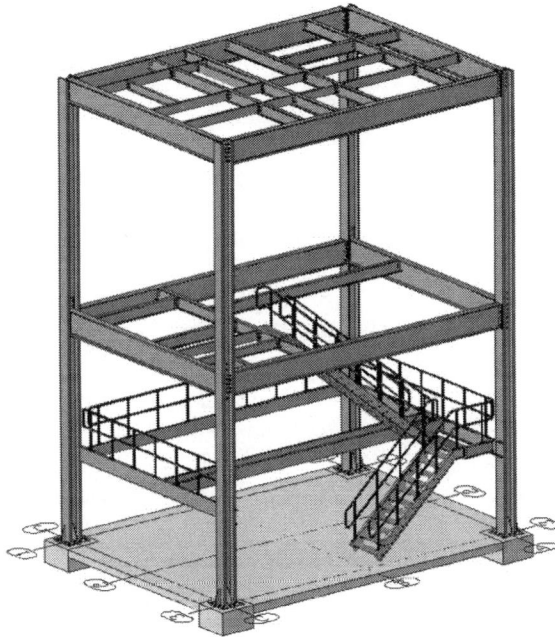

Figure 75 *The Elevator Tower model to insert stairs and handrails*

Bracing Parameters
The following are the bracing parameters. The dimensions not listed below need to be taken as the default value.

Bracing Sections	**AISC 14.1 Angle identical > L6X6X1/2** (for Imperial) or **Australian Equal Angle > 150x150x10 EA** (for Metric).
Member split	**Both diagonals**.
Positioning	Clear the **Back to back/ Toe to Toe** Option

Joists Parameters
The following are the parameters required to insert the joists. **Note that if you are using the Metric units, do not enter the unit string (mm) while typing the dimension values.**

Height	**1'8" or 500 mm**
Height Settings	**Support plates**
Type	**Modified Warren**
Decrease number of fields with	**1**

Stair 1 Parameters

The following are the parameters required for the first stair. **Note that if you are using the Metric units, do not enter the unit string (mm) while typing the dimension values.**

Stair Width	3' or **900 mm**
Stringer offset	2" or **0 mm** (no offset for Metric profiles)
Stringer prof. size	**AISC 14.1 C Channels > C10X30** (for Imperial) or **Australian Parallel Flange Channel > 200 PFC** (for Metric)
Top and Bottom Landings	**AISC 14.1 C Channels > C10X30** (for Imperial) or **Australian Parallel Flange Channel > 200 PFC** (for Metric)
Tread type	**Type 1 > 240-35x2-30x30**
Connection tread to stringer	**Welded-bolted > 1/2 inch** (for Imperial) or **12.00 mm** (for Metric)
Grating	**Variable Grating > Webforge Aluminium > Webforge A253A 25x3**
Tread Width	9" or **230 mm**
Side Height	3 1/4" or **75 mm**
Side Distance	1 1/2" or **35 mm**
Bolts Groups Distance	5" or **125 mm**
Landing Lengths	1' or **310 mm**
Create last tread	**No**
Top Cover Offset from Stringer	1/4" or **5 mm**
Cover Length	1' or **310 mm**

Stair 2 Parameters

The following are the parameters required for the second stair.

Top landing Length	8 7/8" or **290 mm**
Top Landing Offset	1 1/2" or **0 mm** (no offset for Metric profiles)
Top Cover Length	8 7/8" or **290 mm**
Bottom Landing Length	11 5/16" or **288 mm**
Landing Offset	**0**
Bottom Cover Made From	**Grate > Webforge Aluminium > A253A 25x3**
Cover Offset from Stringer	1/8" or **3 mm**
Cover Length	11 5/16' or **310 mm**

Handrail Parameters

The following are the parameters required for the handrails.

Top Handrail Section	**Tube DIN EN 10210-2 > RO42.4X4** (for Imperial) or **Australian Circular Hollow Section - CF C250L0 > 42.4x4.0 CHS** (for Metric)
Distance From Top of Beam	3' 4 1/6" or **1020 mm**

Middle Handrail Section	Tube DIN EN 10210-2 > RO33.7X3.2 (for Imperial) or **Australian Circular Hollow Section - CF C250L0 > 33.7x3.2 CHS** (for Metric)
Sloped Beam Dist from Bottom Handrail	1' 7 1/8" or **485 mm**
Straight Beam Distance from Bottom Handrail	1' 10 1/16" or **560 mm**
Radius of Top Handrail	5 1/2" or **140 mm**
Radius of Middle Handrail	5 1/2" or **140 mm**
Post layout > Calculate dist. for each beam	**No**
Post Connection Type	**Plate with bolts**
Plate thickness	3/8" or **10 mm**
Plate Length	7 7/8" or **200 mm**
Intermediate Bolt Distance	5" or **125 mm**
Bolt Diameter	**1/2 Inch** or **12.00 mm**

Section 1: Inserting and Copying Vertical Bracing

In this section, you will open **C06-Imperial-Frame.dwg** or **C06-Metric-Frame.dwg** file and then insert one vertical bracing. You will then create multiple copies of that bracing using the **Copy** tool.

1. From the **C06 > Struc-BIM** folder, open the **C06-Imperial-Frame.dwg** or **C06-Metric-Frame.dwg** file, based on the preferred units.

   This model is similar to the one you created in the previous chapter.

2. Make sure the **Polar Tracking** is turned on.

3. Zoom to the bottom of the column at the **B1** grid intersection point, as shown in Figure 76.

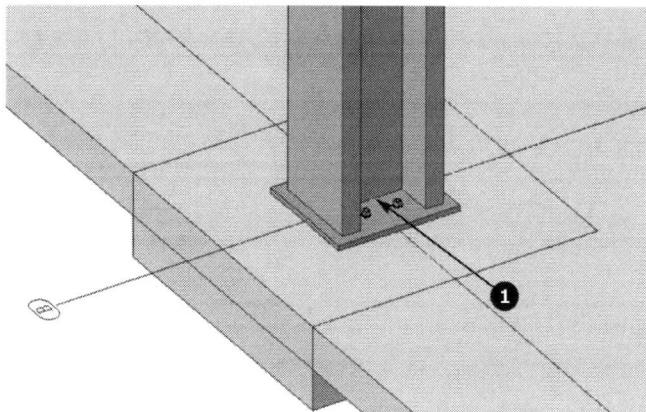

Figure 76 Zooming in to align the UCS

As mentioned earlier in the chapter, the bracing is inserted in the current UCS plane. Therefore, you first need to align the UCS vertically at the center of one of the columns. This is done using the **UCS 3 point** tool from the **Advance Steel Tool Palette**. The origin point of this new UCS will be the point labeled as **1** in Figure 76. For the point along the X-axis, you will track along 270-degrees direction and for point along the Y-axis, you will track vertically along the +Z direction.

4. From the **Advance Steel Tool Palette > UCS** tab, invoke the **UCS 3 points** tool; you are prompted to specify the new origin point.

5. Select the midpoint of the base of the column labeled as **1** in Figure 76; you are prompted to specify a point on positive portion of X-axis.

6. Move the cursor towards the right to make sure you are tracking 270-degrees angle. Once the tracking line is displayed, click anywhere to define the positive X-axis direction; you are prompted to specify a point on positive portion of Y-axis.

7. Move the cursor vertically up to make sure you are tracking along the +Z direction. Once the tracking line is displayed, click anywhere to define the positive Y-axis direction; the UCS is aligned to the midpoint of the column.

   Next, you will insert the vertical bracing using the **Bracing** tool.

8. From the **Home** ribbon tab > **Extended Modeling** ribbon panel, invoke the **Bracing** tool; you are prompted to select the first point.

9. Select the midpoint labeled as **1** in Figure 76 as the first point of the bracing; you are prompted to select the second point.

   The second point of the bracing is the diagonally opposite point, which will be at the top midpoint of the column at the **A1** grid intersection point.

10. Hold down the SHIFT key and the Wheel Mouse Button and orbit the model to a view similar to the one shown in Figure 77.

11. Select the midpoint at the top of the column, which is labeled as **2** in Figure 77 as the second point of the bracing; the bracing with the default value is inserted and the **Advance Steel - Structure element - Bracing [X]** dialog box is displayed with the **Properties** tab active.

12. Activate the **Type & Section** tab of the dialog box.

   In the next chapter, you will insert the bracing joints for which you need to split both the diagonal sections.

13. From the **Member split** list, select **Both diagonals**.

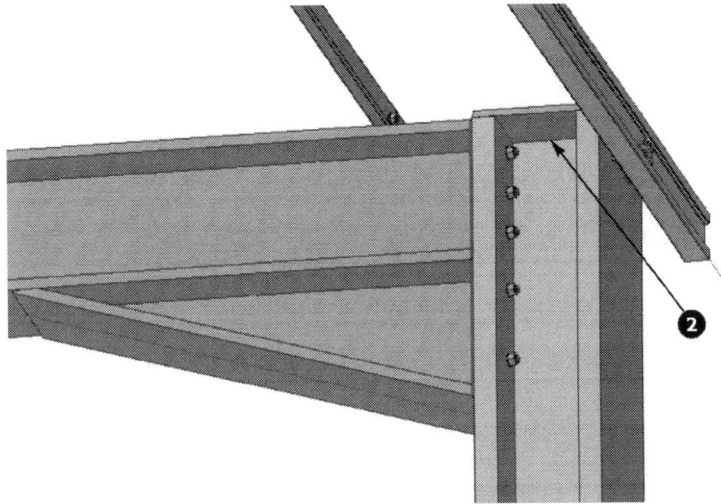

Figure 77 Orbiting to specify the second point

14. From the **Section size** list, select **AISC 14.1 Angle identical > L6X6X1/2** (for Imperial) or **Australian Equal Angle > 150x150x10 EA** (for Metric).

    For inserting the joint between the four diagonals in the next chapter, you need to make some adjustments in the positioning of the diagonals. This is done using the **Positioning** tab.

15. Activate the **Positioning** tab.

16. Clear the **Back to back/ Toe to Toe** tick box.

    Note: Although at this stage, the angle sections appear clashing, but when you add the joint between the four diagonal sections in the next chapter, there will be no clashing.

    The **Geometry** tab of the dialog box allows you to offset the bracings at the four endpoints. However, in this case, you will accept the default values and then use the joint tool in the next chapter to offset the sections.

17. Activate the **Library** tab and save the bracing with the name **Bracing-Frame**.

18. Close the dialog box. The model, after inserting the bracing, is shown in Figure 78.

    Next, you need to copy this bracing at the remaining locations of the frame. To be able to do that, you first need to turn on the **Connection boxes** (for Imperial) or **Joint box** (for Metric) layer.

19. Turn on the **Connection boxes** (for Imperial) or **Joint box** (for Metric) layer; the frames around the bracing and other connections are displayed.

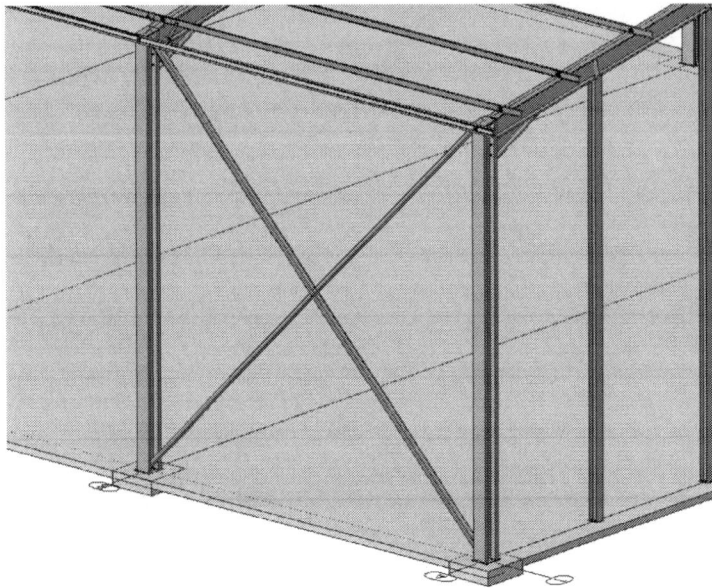

Figure 78 *The model, after inserting the first bracing*

20. Using the **Copy** tool, select the connection box of the bracing and then copy it three times on this side of the frame. You can use the top endpoint of the base plate as the base point and the second point to copy the bracing.

21. Similarly, align the UCS at the bottom midpoint of the column at the **A2** grid intersection point and then insert the bracing from the library.

22. Finally, copy the bracing three more times on this side of the frame to complete the model.

23. Turn off the **Connection boxes** (for Imperial) or **Joint box** (for Metric) layer. The model, after inserting and copying all the bracings, is shown in Figure 79.

Section 2: Inserting and Copying Joists

In this section, you will open **C06-Imperial-Joist.dwg** or **C06-Metric-Joist.dwg** file and then insert one joist. You will then create multiple copies of that joist using the **Copy** tool.

1. From the **C06 > Struc-BIM** folder, open the **C06-Imperial-Joist.dwg** or **C06-Metric-Joist.dwg** file, based on the preferred units.

   This model is similar to the one you created in earlier chapters while inserting the trusses.

2. Zoom to the area shown in Figure 80.

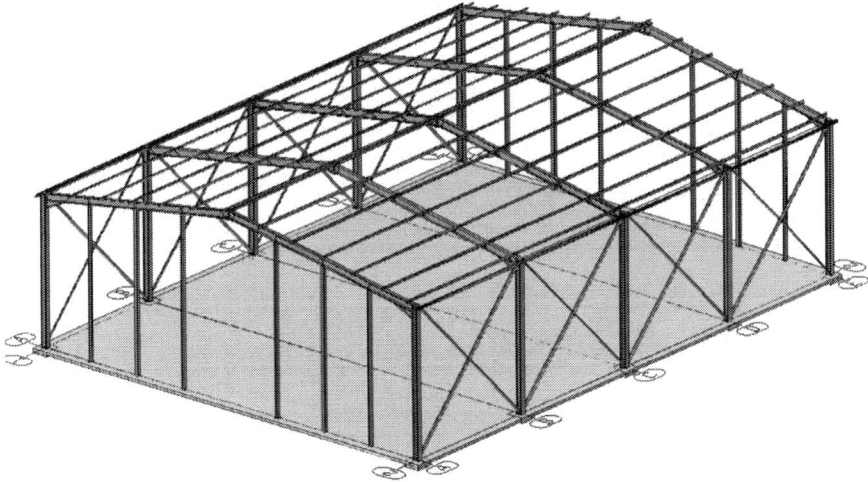

Figure 79 *The completed model after inserting and copying all the bracings*

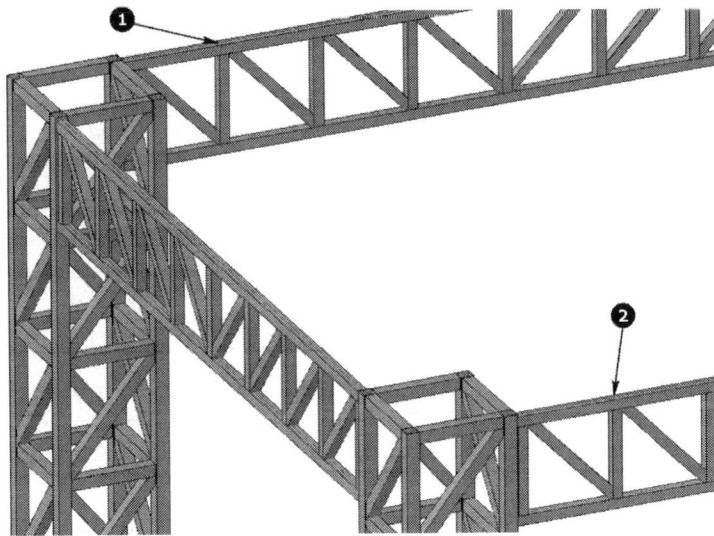

Figure 80 *The zoomed in view for inserting a joist*

The joist will be inserted using the two **Node** points labeled as **1** and **2** in Figure 80. These points represent the top endpoints of the vertical sections of the trusses at the two ends.

3. From the **Home** ribbon tab > **Extended Modeling** ribbon panel, invoke the **Joist** tool; you are prompted to select the start point of the joist.

4. Select the **Node** point labeled as **1** in Figure 80 as the start point of the joist; you are prompted to select the end point of the joist.

5. Select the **Node** point labeled as **2** in Figure 80 as the endpoint of the joist; you are prompted to select a point to define the height and orientation of the joist.

    To define the orientation, you need to move the cursor down the track along the -Z direction.

6. Move the cursor vertically down to make sure you are tracking along the -Z direction. Once the tracking line is displayed, enter **1'8"** or **500**; the joist with default values is created and the **Advance Steel - Joist [X]** dialog box is displayed with the **Properties** tab active.

    By default, the top chords of the joist are aligned with the insertion points. You will now change this to align the support plates with the insertion points.

7. Activate the **Geometry** tab.

    Notice the name in the **Joist name** text box. This is the name that will be used to save this joist in the library. If need be, you can change the name here or while saving it in the library.

8. From the **Height settings** list, select **Support plates**; the height of the joist is adjusted and the support plates are now placed at the insertion points, as encircled in Figure 81.

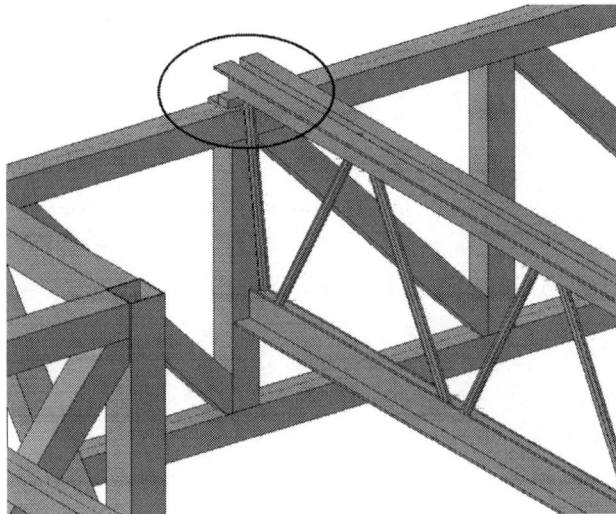

Figure 81 The support plates placed at the joist insertion point

9. From the **Types** list, select **Modified Warren**; the joist is modified and vertical sections are inserted in the joist.

10. In the **Decrease number of fields with** edit box, enter **1**; the end fields of the joist are removed, as shown in Figure 82.

    For the section sizes, joist seat plate thicknesses, and the remaining parameters, you will use the default values.

Figure 82 *The joist modified with a field decreased at the two ends*

11. Save this joist in the library with the name **Joist**.

    Next, you will create multiple copies of this joist at the remaining locations. But first, you need to turn on the **Connection boxes** (for Imperial) or **Joint box** (for Metric) layer.

12. Turn on the **Connection boxes** (for Imperial) or **Joint box** (for Metric) layer; the connection boxes are displayed around the joist and also around other objects.

13. Using the **Copy** tool, copy the joist at the top of the remaining vertical web sections of the trusses on the outside. A total of 14 joists need to be created. You can use the bottom endpoint of these vertical web sections as the base point and to points for copying.

14. Turn off the **Connection boxes** (for Imperial) or **Joint box** (for Metric) layer. The model, after copying all the joists, is shown in Figure 83.

Section 3: Inserting Stairs

In this section, you will open **C06-Imperial-Elev-Tower.dwg** or **C06-Metric-Elev-Tower.dwg** file and then insert two stairs between predefined points. For your convenience, there are points already inserted in this model on the **Points** layer.

1. From the **C06 > Struc-BIM** folder, open the **C06-Imperial-Elev-Tower.dwg** or **C06-Metric-Elev-Tower.dwg** file, based on the preferred units.

    This model looks similar to the one you created in the previous chapter. You will now insert stairs in this model using the start and the end points of the stair. The points to be used as the start and end are on the **Points** layer, which is turned off by default. Therefore, you first need to turn on this layer to display the points.

2. Turn on the **Points** layer; the points created in the model are displayed.

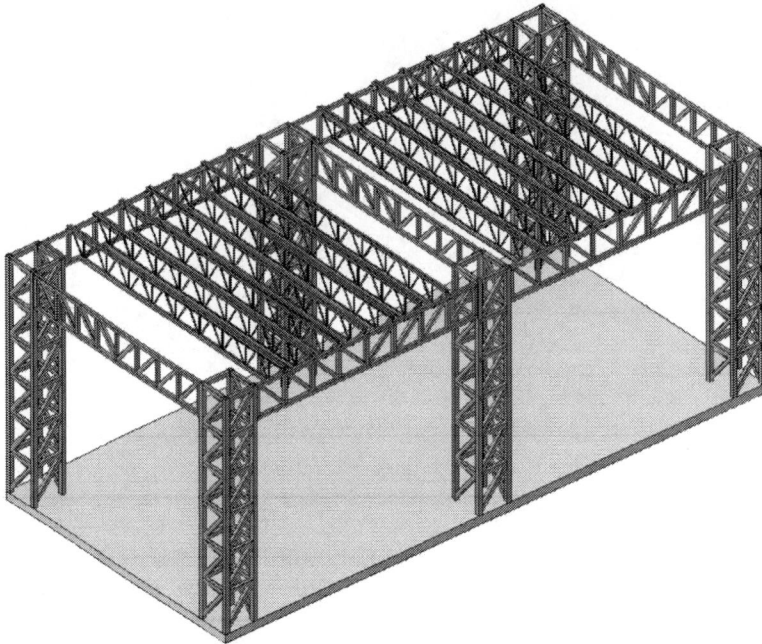

Figure 83 *The model, after inserting and copying the joists*

3. Zoom in to the area shown in Figure 84.

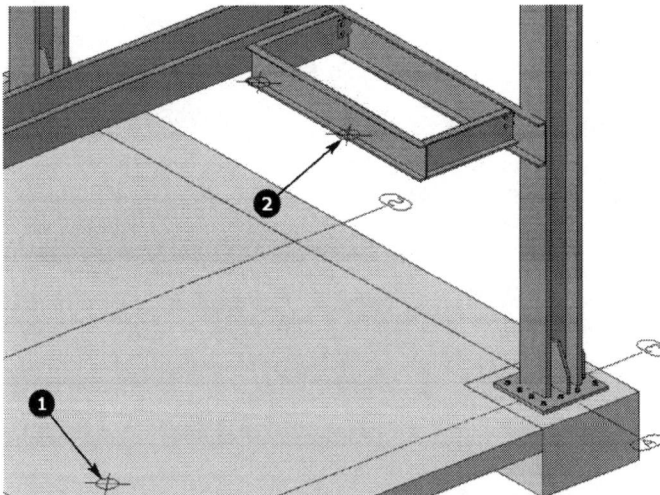

Figure 84 *Zooming in to insert the stair*

4. From the **Home** ribbon tab > **Extended Modeling** ribbon panel, invoke the **Straight stair** tool; you are informed that the start and end point method is the default method to create stair.

5. Press ENTER at this prompt; you are prompted to select the first point of the stair.

6. Select the point labeled as **1** in Figure 84 as the start point of the stair; you are prompted to select the second point of the stair.

7. Select the point labeled as **2** in Figure 84 as the end point of the stair; you are informed that the middle alignment is by default selected to align the stair.

8. Press ENTER at this prompt; a stair with the default values is inserted and the **Advance Steel - Stair [X]** dialog box is displayed.

   First of all, you will change the stringer and landing profiles.

9. Activate the **Distances + Stringer > Stringer** tab.

10. In the **Stringer offset** edit box, enter **2"** or **0** (no offset for the Metric profile).

11. From the **Stringer prof. size front** list, select **AISC 14.1 C Channels > C10X30** (for Imperial) or **Australian Parallel Flange Channel > 200 PFC** (for Metric).

    By default, the **Same for rear** tick box is selected. As a result, the same stringer profile is also defined for the rear stringer.

12. Activate the **Landings > Top landing prof.** tab.

13. Select the **Create front** and **Create rear** tick boxes, if not already selected.

14. From the **Landing prof. size front** list, select **AISC 14.1 C Channels > C10X30** (for Imperial) or **Australian Parallel Flange Channel > 200 PFC** (for Metric).

    By default, the **Same for rear** tick box is selected. As a result, the same top landing profile is also defined for the rear landing, as shown in Figure 85.

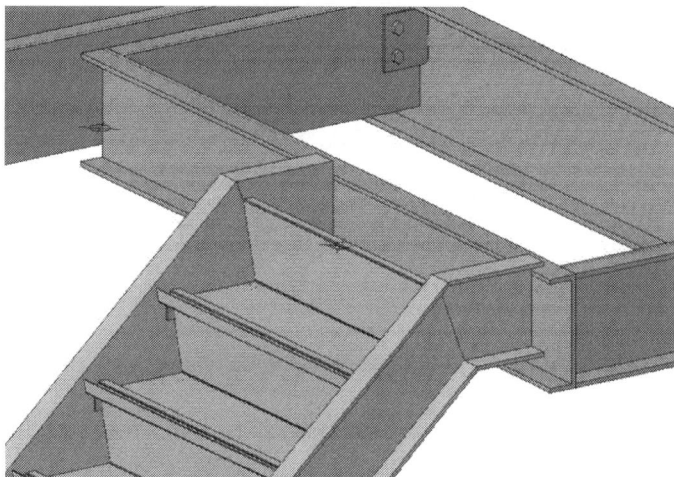

Figure 85 *The profiles defined for the stringers and top landings*

15. Activate the **Landings > Bottom landing prof.** tab.

16. Clear the **Create front** and **Create rear** tick boxes, if selected.

    This ensures the bottom landings are not created. This is because in the next chapter, you can insert the **Stair Anchor Base Plate** joint to the bottom of the stringer.

    Next, you will change the step tread type.

17. Activate the **Step - General > Tread type** tab.

18. From the **Tread type** list, select **Type 1**; the tread type of the stair changes to type 1.

19. From the **Tread size** list, select **240-35x2-30x30**; the tread size is updated.

20. From the **Connection tread to stringer** list, select **welded-bolted**; the treads are now welded to the end plates and the end plates bolted to the stringers. Figure 86 shows a view from below the stairs to show the connection between the treads and the stringers.

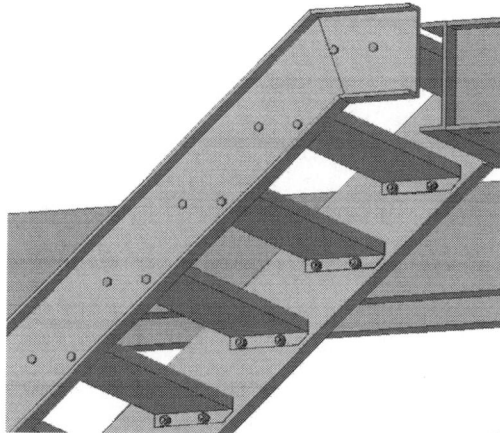

Figure 86 A view showing the connection of the treads to the stringers

21. From the **Bolt diameter** list, select **1/2 inch** (for Imperial) or **12.00 mm** (for Metric).

22. From the **Bolt type** list, select **A325** (for Imperial) or **AS 1252** (for Metric).

23. From the **Bolts/welds location** list, select **site/shop**. This ensures the welding is done on site and the bolted connections are made on site.

What I do

*If I customize the tread parameters, as you did in the above mentioned steps, I prefer saving these settings with the customer name as initials by clicking the **Save As** button. This saves a new tread type in the **Tread size** list and I can then directly select that tread size from the list next time.*

---

24. Click the **Save As** button and save the tread size with your initials as prefix.

    By changing the tread type and size, the width of the stair also changes to a predefined value. You will now change the tread dimensions to ensure you can manually enter the width of the stair.

25. Activate the **Step - General > Tread dimensions 1** tab.

26. From the **Grating** list, select **Variable grating**; the grating changes to a default type.

27. From the **Grating class** list, select **Webforge Aluminium**.

28. From the **Grating size** list, select **Webforge A253A 25x3**; the tread changes to the required grating.

29. In the **Tread width** edit box, enter **9"** or **230**.

    Next, you will change the dimensions of the end plate.

30. Activate the **Step - General > Tread dimensions 2** tab.

31. In the **Side height** edit box, enter **3 1/4"** or **75**; the side plate height updates.

32. In the **Side distance** edit box, enter **1 1/2"** or **35**, the bolt location updates.

33. In the **Bolts groups distance** edit box, enter **5"** or **125**, the spacing between the bolts updates, as shown in Figure 87.

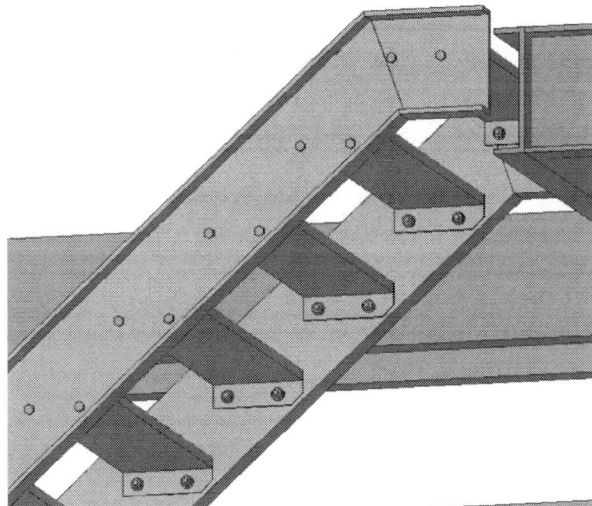

Figure 87 *The zoomed in view of the model, after modifying the side plate parameters*

Next, you will update the width of the stair in the **Distances + stringer** tab.

34. Activate the **Distances + Stringer > Distances + stringer** tab.

35. In the **Width** edit box, enter **3'** or **900**.

    Next, you will modify the landing parameters.

36. Activate the **Landings > Top landing** tab.

37. Select the **Distance from nosing point** tick box.

38. In the **Landing length (front)** and **Landing length (rear)** edit boxes, enter **1'** or **310**.

39. In the **Landing offset** edit box, enter **0** as the value.

    You will now specify the parameters to create the top cover that is placed on top of the landings.

40. Clear the **Create last tread** tick box; the last tread is no more created.

41. Activate the **Landings > Top Cover** tab.

42. From the **Cover made from** list, select **Grate**.

43. In the **Cover thickness** edit box, enter **3/8"** or **10**.

44. From the **Grating class** list, select **Webforge Aluminium**; the top cover changes to the same grating as the treads.

45. From the **Grating size** list, make sure **Webforge A253A 25x3** is selected.

46. Clear the **Cover on top of stringer** tick box; the cover is now placed on top of angle sections welded to the landings. This ensures that later on when you insert the handrails, the post connection plate does not interfere with the grating.

47. In the **Offset from stringer** edit box, enter **1/4"** or **5**.

48. In the **Cover length** edit box, enter **1'** or **310**.

49. Activate the **Landings > Top Cover - Angle** tab.

    You will accept the default angle section size.

50. In the **Offset** edit box, enter **0**; the top cover is modified to suit the requirements, as shown in Figure 88.

    Next, you need to modify the parameters of the bottom landing and the first tread.

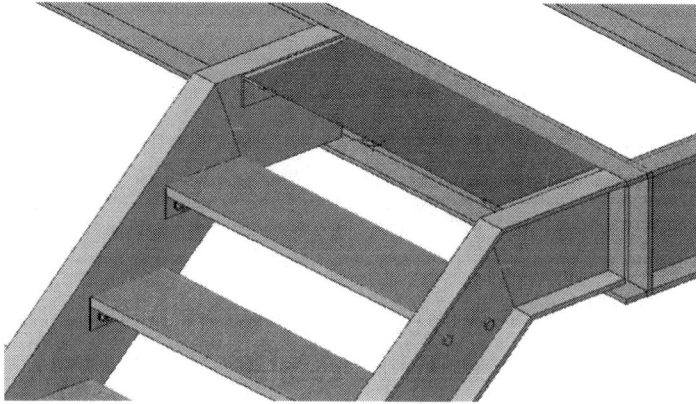

Figure 88 *The top cover modified to suit the requirements*

51. Zoom close to the bottom of the stair.

    Notice that the first tread is not the same as the remaining treads. Therefore, you need to modify this tread to match the remaining treads.

52. Activate the **Step - Bottom > Step size** (for Imperial) or **Tread size** (for Metric) tab.

53. Select the **Same as other steps** tick box; the first tread is modified to match the remaining treads of the stairs.

    This completes all the modifications in the stair. Figure 89 shows the completed Imperial stair. The Metric stair will look a little different at the start of the stair.

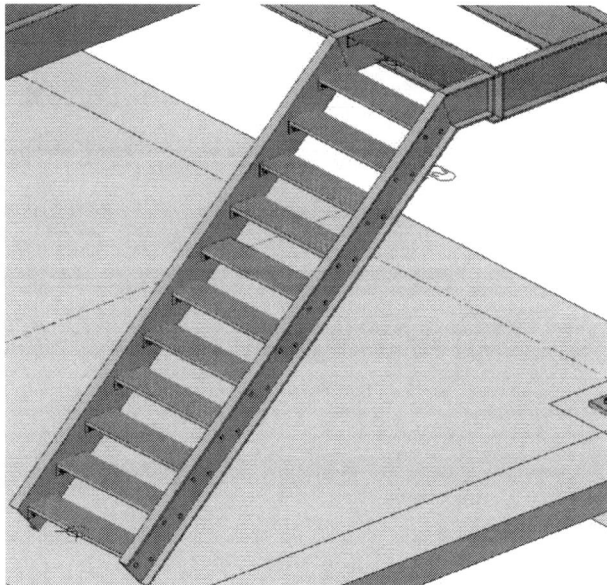

Figure 89 *The completed Imperial stair*

54. Save this stair in the library with the name **Stair-Tower** and close the dialog box.

    Next, you will insert the stairs between the remaining two points. You will use the stair saved in the library for this.

55. Orbit the model to a view similar to the one shown in Figure 90.

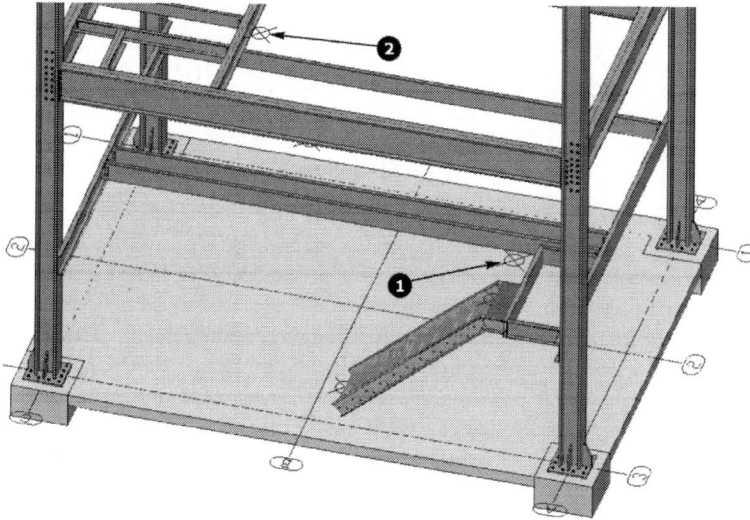

Figure 90 Orbiting the view to create the second stair

56. From the **Home** ribbon tab > **Extended Modeling** ribbon panel, invoke the **Straight stair** tool; you are informed that the start and end point method is the default method to create stair.

57. Press ENTER at this prompt; you are prompted to select the first point of the stair.

58. Select the point labeled as **1** in Figure 90 as the start point of the stair; you are prompted to select the second point of the stair.

59. Select the point labeled as **2** in Figure 90 as the end point of the stair; you informed that the middle alignment is by default selected to align the stair.

60. Press ENTER at this prompt; a stair with the default values is inserted and the **Advance Steel - Stair [X]** dialog box is displayed.

61. Activate the **Properties > Library** tab.

62. Select the stair named **Stair-Tower**.

63. Activate the **Distances, Stringer > Distances, stringer** tab.

64. In the **Width** edit box, enter **3'** or **900**.

65. Activate the **Landings > Top landing** tab.

66. In the **Landing length (front)** edit box, enter **8 7/8"** or **290**.

67. In the **Landing length (rear)** edit box, enter **8 7/8"** or **290**.

68. In the **Landing offset** edit box, enter **1 1/2"** or **0** (no landing offset for the Metric sections).

69. Activate the **Landings > Top Cover** tab.

70. In the **Cover length** edit box, enter **8 7/8"** or **290**.

This modifies the top of the stair, as shown in Figure 91.

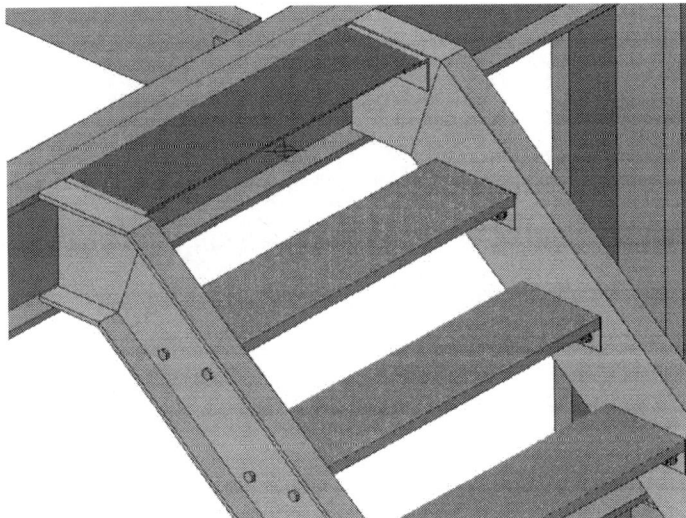

Figure 91 *The top of the stair modified to suit the requirements*

Next, you will modify the bottom of the stair.

71. Activate the **Landings > Bottom landing prof.** tab.

72. Select the **Create front** and **Create rear** tick boxes.

73. From the **Landing prof. size front** list, select **AISC 14.1 C Channels > C10X30** (for Imperial) or **Australian Parallel Flange Channel > 200 PFC** (for Metric).

By default, the **Same for rear** tick box is selected. As a result, the same top landing profile is also defined for the rear landing.

74. Activate the **Landings > Bottom landing** tab.

75. Select the **Distance from nosing point** tick box. This ensures that the values of the bottom cover is taken from the end of the bottom tread so that there is no gap between the cover and the bottom tread. Having a gap between the bottom tread and the bottom cover is a hazard.

76. In the **Landing length (front)** edit box, enter **11 5/16"** or **310**.

77. In the **Landing length (rear)** edit box, enter **11 5/16"** or **310**.

78. Select the **Create first tread** tick box.

79. In the **Landing offset** edit box, enter **0**.

80. Activate the **Landings > Bottom cover** tab.

81. From the **Cover made from** list, select **Grate**.

82. In the **Cover thickness** edit box, enter **3/8"** or **10**.

83. From the **Grating class** list, select **Webforge Aluminium**; the top cover changes to the same grating as the treads.

84. From the **Grating size** list, make sure **Webforge A253A 25x3** is selected.

85. Clear the **Cover on top of stringer** tick box; the cover is placed on top of the angle sections welded to the landings and stringers.

86. In the **Offset from stringer** edit box, enter **1/8"** or **3**.

87. In the **Cover length** edit box, enter **11 5/16'** or **310**; the bottom cover is modified to suit the requirement, as shown in Figure 92.

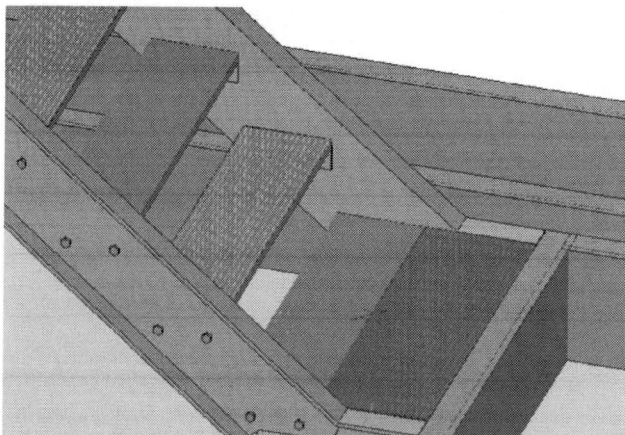

Figure 92 The bottom of the stair modified to suit the requirements

This completes all the modifications in the stairs.

88. Save the file.

Section 4: Inserting and Copying Handrails on the Stairs

In this section, you will insert handrails on both stairs. You will start with the right handrail on the bottom stair and then create the remaining handrails on the stairs using the **Create joint in a joint group** tool.

1. From the **Home** ribbon tab > **Extended Modeling** ribbon panel, invoke the **Hand-railing** tool; you are prompted to select the base beam for railing.

2. Select the beams labeled as **1**, **2** and **3** in Figure 93 as the base beams.

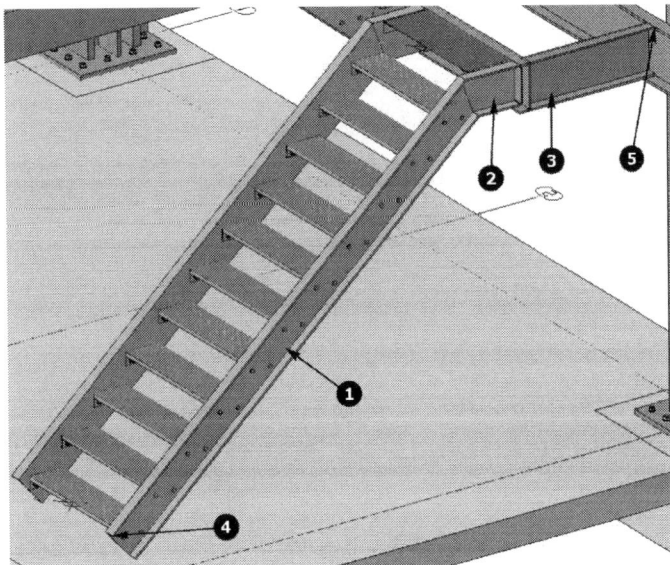

Figure 93 *Making selections for the handrails*

3. Press ENTER; you are prompted to select the start point of the railing.

4. Select the endpoint labeled as **4** as the start point of the railing; you are prompted to select the endpoint of the railing.

   The railing needs to terminate 6" or 150 behind the point labeled as **5**. In Advance Steel, you do not need to use object snap tracking to get that point. Instead, the endpoint can be moved in the dialog box.

5. Select the point labeled as **5** in Figure 93 as the endpoint of the railing.

   After specifying the endpoint of the railing, you are prompted to specify whether you want to select a nosing point relative to the start point.

6. Press ENTER to accept **No** for the previous prompt; the **Advance Steel - Railing [X]** dialog box is displayed and the railing with the default values is created, as shown in Figure 94.

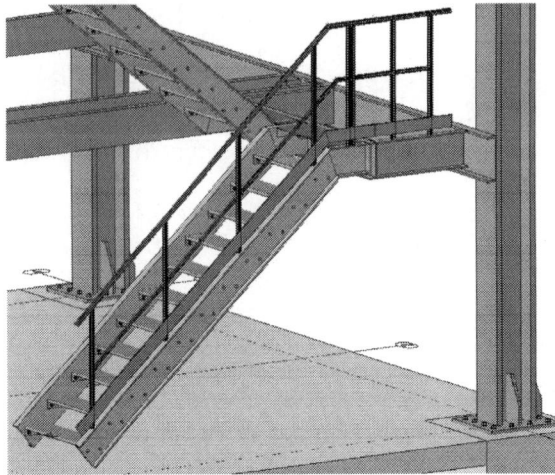

Figure 94 *The handrail with the default values*

7. Activate the **Posts > Post layout** (Imperial) or **Set out of posts** (Metric) tab.

8. In the **Move end base point** edit box, enter **-6"** or **-150**; the endpoint of the railing moves back by the specified value.

9. Activate the **Handrail > Top handrail** tab.

10. From the **Section** list, select **Tube DIN EN 10210-2 > RO42.4X4** (for Imperial) or **Australian Circular Hollow Section - CF C250L0 > 42.4x4.0 CHS** (for Metric).

> **Note**: *If you want, you can select your country standard for the top handrail section. However, ensure that the diameter is around 42.4 mm.*

11 In the **Distance from top of beam** edit box, enter **3' 4 3/16"** or **1020**.

12. In the **Distance from top of sloped beam** edit box, enter **3' 4 3/16"** or **1020**.

13. Activate the **Handrail > Middle handrail horizontal** tab.

14. From the **Section** list, select **Tube DIN EN 10210-2 > RO33.7X3.2** (for Imperial) or **Australian Circular Hollow Section - CF C250L0 > 33.7x3.2 CHS** (for Metric).

15. Make sure in the **Distance from** list, **Bottom** is selected.

16. In the **Distance from bottom handrail** edit box, enter **1' 10 1/16"** or **560**.

    Generally, the distance of the middle handrail on the sloped beam of the stair is less than that on the horizontal beam. Therefore, you need to change this value now.

17. Activate the **Handrail > Middle handrail sloped** tab.

18. In the **Distance from bottom handrail** edit box, enter **1' 7 1/8"** or **485**.

    Next, you will turn off the kickrail as it is not required.

19. Activate the **Handrail > Kickrail** (for Imperial) or **Kick plate** (for Metric) tab.

20. Clear the **Use kick rail** tick box; the kickrail is no more created.

    Next, you will define how the handrails connect at the bends.

21. Activate the **Handrail > Handrail - Handrail** tab.

22. From the **Connection type** list, select **Knee** (for Imperial) or **Radius** (for Metric).

23. In the **Radius of top handrail** edit box, enter **5 1/2"** or **140**.

24. In the **Radius of middle handrail** edit box, enter **5 1/2"** or **140**.

    Next, you will define the connection type at the two ends of the handrail.

25. Activate the **Handrail > End of handrail** (for Imperial) or **End of handrail (start)** tab.

26. From the **Connection type** list, select **Loop**; the handrail loops at the start.

27. Select the **Same joint for both ends** tick box; the handrail loops at the end as well.

28. In the **Radius** edit box, enter **3"** or **75**.

29. In the **Length** edit box, enter **5"** or **125**; the handrail is modified to the required size, as shown in Figure 95.

    Next, you will define the post parameters.

30. Activate the **Posts > Post** tab.

31. From the **Section** list, select **Tube DIN EN 10210-2 > RO42.4X4** (for Imperial) or **Australian Circular Hollow Section - CF C250L0 > 42.4x4.0 CHS** (for Metric).

32. Activate the **Post > Post layout** (for Imperial) or **Set out of posts** (for Metric) tab.

33. In the **Max. distance between posts**, enter **2' 10 1/2"** or **800**.

> *Note: Because of the difference in the values, the number of posts may be different in the Imperial and Metric sections.*

34. Clear the **Calculate dist. for each beam** tick box; the posts are now evenly spread between all the base beams.

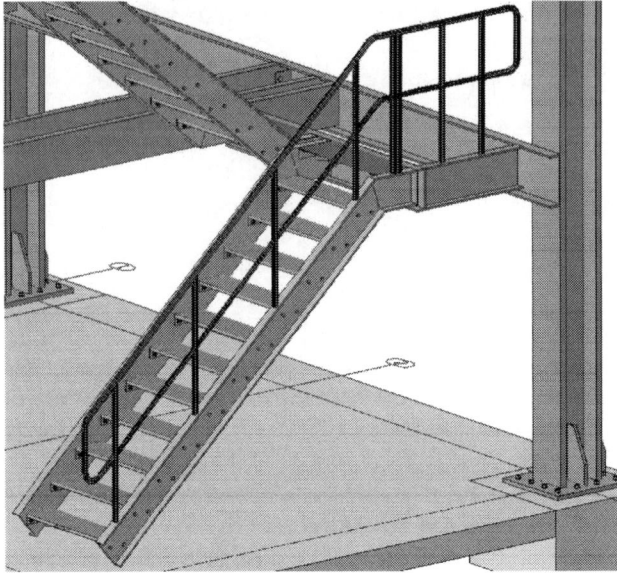

Figure 95 *The handrail modified to the required values*

35. Activate the **Post connection > Post connection** tab (for Imperial) or **Post fixing > Fixing of post** tab (for Metric).

36. From the **Connection type** list, select **Plate with bolts**; plate with bolts are inserted at the bottom of all the posts connecting them to the beams.

37. In the **Plate thickness** edit box, enter **3/8"** or **10**.

38. In the **Plate length** edit box, enter **7 7/8"** or **200**.

39. Activate the **Post connection > Connection parameter** tab (for Imperial) or **Post fixing > Fixing parameters** tab (for Metric).

40. In the **Number of lines** edit box, enter **1**; the connection is modified with 1 bolt on each side of the posts.

41. In the **Intermediate distance** edit box, enter **5"** or **125**.

42. Activate the **Post connection > Slope connection** tab (for Imperial) or **Post fixing > Fixing slope** tab (for Metric).

43. In the **Intermediate distance** edit box, enter **5"** or **125**.

44. Activate the **Post connection > Bolts** tab (for Imperial) or **Post fixing > Fixing bolts** tab (for Metric).

45. From the **Diameter of Bolts** list, select **1/2 Inch** or **12.00 mm**.

46. From the **Standard** list, select **A325** (for Imperial) or **AS 1252** (for Metric).

    This completes all the modifications in the handrails. Figure 96 shows a zoomed in view of the post connections.

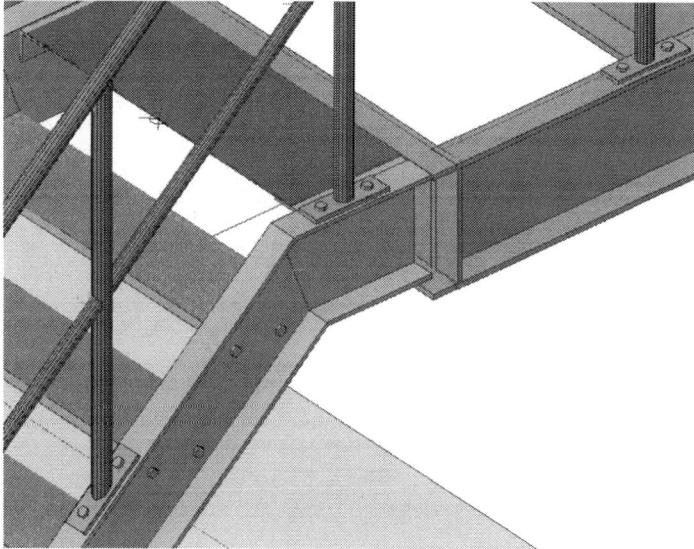

Figure 96 *The post connections modified to suit the requirements*

47. Save this handrail in the library with the name **Tower-Handrail**.

    Next, you will use the **Create joint in a joint group** tool to copy this handrails to the other side of the stair.

48. From the **Advance Steel Tool Palette > Tools** tab, invoke the **Create joint in a joint group** tool; you are prompted to select the connection part.

49. Select one of the handrail sections and then press ENTER; you are prompted to select base beams for the railing.

50. Select the beam labeled as **1** in Figure 97 as the base beam and then press ENTER; you are prompted to select the start point of the railing.

51. Select the endpoint of the beam labeled as **2** as the start point of the railing; you are prompted to select the endpoint of the railing.

52. Select the endpoint labeled as **3** as the endpoint of the railing; you are prompted to specify if you want to select a nosing point relative to the start point.

53. Press ENTER at the previous prompt to accept the default value of **No**; the railing is inserted on the selected base beam and you are again prompted to select a base beam.

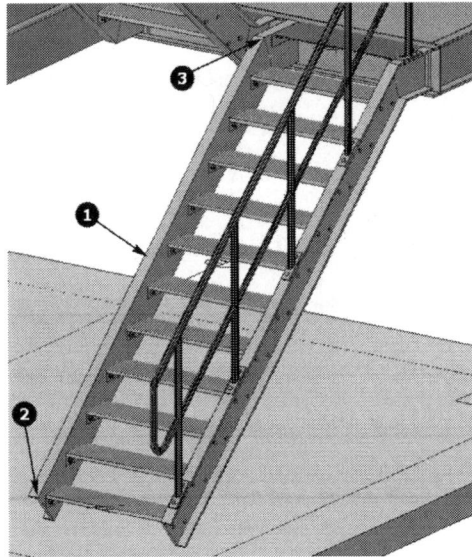

Figure 97 *Selecting the beam and points for the railing*

54. Press ENTER; the **Warning** dialog box is displayed informing you that the slave joint is different from the master joint.

    You need to separate the slave joint from master because the top end of the railing needs to be changed to a different connection.

55. Click **Yes** to separate the joint from the master.

56. Select any section of the new handrails and then right-click in the blank area of the drawing window; a shortcut menu is displayed.

57. Click **Advanced Joint Properties** from the shortcut menu; the **Advance Steel - Railing [X]** dialog box is displayed.

58. Activate the **Handrail > End of handrail** tab (for Imperial) or **End of handrail (start)** (for Metric) tab.

59. Clear the **Same joint for both ends** tick box.

60. Activate the **Handrail > End of handrail (end)** tab.

61. From the **Connection type** list, select **Loop return**; the handrail at the top end is modified.

62. In the **Radius** edit box, enter **3"** or **75**.

63. In the **Height** edit box, enter **1' 8 3/4"** or **530**.

64. In the **Horizontal leg** edit box, enter **1'** or **300**; the handrail is updated, as shown in Figure 98.

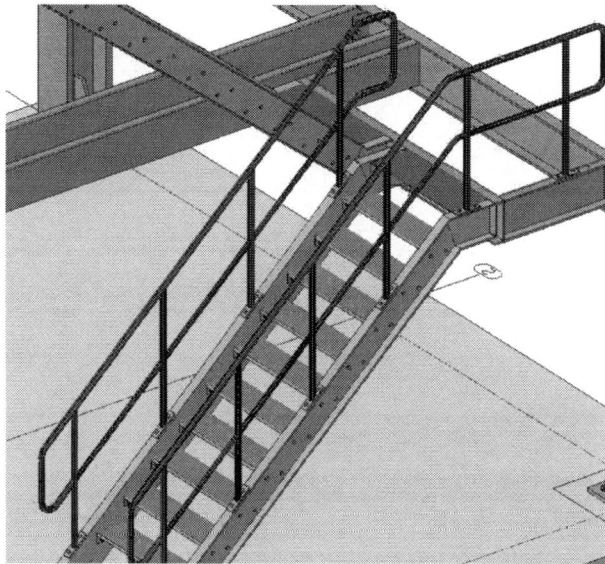

Figure 98 *The second handrail modified to suit the requirement*

65. Activate the **Posts > Post layout** (Imperial) or **Set out of posts** (Metric) tab.

66. In the **Move end base point** edit box, enter **0**; the endpoint of the railing updates.

67. Close the dialog box.

Next, you will use the same method to copy the second handrail to the two sloped beams of the other stair.

68. Using the **Create joint in a joint group** tool, one by one copy the second handrail to the other stair. While selecting the base beams, only select the sloped beam. For the start and endpoint of the railing, select the endpoints at the either ends of the sloped beams. Also, in the **Warning** dialog box, click **No** to retain the associativity of the copied railing with the master railing.

The model, after copying the handrail to the two sloped beams of the second stair, is shown in Figure 99.

Tip: When you create handrails, two small spheres are placed at the start and end points of the stairs. You can move these two spheres to move the end of stairs.

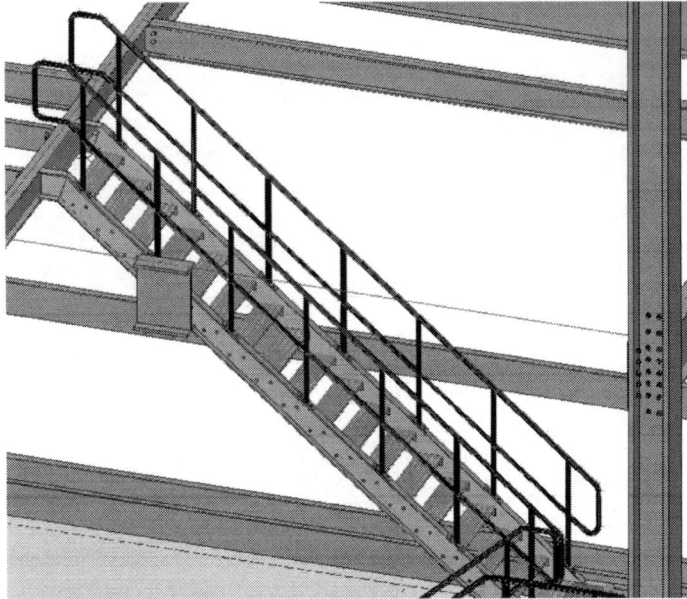

Figure 99 *The handrails copied to the second stair*

Section 5: Creating Handrails to the Perimeter Beams (Optional)

In this section, you will create the handrails on the perimeter beams.

1. Using the **Hand-railing** tool, create the handrails labeled as **1, 2**, and **3** in Figure 100. Use the handrail saved in the library to insert the new handrails. For handrails **1** and **2**, in the **Posts > Post layout** (Imperial) or **Set out of posts** (Metric) tab, move the start and end base points back by a distance of **-1' 8"** or **-510**. For the handrail **3**, move the start and end base points back by **-8"** or **-200**.

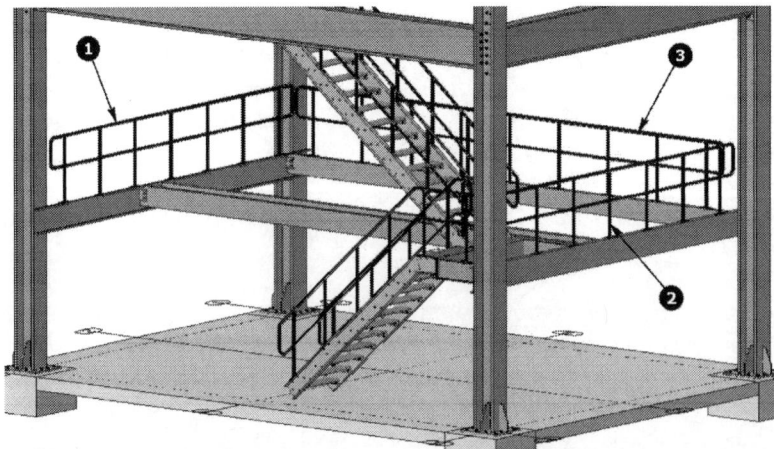

Figure 100 *The handrails copied to the perimeter beams*

Skill Evaluation

Evaluate your skills to see how many questions you can answer correctly. The answers to these questions are given at the end of the book.

1. In Advance Steel, you cannot directly insert bracing. (True/False)

2. While inserting stairs, you can only select channels as the stringer profiles. (True/False)

3. The handrails cannot be copied using the **Create joint in a joint group** tool. (True/False)

4. The handrail posts need to be manually connected to the beams using **Connection vault**. (True/False)

5. Joists are generally used in structural models to support floor or ceilings and to transfer the loads to the structural frame. (True/False)

6. Which tool is used to insert handrails in Advance Steel?

   (A) **Handrails**   (B) **Hand-railing**
   (C) **Rails**   (D) **Railing**

7. While inserting straight stairs, by default they are inserted using which one of the following methods?

   (A) **Start and end point**   (B) **Length and angle**
   (C) **Length and distance**   (D) **Distance and angle**

8. While inserting joists, which two options can be used to define the height settings?

   (A) **Bottom chord**   (B) **Top Chords**
   (C) **Vertical sections**   (D) **Support plates**

9. The top cover of the stairs cannot be made from which of the following items?

   (A) **Plate**   (B) **Angle**
   (C) **Glass**   (D) **Grate**

10. Which one is not a bracing type that can be inserted using the **Bracing** tool?

    (A) **Platform**   (B) **Crossed**
    (C) **Single**   (D) **Inserted**

Class Test Questions

Answer the following questions:

1. Explain briefly the process of inserting the stairs between two points.

2. Explain the process of copying handrails.

3. What is the procedure of inserting bracing?

4. How do you change the end connection of the handrails?

5. Explain briefly how to insert a joist between two points?

Chapter 7 – Inserting the Bracing, Tube, and Stair Joints

The objectives of this chapter are to:

√ *Explain the process of inserting and editing various types of bracing joints*
√ *Explain the process of inserting and editing various types of tube connections*
√ *Explain how to insert various types of stairs and railings joints*

INSERTING THE GENERAL BRACING JOINTS

The **General bracing** category lists various types of joints that can be added to the bracing. It is important to mention here that most of the joints in this category are similar. Therefore, only certain joint types are discussed in detail.

Gusset plate to column and base plate Joint

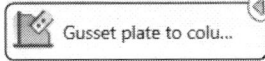

Gusset plate to colu...

This type of joint is used to connect a gusset plate and a bracing section to a column and a base plate or to a column and a secondary beam.

The bracing sections that can be used for this joint include angles, channels, flat bars, and hollow sections. When you invoke this tool, you will be prompted to select the column beam. Once you select the column and press ENTER, you will be prompted to select the secondary beam or the base plate. On making this selection and pressing ENTER, you will be prompted to select the diagonal beam, which is the bracing section. Once you select the bracing section and press ENTER, the **Advance Steel - Gusset plate to column and base plate [X]** dialog box will be displayed with the **Properties** category > **Properties** tab. The tabs in the **Properties** category are similar to those discussed in the earlier chapters. The remaining tabs are discussed next.

*Tip: This is one of the most commonly used bracing connection. Therefore, it is recommended to save the preferred settings of this connection in the **Library** tab of the dialog box for future use.*

General Category > Gusset plate shape Tab

When you click on the **General** category in the left pane of the dialog box, the **Gusset plate shape** tab is the first tab, as shown in Figure 1. The options available in this tab are discussed next.

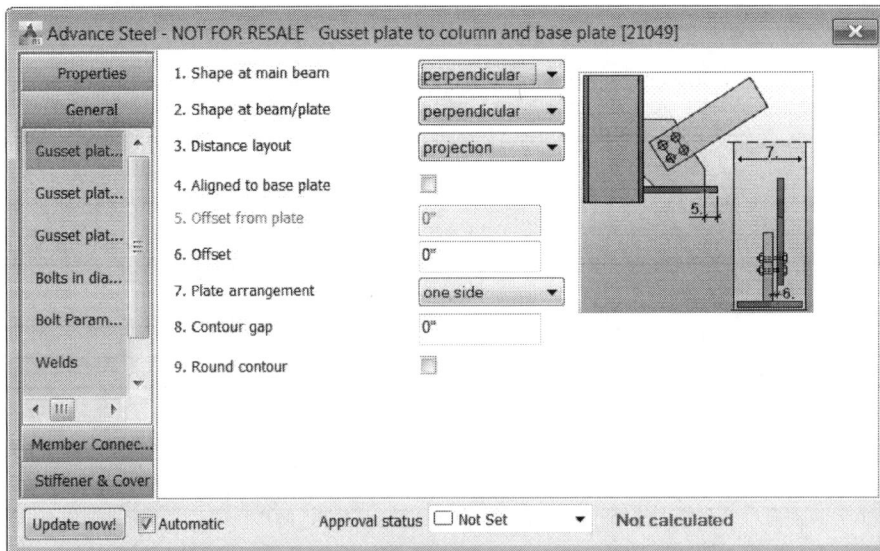

*Figure 1 The **Gusset plate shape** tab of the **General** category*

Shape at main beam

This list allows you to specify the shape of the gusset plate at the main beam. By default, **perpendicular** is selected from this list. As a result, the gusset plate is inserted perpendicular at the main beam, as shown in Figure 2. In this figure, the column is made transparent for better visibility of the gusset plate. If you select **parallel** from this list, the gusset plate will be cut parallel to the bracing member at the main beam, as shown in Figure 3. You can also select **variable** from this list and manually enter the dimensions of the gusset plate.

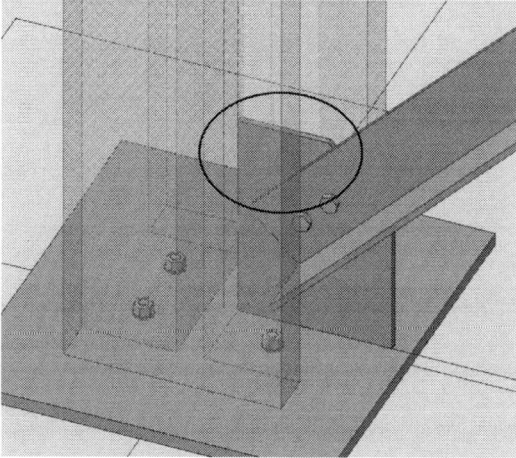

Figure 2 The gusset plate inserted perpendicular at the main beam

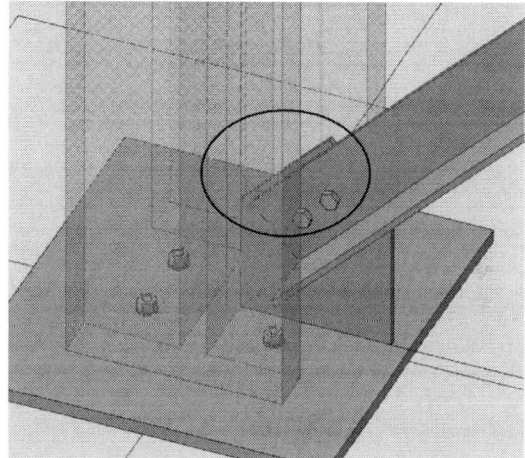

Figure 3 The gusset plate inserted parallel at the main beam

Shape at beam/plate

This list allows you to specify the shape of the gusset plate at the secondary beam or the base plate. The options in this list are similar to those discussed in the **Shape at main beam** list. In Figures 2 and 3 above, the gusset plate is inserted perpendicular to the plate. Figure 4 shows the gusset plate cut parallel to the bracing section at the plate.

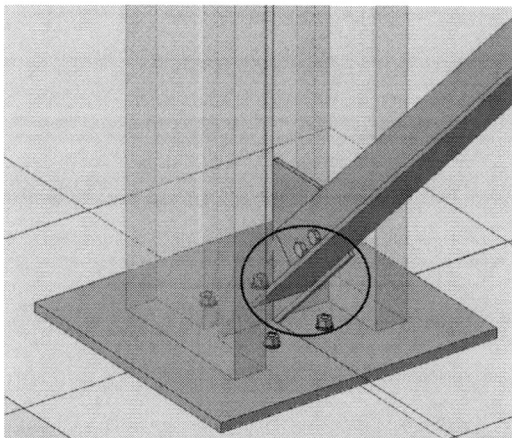

Figure 4 The gusset plate cut parallel to the bracing section

Distance layout

This list is used to specify whether you want to define the layout of the gusset plate using projections or the total distance values.

Align to base plate

While inserting the gusset plate perpendicular to the base plate, selecting this tick box will automatically align the lower width of the gusset plate to the edge of the base plate.

Offset from plate

This edit box is only available when you select the **Align to base plate** tick box. It is used to enter an offset between the gusset plate and the edge of the base plate in this edit box.

Offset

This edit box is used to define an offset between the face of the bracing section and the gusset plate.

Plate arrangement

This list is used to specify the arrangement of the gusset plates around the bracing section.

Contour gap

This edit box is used to specify the value of the contour gap.

Round contour

Select this tick box to make sure the contour is rounded.

General Category > Gusset plate parameters Tab

The **Gusset plate parameters** tab, shown in Figure 5, is used to define the parameters of the gusset plate. The options available in this tab are discussed next.

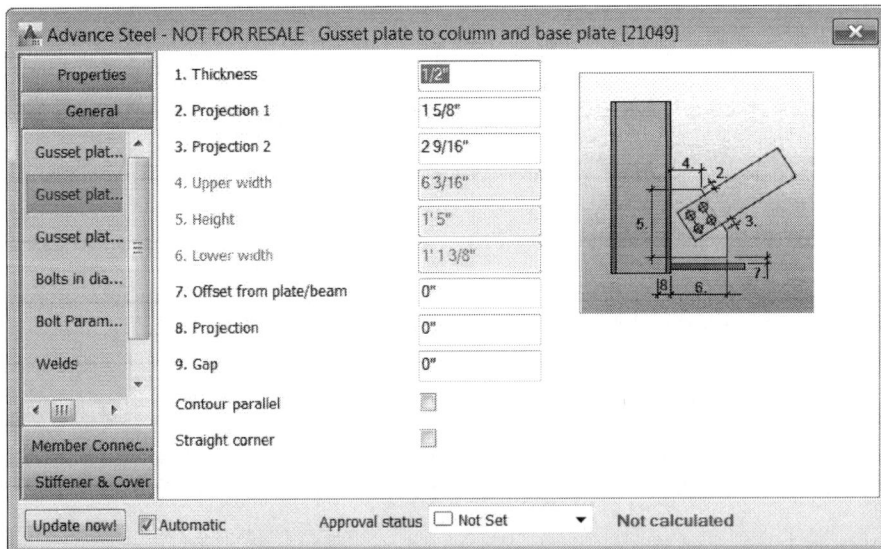

Figure 5 The Gusset plate parameters tab of the General category

Thickness
This edit box is used to specify the thickness of the gusset plate.

Projection 1
This edit box is only available when the **projection** is selected from the **Distance layout** list from the **Gusset plate shape** tab. In that case, you can specify the distance between the bracing section and the top edge of the gusset plate in this edit box. This dimension is labeled as **2** in the preview window on the right side of the dialog box.

Projection 2
This edit box is only available when the **Align to base plate** tick box is not selected from the **Gusset plate shape** tab. In that case, you can specify the distance between the bracing section the and right edge of the gusset plate in this edit box. This dimension is labeled as **3** in the preview window on the right side of the dialog box.

Upper width
This edit box is used to specify the upper width of the gusset plate. This dimension is labeled as **4** in the preview window on the right side of the dialog box.

Height
This edit box is only available when you select **variable** from the **Shape at main beam** list from the **Gusset plate shape** tab. In that case, you can enter the height of the gusset plate in this edit box. This dimension is labeled as **5** in the preview window on the right side of the dialog box.

Lower width
This edit box is only available when you select **variable** from the **Shape at base plate** list from the **Gusset plate shape** tab. In that case, you can enter the lower width of the gusset plate in this edit box. Figure 6 shows the gusset plate with the lower width value manually defined and Figure 7 shows the lower width aligned to the base plate.

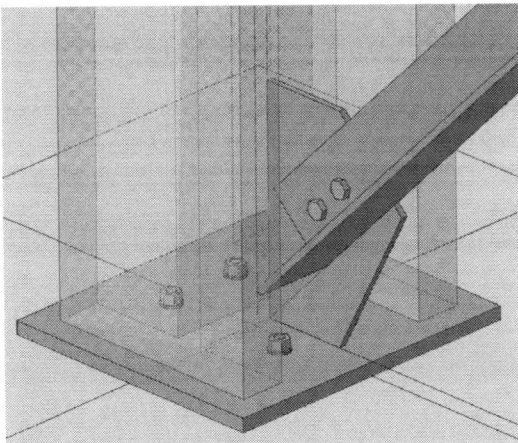

Figure 6 The lower width of the gusset plate entered as a manual value

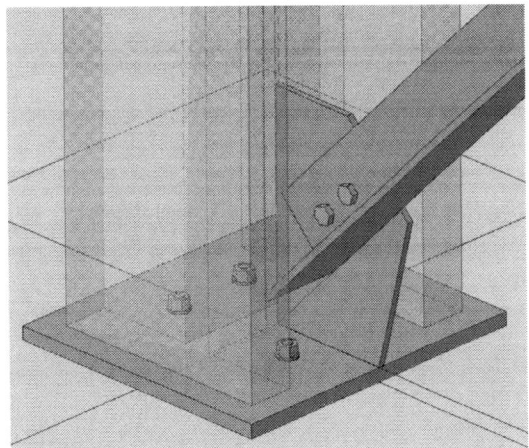

Figure 7 The lower width of the gusset plate aligned to the base plate

Offset from plate/beam

This edit box is used to enter an offset value between the gusset plate and the base plate or the secondary beam.

Projection

This edit box is used to define the penetration of the gusset plate inside the column.

Gap

This edit box is used in combination with the **Projection** edit box and is used to define the gap between the gusset plate and the column opening for the penetration.

Contour parallel

Selecting this tick box will ensure the contour is parallel.

Straight corner

Selecting this tick box will insert the gusset plate with the straight corner. Figure 8 shows the other side view of the gusset plate with this tick box cleared and Figure 9 shows the same gusset plate with this tick box selected.

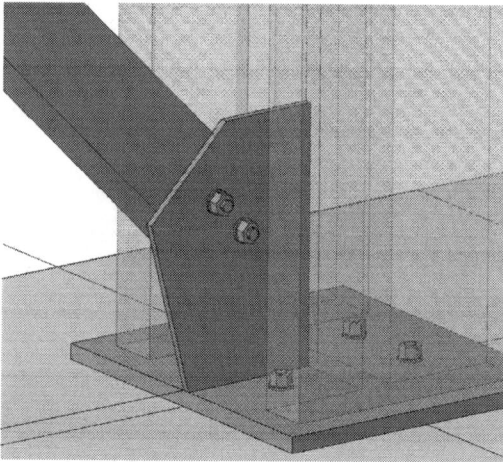

Figure 8 The gusset plate inserted without the straight corner

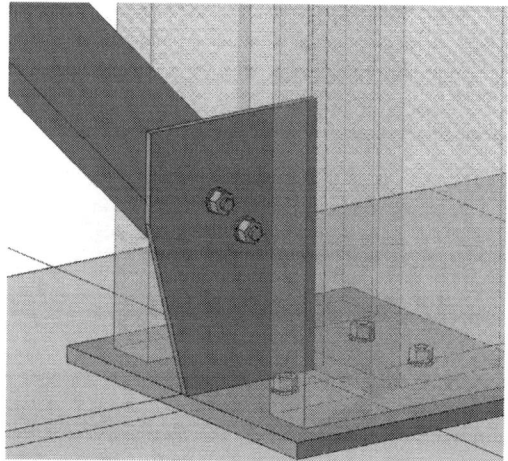

Figure 9 The gusset plate inserted with the straight corner

General Category > Gusset plate contour Tab

The **Gusset plate contour** tab, shown in Figure 10, is used to define the parameters related to the gusset plate contour. You can select the contour from the **Corner type** list. By default, **none** is selected from this list, As a result, no contour is created. Selecting **corner finish** from this list allows you to insert a corner finish. The corner finish type can be selected from the **Corner finish** list and the related parameters can be entered in their respective edit boxes in this tab. Selecting **contour** from the **Corner type** list allows you to insert a contour. You can select the contour type from the **Contour type** list and then enter the contour parameters in their respective edit boxes in this tab.

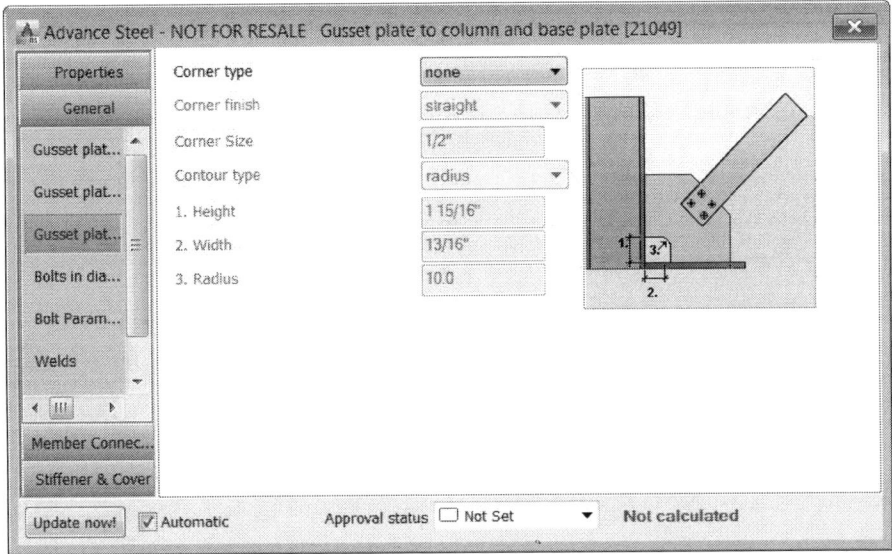

Figure 10   The **Gusset plate contour** tab of the **General** category

General Category > Bolts in diagonal Tab

The **Bolts in diagonal** tab, shown in Figure 11, is used to define the options related to the bolts on the bracing section. These options are discussed next.

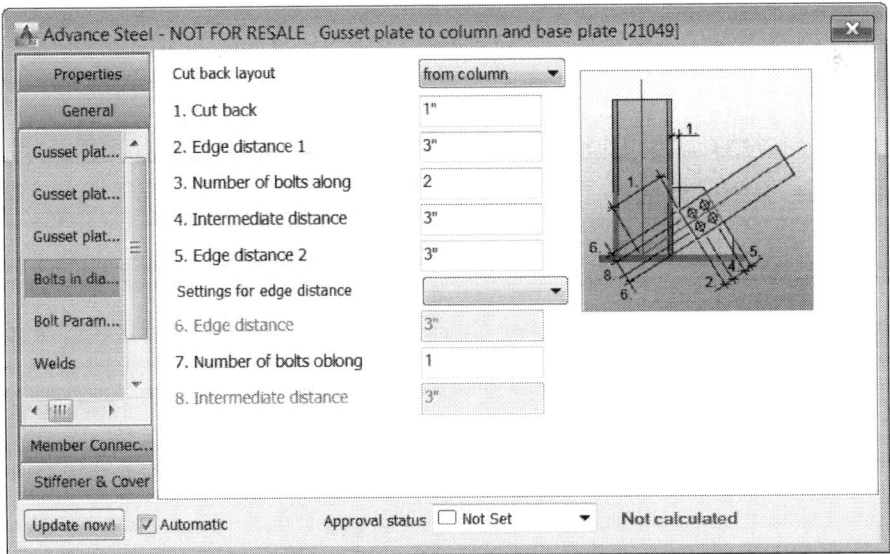

Figure 11   The **Bolts in diagonal** tab of the **General** category

Cut back layout

This list is used to specify whether the bracing cutout will be defined from the column face of the endpoint of the column system axis.

Cut back

This edit box is used to enter cut back value for the bracing section. Figure 12 shows the bracing section with 1" or 25mm cut back value and Figure 13 shows the bracing section with 5" or 125mm cut back value.

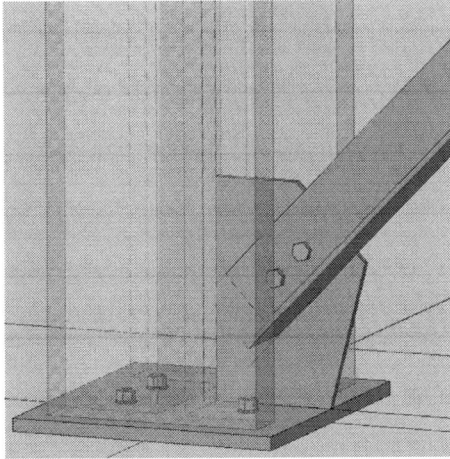

Figure 12   The joint with 1" or 25mm cut back value

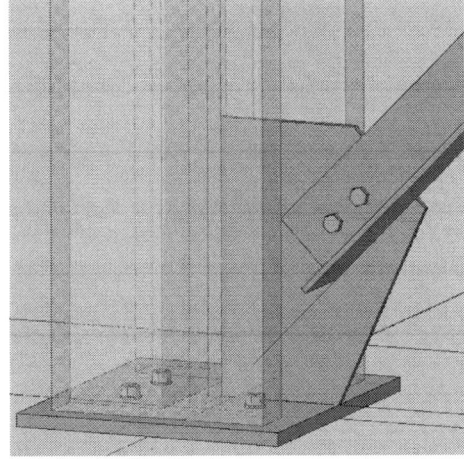

Figure 13   The joint with 5" or 125mm cut back value

Edge distance 1

This edit box is used to enter the distance between the end of the bracing section and the center of the first bolt.

Number of bolts along

This edit box is used to specify the number of bolts to be inserted along the bracing section. Figure 14 shows the joint with 4 bolts along the bracing section.

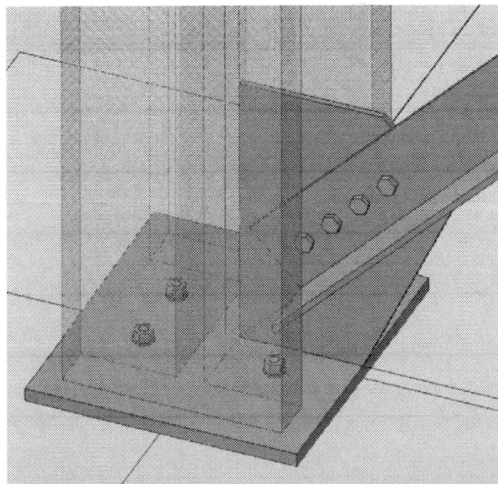

Figure 14 The joint with 4 bolts along the bracing section

Intermediate distance
This edit box is used to enter intermediate distance between the bolt centers along the bracing section.

Edge distance 2
This edit box is used to enter distance between the end of the gusset plate and the center of the last bolt.

Settings for edge distance
This list is used to specify where the bolt centers will be inserted. Selecting **center line** from this list ensures the bolt centers are inserted at the center line of the bracing face. Selecting **gauge line** ensures the bolt centers are inserted at the gauge line. Selecting **edge** from this list will allow you to enter the distance between the bracing edge and the bolt centers.

Number of bolts oblong
This edit box is used to specify the columns of bolts to be inserted in the other direction. Figure 15 shows the bracing joint with 2X2 bolt matrix created by inserting two bolts along the bracing section and two bolt columns inserted in the other direction.

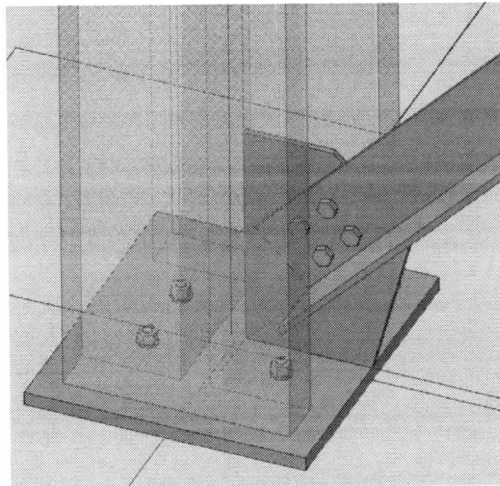

Figure 15 *The bracing joint with 2X2 bolt matrix*

Intermediate distance
This edit box is used to enter the intermediate distance between the bolt centers in the other direction.

The remaining tabs and categories are similar to those discussed in earlier chapters.

Gusset plate at one diagonal Joint

Gusset plate at o...

This type of joint is similar to the previous joint, with the difference that in this joint you do not need to select the base plate or the secondary beam. Figure 16 shows this type of joint.

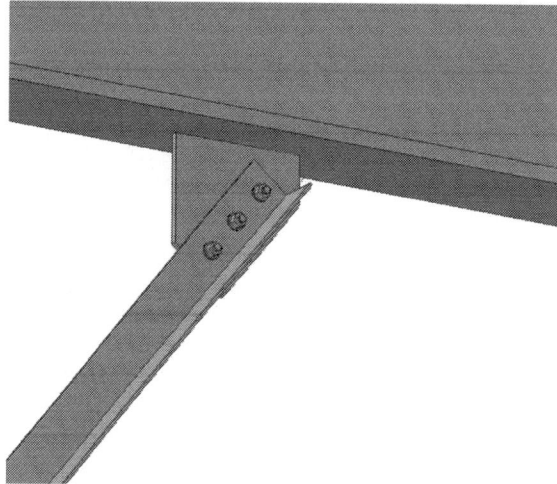

Figure 16   *The **Gusset plate at one diagonal** joint*

Gusset plate for 2 diagonals Joint

Gusset plate for...

This type of joint is used to connect two bracing sections to a beam using a gusset plate. Figure 17 shows this type of joint.

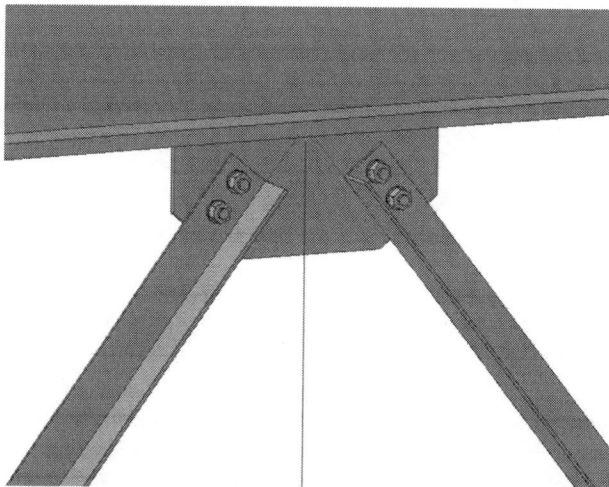

Figure 17   *The **Gusset plate for 2 diagonals** joint*

> *Tip*: In some joints, such as the **Gusset plate for 2 diagonals** joint, do not provide the option to invert the bolts. In that case, you can double-click on the bolts and then select the **Inverted** tick box from the **Definition** tab of the dialog box.

Four diagonals - Middle gusset plate Joint

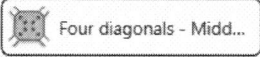

| Four diagonals - Midd... |

This type of joint is used to connect four diagonal bracing sections using a gusset plate at the center. When you invoke this tool, you will be prompted to select the four diagonals one by one. Remember that you need to press ENTER after selecting every diagonal section. Once you press ENTER after selecting the fourth diagonal, the **Advance Steel - Middle gusset - 4 diagonals** dialog box will be displayed. The options in this dialog box are similar to those discussed earlier in this chapter. Figure 18 shows this type of joint.

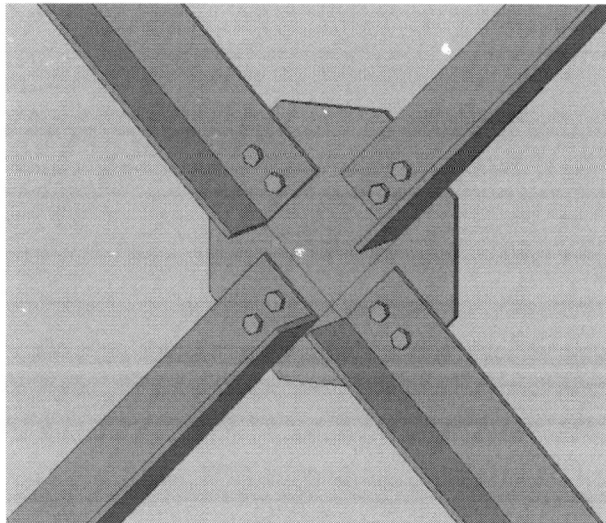

Figure 18 The *Four diagonals - Middle gusset plate* joint

INSERTING TUBE JOINTS

The **Tube connections** category lists various joints that can be added to the hollow section, such as circular hollow sections, rectangular hollow sections, and so on. Note that most of the joints in this category are similar. Therefore, only certain joint types are discussed in detail.

Tube connection with sandwich plate - additional objects Joint

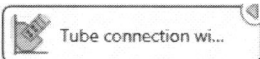

| Tube connection wi... |

This type of joint is used to connect a tube section to a column and a base plate or to a column and a secondary beam using a gusset plate. The gusset plate, in turn, is connected to the bracing section using sandwich plates. Figure 19 shows this type of joint and various items in it. These items are explained below the figure.

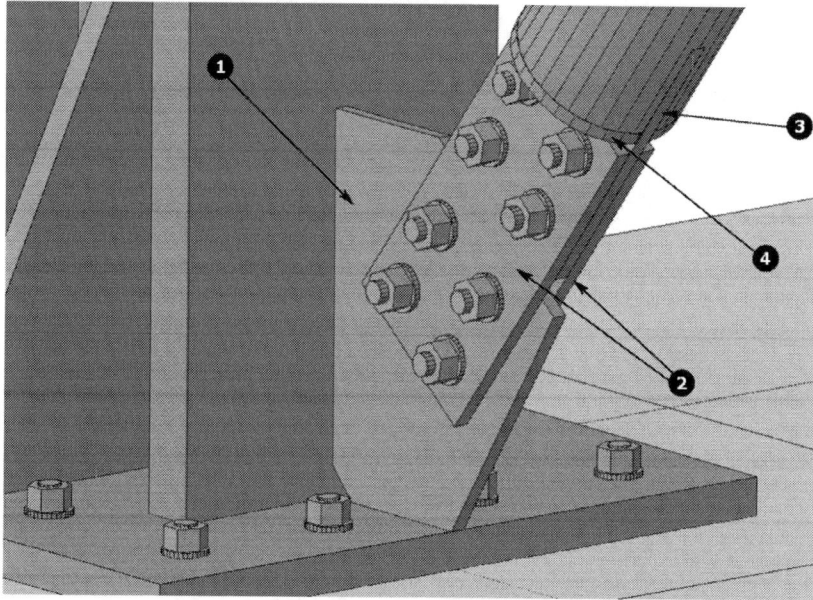

*Figure 19 The **Tube connection with sandwich plate - additional object** joint*

Item 1	**Gusset Plate**
Item 2	**Sandwich Plates**
Item 3	**Tab Plate**
Item 4	**Cover Plate**

When you invoke this tool, you will be prompted to select the column beam. Once you select the column and press ENTER, you will be prompted to select the secondary beam or the base plate. On making this selection and pressing ENTER, you will be prompted to select the diagonal beam, which is the bracing section. Once you select the bracing section and press ENTER, the **Advance Steel - Tube connection with sandwich plates [X]** dialog box will be displayed. Most of the categories and tabs in this dialog box are similar to those discussed earlier in this chapter. The remaining tabs are discussed next.

General Category > Sandwich plates Tab

The **Sandwich plates** tab, shown in Figure 20, is used to specify the options related to the connection between the sandwich plates and the tab plate. These options are discussed next.

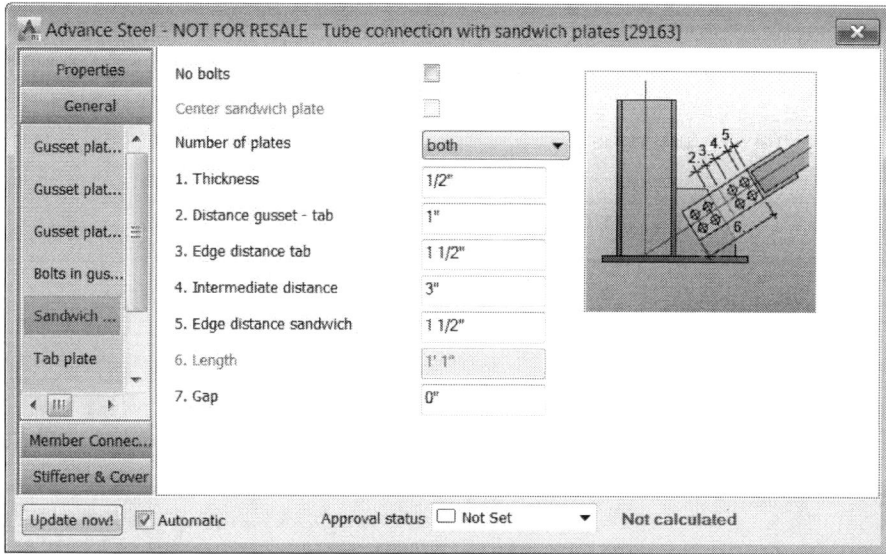

*Figure 20 The **Sandwich plates** tab of the **General** category*

No bolts

Selecting this tick box ensures no bolts are inserted between the sandwich plates and tab plates. Figure 21 shows the sandwich plates with this tick box cleared and Figure 22 shows the sandwich plates with this tick box selected.

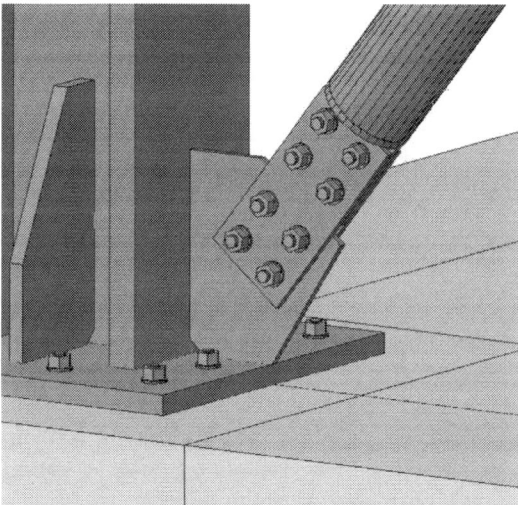

Figure 21 Sandwich plates with bolts inserted

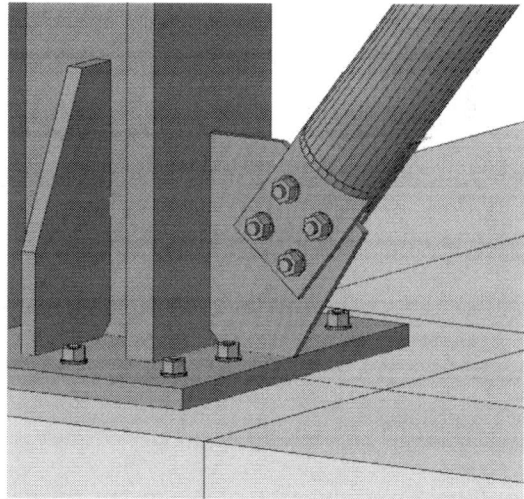

Figure 22 Sandwich plates with bolts not inserted

Center sandwich plate

This tick box is selected to ensure the sandwich plate is centered.

Number of plates

This list is used to specify whether the sandwich plates are to be inserted on the front, back, or both sides of the gusset plate. You can also select **none** from this list to ensure that the sandwich plates are not inserted.

Thickness

Enter the thickness of the sandwich plate in this edit box.

Distance gusset - tab

This edit box is used to specify the distance between the gusset plate and the tab plate.

Edge distance tab

This edit box is used to specify the distance between the edge of the tab plate and the center of the first bolt column connecting the sandwich and tab plates.

Intermediate distance

This edit box is used to specify the intermediate distance between the center of the bolts connecting the sandwich plate to the tab plate.

Edge distance sandwich

This edit box is used to specify the distance between the edge of the sandwich plate and the center of the last bolt column connecting the sandwich and tab plates.

Length

This edit box is used to specify the length of the sandwich plates.

Gap

This edit box is used to specify the gap between the sandwich plates and the tab plate.

General Category > Tab plate Tab

The **Tab plate** tab, shown in Figure 23, is used to specify the options related to tab plate. These options are discussed next.

Create plate

This tick box is only available when you select the **No bolts** tick box from the **Sandwich plate** tab. In that case, clearing this tick box will not create the tab plate.

Type

This list is used to specify the type of the tab plate. By default, **Pipe slotted** is selected from this list. As a result, a slot is created in the hollow section and the tab plate is welded to that slot. If you select **tab plate slotted**, the tab plate will terminate on the end face of the hollow section. If you select **tab plate inside**, the tab plate will be welded on the inside of the hollow section.

Thickness

This edit box is used to specify the thickness of the tab plate.

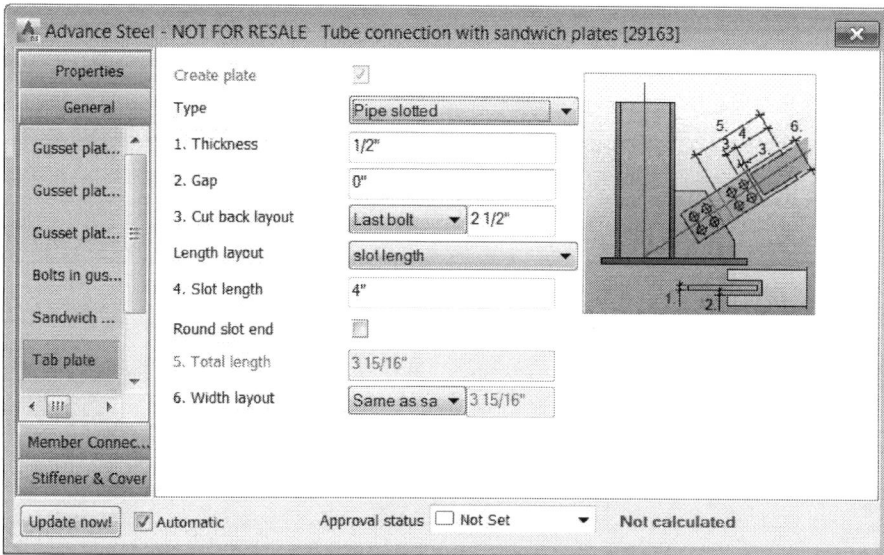

*Figure 23 The **Tab plate** tab of the **General** category*

Gap

This edit box is used to specify the gap between the tab plate and the hollow section. Figure 24 shows the tab plate with a 1/2" or 12mm gap.

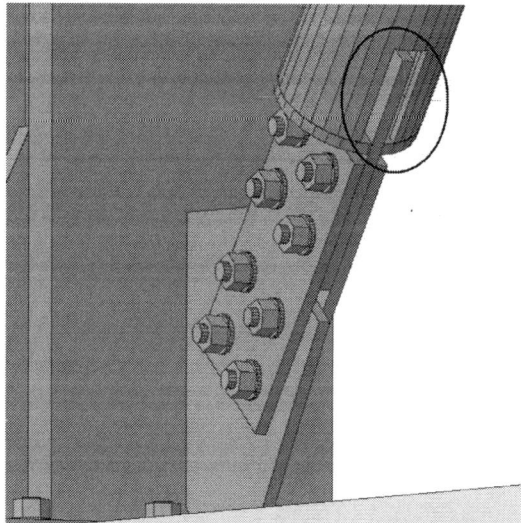

Figure 24 The tab plate with a gap of 1/2" or 12mm

Cut back layout

This list is used to specify the cut back layout method for the tab plate. By default, the **Last bolt** option is selected. As a result, the tab plate cut back is measured from the center of the last bolt on the sandwich plate. The distance value is entered in the edit box on the right of this list. You can also select **Plate** from this list to measure the cut back distance from the end of the sandwich plate.

Length layout

This list is used to specify how the length of the tab plate will be defined. Selecting **slot length** from this list allows you to define the length of the tab plate based on the slot length on the hollow section. The slot length is defined in the **Slot length** edit box below this list. Alternatively, you can select **total** from this list and then enter the total length of the tab plate in the **Total length** edit box.

Round slot end

Select this tick box to round the end of the slot on the hollow section.

Width layout

This list is used to specify how the width of the tab plate will be defined. By default, **same as sandwich** is selected from this list. As a result, the width of the tab plate is the same as the sandwich plate. Alternatively, you can select **total** from this list and enter the total width of the tab plate in the edit box on the right of this list.

The remaining tabs and categories are similar to those discussed in earlier chapters.

Tube connection with sandwich plates Joint

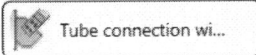

Tube connection wi...

This type of joint is similar to the previous joint, with the difference that in this joint you do not need to select the base plate or the secondary beam. Figure 25 shows this type of joint at the bottom face of a beam.

*Figure 25 The **Tube connection with sandwich plate** joint*

Tube connection with sandwich plates - 2 diagonals Joint

Tube connection wi...

This type of joint is used to connect two hollow sections to a beam using a gusset plate. Figure 26 shows this type of joint at the bottom face of a beam.

*Figure 26 The **Tube connection with sandwich plate - 2 diagonals** joint*

INSERTING MISCELLANEOUS JOINTS

The **Miscellaneous** category lists a number of miscellaneous joints that can be added to stairs, handrail posts, joist seats, and so on. Some of these joints are discussed next.

Stair Anchor Base Plate Joint

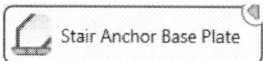

Stair Anchor Base Plate

This type of joint is used to create a base plate at the end of the selected stringer of a stair. It is important to note that for this joint to work properly, the UCS needs to be set to the World position. This joint also provides you an option of creating a vertical extension to connect to the base slab. When you invoke this tool, you will be prompted to select a stringer. On selecting the stringer and pressing ENTER, you will be prompted to specify whether or not you want to select a reference point. On specifying the option in this prompt, the **Advance Steel - Stair anchor base plate [X]** dialog box will be displayed and the joint with the default values will be created. Some of the categories and tabs in this dialog box are similar to those discussed in earlier chapters. The rest of the tabs are discussed next.

General Category > Vertical extension Tab

The **Vertical extension** tab, shown in Figure 27, is used to specify the options related to creating a vertical profile at the end of the stringer. These options are discussed next.

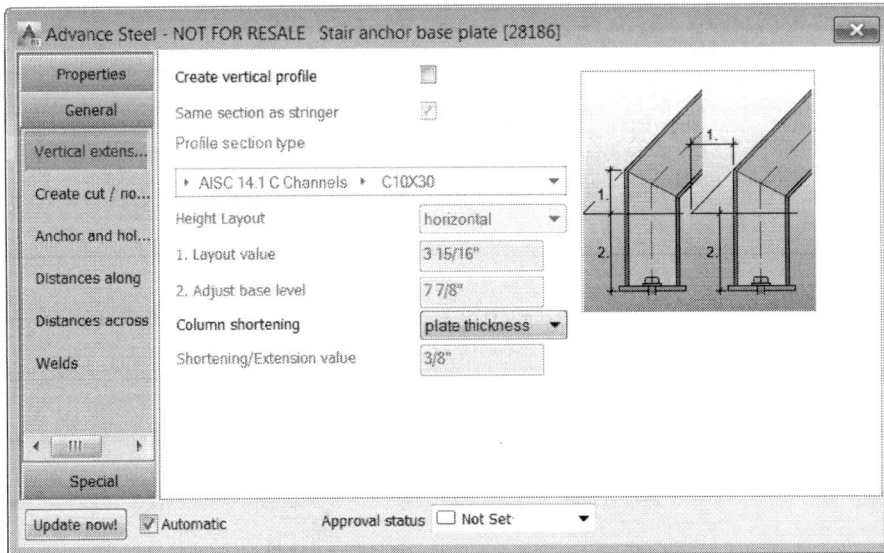

Figure 27 *The* **Vertical extension** *tab of the* **General** *category*

Create vertical profile

This tick box is used to specify whether or not the vertical profile will be created at the end of the stringer. By default, this tick box is not selected. As a result, the vertical profile is not created, as shown in Figure 28. If you select this tick box, the vertical profile will be created, as shown in Figure 29.

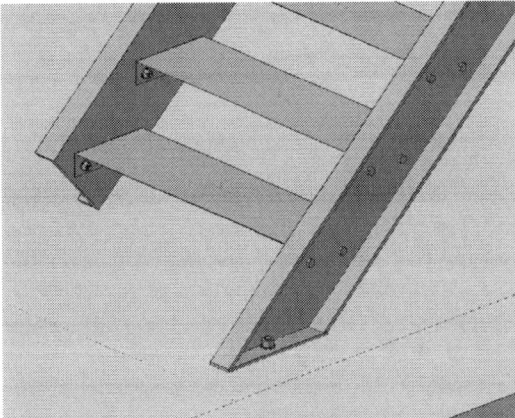

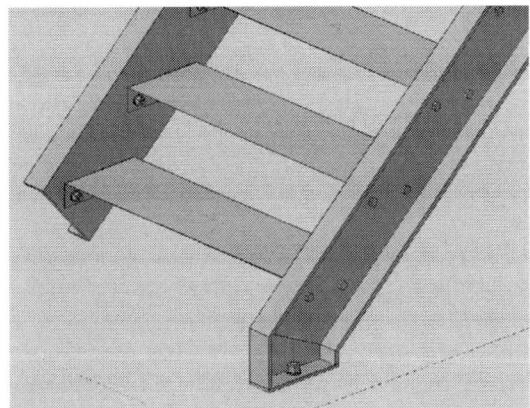

Figure 28 *No vertical profile created at the end of the stringer*

Figure 29 *Vertical profile created at the end of the stringer*

Same section as stringer

Select this tick box to insert the vertical profile with the same section as the stringer. If you clear this tick box, you can select the section for the vertical profile using the **Profile section type** list available below this tick box.

Height Layout

This list is used to specify whether the height layout of the vertical profile will be measured horizontally or vertically from the stringer. The layout value can be entered in the **Layout value** edit box available below this list. Figure 30 shows the vertical profile with 6" or 150mm as the vertical layout value and Figure 31 shows the vertical profile with 9" or 230mm as the vertical layout value.

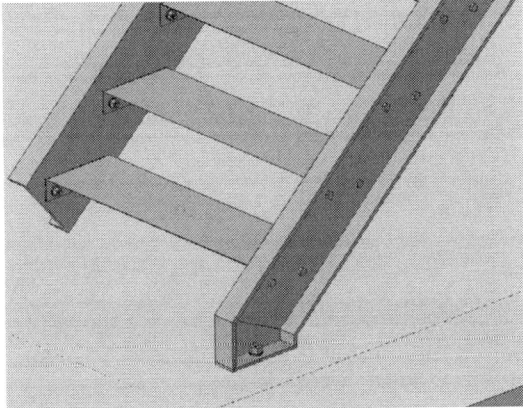

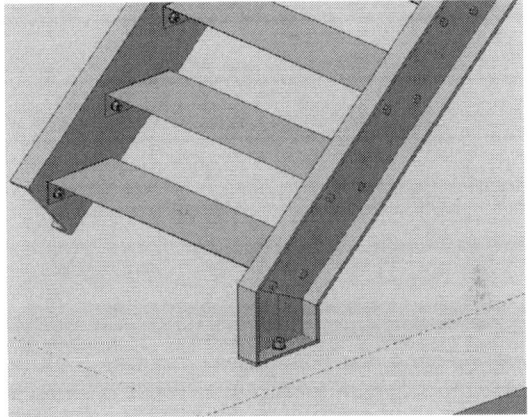

Figure 30 Vertical layout value set to 6" or 150mm   *Figure 31* Vertical layout value set to 9" or 230mm

Adjust base level

This edit box is used to specify the height of the vertical profile from the end of the height layout to the bottom of the base plate.

*Tip: If the stair you have created needs to be connected to the ground slab, setting the **Adjust base level** value to **0** will automatically align the base plate on top of the ground slab.*

Column shortening

This list is only available if you clear the **Create vertical profile** tick box at the top of this tab. In that case, you can select the column shortening method from this list. By default, **none** is selected from this list. As a result, the column is not shortened. Selecting **plate thickness** from this list will shorten the column by the value equal to the plate thickness. You can also select **value** from this list and then enter the shortening value in the **Shortening/Extension value** edit box available below this list.

General Category > Create cut / nose plate Tab

The options in the **Create cut / nose plate** tab, shown in Figure 32, are only available if you clear the **Create vertical profile** tick box at the top of the **Vertical extension** tab. The options in this tab are used to specify the parameters related to the stringer cut and the nose plate. These options are discussed next.

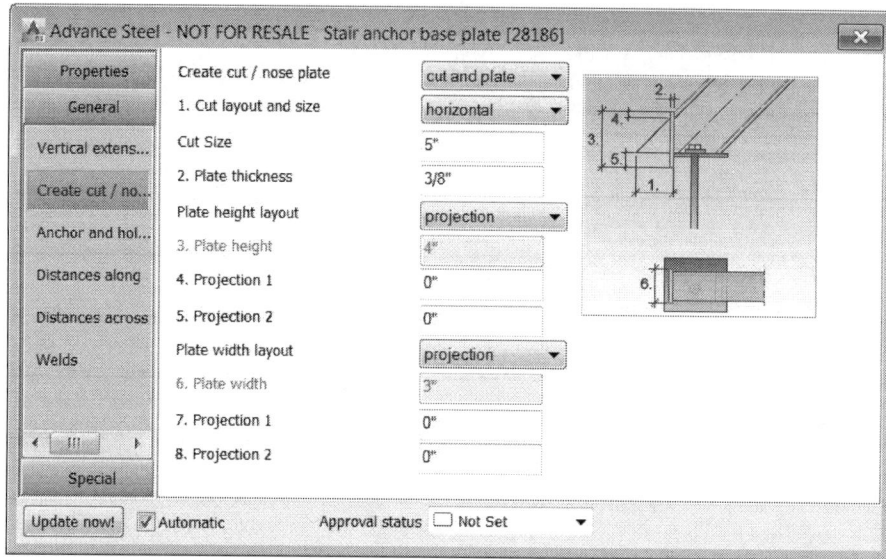

*Figure 32 The **Create cut / nose plate** tab of the **General** category*

Create cut / nose plate List

This list is used to specify whether the cut and the nose plate will be created or not. If you select **cut only** from this list, only the stringer cut will be created, as shown in Figure 33. However, if you select **cut and plate** from this list, the nose plate will also be created, as shown in Figure 34.

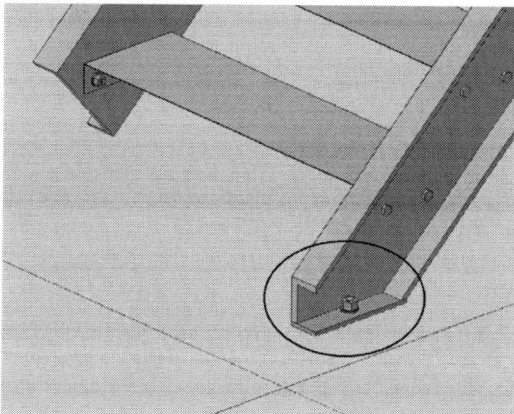

Figure 33 Cut only created on the stringer

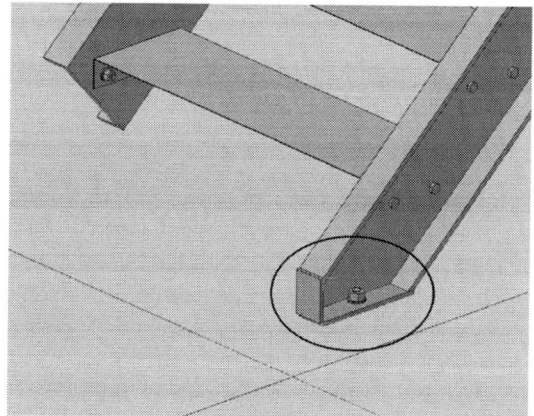

Figure 34 Cut and nose plate created

Cut size

This edit box is used to specify the size of the cut.

Plate thickness

This edit box is used to specify the thickness of the plate.

Plate height layout

This list is used to specify the layout for the plate height. By default, **projection** is selected from this list. As a result, the plate height is defined in terms of the projection values entered in the edit boxes available below this list. If you select **total**, you can specify the total height of the plate in the edit box available below this list.

Plate width layout

This list is used to specify the layout for the plate width. By default, **projection** is selected from this list. As a result, the plate width is defined in terms of the projection values entered in the edit boxes available below this list. If you select **total**, you can specify the total width of the plate in the edit box available below this list.

The remaining tabs and categories in this joint are the same as those discussed in the earlier chapters.

Stair Anchor Angle Joint

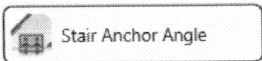

This type of joint is used to connect the selected stringer of the stair to the base using one or two angles. The angle can be welded to the stringer or bolted. You can specify whether or not you want to create the vertical profile. You can also specify whether the angle will be created on one side of the stringer or both sides. Figure 35 shows this type of joint without the vertical profile and Figure 36 shows this type of joint with the vertical profile.

Figure 35 The joint without the vertical profile

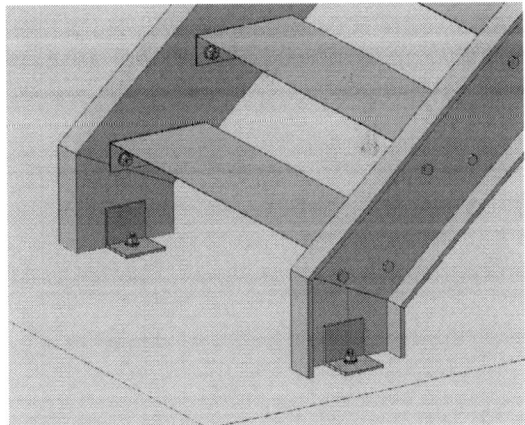

Figure 36 The joint with the vertical profile

Railing joint handrail Joint

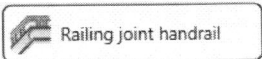

This type of joint is used to connect two handrailings together. You can define the connection to be miter cut or knee joint. Note that if the handrailing you are connecting has more than one rails, such as the top handrail and the middle handrail then they need to be connected individually using two separate joints. Figure 37 shows the two separate handrailings and Figure 38 shows the handrails connected together. Note that in this figure, the column visibility is changed to symbol.

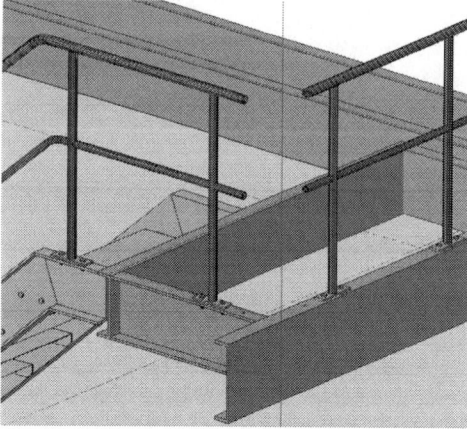

Figure 37 *Before connecting the handrails*

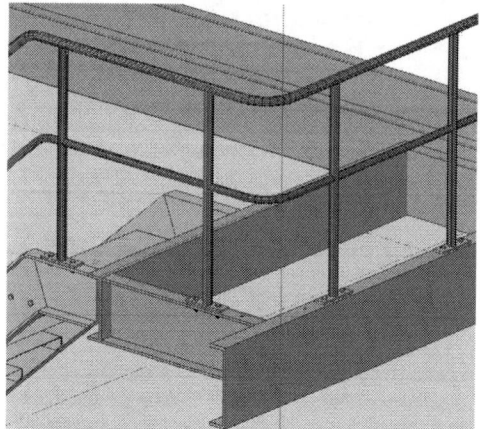

Figure 38 *After connecting the handrails*

Hands-on Tutorial (STRUC/ BIM)	In this tutorial, you will complete the following tasks: 1. Open the file containing the frames. 2. Insert vertical bracing joint and then create multiple copies of it, as shown in Figures 39 and 40. 3. Create the four diagonals bracing joint, as shown in Figure 41. 4. Open the elevator tower file. 5. Insert the tube joints, as shown in Figures 42 and 43. 6. Insert stairs anchor joints, as shown in Figure 44. 7. Insert the shear plate joint between the stair top landing and the beam, as shown in Figure 45.

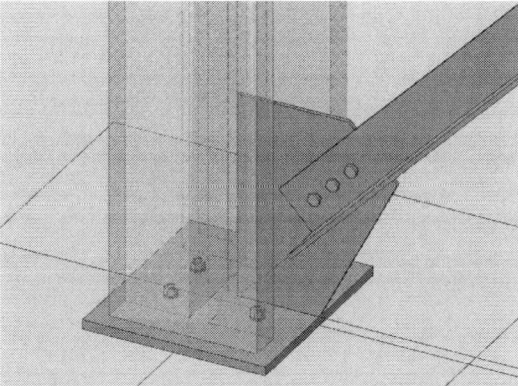

Figure 39 *Bracing joint at the column base*

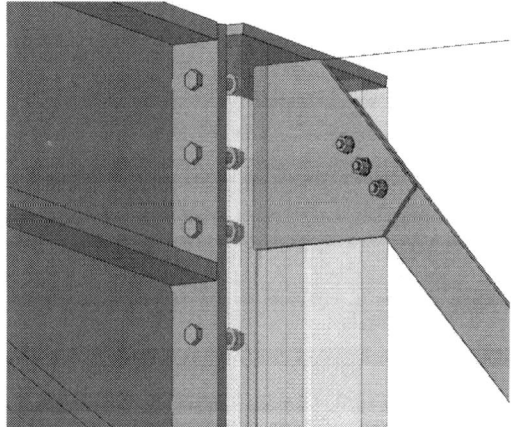

Figure 40 *Bracing joint at the top of the column*

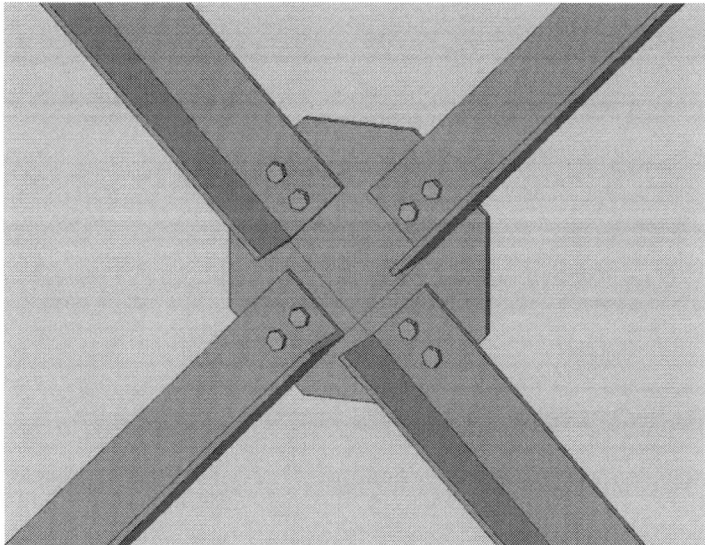

Figure 41 *Four diagonal bracing joint*

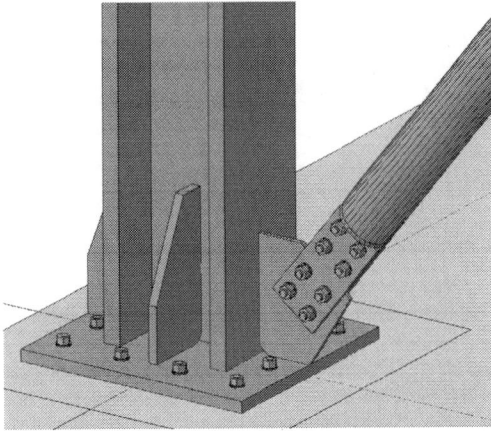

Figure 42 *Tube joint at the column base*

Figure 43 *Tube joint connecting two round bars to the beam*

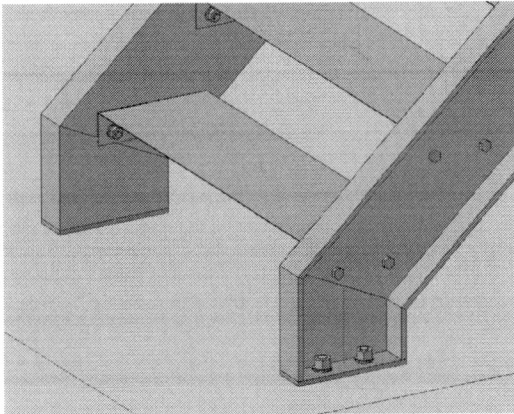

Figure 44 *Stair anchor base joint*

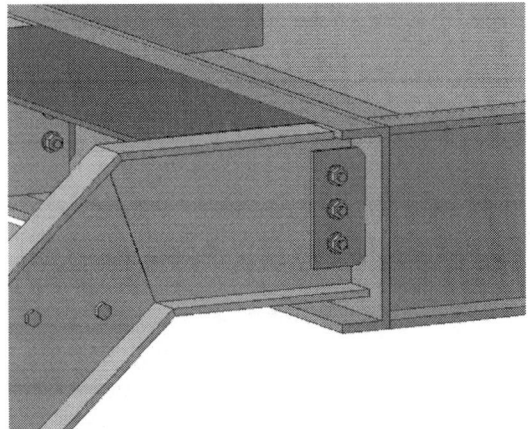

Figure 45 *Stair-beam shear plate joint*

Bottom Bracing Joint Parameters

The following are the bracing joint parameters. The dimensions are provided in both Imperial and Metric units. The dimensions not listed below need to be taken as the default value. **Note that if you are using the Metric units, do not enter the unit string (mm) while typing the dimension values.**

Shape at beam/plate	**Variable**
Align to base plate	**Yes**
Gusset Plate Thickness	**1/2" or 12 mm**
Projections 1 and 2	**1" or 25 mm**
Corner Finish	**Straight**
Corner Size	**1" or 25 mm**
Cut Back from Column	**4" or 100 mm**
Edge Distance 1	**2" or 50 mm**

Number of Bolts Along	3
Intermediate distance	3" or 75 mm
Edge Distance 2	2" or 50 mm
Bolt Diameter	3/4" or 20 mm

Top Bracing Joint Parameters

Most of the parameters of this joint are the same as the bottom bracing joint. The values that are different are listed below.

Align to base plate	No
Lower Width	8 1/2" or 215 mm
Cut Back	8" or 200 mm

Four diagonals - Middle gusset plate Joint Parameter

The following are the parameters required to insert this joint.

Plate Thickness	1/2" or 12 mm
Gap Between Diagonals	2" or 50 mm
Projection	1" or 25 mm
Bolts Diameter	3/4" or 20 mm
Distance diagonal - bolt	1 1/2" or 35 mm
Number of Lines of Bolts	2
Intermediate Distance	3" or 75 mm
Distance bolt - gusset	1 1/2" or 35 mm
Settings for the Edge distance	Center line
Edge distance	0
Number of Columns of Bolts	1

Tube Connection with Column and Gusset Plate Joint Parameters

The following are the parameters required for the tube connection with column and gusset plate.

Gusset Plate Thickness	3/4" or 20 mm
Projections	1/2" or 12 mm
Gusset Plate Corner	1" or 25 mm
Cut back from Column	4" or 100 mm
Edge Distance Sandwich	1 1/2" or 35 mm
Number of Bolts Along	2
Intermediate Distance	3" or 75 mm
Edge Distance Gusset	1 1/2" or 35 mm
Edge Distance	1 1/2" or 35 mm
Number of Bolts Oblong	2
Intermediate Distance	3" or 75 mm
Sandwich Plate Thickness	3/4" or 20 mm
Distance Gusset - Tab	1" or 25 mm
Edge Distance Tab	1 1/2" or 35 mm
Intermediate Distance	3" or 75 mm
Edge Distance Sandwich	1 1/2" or 35 mm

Tab Plate	**Pipe Slotted**
Thickness	**3/4" or 20 mm**
Gap	**0**
Cut Back Last bolt	**3" or 75 mm**
Slot Length	**4" or 100 mm**
Width Layout	**Same as Sandwich**
Bolts Diameter	**3/4" or 20.00 mm**
Cover Plate	**Cover Slotted**
Thickness	**3/4" or 20 mm**
Projection	**0**

Tube Connection with Two Diagonals Joint Parameters

The following are the parameters required for this joint.

Gusset Plate Thickness	**3/4" or 20 mm**
Cut back from Column	**2" or 50 mm**
Edge distance sandwich	**1 1/2" or 35 mm**
Number of Bolts Along	**2**
Intermediate Distance	**3" or 75 mm**
Edge Distance Gusset	**3" or 75 mm**
Edge Distance	**1 1/2" or 35 mm**
Number of Bolts Oblong	**2**
Intermediate Distance	**4" or 100 mm**
Gusset Bolts Diameter	**3/4" or 20.00 mm**
Sandwich diagonal 1	**No bolt Cleared**
Number of Plates	**Both**
Thickness	**3/4" or 20 mm**
Distance Gusset - Tab	**1" or 25 mm**
Edge Distance Tab	**3" or 75 mm**
Intermediate Distance	**3" or 75 mm**
Edge Distance Sandwich	**3" or 75 mm**
Tab Plate Type	**Pipe Slotted**
Thickness	**3/4" or 20 mm**
Gap	**0**
Cut back from Last bolt	**3" or 75 mm**
Length layout from Slot Length	**4" or 100 mm**
Width Layout	**Same as Sandwich**
Cover Plate	**Cover Slotted**
Thickness	**3/4" or 20 mm**
Projection	**0**

Stair Anchor Base Plate Joint Parameters

The following are the parameters required for this joint.

Create Vertical Profile	**Yes**
Height Layout	**Horizontal**
Layout Value	**9" or 225**
Adjust Base Level	**0**
Anchor type	**US Normal Anchors** (for Imperial) or **HILITI HAS** (for Metric)
Anchor Diameter	**5/8 inch** (for US Normal Anchors) or **16.00 mm** (For HILITI HAS Anchors)
Anchor Length	**8"** (for US Normal Anchors) or **7 1/2"** (For HILITI HAS Anchors)
Plate Thickness	**3/8" or 10 mm**
Plate Distance Layout	**Projections**
Projections	**0**
Bolt Reference From	**Centered**
Number of Bolts	**2**
Intermediate Distance	**4" or 100 mm**
Distance Across Bolt	
Reference From	**Plate Center**
Number of Bolts	**1**
Offset from Center	**0**

Section 1: Inserting and Copying Bracing Joints

In this section, you will open the **C07-Imperial-Frame.dwg** or the **C07-Metric-Frame.dwg** file and then insert bracing joints. You will then create multiple copies of those joints in the model.

1. From the **C07 > Struc-BIM** folder, open the **C07-Imperial-Frame.dwg** or **C07-Metric-Frame.dwg** file based on the preferred units.

   This model is similar to the one you created in the previous chapter.

2. Zoom to the base of the column at the **B1** grid intersection point, as shown in Figure 46. This figure also shows the selection sequence to create the joint.

3. From the **Connection vault** palette > **General bracing** category, invoke the **Gusset plate to column and base plate** joint tool, which is the first button at the top; you are prompted to select the column beam.

4. Select the column at the **B1** grid intersection point labeled as **1** in Figure 46 and press ENTER; you are prompted to select the secondary beam/base plate.

5. Select the base plate labeled as **2** in Figure 46 and press ENTER; you are prompted to select the diagonal beam.

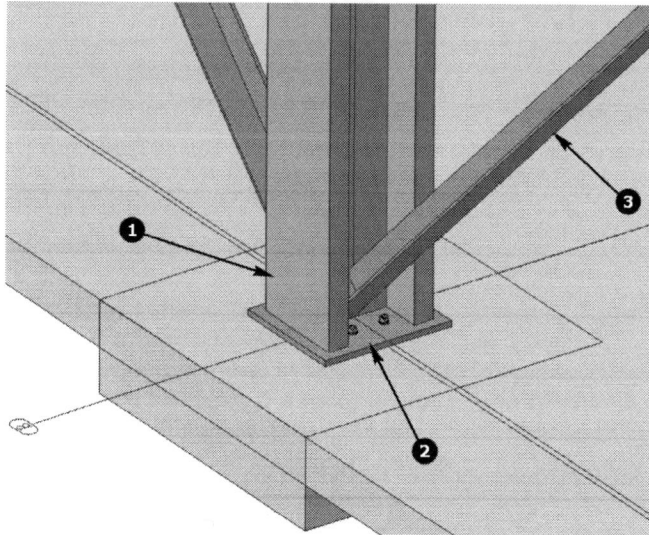

Figure 46 *The selections for the bracing joint*

6. Select the bracing member labeled as **3** in Figure 46 and press ENTER; the joint with the default values is created, as shown in Figure 47, and the **Advance Steel - Gusset plate to column and base plate [X]** dialog box is displayed.

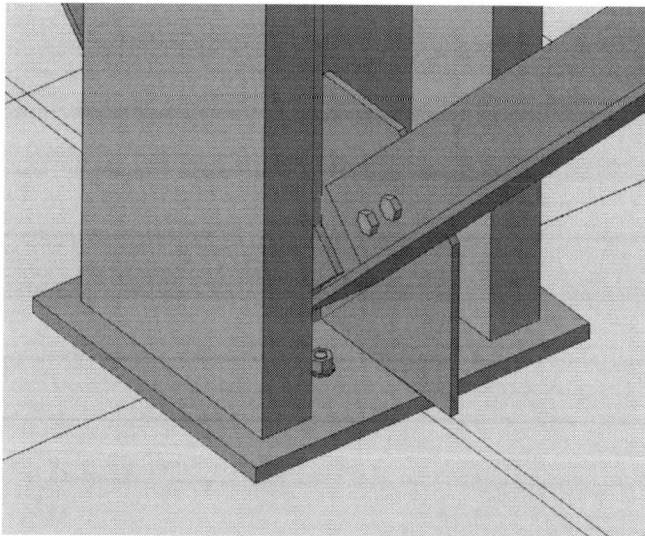

Figure 47 *The joint with the default values*

Note: *The default joint you see in your model may be different from the one shown in Figure 47. Once you change the parameters, the joint will match the one in the book.*

You will now modify the joint parameters to suit your requirements.

7. From the left pane of the dialog box, select the **General** category; the **Gusset plate shape** tab is activated as the default tab.

8. From the **Shape at beam/plate** list, select **variable**.

9. Select the **Align to base** (Imperial) or **Align to base plate** (Metric) tick box.

10. Activate the **General** category > **Gusset plate parameter** tab.

11. In the **Thickness** edit box, enter **1/2"** or **12**.

12. In the **Projection 1** edit box, enter **1"** (for Imperial) or **25** (for Metric).

13. In the **Projection 2** edit box, enter **1"** (for Imperial) or **25** (for Metric).

14. Activate the **General** category > **Gusset plate contour** tab.

15. From the **Corner type** list, select **corner finish**.

16. From the **Corner finish** list, select **straight**.

17. In the **Corner size** edit box, enter **1"** or **25**.

18. Activate the **General** category > **Bolts in diagonal** tab.

19. From the **Cut back layout** list, make sure **from column** is selected.

20. In the **Cut back** edit box, enter **4"** or **100**.

21. In the **Edge distance 1** edit box, enter **2"** or **50**.

22. In the **Number of bolts along** edit box, enter **3**; the number of bolts along the bracing change to 3 and the gusset plate is modified.

23. In the **Intermediate distance** edit box, enter **3"** or **75**.

24. In the **Edge distance 2** edit box, enter **2"** or **50**.

25. From the **Settings for edge distance** list, select **center line**.

26. In the **Number of bolts oblong** edit box, enter **1**.

27. Activate the **General** category > **Bolt Parameters** tab.

28. From the **Diameter** list, select **3/4"** or **20.00 mm**.

29. From the **Bolt Type** list, select **A325** (for Imperial) or **AS 1252** (for Metric).

This completes the editing of all the parameters of the joint. The joint, after updating all the parameters, is shown in Figure 48.

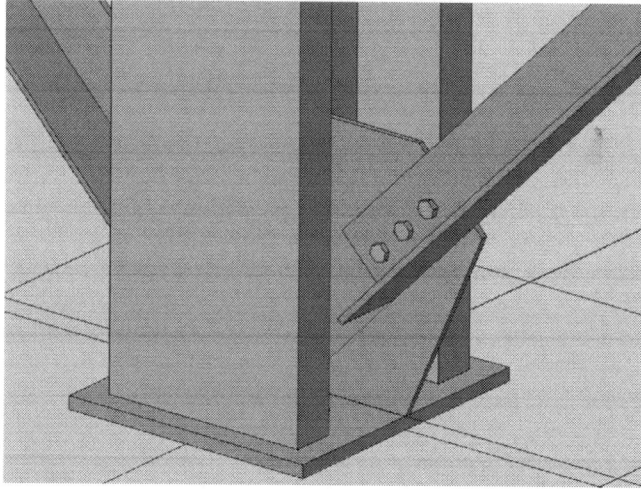

Figure 48 *The bracing joint after updating the parameters*

30. Save this joint in the library with the name **Frame-Bracing-Base**.

Next, you need to add a similar joint to the bracing at the top of the column. You will use the joint saved in the library and then modify the parameters to suit the requirements.

31. Zoom to the top of the column to which you added the joint in the earlier steps, as shown in Figure 49. This figure also shows the selection sequence to create the joint.

> *Tip: For clarity of the joint, the visibility of the eaves beam to the right of the column and the eaves beam joint is turned off using the **Advance Steel Tool Palette > Quick views > Selected objects off** tool. You will learn more about this tool in later chapters.*

32. From the **General bracing** category, invoke the **Gusset plate to column and base plate** joint tool, which is the first button at the top; you are prompted to select the column beam.

33. Select the column labeled as **1** in Figure 49 and press ENTER; you are prompted to select secondary beam/base plate.

34. Select the cover plate labeled as **2** in Figure 49 and press ENTER; you are prompted to select the diagonal beam.

35. Select the bracing member labeled as **3** in Figure 49 and press ENTER; the joint with the default values is created, and the **Advance Steel - Gusset plate to column and base plate [X]** dialog box is displayed.

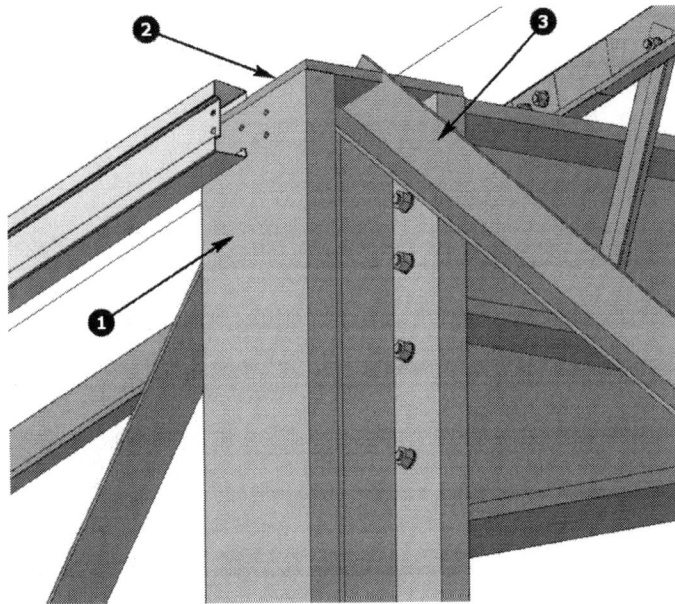

Figure 49 *The selections for the bracing joint*

36. From the **Library** tab, select the joint that you saved earlier the name **Frame-Bracing-Base**; the joint is modified to the parameters of the saved joint.

    Next, you need to modify some of these parameters.

37. Activate the **General** category > **Gusset plate shape** tab and make sure **variable** is selected from the **Shape at beam/plate** list.

38. Clear the **Align to base plate** tick box.

39. Activate the **General** category > **Gusset plate parameter** tab.

40. In the **Projection 2** edit box, enter **0**.

41. In the **Lower width** edit box, enter **8 1/2"** or **215**.

42. Activate the **General** category > **Bolts in diagonal** tab.

43. In the **Cut back** edit box, enter **8"** or **200**.

    This completes the modification of all the parameter of the joint. The updated joint is shown in Figure 50.

44. Save this joint in the library with the name **Frame-Bracing-Top**.

    Next, you will insert the four diagonal bracing joint.

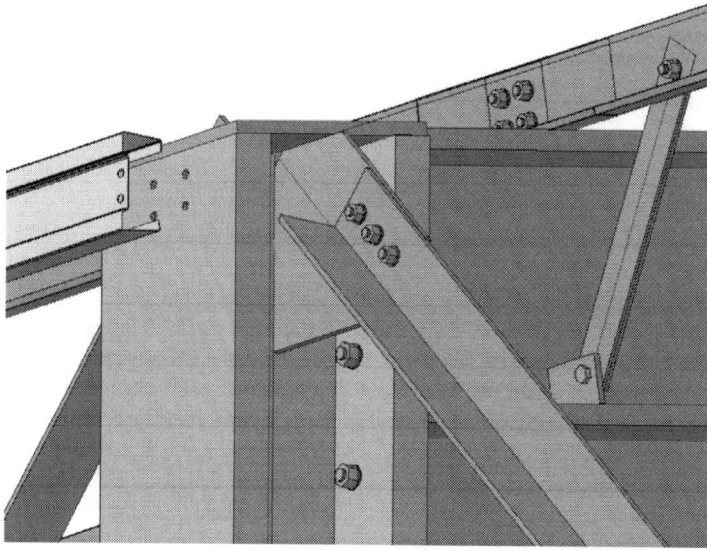

Figure 50 *The joint, after modifying the parameters*

45. Zoom close to the intersection of the four diagonal members at the center of the bracing, as shown in Figure 51. This figure also shows the selection sequence to create the joint.

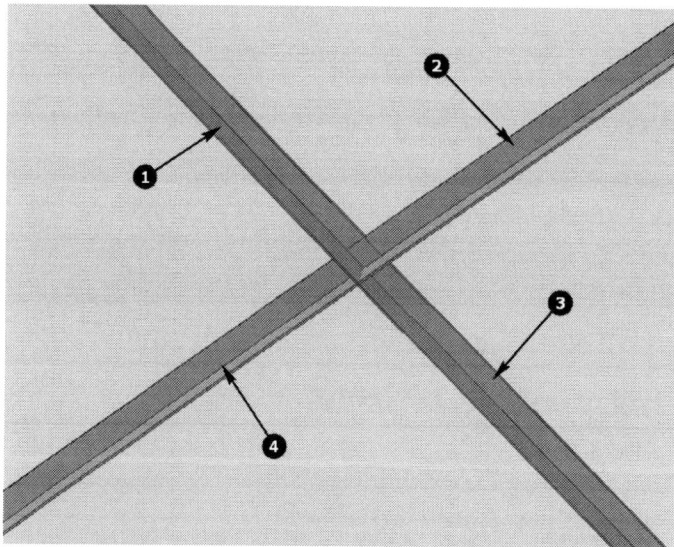

Figure 51 *The four bracing members to insert the joint*

46. From the **General bracing** category, invoke the **Four diagonals - Middle gusset plate** joint tool; you are prompted to select diagonal 1.

47. Select the diagonal labeled as **1** in Figure 51 and press ENTER; you are prompted to select diagonal 2.

Note: *It is important to note that the selections need to be made either in the clockwise or counterclockwise direction.*

48. Select the diagonal labeled as **2** in Figure 51 and press ENTER; you are prompted to select diagonal 3.

49. Select the diagonal labeled as **3** in Figure 51 and press ENTER; you are prompted to select diagonal 4.

50. Select the diagonal labeled as **4** in Figure 51 and press ENTER; the joint with the default values is created and the **Advance Steel - Middle gusset - 4 diagonals [X]** dialog box is displayed.

    You will now edit the parameters of this joint to suit the requirements.

51. Activate the **Gusset plate** tab.

52. In the **Plate thickness** edit box, enter **1/2"** or **12**.

53. In the **Settings for gap** list, make sure **gap between diagonals** is selected.

54. In the **Gap between diagonals** edit box, enter **2"** or **50**.

55. Activate the **Projection of the gusset** tab.

56. Make sure the **Same Projection for all corners** tick box is selected.

57. In the **Projection** edit box, enter **1"** or **25**.

58. Activate the **Bolts** tab.

59. From the **Diameter** list, select **3/4"** or **20.00 mm**.

60. From the **Bolt Type** list, select **A325** (for Imperial) or **AS 1252** (for Metric).

61. Make sure the **Same disposition of bolts for all** tick box is selected. This will ensure that the same bolt parameters are defined for all the four diagonals.

62. Activate the **Bolt distances - diagonal 1** tab.

63. From the **Connection type** list, make sure **bolts** is selected.

64. In the **Distance diagonal - bolt** edit box, enter **1 1/2"** or **35**.

65. In the **Number of lines of bolts** edit box, enter **2**.

66. In the **Intermediate distance** edit box, enter **3"** or **75**.

67. In the **Distance bolt - gusset** edit box, enter **1 1/2"** or **35**.

68. Clear the **Place bolts on gauge line** tick box.

69. From the **Settings for the Edge distance** list, select **center line** (for Imperial) or **Centerline** (for Metric).

70. In the **Edge distance** edit box, enter **0**.

71. In the **Number of columns of bolts** edit box, enter **1**.

   This completes the modifications of all the parameters of this joint. The updated joint is shown in Figure 52.

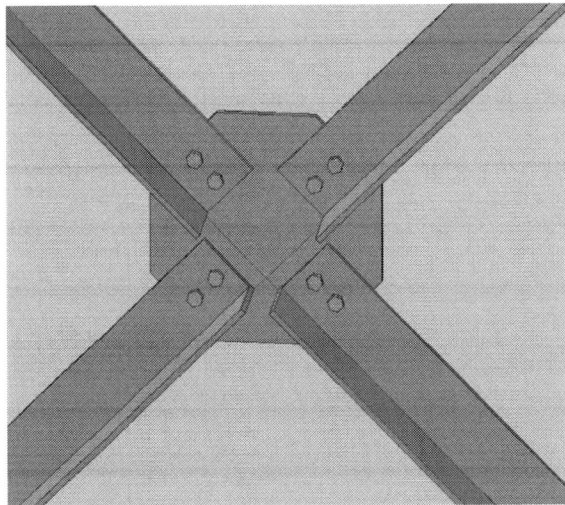

Figure 52 *The four diagonal bracing joint*

72. Save this joint in the library with the name **Frame-Bracing-Diagonals**.

   Finally, you will copy all these three bracing joints. You can use the **Create joint in the joint group, multiple** tool or the **Create joint in the joint group** tool to copy the joints.

73. Using the **Create joint in the joint group, multiple** tool or the **Create joint in the joint group** tool, one by one copy the three bracing joints to all the remaining locations. The zoomed in view of the model, after copying all the joints, is shown in Figure 53.

What I do

*In cases like copying the four diagonal bracing joint, I prefer using the **Create joint in a joint** group tool rather than the **Create joint in a joint group, multiple** tool. This ensures I do not need to zoom and pan all around the model to make my selections. I can make all the selections on one set of diagonals to copy the joint and then move to the next set.*

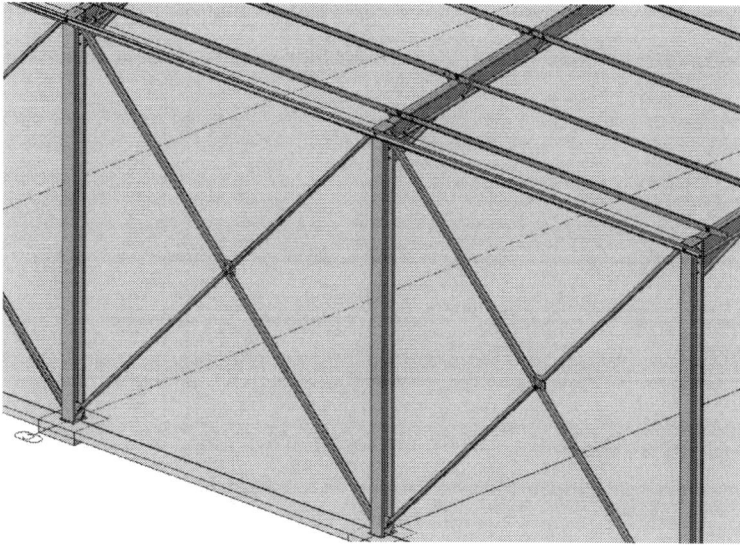

Figure 53 *The zoomed in view of the model after copying the joints*

Section 2: Inserting and Copying Tube Joints

In this section, you will open **C07-Imperial-Elev-Tower.dwg** or **C07-Metric-Elev-Tower.dwg** file and then insert tube joints. You will then create multiple copies of those joints in the model.

1. From the **C07 > Struc-BIM** folder, open the **C07-Imperial-Elev-Tower.dwg** or **C07-Metric-Elev-Tower.dwg** file, based on the preferred units.

   This model is similar to the one you created in the previous chapter. However, notice that there are circular hollow sections inserted around the model. Note that these sections are not inserted as bracing, but as beams. However, you can also use the **Bracing** tool to insert the sections as single bracing.

2. Zoom to the base of the column at the **C3** grid intersection point, as shown in Figure 54. This figure also shows the selection sequence to create the tube joint.

3. From the **Connection vault** palette > **Tube connections** category, invoke the **Tube connection with sandwich plates - additional object** joint tool, which is the first button at the top; you are prompted to select the column beam.

4. Select the column labeled as **1** in Figure 54 and press ENTER; you are prompted to select secondary beam/base plate.

5. Select the base plate labeled as **2** in Figure 54 and press ENTER; you are prompted to select the diagonal beam.

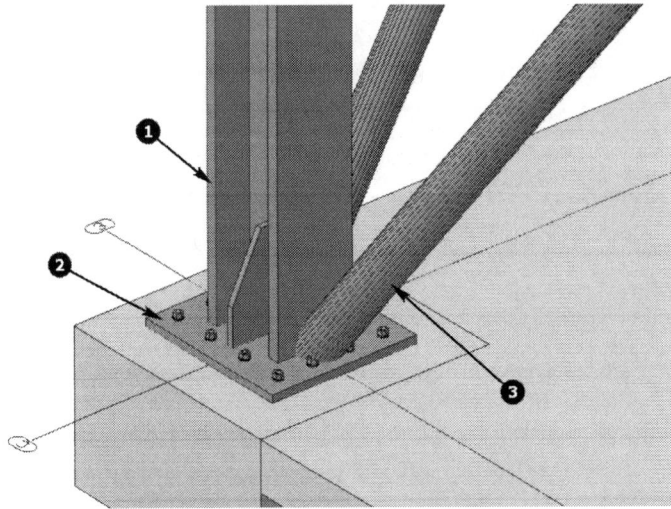

Figure 54 *The selection for the tube joint*

6. Select the tube member labeled as **3** in Figure 54 and press ENTER; the joint with the default values is created, as shown in Figure 55, and the **Advance Steel - Tube connection with sandwich plates [X]** dialog box is displayed.

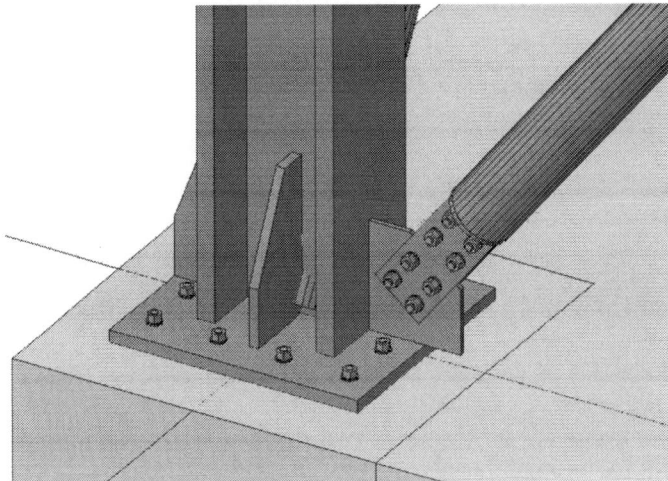

Figure 55 *The joint with the default values*

Note: *The default connection that is created in your case may be different from the one shown in Figure 55. However, once the parameters are modified, you will get the right results.*

Next, you will modify the joint parameters. Although this model only shows two levels of the Elevation Tower, in the real world, it has 6-8 levels. Therefore, the tube joints that you need to create are going to be heavy duty to be able to withstand all the loads.

7. From the left pane of the dialog box, select the **General** category; the **Gusset plate shape** tab is activated as the default tab.

8. From the **Shape at beam/plate** list, select **variable**.

9. Select the **Align to base** tick box.

10. Activate the **General** category > **Gusset plate parameter** tab.

11. In the **Thickness** edit box, enter **3/4"** or **20**.

12. In the **Projection 1** edit box, enter **1/2"** or **12**.

13. In the **Projection 2** edit box, enter **1/2"** or **12**.

14. Activate the **General** category > **Gusset plate contour** tab.

15. From the **Corner type** list, select **corner finish**.

16. From the **Corner finish** list, select **straight**.

17. In the **Corner size** edit box, enter **1"** or **25**.

18. Activate the **General** category > **Bolts in gusset plate** tab.

    This tab allows you to modify the bolt parameters in the gusset plate.

19. From the **Cut back layout** list, make sure **from column** is selected.

20. In the **Cut back** edit box, enter **4"** or **100**.

21. In the **Edge distance sandwich** edit box, enter **1 1/2"** or **35**.

22. In the **Number of bolts along** edit box, enter **2**.

23. In the **Intermediate distance** edit box, enter **3"** or **75**.

24. In the **Edge distance gusset** edit box, enter **1 1/2"** or **35**.

25. In the **Edge distance** edit box, enter **1 1/2"** or **35**.

26. In the **Number of bolts oblong** edit box, enter **2**.

27. In the **Intermediate distance** edit box, enter **3"** or **75**.

    Next, you will modify the bolt parameters in the sandwich plate.

28. Activate the **General** category > **Sandwich plates** tab.

29. Make sure the **No bolts** tick box is cleared. This will ensure you are inserting bolts on the sandwich plate.

30. From the **Number of plates** list, select **both**.

31. In the **Thickness** edit box, enter **3/4"** or **20**.

32. In the **Distance gusset - tab** edit box, enter **1"** or **25**.

33. In the **Edge distance tab** edit box, enter **1 1/2"** or **35**.

34. In the **Intermediate distance** edit box, enter **3"** or **75**.

35. In the **Edge distance sandwich** edit box, enter **1 1/2"** or **35**.

36. Activate the **General** category > **Tab plate** tab.

37. From the **Type** list, make sure **pipe slotted** is selected.

38. In the **Thickness** edit box, enter **3/4"** or **20**.

39. Make sure the value of the **Gap** edit box is set to **0**.

40. From the **Cut back layout** list, select **Last bolt**.

41. In the edit box on the right of the **Cut back layout** list, enter **3"** or **75**.

42. From the **Length layout** list, select **slot length**.

43. In the **Slot length** edit box, enter **4"** or **100**.

44. Make sure from the **Width layout** list, **same as sandwich** is selected.

45. Activate the **General** category > **Bolts & Welds** tab.

46. From the **Diameter** list, select **3/4"** or **20.00 mm**.

47. From the **Bolt Type** list, select **A325** (for Imperial) or **AS 1252** (for Metric).

48. Activate the **Stiffener & Cover** category > **Cover plate** tab.

49. From the **Cover plate** list, make sure **cover slotted** is selected.

50. In the **Thickness** edit box, enter **3/4"** or **20**.

51. From the layout list, make sure **projection** is selected.

52. In the **Projection** edit box, enter **0**.

This completes the editing of all the parameters of the joint. The joint, after updating all the parameters, is shown in Figure 56.

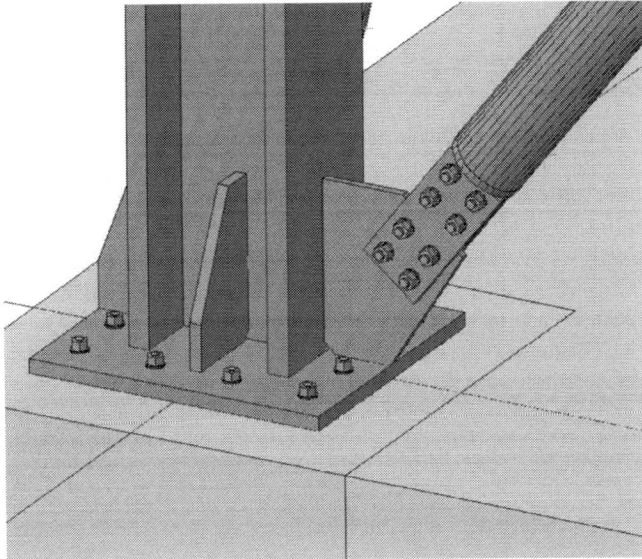

Figure 56 *The joint, after modifying the parameters*

53. Save this joint in the library with the name **Tower-Tube-Flange**.

Next, you will create a similar joint on the web side tube and the same column that you selected for the previous joint.

54. Orbit the model so you can see the web side tube and the column at the **C3** grid intersection point, as shown in Figure 57. This figure also shows the selection sequence to create the tube joint.

55. From the **Tube connections** category, invoke the **Tube connection with sandwich plates - additional object** joint tool; you are prompted to select the column beam.

56. Select the column labeled as **1** in Figure 57 and press ENTER; you are prompted to select secondary beam/base plate.

57. Select the base plate labeled as **2** in Figure 57 and press ENTER; you are prompted to select the diagonal beam.

58. Select the tube member labeled as **3** in Figure 57 and press ENTER; the joint with the default values is created and the **Advance Steel - Tube connection with sandwich plates [X]** dialog box is displayed.

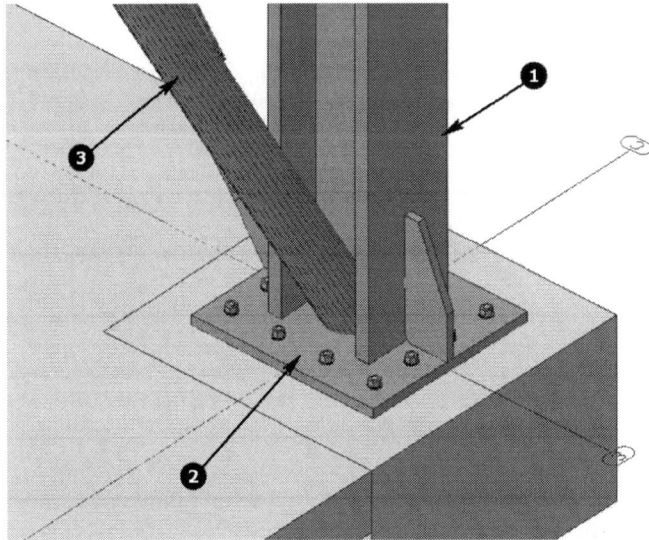

Figure 57 *The selections to be made for the joint on the web side*

59. From the **Library** tab, select **Tower-Tube-Flange**, the joint you saved in the earlier steps; the joint is modified to match the saved values.

   Next, you will modify some of the joint parameters to suit your requirements.

60. Activate the **General** category > **Gusset plate shape** tab.

61. Clear the **Align to base plate** tick box. This ensures the gusset plate of the joint is not aligned to the base plate of the column.

62. Activate the **General** category > **Gusset plate parameter** tab.

63. In the **Lower width** edit box, enter **8"** or **200**.

64. Activate the **General** category > **Bolts in gusset plate** tab.

65. From the **Cut back layout** list, select **from system end**. This ensures the values defined in this tab are measured from the end of the system axis of the column and not the face of the column.

66. In the **Cut back** edit box, enter **10"** or **250**.

   This completes the editing of all the parameters of the joint. The joint, after updating all the parameters, is shown in Figure 58.

67. Save this joint in the library with the name **Tower-Tube-Web**.

   Next, you will create a joint between the beam and the two circular hollow sections.

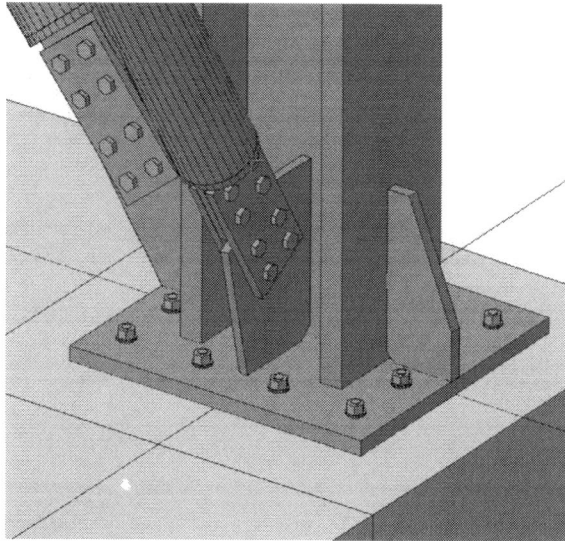

Figure 58 *The joint, after modifying the parameters*

68. Zoom to the area shown in Figure 59. This figure also shows the selection sequence to create the tube joint.

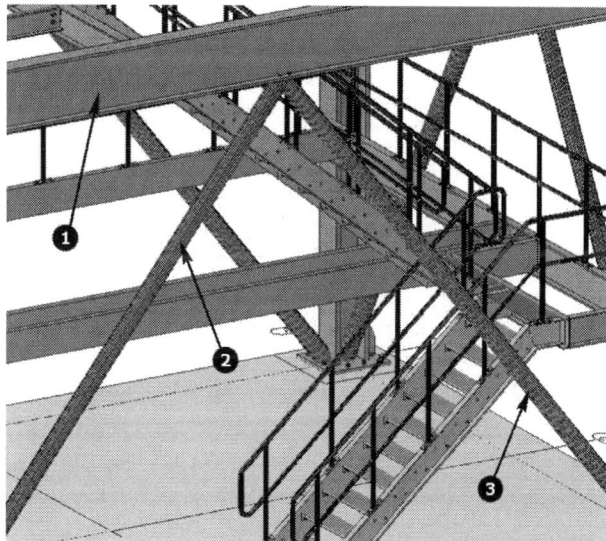

Figure 59 *The selections to be made for the tube joint*

69. From the **Tube connections** category, invoke the **Tube connection with sandwich plates - 2 diagonals** joint tool; you are prompted to select the main beam.

70. Select the beam labeled as **1** in Figure 59 and press ENTER; you are prompted to select the first attached diagonal.

71. Select the tube member labeled as **2** in Figure 59 and press ENTER; you are prompted to select the second attached diagonal.

72. Select the tube member labeled as **3** in Figure 59 and press ENTER; the joint with the default values is created, as shown in Figure 60, and the **Advance Steel - Tube connection with sandwich plates - 2 diagonals [X]** dialog box is displayed.

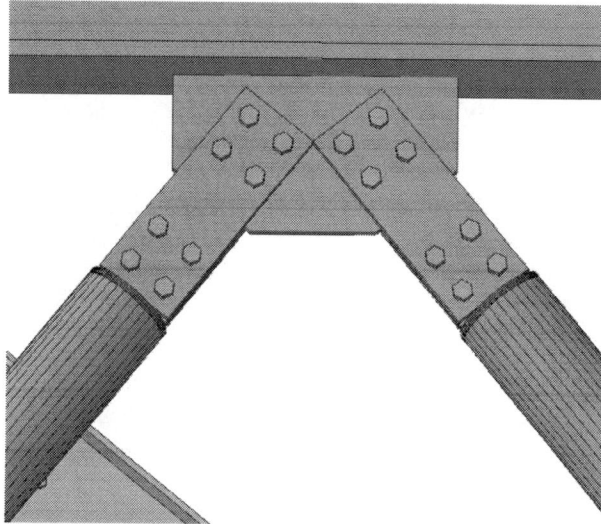

Figure 60 The joint with the default values

Next, you will modify the parameters of this joint to suit your requirements.

73. Activate the **Gusset** category in the dialog box; the **Gusset Shape and Width** tab is active by default.

74. In the **Thickness** edit box, enter **3/4"** or **20**.

75. Activate the **Gusset** category > **Gusset bolt diagonal 1** tab.

76. From the **Cut back layout** list, make sure **from column** is selected.

77. In the **Cut back** edit box, enter **2"** or **50**.

78. Make sure the value in the **Edge distance sandwich** edit box is set to **1 1/2"** or **35**.

79. In the **Number of bolts along** edit box, enter **2**.

80. In the **Intermediate distance** edit box, enter **3"** or **75**.

81. In the **Edge distance gusset** edit box, enter **3"** or **75**.

82. In the **Edge distance** edit box, enter **1 1/2"** or **50**.

83. In the **Number of bolts oblong** edit box, enter **2**.

84. In the **Intermediate distance** edit box, enter **3"** or **75**.

By default, in the **Gusset bolt diagonal 2** tab, the **Same as other side** tick box is selected. As a result, the first diagonal bolt values are used for the second diagonal as well.

85. Activate the **Gusset** category > **Bolts parameter** tab.

86. From the **Diameter** list, select **3/4"** or **20.00 mm**.

87. From the **Bolt Type** list, select **A325** (for Imperial) or **AS 1252** (for Metric).

88. Activate the **Sandwich & Tabs** category > **Sandwich diagonal 1** tab.

89. Make sure the **No bolts** tick box is cleared. This will ensure you are inserting bolts on the sandwich plate.

90. From the **Number of plates** list, select **both**.

91. In the **Thickness** edit box, enter **3/4"** or **20**.

92. In the **Distance gusset - tab** edit box, enter **1"** or **25**.

93. In the **Edge distance tab** edit box, enter **3"** or **75**.

94. In the **Intermediate distance** edit box, enter **3"** or **75**.

95. In the **Edge distance sandwich** edit box, enter **3"** or **75**.

By default, in the **Sandwich diagonal 2** tab, the **Same as other side** tick box selected. As a result, the first sandwich diagonal values are used for the second diagonal as well.

96. Activate the **Sandwich & Tabs** category > **Tab plate diagonal 1** tab.

97. From the **Type** list, make sure **pipe slotted** is selected.

98. In the **Thickness** edit box, enter **3/4"** or **20**.

99. Make sure the value of the **Gap** edit box is set to **0**.

100. From the **Cut back layout** list, select **Last bolt**.

101. In the edit box on the right of the **Cut back layout** list, enter **3"** or **75**.

102. From the **Length layout** list, select **slot length**.

103. In the **Slot length** edit box, enter **4"** or **100**.

104. Make sure from the **Width layout** list, **same as sandwich** is selected.

   By default, in the **Tab plate diagonal 2** tab, the **Same as other side** tick box is selected. As a result, the first tab plate values are used for the second tab plate as well.

105. Activate the **Cover & Stiffener** category > **Cover plate diagonal 1** tab.

106. From the **Cover plate** list, make sure **cover slotted** is selected.

107. In the **Thickness** edit box, enter **3/4"** or **20**.

108. From the layout list, make sure **projection** is selected.

109. In the **Projection** edit box, enter **0**.

   This completes the modifications of all the parameters of this joint. The joint, after modifying the parameters is shown in Figure 61.

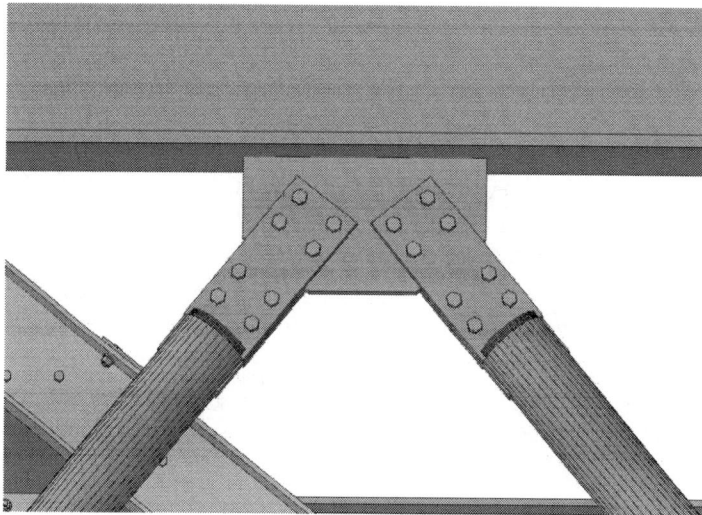

Figure 61 The joint, after modifying the parameters

110. Save this joint in the library with the name **Tower-Beam-2 Tubes**.

   Next, you will create a similar joint on the beam and tubes shown in Figure 62.

111. Zoom to the area shown in Figure 62. This figure also shows the selections to be made for the joint.

112. From the **Tube connections** category, invoke the **Tube connection with sandwich plates - 2 diagonals** joint tool; you are prompted to select the main beam.

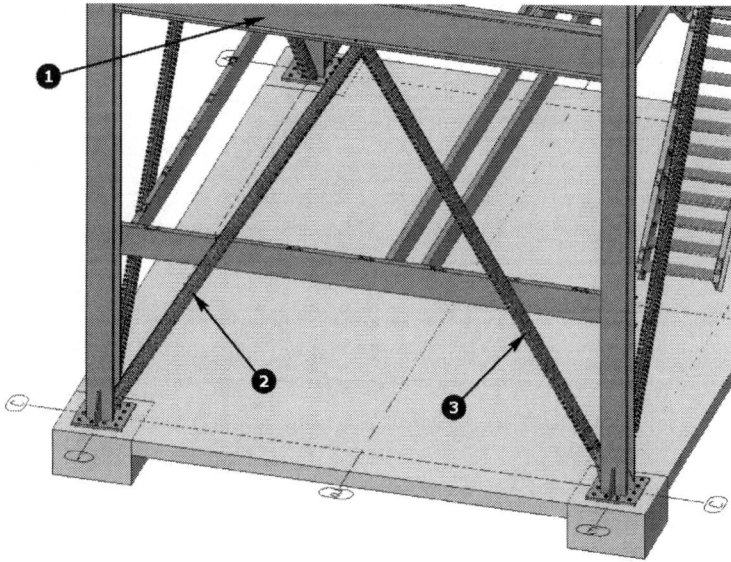

Figure 62 *The selections to be made for the joint*

113. Select the beam labeled as **1** in Figure 62 and press ENTER; you are prompted to select the first attached diagonal.

114. Select the tube member labeled as **2** in Figure 62 and press ENTER; you are prompted to select the second attached diagonal.

115. Select the tube member labeled as **3** in Figure 62 and press ENTER; the joint with the default values is created and the **Advance Steel - Tube connection with sandwich plates - 2 diagonals [X]** dialog box is displayed.

116. From the **Library** tab, select **Tower-Beam-2 Tubes**.

    With this joint, you need to only change the **Cut back** value.

117. Activate the **Gusset** category > **Gusset bolts diagonal 1** tab.

118. In the **Cut back** edit box, enter **4"** or **100**; the joint updates, as shown in Figure 63.

    Finally, you will copy all the tube joints at the remaining locations in the model.

119. Using the **Create joint in a joint group, multiple** or **Create joint in a joint group** tool, copy the tube joints created in the earlier steps to the remaining locations. The model, after copying all the joints, is shown in Figure 64.

120. Save the file.

Figure 63 *The joint after modifying the **Cut back** value*

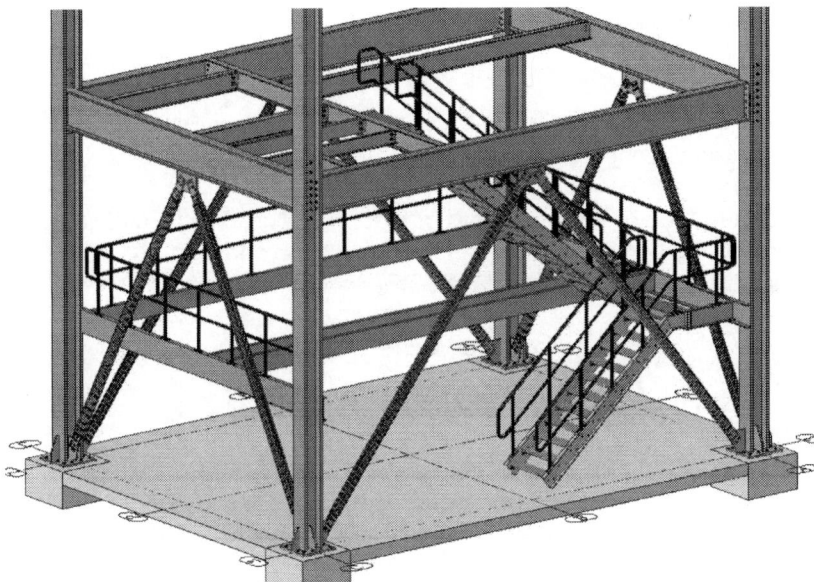

Figure 64 *The model after copying all the tube joints*

Section 3: Inserting and Copying Stair Anchor Joints

In this section, you will insert the stair anchor joint on one of the stringers of the stair at the ground level and then copy it to the other stringer. You will then create the stair end plate joint at the top landing of the same stair and then copy it to the other landing. You will also copy the same joint to landings of the stair at level 1.

1. Zoom to the base of the stair at the ground level, as shown in Figure 65. This figure also shows the stringer to be selected to create the joint.

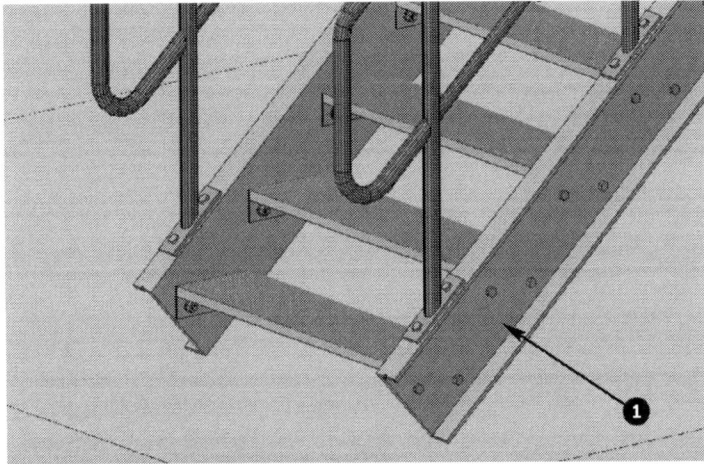

Figure 65 The stair to insert the anchor joint

> **Note**: If you are using the Metric file, the first step will appear outside the stringer. This will be fixed using the parameters defined in the anchor joint you are creating.

2. From the **Connection vault** palette > **Miscellaneous** category, invoke the **Stair Anchor Base Plate** joint tool, which is the first button at the top; you are prompted to select the stringer.

3. Select the stringer labeled as **1** in Figure 65 and press ENTER; you are prompted to confirm if you want to select a reference point.

4. Press ENTER at this prompt to accept the default value of **No**; the joint with the default value is created, as shown in Figure 66, and the **Advance Steel - Stair anchor base plate [X]** dialog box is displayed.

> **Note**: The default connection that is created in your case may be different from the one shown in Figure 66. However, once the parameters are modified, you will get the right results.

Next, you will modify the parameters of this joint to suit your requirements.

5. Activate the **General** category in the dialog box; the **Vertical extension** is the default tab active.

6. Select the **Create vertical profile** tick box; the vertical section is added to the stringer.

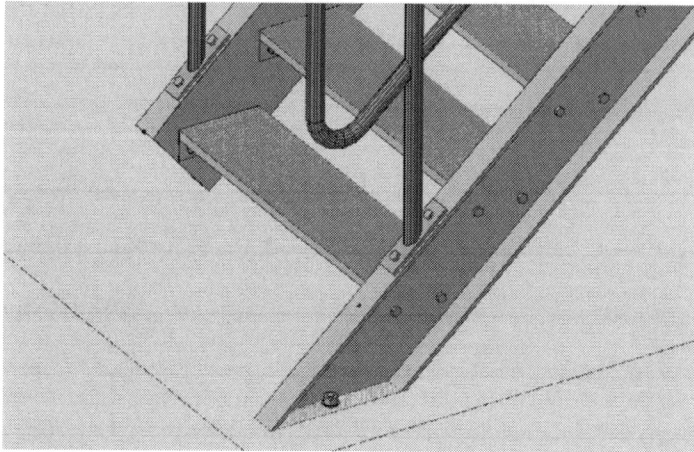

Figure 66 *The stair anchor base plate joint with the default values*

7. Make sure the **Same section as stringer** tick box is selected. This ensures the vertical profile has the same section as the stringer.

8. From the **Height layout** list, make sure **horizontal** is selected.

9. In the **Layout value** edit box, enter **9"** or **190**; the height of the vertical profile is increased.

10. In the **Adjust base level** edit box, enter **0**; the base level is automatically adjusted to the top of the ground slab.

11. Activate the **General** category > **Anchor and holes** tab.

    You first need to set the anchor type before you select the anchor diameter. This is because the anchor diameters are dependent on the anchor type.

12. From the **Anchor type** list, select **US Normal Anchors** (for Imperial) or **HILITI HAS** (for Metric).

13. From the **Anchor diameter** list, select **5/8 inch** (for US Normal Anchors) or **16.00 mm** (For HILITI HAS Anchors).

14. From the **Anchor length** list, select **8"** (for US Normal Anchors) or **190.0** (For HILITI HAS Anchors).

    Next, you will change the number of anchors to 2 and modify the intermediate distance between the anchors.

15. Activate the **General** category > **Distance along** tab.

16. In the **Plate thickness** edit box, enter **3/8"** or **10**.

17. Make sure from the **Plate distance layout** list, **projections** is selected.

18. Make sure the **Projection 1** and **Projection 2** edit box values are set to **0**.

19. From the **Bolt reference from** list, select **centered**, if not already selected.

20. In the **Number of bolts** edit box, enter **2**.

21. In the **Intermediate distance** edit box, enter **4"** or **80**.

22. Activate the **General** category > **Distance across** tab.

23. Make sure **projections** is selected from the **Plate distance layout** list.

24. Make sure the two projection values are set to **0** in the **Projection 1** and **Projection 2** edit boxes.

25. From the **Bolt reference from** list, select **plate center**, if not already selected.

26. In the **Number of bolts** edit box, enter **1** as the value, if not already set to that.

27. In the **Offset from center** edit box, enter **0**.

This completes the modifications of all the parameters of this joint. The joint, after modifying the parameters, is shown in Figure 67.

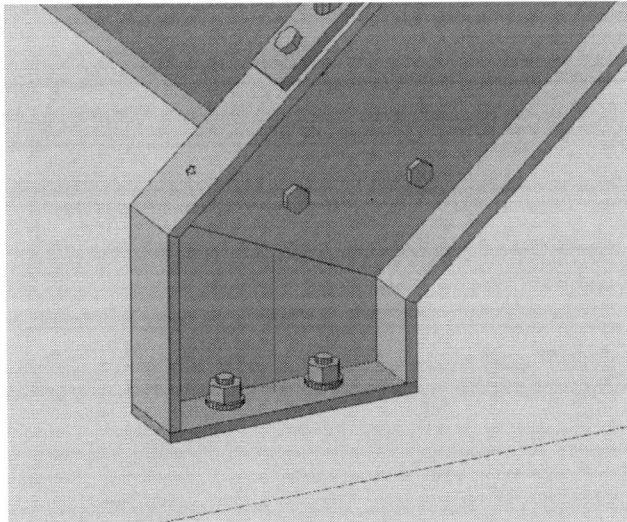

Figure 67 The stair anchor base plate joint after modifying the parameters

28. Save this joint in the library with the name **Tower-Stair-Anchor**.

29. Copy this joint to the other stringer. The zoomed in view of the model, after copying the joint, is shown in Figure 68.

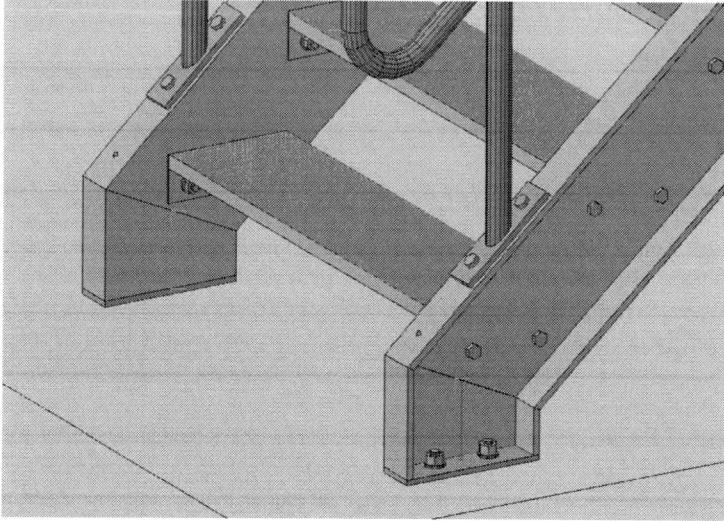

Figure 68 *The model after copying the stair anchor joint to the other stringer*

30. Save the file and then close it.

Section 4: Creating Shear Plate Joints between Stair Landings and Beams (Optional)

1. If time permits, create shear plate joints between the landings of the two stairs and the beams to which the landings are connecting. You can specify your own dimensions. The joints should look similar to the one shown in Figure 45 at the start of this tutorial.

Skill Evaluation

Evaluate your skills to see how many questions you can answer correctly. The answers to these questions are given at the end of the book.

1. In Advance Steel, there is a special category of joints for bracing. (True/False)

2. Tube connections cannot be inserted in Advance Steel. (True/False)

3. Stair anchors are added as part of inserting stairs. (True/False)

4. The **Stair Anchor Angle** is not a type of joint. (True/False)

5. Two separate handrails cannot be connected together in Advance Steel. (True/False)

6. Which joint is used to connect a bracing to a column and a base plate using a gusset plate?

   (A) **Gusset plate to column and base plate**　(B) **Gusset Bracing**
   (C) **Gusset plate with base plate**　　　　　　(D) **None**

7. Which joint is used to connect the selected stringer of the stair to the base using one or two angles?

   (A) **Stair Angle Joint**　　　　　　(B) **Angle Joint**
   (C) **Gusset Angle Joint**　　　　　(D) **Stair Anchor Angle Joint**

8. Which type of joint is used to connect two handrailings together?

   (A) **Railing Joint**　　　　　　　　　(B) **Handrail Joint**
   (C) **Railing joint handrail Joint**　　(D) **None**

9. Which type of joint is used to connect two hollow sections to a beam using a gusset plate?

   (A) **Tube connection with sandwich plates**
   (B) **Sandwich plates - 2 diagonals**
   (C) **Tube - 2 diagonals**
   (D) **Tube connection with sandwich plates - 2 diagonals**

10. The joints meant for hollow sections (tubes) are available in which category?

    (A) **Tube connections**　　　　　　(B) **Round Member Connections**
    (C) **Hollow Member Connections**　(D) **None**

Class Test Questions

Answer the following questions:

1. Explain briefly the process of connecting two separate handrails together.

2. Explain the process of inserting a connection between a beam and two diagonal tubes.

3. What is the procedure of inserting a stair anchor base plate joint?

4. How do you connect a stair landing to a beam?

5. Explain briefly how to connect a tube to a column and a base plate with a gusset plate?

Chapter 8 – Inserting Plates and Gratings, and Controlling Object Visibility

The objectives of this chapter are to:

√   *Explain the process of creating various types of flat plates*
√   *Explain the process of creating various types of folded plates*
√   *Explain the process of checking the unfolding of the folded plates*
√   *Explain the process of creating various types of gratings*

USE OF PLATES IN ADVANCE STEEL

Plates are an integral part of most of the structural models created in Advance Steel. They are either added automatically as part of inserting connections or can be manually added to complete the model. In Advance Steel, you can create the following two types of plates:

> **Flat Plates**: *Plates in the same plane such as rectangular or polygonal plates*
> **Folded Plates**: *To show the folded sheet metal objects such as chutes and hoppers*

Both these types of plates are discussed next.

CREATING FLAT PLATES

In Advance Steel, you can create various types of flat plates. You can also convert a sketch into a plate. These methods are discussed next.

Creating Rectangular Plates by Specifying its Center Point

Home Ribbon Tab > Objects Ribbon Panel (Expanded) > Rectangular plate, center
Objects Ribbon Tab > Plates Ribbon Panel > Rectangular plate, center

This tool is used to create a rectangular plate by specifying its center point. When you invoke this tool, you will be prompted to specify the center point of the plate to be created. On specifying the center point, the plate with the default values will be created, as shown in Figure 1, and the **Advance Steel - Plate [X]** dialog box will be displayed, as shown in Figure 2.

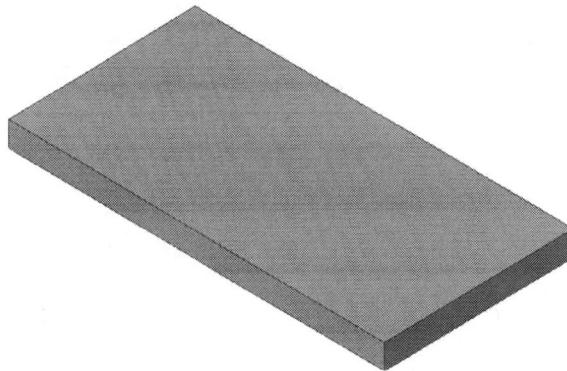

Figure 1 A default flat plate

The options available in the **Advance Steel - Plate [X]** dialog box are similar to those in the other dialog boxes discussed earlier in this book.

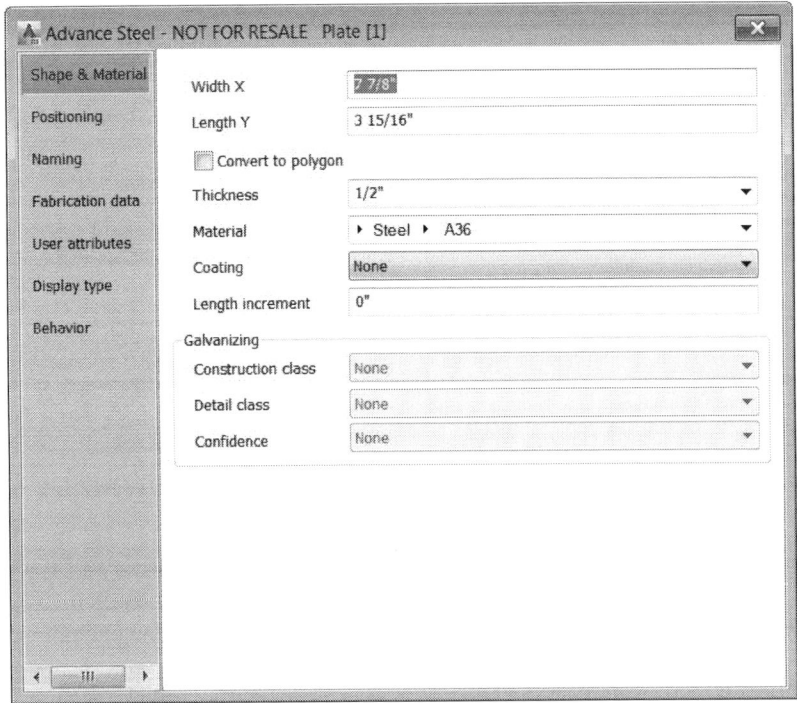

*Figure 2 The **Advance Steel - Plate [X]** dialog box*

Creating Rectangular Plates by Specifying Two Corner Points

Objects Ribbon Tab > Plates Ribbon Panel > Rectangular plate, 2 points

This tool is used to create a rectangular plate by specifying its two diagonally opposite corner points. When you invoke this tool, you will be prompted to specify the start point of the plate's diagonal line. On specifying this point, you will be prompted to specify the end point of the plate's diagonal line. On specifying this point, the plate with the specified values will be created and the **Advance Steel - Plate [X]** dialog box will be displayed.

*Tip: When you are prompted to specify the second point of the plate's diagonal, you can enter the length and width of the plate, separated by a comma (,). For example, if you need to create a plate of 10" length and 4" width, you can specify **10",4"** as the value in the second prompt.*

Creating Rectangular Plates by Specifying Three Corner Points

Objects Ribbon Tab > Plates Ribbon Panel > Rectangular plate, 3 points

This tool is used to create a rectangular plate by specifying its three corner points. When you invoke this tool, you will be prompted to specify the first point of the plate's contour. On specifying this point, you will be prompted to specify the second point to define the plate's X direction and dimension. This point defines the length of the plate and the direction of the first edge. On specifying this point, you will be prompted to specify the third point to define the plate's plane and Y dimension. This point defines the width of the plate and the direction of the second edge. On specifying this point, the plate with the specified values will be created and the **Advance Steel - Plate [X]** dialog box will be displayed.

Creating Polygonal Plates

Home Ribbon Tab > Objects Ribbon Panel (Expanded) > Polygonal plate
Objects Ribbon Tab > Plates Ribbon Panel > Polygonal plate

This tool is used to create a polygonal plate by specifying points that define its contour. When you invoke this tool, you will be prompted to specify the points that define the contour of the plate. After specifying all the required points, press ENTER to create the plate. Figure 3 shows a polygonal plate created by specifying the six polygonal points.

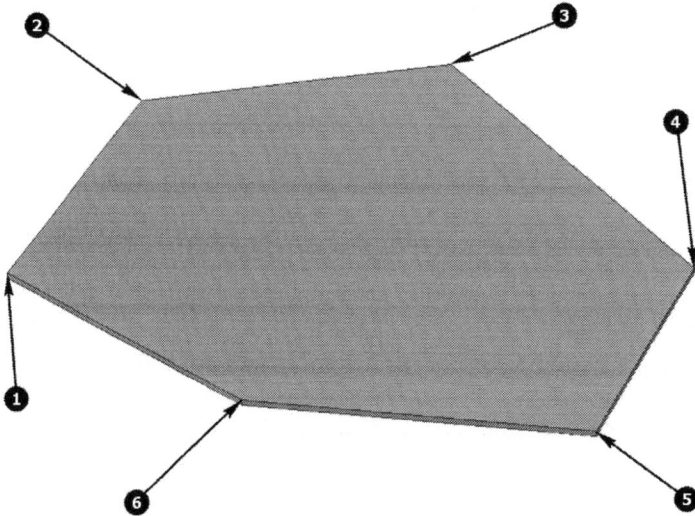

Figure 3 A polygonal plate created using six points

Creating a Plate using a Polyline

Objects Ribbon Tab > Plates Ribbon Panel > Plate at poly line

This tool is used to create a plate using the shape of an existing closed polyline. When you invoke this tool, you will be prompted to select a polyline. After you have selected the polyline, press ENTER; a plate will be created with the same shape as that of the polyline and the **Advance Steel - Plate [X]** dialog box will be displayed. Figure 4 shows a polyline and Figure 5 shows a plate created using the polyline.

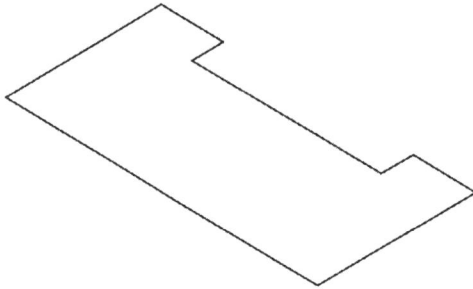

Figure 4 *The polyline to create the plate*

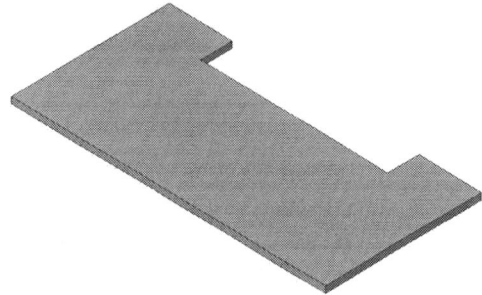

Figure 5 *The plate created using the polyline*

Note: *The polyline used to create the plate is not deleted when the plate is created. You can turn off the layer of the polyline or delete it, if you do not require it.*

Creating a Polyline using a Plate

> **Objects Ribbon Tab > Plates Ribbon Panel > Plate to poly line**

This tool is used to convert a plate into a polyline. When you invoke this tool, you will be prompted to select a plate. You can select one or more plates to convert into polylines. On selecting all the plates, press ENTER. You will then be prompted to specify if you want to delete the plates after creating polylines. You can select **Yes** at this prompt to delete the plates and only retain polylines or select **No** to retain the plates and also create the polyline.

CREATING FOLDED PLATES

As mentioned earlier, the folded plates are used to represent lofted sheet metal parts such as chutes and hoppers. However, it is important to mention here that Advance Steel is not a sheet metal software. Features such as corner relief are not created and the only bend relief that is created is the tear relief. As a result, you can use folded plates mainly for representations.

What I do

If my model requires complex folded plates that require detail drawings and DXF files for laser/plasma cutting, I prefer using Autodesk Inventor software to model those plates. This software is an extremely robust sheet metal software and is used to create complex sheet metal components with a lot of ease. Once the complex sheet metal part is created, I can create flat patterns of those and then create DXF files for laser/plasma cutting straight from the flat pattern. To use those sheet metal parts in Advance Steel, I export them from Autodesk Inventor in a DWG format that I can insert into my Advance Steel project. You will learn more about this later in this chapter.

Creating Folded Plates Without Position Adjustment

Objects Ribbon Tab > Plates Ribbon Panel > Create folded plate - without position adjustment

This tool is used to join two existing plates together to create a folded plate. When you invoke this tool, you are prompted to identify the plate to connect to. This is the plate that remains stationary. On selecting this plate, you will be prompted to select the plate to be connected. It is important to note that while using this tool, if the two plates are positioned away from each other, their positions will not be adjusted. Instead, their size will be increased on joining. On selecting the second plate, the two plates will be joined and the **Advance Steel - Folded Plate Relation** dialog box will be displayed, as shown in Figure 6. This dialog box allows you to specify the angle of fold, the justification of fold, the fold radius, and the incremental angle for grip editing of the fold.

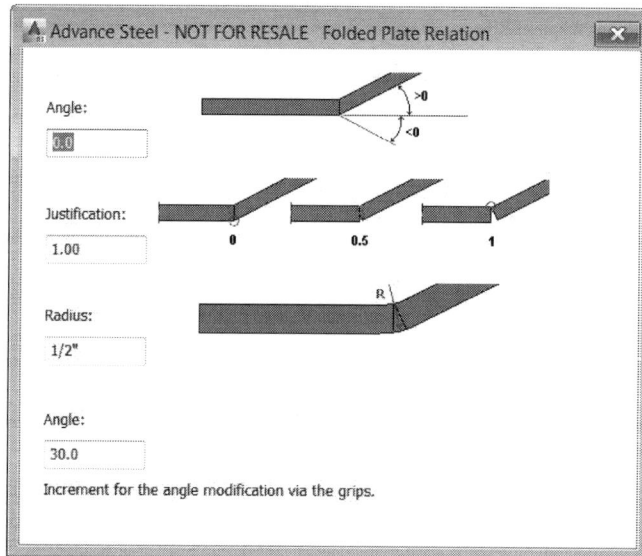

*Figure 6 The **Advance Steel - Folded Plate Relation** dialog box*

Figure 7 shows two plates to join as folded plates and Figure 8 shows the folded plate created with a fold angle of 90-degrees. Note that because the length of the second plate is smaller than that of the first plate, the tear bend relief is created. As mentioned earlier, this is the only type of bend relief that can be created in Advance Steel. Also, notice that the width of the two plates are increased after joining.

Tip: To edit the fold using grips, select the bend created between the two plates; a grip point will be displayed on the plate that was folded. Use that grip point to edit the fold in the increments of the fold grip angle increment defined in the **Advance Steel - Folded Plate Relation** dialog box.

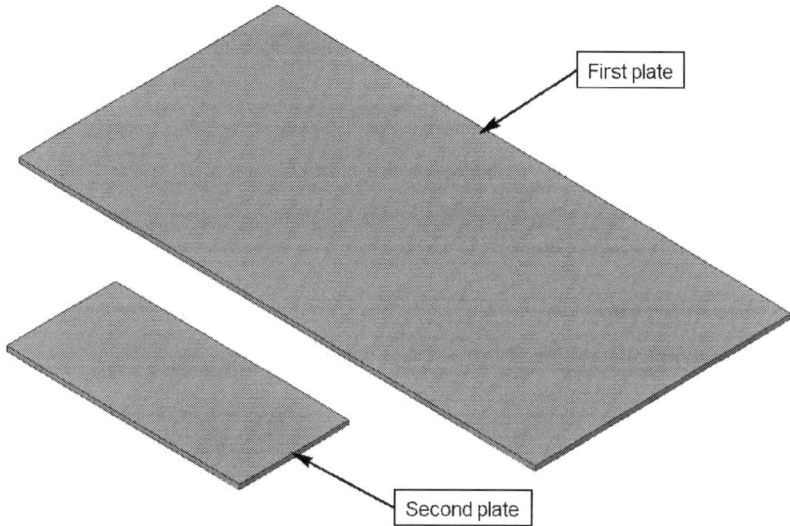

Figure 7 Two plates to be joined to create a folded plate

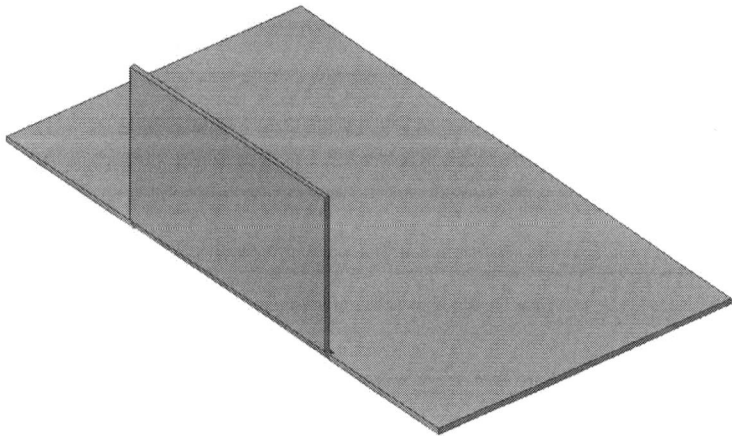

Figure 8 The folded plate created after joining the two plates

Creating Folded Plates With Position Adjustment

Objects Ribbon Tab > Plates Ribbon Panel > Create folded plate - with position adjustment

This tool is used to join two existing plates together to create a folded plate by adjusting the position of the second plate. When you invoke this tool, you are prompted to identify the plate to connect near the edge. This is the plate that remains stationary. The second plate is connected to the edge near which you click to select the first plate. On selecting this plate, you will be prompted to select the plate to be connected near an edge. On selecting the second plate, it will be moved to the first plate and you will be prompted to specify the fold angle. On specifying the fold angle, the **Advance Steel - Folded Plate Relation** dialog box will be displayed. Figure 9 shows the two plates to be joined to create a folded plate and Figure 10 shows the folded plate created. Notice that in Figure 10, the sizes of the two joined plates is the same as the original plate sizes.

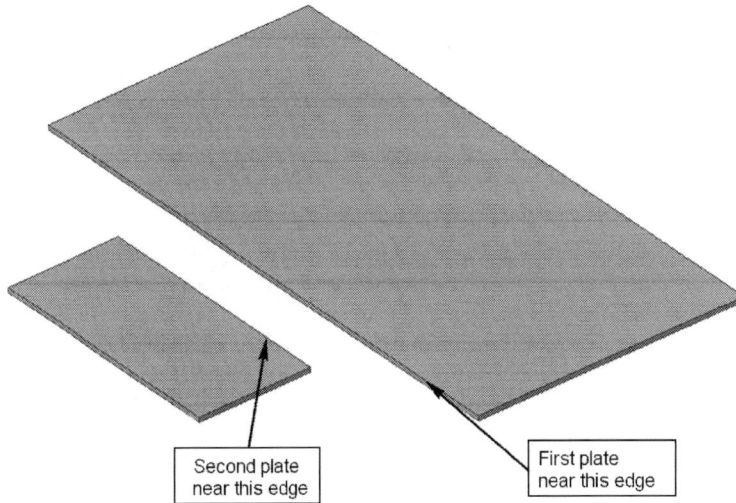

Second plate near this edge

First plate near this edge

Figure 9 *Two plates to be joined to create a folded plate*

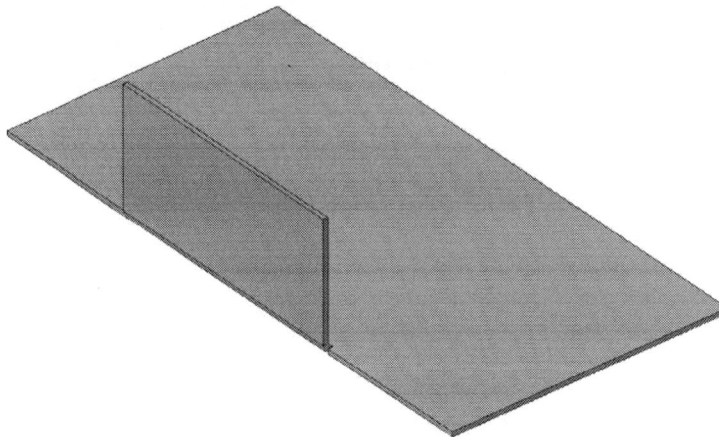

Figure 10 *The folded plate created after joining the two plates*

Creating Conical Folded Plates

> **Objects Ribbon Tab > Plates Ribbon Panel > Create conical folded plate**

This tool is used to create a folded plate by blending two closed contour profiles or two beam ends. Note that the contour profiles need to be closed polylines for this tool to work.

When you invoke this tool, you will be prompted to specify the start shape type. You can select either the **Contour** option or the **Beam** option. Once you select the contour or the beam end and press ENTER, you will be prompted to select the end shape type. For the end shape also, you can either select the **Contour** option or the **Beam** option. Once you select the shape and press ENTER, the **Advance Steel - Conical Folded Plate** dialog box is displayed, as shown in Figure 11.

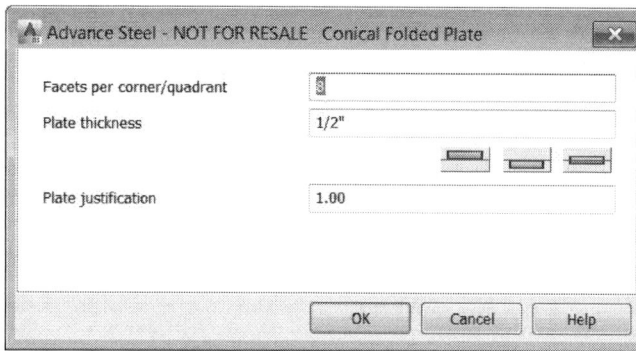

*Figure 11 The **Advance Steel - Conical Folded Plate** dialog box*

In this dialog box, you can specify the number of facets per corner, the plate thickness, and the plate justification. Figure 12 shows the two contours drawn with Z offset to create the conical folded plate and Figure 13 shows the resulting conical plate with 8 facets per quadrant.

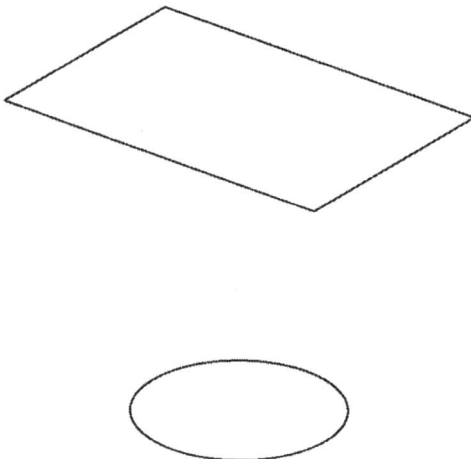

Figure 12 The two contours to create the plate

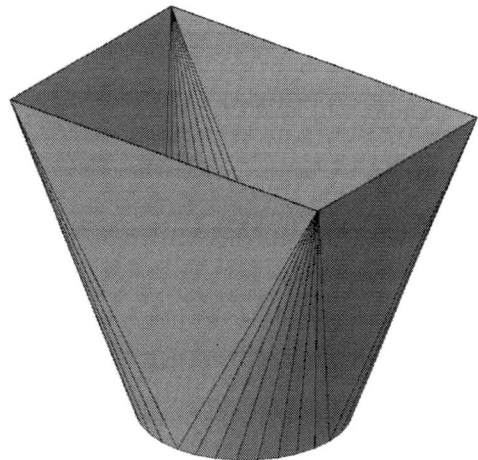

Figure 13 The resulting conical folded plate

Figure 14 shows two beams to create the conical folded plate and Figure 15 shows the resulting plate between the two beams.

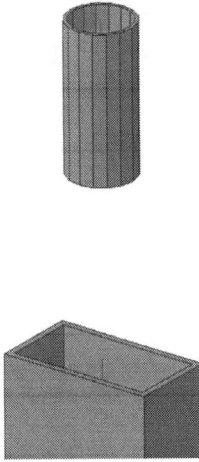

Figure 14 *The two beams to create the plate*

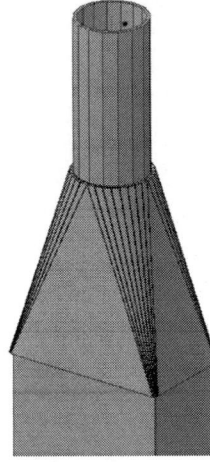

Figure 15 *The resulting conical folded plate*

Creating Twisted Folded Plates

> **Objects Ribbon Tab > Plates Ribbon Panel > Create twisted folded plate**

This tool is used to create a twisted folded plate by blending two sketch elements. The sketch elements that can be selected to create this plate include lines, arcs, polylines, and splines. When you invoke this tool, you will be prompted to select the first entity near the end point from where the plate creation needs to start. Once you select the first sketch entity, you will be prompted to select the second entity. On selecting the second entity, the **Advance Steel - Twisted Folded Plate** dialog box will be displayed, as shown in Figure 16.

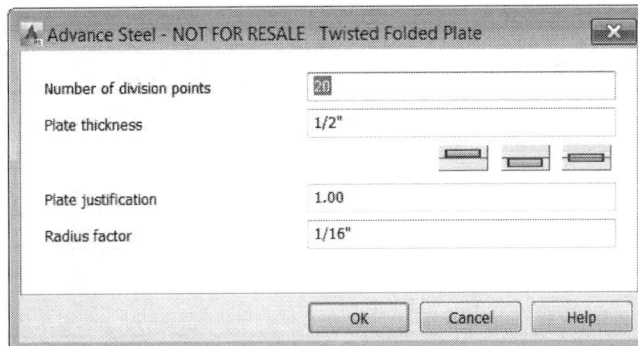

Figure 16 *The **Advance Steel - Twisted Folded Plate** dialog box*

In this dialog box, you can specify the number of division points to create the plate, the plate thickness, the plate justification, and the radius factor. Figure 17 shows the two arcs drawn with Z offset to create the folded plate and Figure 18 shows the resulting folded plate with the radius factor of 1". Note that to create the plate, both the arcs were selected close to their left end points.

Figure 17 The two arcs to create the plate

Figure 18 The resulting twisted folded plate

Creating Circular Plate at the Origin

Objects Ribbon Tab > Plates Ribbon Panel > Circular plate

This tool is used to create a circular plate at the origin of the current UCS. When you invoke this tool, the **Create circular plate** dialog box will be displayed, as shown in Figure 19.

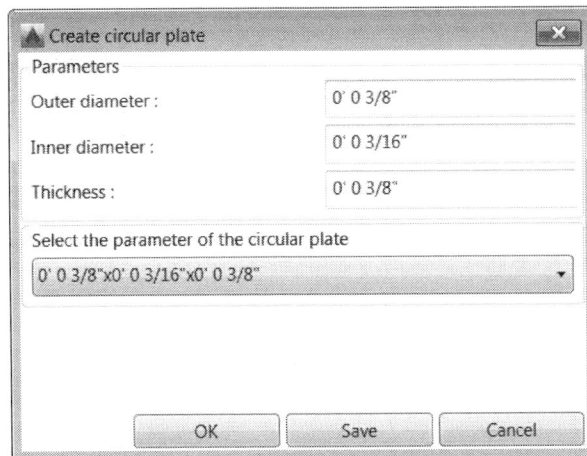

*Figure 19 The **Create circular plate** dialog box*

In this dialog box, you can enter the outer diameter, inner diameter, and the thickness of the plate. By specifying the inner diameter, you can create a hole at the center of this plate. If you do not require a hole at the center, you can enter 0 as the value of the inner diameter. After entering the values in this dialog box, you can press the **Save** button to save the specified value in the list available in the lower half of this dialog box. This allows you to select the same values while creating a new plate next time. When you click **OK** in the dialog box, a new plate will be created at the origin of the current UCS. The resulting plate will show you features in the four quadrants of the circular plate, as shown in Figure 20. You can double-click on these features to edit them.

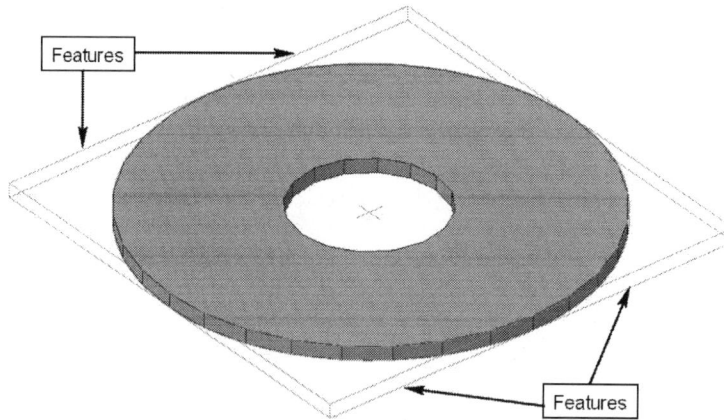

Figure 20 *The circular plate with features*

Tip: You may need to zoom to the extents of the drawing to view the plate. This can be done by double-clicking on the wheel button of the mouse.

Checking the Plate Unfolding

Home Ribbon Tab > Objects Ribbon Panel (Expanded) > Check unfolding
Objects Ribbon Tab > Plates Ribbon Panel > Check unfolding

Checking unfolding is an important part of the process of creating folded plates. It allows you to make sure the plate can actually be unfolded for laser or plasma cutting. In Advance Steel, you can use the **Check unfolding** tool to check the unfolding of the folded plates. When you invoke this tool, you will be prompted to select the folded plate. When you select the plate, you will be prompted to specify whether or not you want to display the unfolded representation. If you select **Yes** at this prompt, the unfolded representation of the plate will be displayed. Figure 21 shows a folded plate and its unfolded representation.

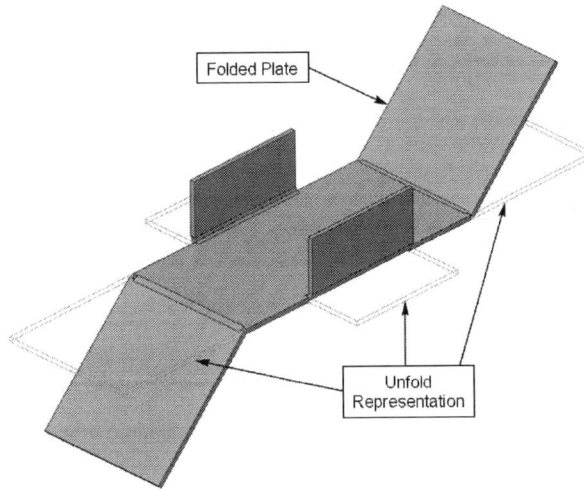

Figure 21 *The folded plate and the unfolded representation*

Setting Folded Plate Main Object

Home Ribbon Tab > Objects Ribbon Panel (Expanded) > Set folded plate main object
Objects Ribbon Tab > Plates Ribbon Panel > Set folded plate main object

When viewing the unfolded representation of the folded plate, the main plate remains stationary and the remaining folded plate unfolds around it. However, in some cases, you may want to change the main plate. This can be done by using the **Set folded plate main object** tool. When you invoke this tool, you will be prompted to select the new main plate. You can click on any face of the folded plate to use that as the main plate. Figure 22 shows the folded and unfolded representation of a conical plate showing the default main plate. Figure 23 shows the same conical folded plate, but with a different main plate and the resulting unfold representation.

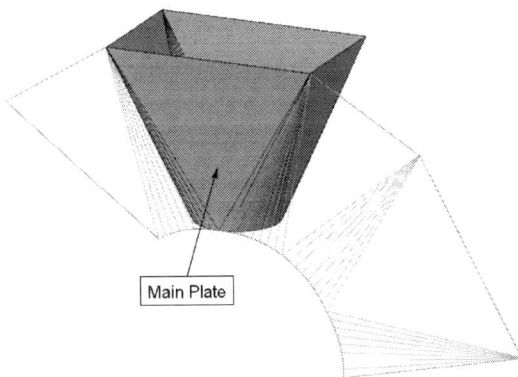

Figure 22 *The main plate of a conical folded plate and the unfold representation*

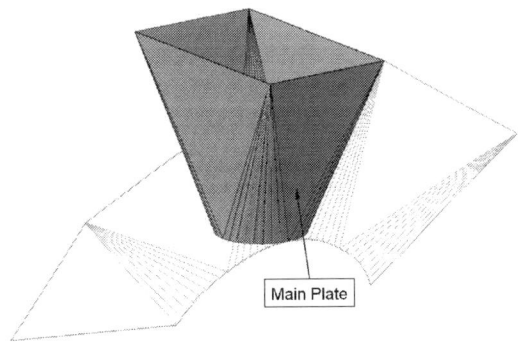

Figure 23 *The conical plate with a different face as the main plate and the unfold representation*

EDITING FLAT AND FOLDED PLATES

Advance Steel provides a number of tools to edit flat or folded plates. These tools are discussed next.

Splitting Plates Using Two Points

> **Objects Ribbon Tab > Plates Ribbon Panel > Split plates by 2 points**

This tool is used to split the selected flat or folded plate by specifying two points on the plate. When you invoke this tool, you will be prompted to select a plate. After selecting the plate and pressing ENTER, you will be prompted to specify the first point of the split line or define the gap value. By defining the gap value, you can specify the gap between the two resulting split plates. After specifying the gap and the first point of split line, you will be prompted to specify the second point of the split line. After you specify the second point, the selected plate will be split. Figure 24 shows a flat plate with the two points used as the points of split line and Figure 25 shows the resulting split plates with a gap.

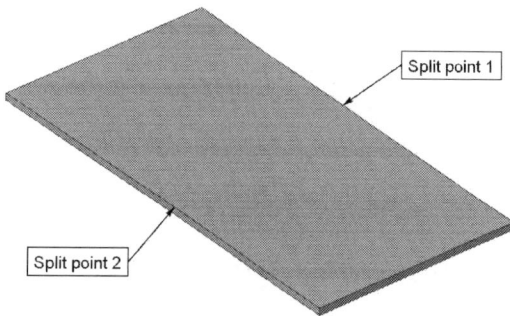

Figure 24   A flat plate with the two split points   *Figure 25   The two resulting split plates*

Splitting Plates Using a Line

> **Objects Ribbon Tab > Plates Ribbon Panel > Split plates at lines**

This tool is used to split the selected flat or folded plate by specifying one or more lines. When you invoke this tool, you will be prompted to select a plate. After selecting the plate and pressing ENTER, you will be prompted to select a line or define the gap value. After specifying the gap and the split line, you will again be prompted to select the split line. Once you have selected all the split lines, press ENTER; the selected flat or folded plate will be split. Figure 26 shows a conical folded plate with the split line and Figure 27 shows the plate after splitting.

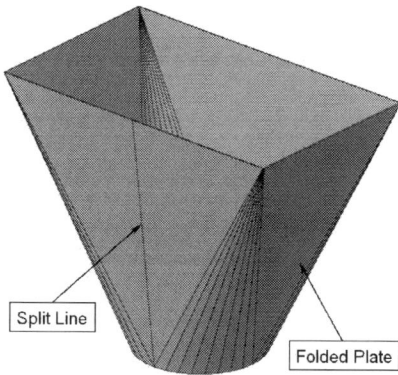

Figure 26 *A folded plate with the split line*

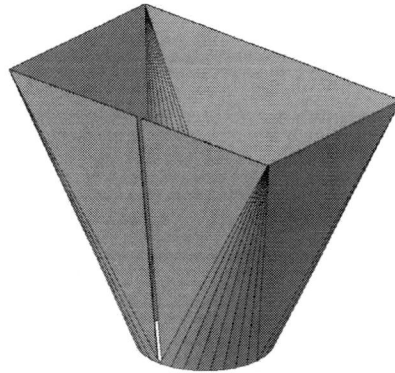

Figure 27 *The resulting split in the folded plate*

Merging Plates

Objects Ribbon Tab > Plates Ribbon Panel > Merge plates

This tool is used to merge two plates into a single plate. It is important to note here that if there is a gap between the plates, then they cannot be merged. For the plates to be merged, you first need to move them so that there is no gap between them. When you invoke this tool, you will be prompted to select the plates to be merged. Select all the plates that you want to merge and then press ENTER; the selected plates will be merged. Figure 28 shows two separate plates that need to be merged and Figure 29 shows the merged plate created using the **Merge plates** tool.

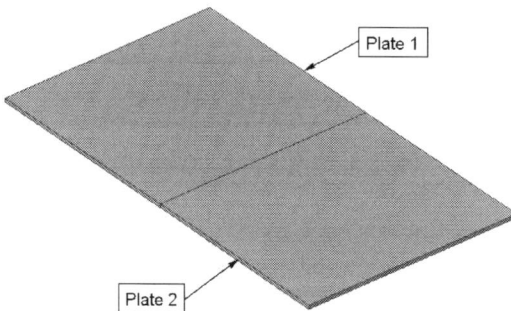

Figure 28 *Two plates before merging*

Figure 29 *The resulting merged plate*

Shrinking and Expanding Poly Plates

Objects Ribbon Tab > Plates Ribbon Panel > Shrink / expand poly plate

This tool is used to shrink or expand the selected plates by a specified value. When you invoke this tool, you will be prompted to select plates to shrink. After selecting all the plates, press ENTER; you will be prompted to specify the offset value. At this prompt, you can specify a positive offset value to expand the plates by that value. Alternatively, you can specify a negative offset value to shrink the plates by that value.

INSERTING GRATINGS

Similar to plates, gratings are also an important part of the structural model required in the plant and mining industry. In Advance Steel, you can create a standard grating that gets its values from the standard library or create variable grating where you can enter its size. You can also convert a sketch into a grating. All these methods of creating gratings are discussed next.

Inserting Standard Grating

> **Home Ribbon Tab > Objects Ribbon Panel > Standard Grating**
> **Objects Ribbon Tab > Grating Ribbon Panel > Standard Grating**

This tool is used to insert a standard grating with its size selected from the standard list of available sizes. When you invoke this tool, you will be prompted to specify the center point of the grating. When you specify the point, a standard grating will be inserted with the specified point as its center and the **Advance Steel - Standard grating** dialog box will be displayed with the **Shape & Connector** tab active, as shown in Figure 30. You can select the class of the grating to be inserted from the **Grating class** list. Depending on the grating class selected, the sizes available in the **Grating size** list will change. You can select the required size from this list.

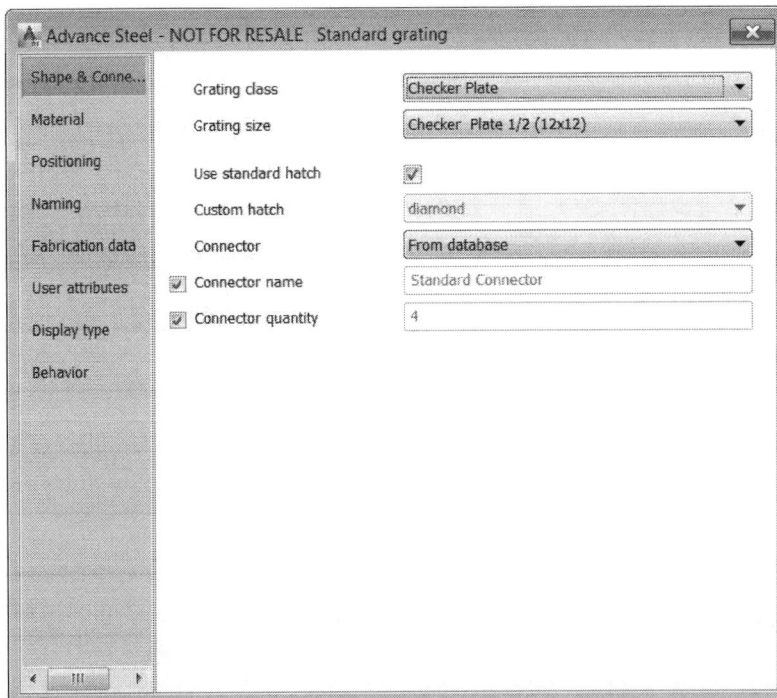

*Figure 30 The **Shape & Connector** tab of the **Advance Steel - Standard grating** dialog box*

Note: *The options in the remaining tabs of the **Advance Steel - Standard grating** dialog box are the same as those discussed in earlier chapters.*

Inserting Rectangular Variable Grating

> **Home Ribbon Tab > Objects Ribbon Panel > Variable Grating, rectangular**
> **Objects Ribbon Tab > Grating Ribbon Panel > Variable Grating, rectangular**

This tool is used to insert a rectangular grating by specifying its two diagonally opposite corners. When you invoke this tool, you will be prompted to specify the start point of the grating's diagonal line. When you specify the point, you will be prompted to specify the endpoint of the grating's diagonal line. On specifying the endpoint, the grating will be inserted and the **Advance Steel - Variable grating** dialog box will be displayed with the **Shape & Connector** tab active, similar to the one shown in Figure 30. However, because you are inserting variable grating, in this case, you will also see the **Grating length** and **Grating width** edit boxes in the **Shape & Connector** tab to enter the exact values of the length and width of the grating.

Inserting Bar Grating

> **Home Ribbon Tab > Objects Ribbon Panel > Bar Grating**
> **Objects Ribbon Tab > Grating Ribbon Panel > Bar Grating**

This tool is used to insert a bar grating by specifying two points that define the length of the grating. When you invoke this tool, you will be prompted to specify the start point of the grating. When you specify the point, you will be prompted to specify the endpoint of the grating. On specifying the endpoint, the bar grating will be inserted and the **Advance Steel - Bar grating** dialog box will be displayed with the **Shape & Connector** tab active, as shown in Figure 31. You can select the series of the bar grating to be inserted from the **Grating series** list. The bearing spacing and the cross bar spacing can be selected from the **Bearing/Cross bar spacing** lists and the number of bearing bars and their widths can be selected from the **Bearing bars/width** lists.

Inserting Polygonal Variable Grating

> **Home Ribbon Tab > Objects Ribbon Panel (Expanded) > Variable Grating, polygonal**
> **Objects Ribbon Tab > Grating Ribbon Panel > Variable Grating, polygonal**

This tool is used to insert a polygonal grating by specifying its corners. When you invoke this tool, you will be prompted to specify the corners that define the contour of the grating. Once you have specified all the points, press ENTER; the grating will be inserted and the **Advance Steel - Variable grating** dialog box will be displayed with the **Shape & Connector** tab active, similar to the one shown in Figure 30.

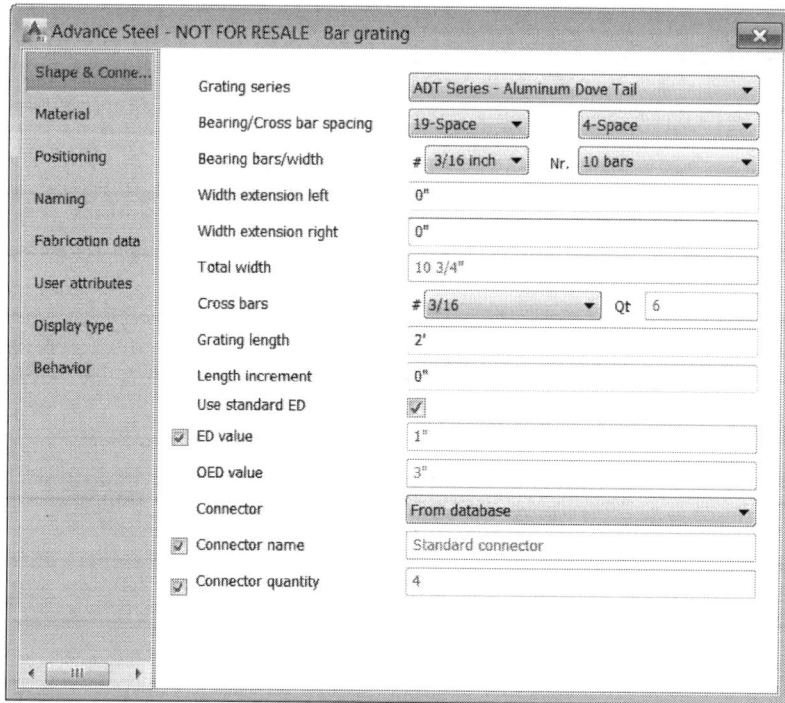

Figure 31 The **Shape & Connector** tab of the **Advance Steel - Bar grating** dialog box

Inserting Grating at a Polyline

Home Ribbon Tab > Objects Ribbon Panel (Expanded) > Grating at Polyline
Objects Ribbon Tab > Grating Ribbon Panel > Grating at Polyline

This tool is used to insert grating at a selected closed polyline. When you invoke this tool, you will be prompted to select the polyline. After you select the polyline and press ENTER, the grating of the shape of the polyline will be inserted and the **Advance Steel - Variable grating** dialog box will be displayed with the **Shape & Connector** tab active, similar to the one shown in Figure 30.

CREATING PLATE AND GRATING FEATURES

The **Features** tab of the **Advance Steel Tool Palette** provides a number of tools to create plate and grating features, as shown in Figure 32. These tools are discussed next.

Rectangular contour, center

This tool is used to create a rectangular contour cut on the selected flat or folded plate or a grating by specifying the center point of the contour. When you invoke this tool, you will be prompted to select a plate to be modified near a corner. Once you select the plate, you will be prompted to specify the center of the plate cope to be created. This is the point where the center of the rectangular cut will be located. Once you specify this point, a rectangular contour cut will be created on the selected plate and the **Advance Steel - Plate**

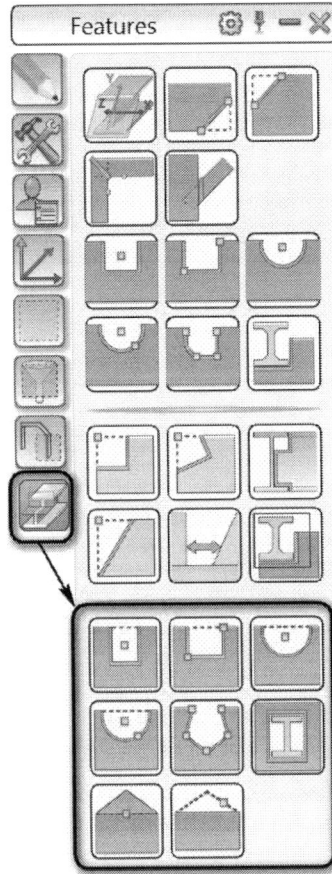

*Figure 32 The **Plate Feature** tools*

contour dialog box will be displayed. Most of the options in this dialog box are similar to those discussed in earlier chapters. Figure 33 shows a plate and the point defined as the center of the cope and Figure 34 shows the resulting cut feature. In this feature, the radius is added to the cut using the **Corner finish** tab of the **Advance Steel - Plate contour** dialog box.

Center of the cope

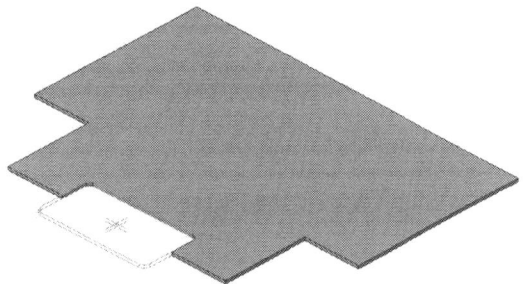

Figure 33 The location of the cope center on the plate

Figure 34 The flat plate with the rectangular cope feature created

Figure 35 shows a folded plate and the location of the cope feature center and Figure 36 shows the resulting cope feature created.

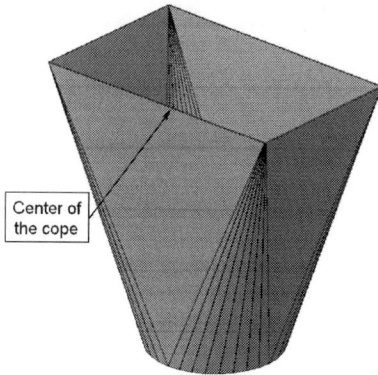

Figure 35   *The location of the cope center on the folded plate*

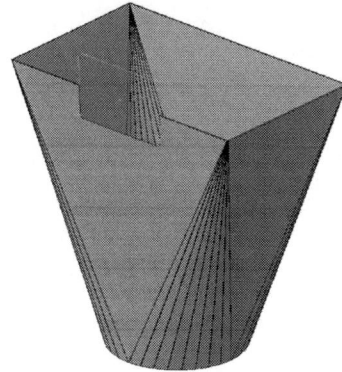

Figure 36   *The folded plate with the rectangular cope feature created*

Tip: You can use the **Positioning** tab of the **Advance Steel - Plate contour** dialog box to change the position of the cope feature on the plate.

To turn off the visibility of the profile of the contour, double-click on the plate to display the plate dialog box. Then in the **Display type** tab, click once on **Features** and then click **Standard** again.

Rectangular contour, 2 points

This tool is used to create a rectangular contour cut on the selected flat or folded plate or the selected grating by specifying two diagonally opposite points of the contour. When you invoke this tool, you will be prompted to select a plate to be modified near a corner. Once you select the plate, you will be prompted to specify two diagonal points for the rectangle inner contour. Once you specify the two points, a rectangular contour cut will be created on the selected plate and the **Advance Steel - Plate contour** dialog box will be displayed.

Circular contour, center

This tool is used to create a circular contour cut on the selected flat or folded plate or the selected grating by specifying the center point of the contour. When you invoke this tool, you will be prompted to select a plate to be modified near a corner. Once you select the plate, you will be prompted to specify the center of the plate cope to be created. This is the point where the center of the circular cut will be located. Once you specify this point, a circular contour cut will be created on the selected plate and the **Advance Steel - Plate contour** dialog box will be displayed. Using this dialog box, you can modify the radius, positioning, and the contour offset for the cut. Figure 37 shows a flat plate and the midpoint to locate the center of the contour and Figure 38 shows the plate with the circular contour cut feature.

Figure 37 *The location of the cope center on the flat plate*

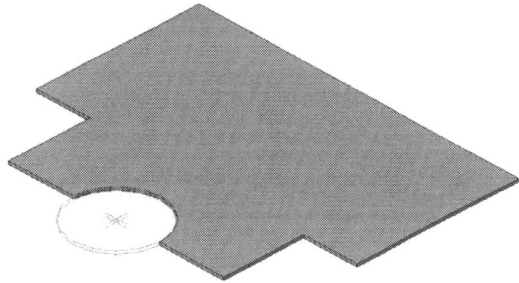

Figure 38 *The flat plate with the circular cope feature created*

Circular contour, 2 points

This tool is used to create a circular contour cut on the selected flat or folded plate or the selected grating by specifying the center point and the radius of the contour cut. When you invoke this tool, you will be prompted to select a plate to be modified near a corner. Once you select the plate, you will be prompted to specify the center of the plate cope to be created. Once you specify the center point, you will be prompted to specify the radius of the cope. On specifying the radius, a circular contour cut will be created on the selected plate and the **Advance Steel - Plate contour** dialog box will be displayed.

Polygonal contour

This tool is used to create a polygonal contour cut on the selected flat or folded plate or the selected grating by specifying the endpoints of the segments of the polygonal cut or by selecting an existing closed polyline. You can specify line or arc segments for the cut. However, you need to make sure that the polyline segments you are specifying are closed. For your convenience, the prompt sequence provides you with the **Close** option to close the polyline you are defining. When you invoke this tool, you will be prompted to select a plate to be modified near a corner. Once you select the plate, you will be prompted to specify the start point of the polyline or select an existing polyline. On specifying the polyline segments or selecting a polyline and pressing ENTER, a polyline contour cut will be created on the selected plate and the **Advance Steel - Plate contour** dialog box will be displayed. You can use this dialog box to define the contour offset for the cut and the corner finishes at each corner of the polyline cut. Figure 39 shows a plate and a closed polyline to create the cut and Figure 40 shows the plate after creating the cut.

Element contour

This tool is used to create a contour cut on the selected flat or folded plate or the selected grating by specifying the shape of an element, such as a beam, a column, and so on. It is important to note that the cut shape that is created can be the exact cut or the casing

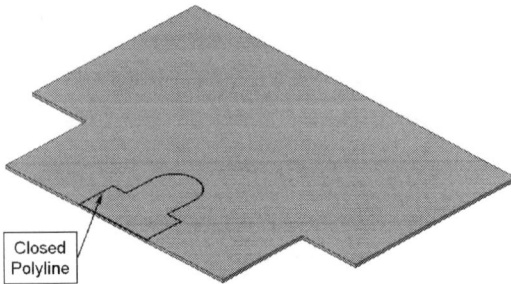

Figure 39 *The flat plate and the polyline to be used for the cut*

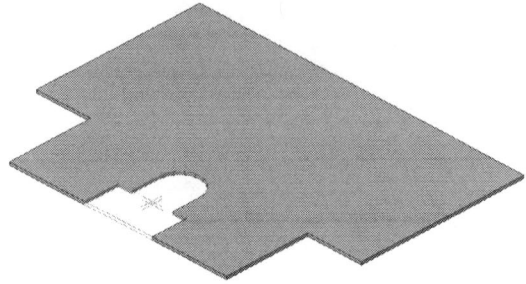

Figure 40 *The flat plate with the polyline cope feature created*

cut, depending on the shape type active in the **Objects** ribbon tab > **Switch** ribbon panel, as shown in Figure 39. By default, the **Exact cross section** is the active shape, as shown in Figure 41. If you click this button, the button switches to **Casing cross section**, as shown in Figure 42.

Figure 41 *The Switch ribbon panel with the Exact cross section shape active*

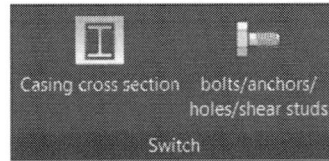

Figure 42 *The Switch ribbon panel with the Casing cross section shape active*

The process for creating the element contour using both these shape types are discussed next.

Process for Creating Exact Cross Section Element Contour

The following are the steps required to create the element contour with the exact cross section shape.

1. From the **Objects** ribbon tab > **Switch** ribbon panel, make sure **Exact cross section** shape is active.

2. From the **Advance Steel Tool Palette** > **Features** tab, invoke the **Element contour** tool; you will be prompted to identify the plate to be modified near a corner.

3. Select the plate; you will be prompted to identify the object to cut to.

4. Select the steel section to cut to; the element contour cut is created and the **Advance Steel - Contour processing** dialog box is displayed using which you can define the gap offset, the size of the contour cut on the two sides, make it a straight cut, and define corner finishes at each corner of the cut.

Figure 43 shows a plate and a column to create the element contour cut and Figure 44 shows the plate after creating the exact element contour cut. In Figure 44, the visibility of the column is turned off for better visibility of the cut.

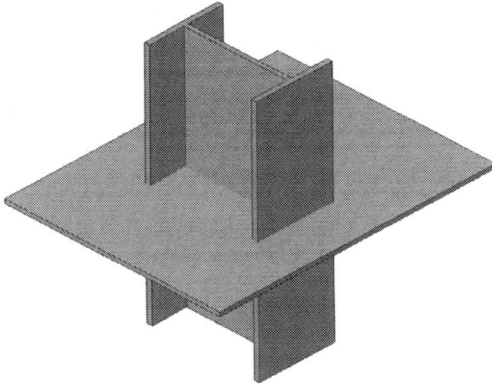

Figure 43 The plate and the column to be used for the cut

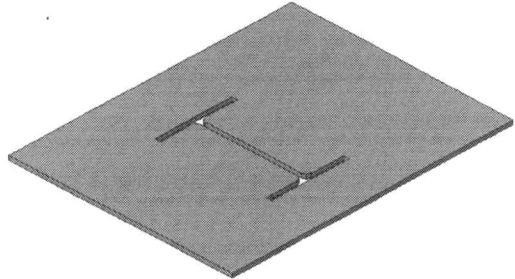

Figure 44 The flat plate after creating the exact cut using the column

Process for Creating Casing Cross Section Element Contour

The following are the steps required to create the element contour with the casing cross section shape.

1. From the **Objects** ribbon tab > **Switch** ribbon panel, make sure **Casing cross section** shape is active.

2. From the **Advance Steel Tool Palette** > **Features** tab, invoke the **Element contour** tool; you will be prompted to identify the plate to be modified near a corner.

3. Select the plate; you will be prompted to identify the object to cut to.

4. Select the steel section to cut to; the element contour cut is created and the **Advance Steel - Contour processing** dialog box is displayed using which you can define the gap offset, the size of the contour cut on the two sides, make it a straight cut, and define corner finishes at each corner of the cut.

    Figure 45 shows a plate and a column to create the element contour cut and Figure 46 shows the plate after creating the exact element contour cut. In Figure 46, the visibility of the column is turned off for better visibility of the cut.

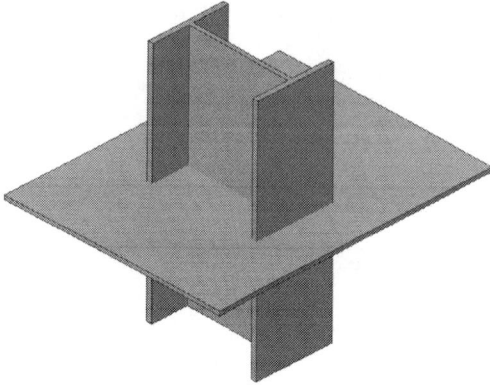

Figure 45 *The plate and the column to be used for the cut*

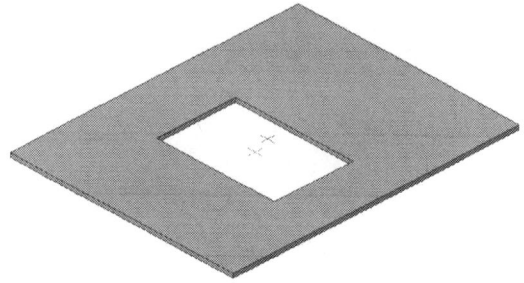

Figure 46 *The flat plate after creating the casing cut using the column*

Inserting a Corner on a Poly Plate

This tool is used to insert a corner on the selected edge of a poly plate. When you invoke this tool, you will be prompted to select a plate near the edge that will be split. The selected edge will be split in half and the midpoint of the new edge can be used to stretch the poly plate. Figure 47 shows a poly plate showing the default grip points and the edge to be used to select the plate and Figure 48 shows the grip points after inserting the corner on the selected edge.

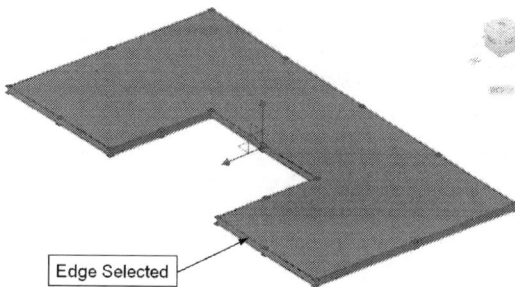

Edge Selected

Figure 47 *The poly plate with default grip points and the edge used to select the plate*

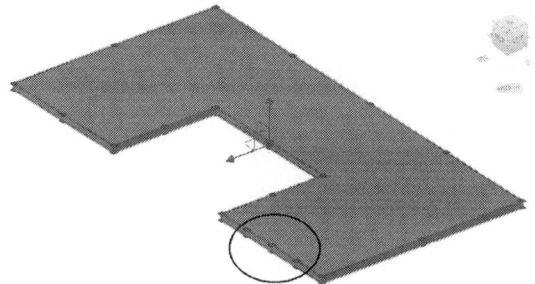

Figure 48 *The grip points on the poly plate after inserting a corner on the selected edge*

Removing a Corner from the Poly Plate

This tool is used to remove the selected corner from the poly plate. When you invoke this tool, you will be prompted to select a plate near the corner that you want to be deleted. Figure 49 shows a poly plate and the corner to be deleted and Figure 50 shows the poly plate after deleting the corner.

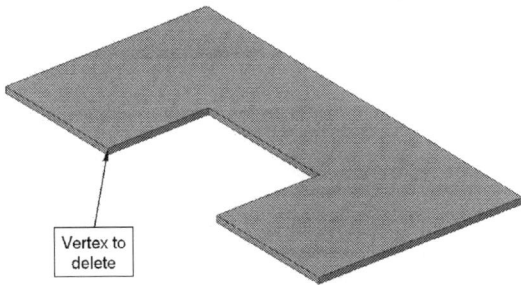

Figure 49 *The poly plate and the vertex selected to be deleted*

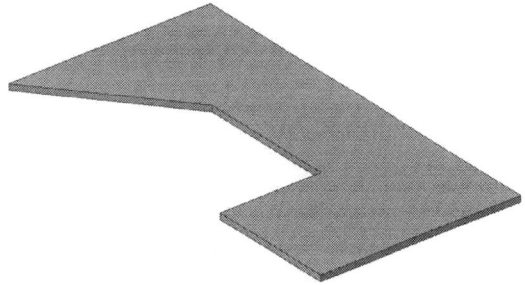

Figure 50 *The poly plate, after deleting the selected vertex*

CONTROLLING OBJECT VISIBILITY

The **Quick views** tab of the **Advance Steel Tool Palette**, shown in Figure 51, provides various tools to control the visibility of the Advance Steel objects. These tools are extremely useful and can help you control the visibility of various objects, thus making it easier to work with large models. Some of these tools are discussed next.

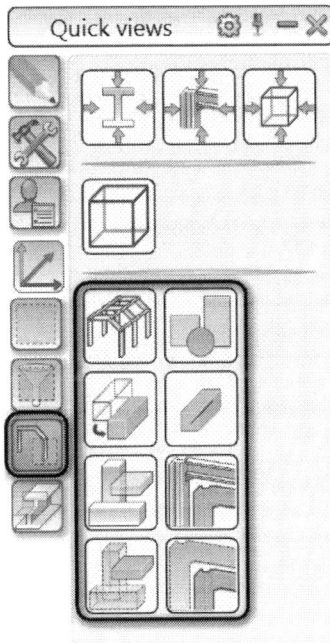

Figure 51 *The **Quick views** tools*

What I do

*I am often asked why I prefer to use the **Quick views** tools and not the **Layer Isolate** tool to turn off the visibility of the objects that are not required. The reason for this is that while working with Advance Steel, the **Standard** layer is the active layer by default. As a result, any element that you insert will be placed on its respective layer. For example, if you insert beams, they will be placed on the **Beams** layer. Similarly, if you insert gratings, they will be placed on the **Gratings** layer. However, when you isolate a layer, it will become the current layer and all the other layers will be turned off. As a result, all the new objects you create will be placed on the current layer and not on their respective layers. However, when you use the **Quick views** tools, only the objects are turned off and not their layers. Also, the **Standard** layer is still the current layer, which allows all the new objects to be placed on their respective layers.*

Turning off the Visibility of the Selected Objects

The **Selected objects off** tool is used to turn off the visibility of the selected objects. When you invoke this tool, you will be prompted to select objects. You can select all the objects that you want to turn off and then press ENTER; all the selected objects will be turned off.

*Tip: The **Selected objects off** tool is extremely useful when combined with the **Select Similar** option, which is available when you select an object in the drawing window and then right-click. For example, if you want to turn off the visibility of all the railings, you can select one of the railings and then right-click and click **Select Similar** to select all the railings. Next, click the **Selected objects off** tool to turn off the visibility of all the railings.*

Turning on the Visibility of All the Objects

The **All visible** tool is used to turn on the visibility of all the objects. This tool is generally used when you used one of the other **Quick views** tools to turn off the visibility of some of the objects.

Showing Only the Selected Objects

The **Show only selected objects** tool is used to turn off the visibility of all the unselected objects. When you invoke this tool, you will be prompted to select objects. You can select all the objects that you want to retain and then press ENTER; all the unselected objects will be turned off. This tool is also extremely useful when combined with the **Select Similar** option.

Turning off the Visibility of the Selected Assemblies

For Advance Steel, assemblies are the items that are welded together. The **Selected assemblies off** tool allows you to turn off the selected objects and all other objects that are welded to them.

Showing Only the Selected Assemblies

The **Show only selected assemblies** tool allows you to retain the visibility of the selected assemblies and turn everything else off. When you invoke this tool, you will be prompted to select objects. Once you select objects and press ENTER, only those selected objects and any other object welded to them will remain turned on and rest everything will be turned off.

Cycling through the Display Type of the Objects

The **Change presentation type** tool is used to cycle through the display type of the selected objects. When you invoke this tool, you will be prompted to select objects. Once you select the objects and press ENTER, their display type will be changed to the next display type available in the **Display type** tab of the dialog box of those objects.

Restoring the Standard Display Type of the Objects

The **Standard Presentation** tool is used to restore the **Standard** display type of the selected objects. When you invoke this tool, you will be prompted to select objects. Once you select the objects and press ENTER, the display type of all the selected objects will be restored to **Standard**.

Hands-on Tutorial (STRUC/ BIM)	*In this tutorial, you will complete the following tasks:*

In this tutorial, you will complete the following tasks:
1. *Open the Elevation Tower file.*
2. *Change the UCS and insert a variable grating at the mezzanine level of the Elevation Tower, as shown in Figure 52.*
3. *Change the UCS and insert a variable grating at level 1 of the Elevation Tower, as shown in Figure 53.*
4. *Insert plates at level 1, as shown in Figure 53.*
5. *Create various grating and plate features to complete the model.*

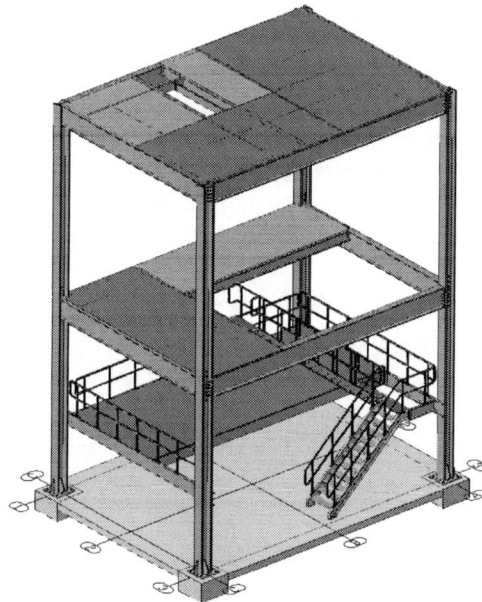

Figure 52 *The completed model for the tutorial*

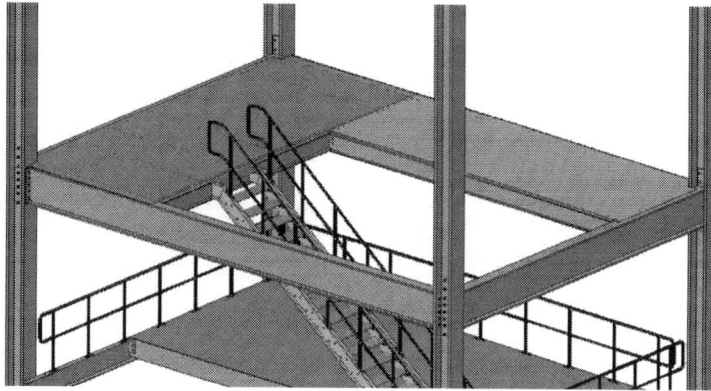

Figure 53 *The gratings and plate to be inserted at level 1*

Section 1: Inserting Grating at the Mezzanine Level

In this section, you will open **C08-Imperial-Elevation-Tower.dwg** or **C08-Metric-Elevation-Tower.dwg** file and then insert align the UCS at the mezzanine level. You will then insert variable rectangular grating at that level.

1. From the **C08 > Struc-BIM** folder, open the **C08-Imperial-Elevation-Tower.dwg** or **C08-Metric-Elevation-Tower.dwg** file, based on the preferred units.

    Before you align the UCS and insert the grating, it is better to turn off the visibility of all the objects, except the beams and columns. This will be done using the **Show only selected objects** tool on the **Quick views** tab of the **Advance Steel Tool Palette**. To make sure you select all beams and columns, you will use the **Select Similar** option and then invoke this tool.

2. Select one of the columns.

3. Right-click in the blank area of the drawing window and click **Select Similar**; all the beams, columns, and the bracing circular hollow sections are selected.

    Because you do not require the bracing circular hollow sections while inserting the grating, you need to deselect them.

4. Hold down the SHIFT key and one by one click on the eight circular hollow sections to deselect them.

5. From the **Advance Steel Tool Palette > Quick views** tab, invoke the **Show only selected objects** tool; the visibility of all the objects, except beams and columns, is turned off.

6. Press the ESC key to deselect the beams and columns. The model, with only beams and columns turned on, is shown in Figure 54.

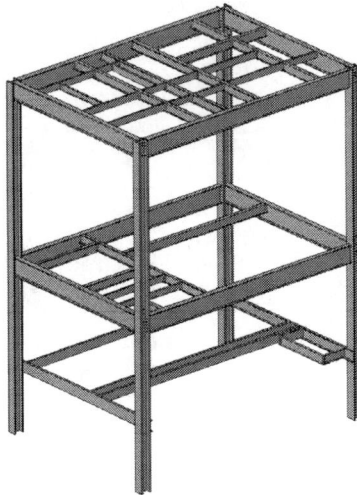

Figure 54 *The model, after isolating the layer of the beams*

Next, you need to align the UCS at the mezzanine level. For this, you will use the **UCS 3 points** tool from the **Advance Steel Tool Palette > UCS** tab. But first, it is recommended that you zoom closer to the area where you will align the UCS.

7. Zoom to the area shown in Figure 55. This figure also shows the vertices to be used to align the UCS.

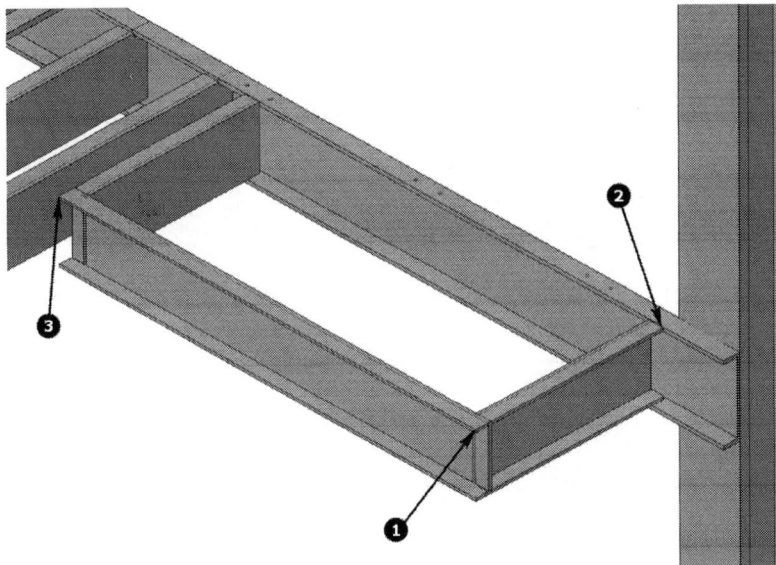

Figure 55 *Zoomed in view of the vertices to be selected to align the UCS*

8. From the **Advance Steel Tool Palette > UCS** tab, invoke the **UCS 3 points** tool; you are prompted to specify the new origin point.

9. Select the endpoint labeled as **1** in Figure 55; you are prompted to specify the point on the positive portion of the X-axis.

10. Select the endpoint labeled as **2** in Figure 55; you are prompted to specify the point on the positive-Y portion of the UCS XY plane.

11. Select the endpoint labeled as **3** in Figure 55; the UCS is aligned.

    You will now use the **Variable Grating, rectangular** tool to insert the grating on the mezzanine level using two corner points.

12. From the **Home** ribbon tab > **Objects** ribbon panel, invoke the **Variable Grating, rectangular** tool; you are prompted to define the start point of the grating's diagonal line.

13. Zoom close to the area shown in Figure 56 and select the endpoint of the channel labeled as **1** in this figure as the start point of the grating's diagonal; you are prompted to define the endpoint of the grating's diagonal line.

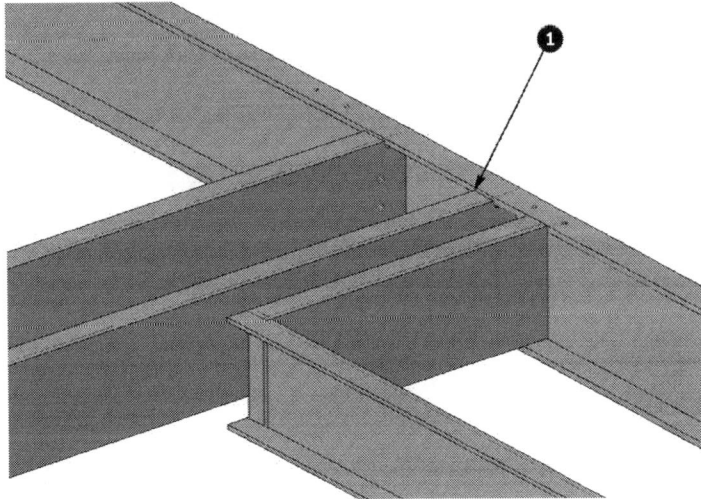

Figure 56 Specifying the start point of the grating's diagonal line

14. Zoom close to the area on the diagonally opposite side, as shown in Figure 57, and select the endpoint of the channel labeled as **2** in this figure as the endpoint of the grating's diagonal line; the grating is created and the **Advance Steel - Variable grating** dialog box is displayed with the **Shape & Connector** tab active.

    Next, you will change the class, size, and position of the grating.

15. From the **Grating class** list, select **Checker Plate** (for Imperial) or **Webforge Aluminium** (for Metric); the grating is changed in the model.

16. From the **Grating name** list, select **Checker Plate 1/2"** (for Imperial) or **Webforge A253A 25x3** (for Metric).

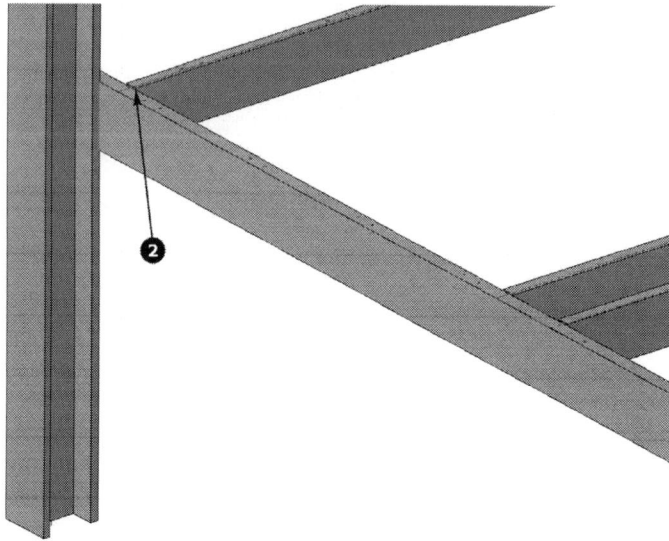

Figure 57 *Specifying the endpoint of the grating's diagonal line*

17. Activate the **Positioning** tab.

    Next, you need to make sure the grating is placed on top of the beams. For this, you will change the justification of the grating.

18. Click the first button in the **Justification** area; the value in the **Justification** edit box changes to **1.00**.

    While inserting the grating, you selected the endpoints of the channels. Therefore, you now need to edit the length and width of the grating so that it sits partially on the channels at the four ends.

19. Activate the **Shape & Connector** tab to return back to this tab.

20. In the **Grating length** edit box, enter **28'6"** or **8720**.

21. In the **Grating width** edit box, enter **12'2"** or **3650**.

    By default, the grating is inserted with hatching. In a large model, this slows down the performance of the drawing. Therefore, it is recommended to turn off the display of the hatches in the gratings.

22. Activate the **Display type** tab.

23. Select the **Exact** radio button; the display of hatching is turned off.

24. Close the dialog box. The zoomed in view of the model, after inserting the grating, is shown in Figure 58.

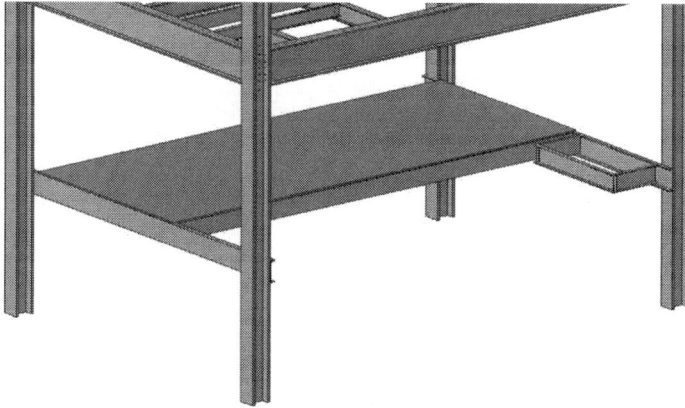

Figure 58 *The model, after inserting the grating*

Section 2: Inserting Grating and Plates at Level 1

In this section, you will align the UCS at level 1 and then insert variable rectangular grating at that level. You will then insert plates at this level.

1. Orbit and zoom close to the area of level 1 shown in Figure 59. This figure also shows the vertices to be used to align the UCS.

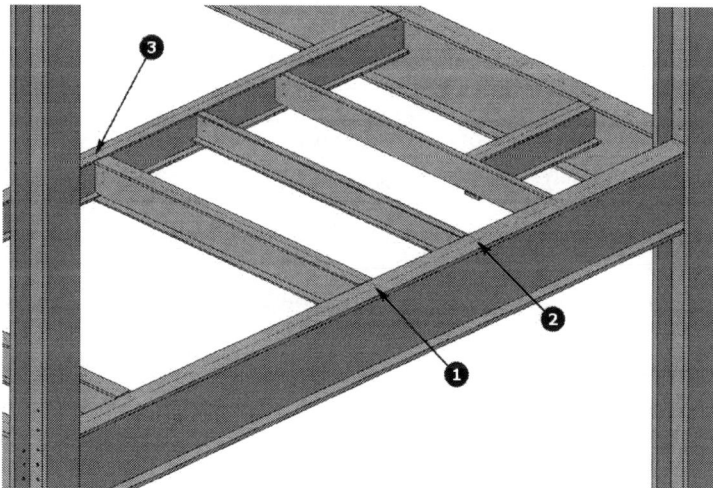

Figure 59 *Zooming in to the area to align the UCS*

2. Using the **UCS 3 points** tool, align the UCS to level 1. Use the node point labeled as **1** in Figure 58 as the origin point, node point labeled as **2** as the point along the X-axis, and the point labeled as **3** as the point along the Y-axis.

   Next, you will insert a rectangular variable grating using two corner points.

3. From the **Home** ribbon tab > **Objects** ribbon panel, invoke the **Variable Grating, rectangular** tool; you are prompted to define the start point of the grating's diagonal line.

4. Orbit and zoom close to the area shown in Figure 60 and select the node point of the beam labeled as **1** in this figure as the start point of the grating's diagonal; you are prompted to define the endpoint of the grating's diagonal line.

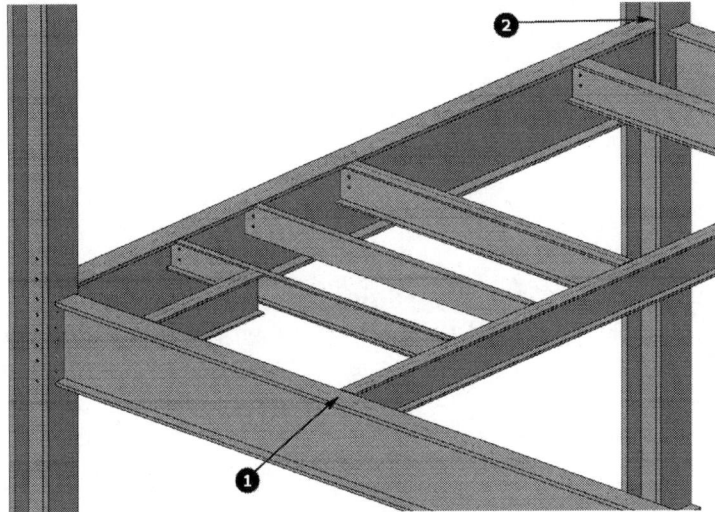

Figure 60 *The points to be used to insert the grating*

5. Select the node point of the beam labeled as **2** in Figure 60 as the endpoint of the grating's diagonal line; the grating is created and the **Advance Steel - Variable grating** dialog box is displayed with the **Shape & Connector** tab active.

   You will now modify the parameters of this grating.

6. From the **Grating class** list, make sure **Checker Plate** (for Imperial) or **Webforge Aluminium** (for Metric) is selected.

7. From the **Grating name** list, make sure **Checker Plate 1/2"** (for Imperial) or **Webforge A253A 25x3** (for Metric) is selected.

8. In the **Grating length** edit box, enter **21' 10"** or **6620**.

9. Activate the **Positioning** tab.

10. Click the first button in the **Justification** area; the value in the **Justification** edit box changes to **1.00**.

11. Activate the **Display type** tab.

12. Select the **Exact** radio button; the display of hatching is turned off.

13. Close the dialog box. The zoomed in view of the model, after inserting the grating, is shown in Figure 61.

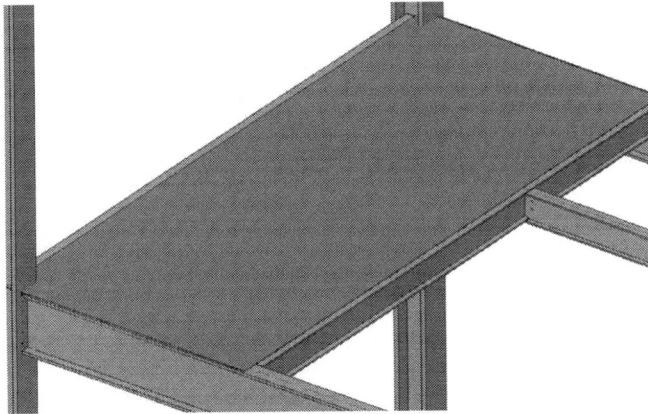

Figure 61   *The model, after inserting the grating*

Next, you will insert a rectangular plate at the same level.

14. From the **Home** ribbon tab > **Objects** ribbon panel, invoke the **Rectangular plate, 2 points** tool; you are prompted to define the start point of the plate's diagonal line.

15. Select the node point labeled as **1** in Figure 62 as the start point of the plate; you are prompted to define the endpoint of the plate's diagonal line.

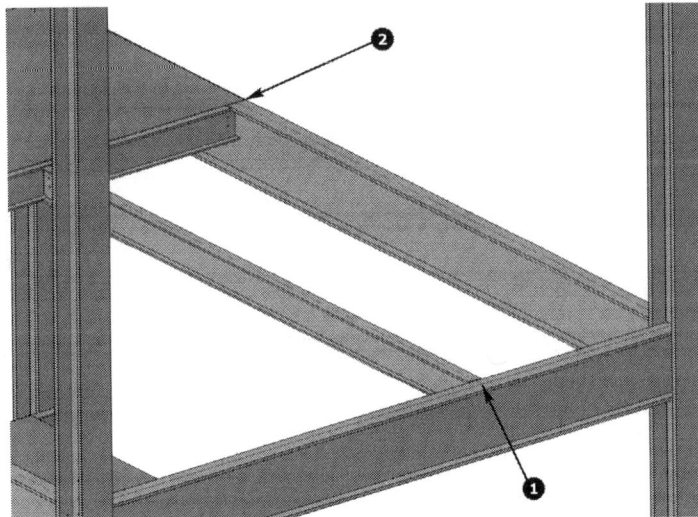

Figure 62   *Selecting the points to create the rectangular plate*

16. Select the node point labeled as **2** in Figure 63 as the endpoint of the plate; the plate is created and the **Advance Steel - Plate [X]** dialog box is displayed with the **Shape & Material** tab active.

You now need to modify the thickness of the plate so this thickness is the same as that of the grating.

17. In the **Thickness** edit box, enter **1/2"** or **25**.

18. Activate the **Positioning** tab.

19. Make sure the **Justification** edit box in the **Justification** area shows the value of **1.00**. If not, click the first button in this area.

20. Close the dialog box. The zoomed in view of the model, after inserting the plate, is shown in Figure 63.

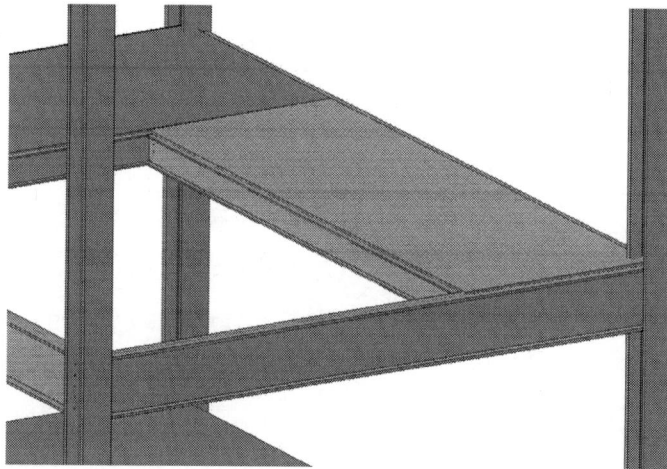

Figure 63   *The model, after inserting the plate*

Section 3: Creating Grating and Plate Features

In this section, you will create various grating and plate features so that they are not intersecting with the columns and the base plates of the handrail posts. However, you first need to turn on the visibility of all the objects.

1. From the **Advance Steel Tool Palette > Quick views** tab, invoke the **All visible** tool; the visibility of all the objects is turned on.

2. From the **Advance Steel Tool Palette > Features** tab, invoke the **Element contour** tool, as shown in Figure 64; you are prompted to identify the plate to be modified near a corner (reference system).

To make the selections, it is better to zoom close to the area where the feature will be created.

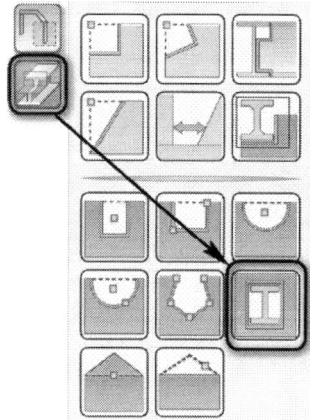

Figure 64 *The* *Element contour* *tool*

3. Zoom close to the area, as shown in Figure 65, and select the grating labeled as **1** in this figure; you are prompted to identify the object to cut to.

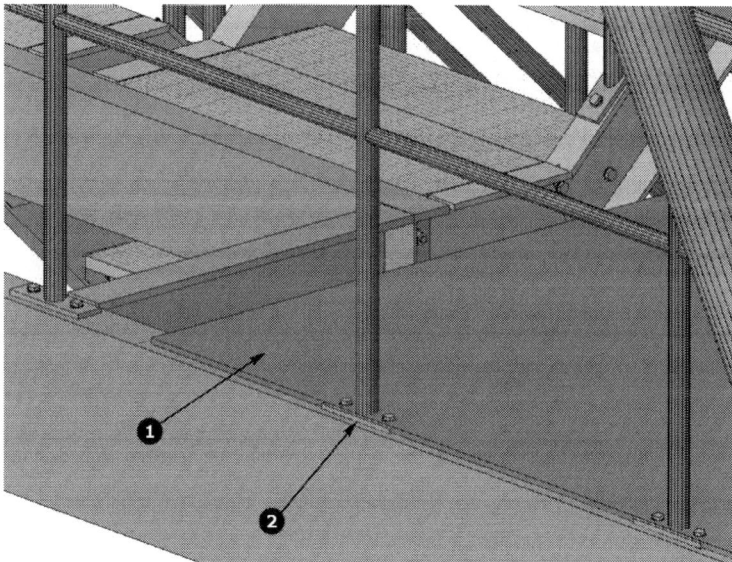

Figure 65 *The grating and the plate to create features*

4. Select the plate labeled as **2** in Figure 65; the feature is created and the **Advance Steel - Contour processing** dialog box is displayed with the **Contour** tab active.

5. In the edit box available in the **Gap width** area, enter **1/2"** or **12** as the value; the element contour feature is modified to create a 1/2" or 12 mm gap around the plate.

   Notice that because the thickness of grating is more than that of the post base plate, the feature does not cut through the grating. You will fix this next.

6. In the **Boundary** area, clear the **Side 1** tick box; the feature cuts through the grating, as shown in Figure 66.

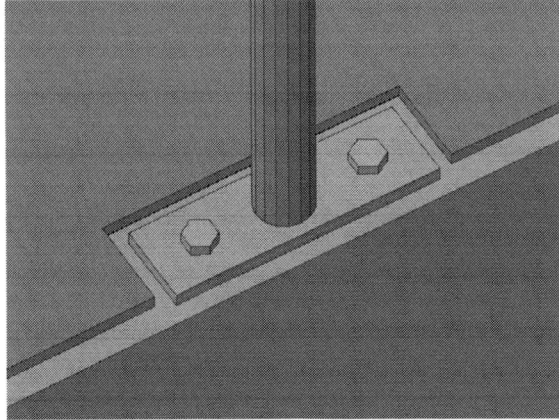

Figure 66 The zoomed in view of the element contour feature

Next, you need to create three more instances of this feature around the remaining plates of this handrail. To create the remaining instances, you will use the **Advance copy** tool available on the **Advance Steel Tool Palette > Tools** tab.

7. From the **Advance Steel Tool Palette > Tools** tab, invoke the **Advance copy** tool, as shown in Figure 67; the **Transform elements** dialog box is displayed.

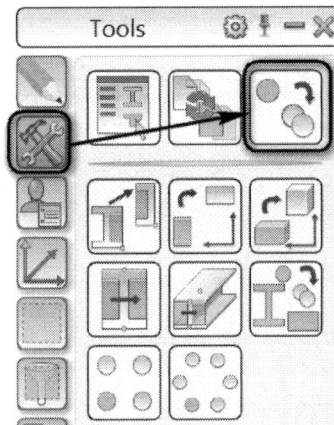

*Figure 67 The **Advance copy** tool*

8. Click the **Select objects** button at the top in this dialog box; you are returned to the drawing window and are prompted to select objects.

While creating the element contour feature, a rectangle was drawn in the middle of the plate around which the feature was created. You need to select this rectangle as the feature to copy.

9. Zoom close to the plate around which the element contour feature was created, as shown in Figure 66, and then select the rectangle that represents the contour cut feature.

10. Press ENTER to return to the dialog box.

11. Under the **Distance** radio button, clear the **Y** and **Z** tick boxes.

12. In the edit box available on the right of the **X** tick box, enter **-2' 11 1/2"** or **-908** as the value. This is the spacing between the railing posts.

13. In the **Number of copies** edit box, enter **3**.

Before you accept the feature, it is better to preview it to make sure all the values are correct.

14. From the top of the dialog box, click the **Preview** button; the **Preview** dialog box is displayed near the top left of the Advance Steel window and the preview of the feature is displayed in the drawing window.

15. Click **OK** in the **Preview** dialog box to accept the feature. The zoomed in view of the model, after creating the **Advance copy** feature, is shown in Figure 68.

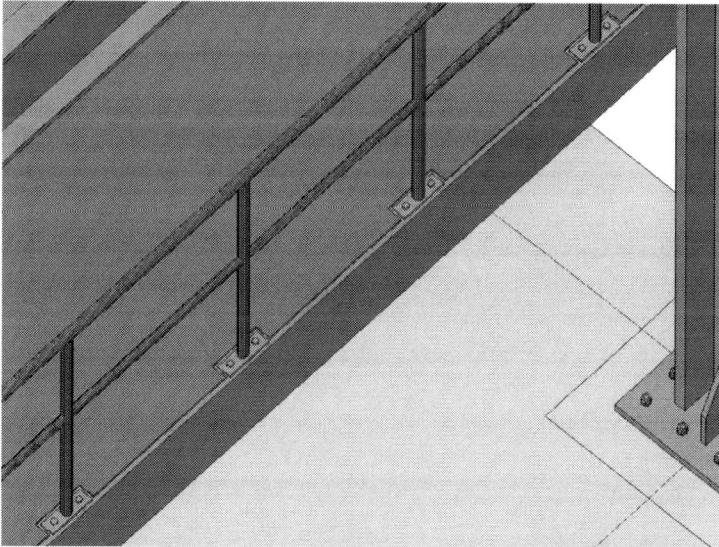

Figure 68 *Zoomed in view of the element contour feature*

16. Similarly, create the element contour features on the remaining two sides of the grating as well. You can create the first feature using the **Element contour** tool and the remaining features along that side using the **Advance copy** tool. To find the spacing for copying the feature, you can measure the distance between the corners of the plates. Also, note that because of the direction of the handrail base plates, you may need to change the distance direction from X to Y in some cases.

Next, you need to create the cut features between the plate and grating on level 2 and the columns at the **A1**, **C1**, and **C3** grid intersection points. However, before you do that, it is important to make sure you switch to the **Casing cross section** type.

17. From the **Objects** ribbon tab > **Switch** ribbon panel, click the **Exact cross section** button; the **Casing cross section** is now displayed in the ribbon panel.

    You are now ready to create the element contour feature between the column and the plate. However, for the ease of selection, it is better to zoom in the area where the feature will be created.

18. Zoom close to the area shown in Figure 69. This figure also shows the plate and the column to be selected to create the element contour feature.

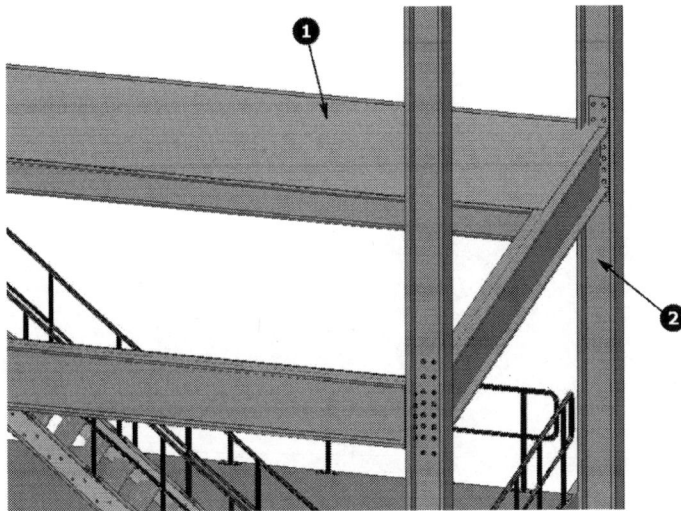

Figure 69 *The plate and the column to create features*

19. From the **Advance Steel Tool Palette > Features** tab, invoke the **Element contour** tool; you are prompted to identify the plate to be modified near a corner (reference system).

20. Select the plate labeled as **1** in Figure 69; you are prompted to identify the object to cut to.

21. Select the column labeled as **2** in Figure 69; the feature is created and the **Advance Steel - Contour processing** dialog box is displayed with the **Contour** tab active.

22. In the edit box available in the **Gap width** area, enter **1/2"** or **12** as the value; the element contour feature is modified to create a 1/2" or 12 mm gap around the plate.

23. Close the dialog box. The zoomed in view of the element contour feature is shown in Figure 70.

24. Similarly, create the same feature between the grating on this level and the two columns it is intersecting with.

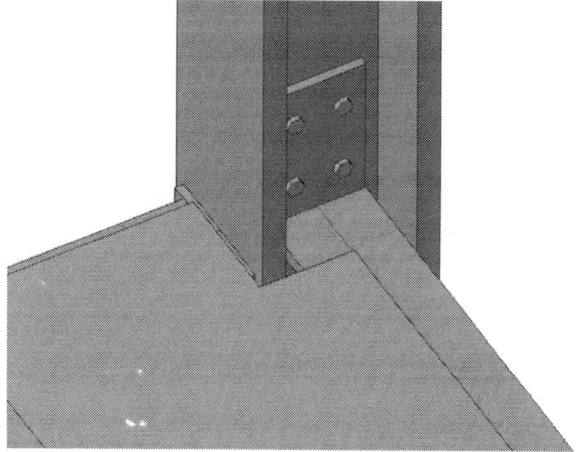

Figure 70 The element contour feature created around the column

Section 4: Inserting the Gratings and Plates at the Top Level (Optional)

In this section, you define the UCS on the top level of the tower and then insert the two gratings.

1. Align the UCS at the top level of the Elevation Tower.

2. Using the **Variable grating, rectangular** tool, insert the two gratings, as shown in Figure 71.

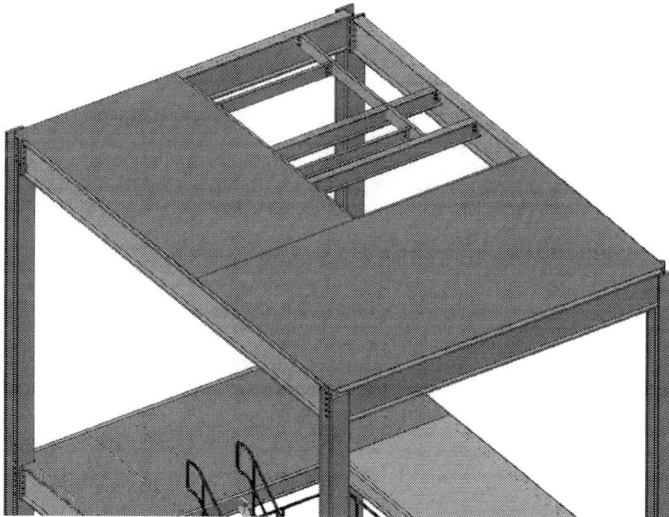

Figure 71 The model, after inserting the two gratings at the top level

3. Using the **Rectangular plate, 2 points** tool, insert the two plates, as shown in Figure 72.

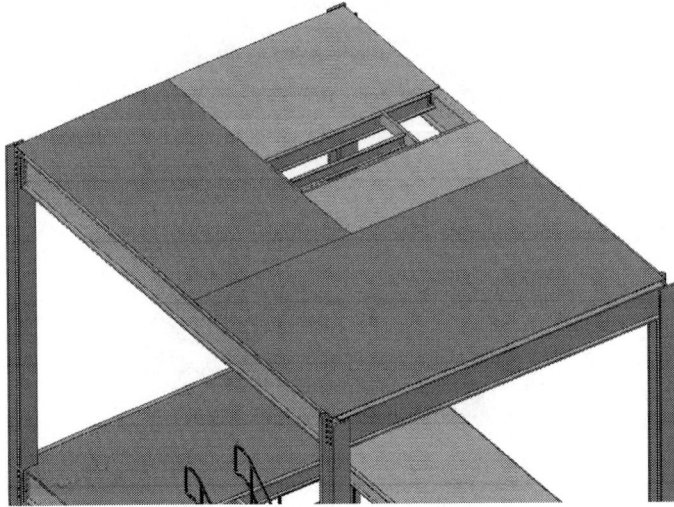

Figure 72 *The model, after inserting the two plates at the top level*

4. Save and close the file.

Skill Evaluation

Evaluate your skills to see how many questions you can answer correctly. The answers to these questions are given at the end of the book.

1. In Advance Steel, both plates and gratings can only be created while inserting stairs and railings. (True/False)

2. In Advance Steel, you can also created folded gratings. (True/False)

3. You can create a polyline using a plate as well. (True/False)

4. Advance Steel allows you to insert the bar grating as well. (True/False)

5. In Advance Steel, you can turn on or off the visibility of the selected objects. (True/False)

6. Which tool is used to create a plate using the shape of an existing closed polyline?

    (A) **Plate at poly line**   (B) **Polyline To Plate**
    (C) **P2P**   (D) **Convert to Plate**

7. Which tool is used to merge two plates into a single plate?

    (A) **Plate Joint**   (B) **Combine plates**
    (C) **Merge plates**   (D) **Joint plates**

8. Which tab of the **Advance Steel Tool Palette** provides the tools to control the visibility of the Advance Steel objects?

    (A) **Visibility**   (B) **Quick views**
    (C) **Display**   (D) **Views**

9. Which tool is used to create a rectangular contour cut on the selected flat or folded plate or a grating by specifying the center point of the contour?

    (A) **Center cut**   (B) **Rectangular cut**
    (C) **Center contour**   (D) **Rectangular contour, center**

10. The **Element contour** tool creates the cuts on plates and gratings depending on which two cross section types?

    (A) **Casing cross section**   (B) **Planar cross section**
    (C) **Hollow cross section**   (D) **Exact cross section**

Class Test Questions

Answer the following questions:

1. Explain briefly the types of plates that can be created in Advance Steel.

2. What does the **Create folded plate - without position adjustment** tool do?

3. Explain briefly the process of creating a lofted plate between a circle and a rectangle.

4. How do you turn on the visibility of all the objects using quick views?

5. Explain briefly how to create a cut in the grating around the base plate of handrail post.

Chapter 9 – Extended Modeling and Productivity Tools

The objectives of this chapter are to:

√ *Explain the process of creating cage ladder*
√ *Explain the process of creating spiral stair*
√ *Explain various selection tools available in the **Advance Steel Tool Palette***
√ *Explain how to create miter beam cuts*
√ *Explain how to cut beams at an object*
√ *Explain the process of manually connecting objects using bolted connections and welds*

CREATING CAGE LADDERS

Home Ribbon Tab > Extended Modeling Ribbon Panel > Cage ladder
Extended Modeling Ribbon Tab > Structural Elements Ribbon Panel > Cage ladder

Cage ladders are an important part of a large number of structural models. Therefore, Autodesk Advance Steel provides the **Cage ladder** tool to directly create the cage ladders. It is important for you to note that the cage ladder is created with its top at the current UCS position. Also, the cage is always created in the +XY quadrant. So in order to get the orientation of the cage right, in some cases, you may have to rotate the UCS around the Z axis.

You can also use this tool to create simple ladders without the cages. Figure 1 shows a simple ladder created using this tool and Figure 2 shows the cage ladder created using the same tool.

Figure 1 A simple ladder

Figure 2 A cage ladder

When you invoke this tool, you will be prompted to select the start point of the ladder. Once you specify the start point of the ladder, you will be prompted to select the point to define the height of the ladder. At this prompt, you can also move the cursor and specify the height of the ladder. However, if you are not sure about the height, you can specify some value close to the height and then later on override the height in the dialog box. Once you specify the height, the **Advance Steel - Cage ladder [X]** dialog box will be displayed, as shown in Figure 3. The tabs in the **Properties** category in this dialog box are similar to those discussed in earlier chapters. The remaining categories and tabs are discussed next.

Ladder Category > Ladder Tab

The options available in the **Ladder** tab, shown in Figure 4, are used to specify the size of the ladder. These options are discussed next.

Height

This option is used to override the height value that was defined while creating the ladder.

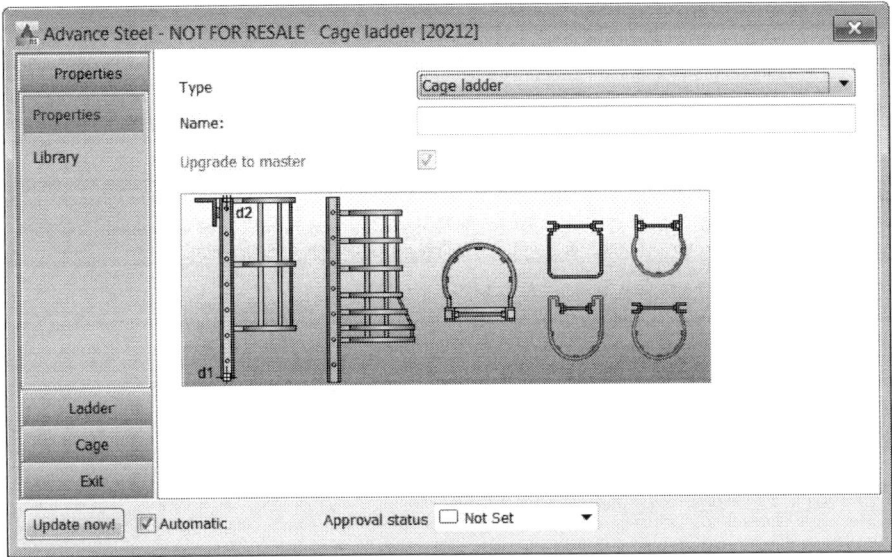

Figure 3 *The **Properties** tab of the **Properties** category*

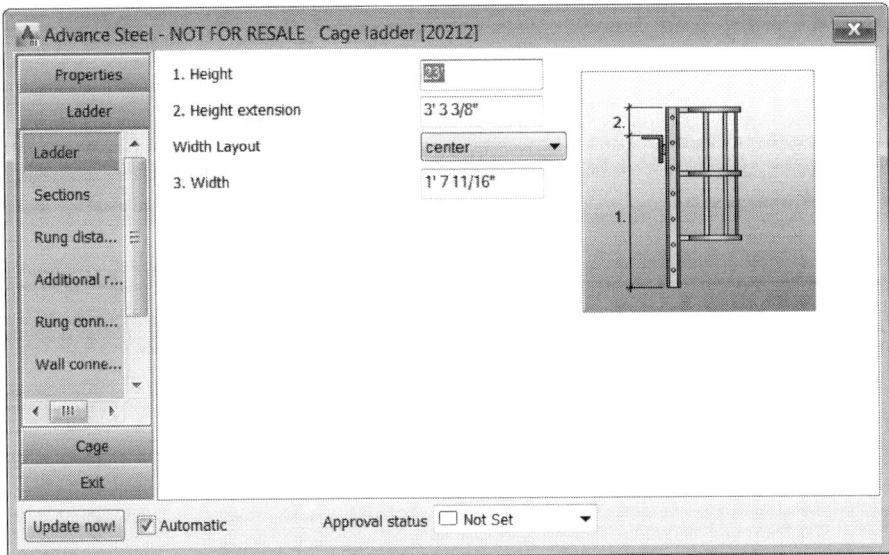

Figure 4 *The **Ladder** tab of the **Ladder** category*

Height extension
This option is used to specify the extension of the ladder at the top.

Width Layout
This list is used to specify whether width of the ladder will be measured from the inside, outside, or the center of the stringers.

Width

This edit box is used to specify the width of the ladder.

TAKE A NOTE **Note:** *Sometimes when you enter the width value, the connection box of the ladder will turn Red. This implies that the other values of the ladder do not comply with the width you have specified. In that case, you will have to go and change the other values of the cage and the ladder as well to suit the width value.*

Ladder Category > Sections Tab

The options available in the **Sections** tab, shown in Figure 5, are used to specify the sections type of the ladder stringer and ladder rung. These options are similar to those discussed in the earlier chapters.

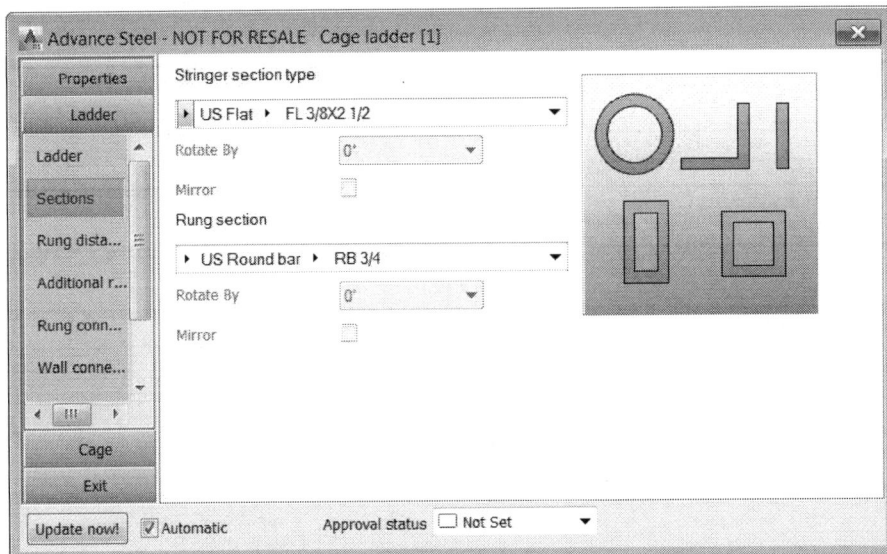

Figure 5 The Sections tab of the Ladder category

Ladder Category > Rung distances Tab

The options available in the **Rung distances** tab, shown in Figure 6, are used to specify the rung distances. These options are discussed next.

Distance layout

This list is used to specify whether the rung distances will be specified from the bottom or from the top.

Position

This list is used to specify whether the rung position will be specified from the top, middle, or bottom.

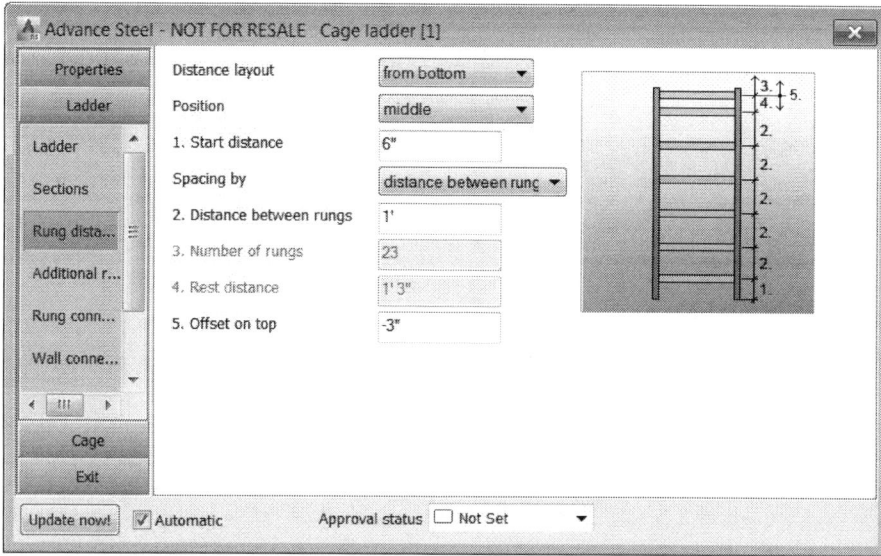

*Figure 6 The **Rung distances** tab of the **Ladder** category*

Start distance

This edit box allows you to specify the start distance of the first rung. Figure 7 shows the zoomed in view of the bottom of the ladder with the first rung start distance of 6" or 150mm and Figure 8 shows the same ladder with the first rung distance of 1' or 300mm.

Figure 7 The start rung at 6" or 150mm

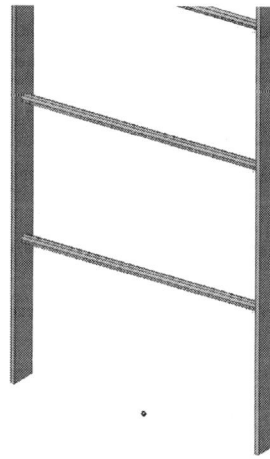

Figure 8 The start rung at 1' or 300mm

Spacing by

This list is used to specify whether the spacing of the rungs will be defined by the distance between the rungs or the total number of rungs. By default, the **distance between rungs** is selected from this list. As a result, you can specify the distance between the rungs in the **Distance**

between rungs edit box available below this list. However, if you select **number of rungs**, you can specify the total number of rungs in the **Number of rungs** edit box, also available below this list.

Rest distance
This edit box is used to specify the rest distance.

Offset on top
This edit box is used to offset the top rung, if required. The positive value will offset the last rung upwards and the negative value will offset it downwards.

Ladder Category > Additional rungs Tab
The options available in the **Additional rungs** tab, shown in Figure 9, are used to specify whether or not you want to create additional rungs. The additional rungs are generally required if you have extended the ladder using the **Height extension** edit box in the **Ladder** tab. These options are discussed next.

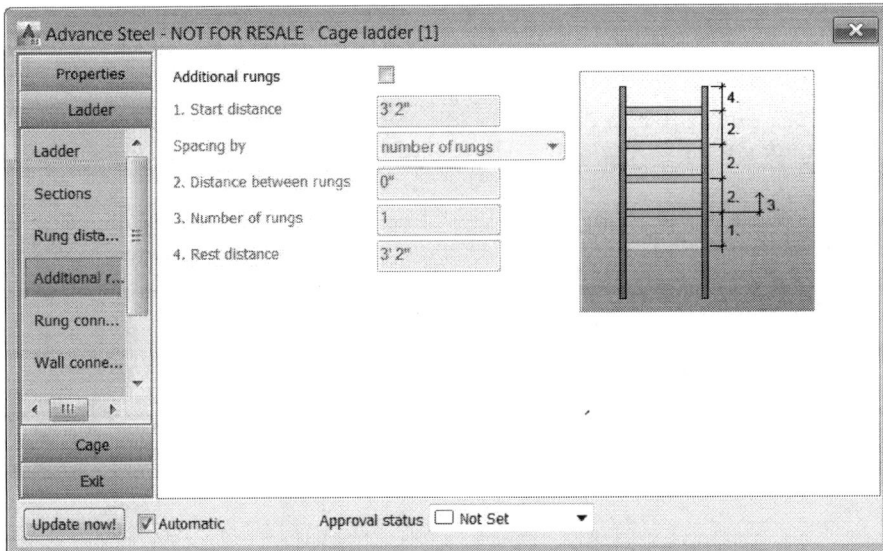

Figure 9   *The **Additional rungs** tab of the **Ladder** category*

Additional rungs
This tick box is used to specify whether or not the additional rungs will be created. By default, this tick box is not selected. As a result, the options in this tab are disabled. If you select this tick box, these options will be enabled.

Start distance
This edit box is used to specify the start distance of the additional rung. Note that this value is calculated from the last rung of the ladder.

Spacing by

This list is used to specify whether the spacing of the rungs will be defined by the distance between the rungs or the total number of rungs. Depending on the option you select from this list, you can specify the total number of rungs or the spacing in the edit boxes available below this list.

Ladder Category > Rung connection Tab

The options available in the **Rung connection** tab, shown in Figure 10, are used to specify the connection type between the rungs and the stringers. The weld options in this tab are similar to those discussed in earlier chapters. The remaining options are discussed next.

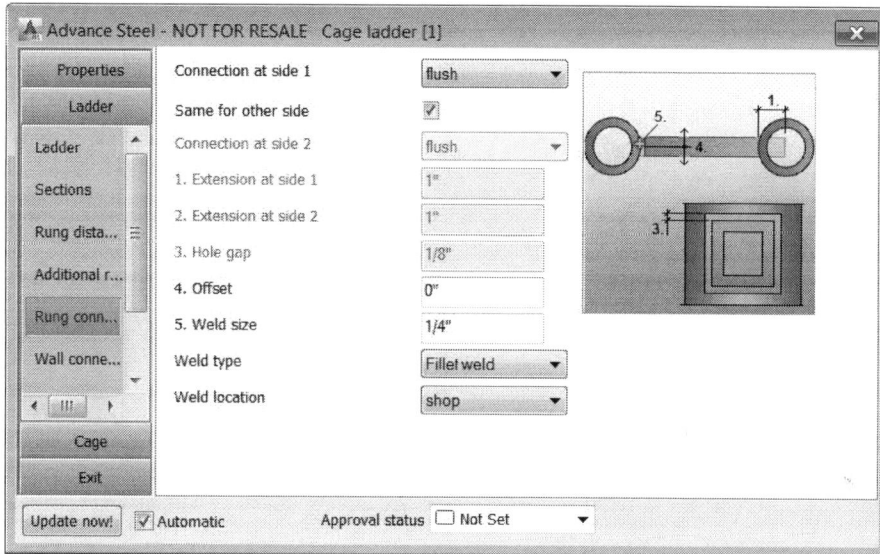

*Figure 10 The **Rung connection** tab of the **Ladder** category*

Connection at side 1

The options available in this list are discussed next.

flush

This option ensures the rungs are connected flush with the inside of the stringers. Figure 11 shows the zoomed in view of the rungs flush with the stringers.

straight

This option ensures the rungs are connected straight to the stringers.

extended

This option extends the rungs outside the stringers. The extension value can be defined in the **Extension at side 1** and **Extension at side 2** edit boxes. When using this option, you can also specify the hole gap for the stringer in the **Hole gap** edit box. Figure 12 shows the extended rungs with the hole gaps.

Figure 11 *The rungs connected flush with the stringers*

Figure 12 *The rungs extended on both sides with a hole gap value defined for the stringer*

Same for other side
This tick box is selected to ensure the same connection is defined with the other stringer as well. By default, this tick box is selected.

Offset
The edit box is used to offset the rungs from the middle of the stringers.

Ladder Category > Wall connection Tab
The options available in the **Wall connection** tab, shown in Figure 13, are used to connect the ladder to the wall. These options are discussed next.

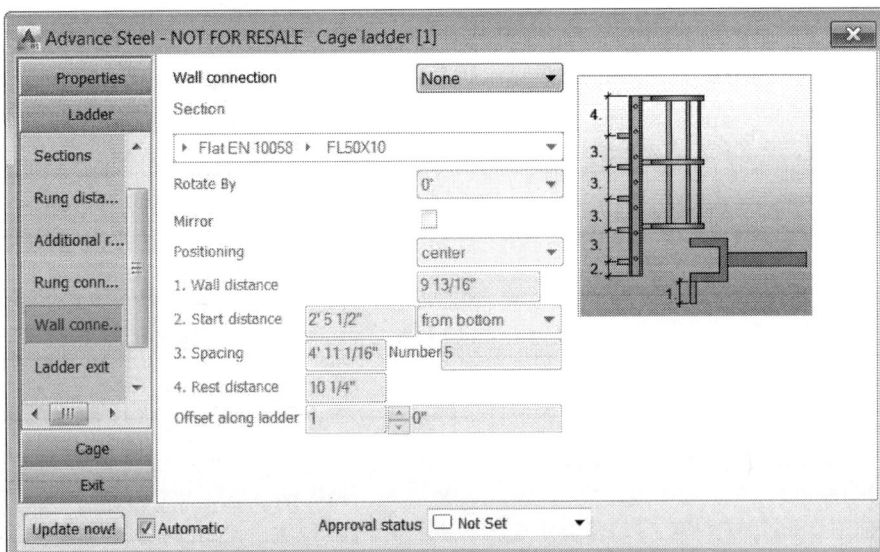

Figure 13 *The **Wall connection** tab of the **Ladder** category*

Wall connection

This list is used specify the wall connection type. By default, **None** is selected from this list. As a result, the options in this tab are disabled. The remaining options in this list are discussed next.

Section

This option is selected to use a steel section to connect the ladder with the wall. You can select the section type and size from the **Section** list. The distance of the ladder from the wall can be specified in the **Wall distance** edit box. Figure 14 shows a flat section used to connect the ladder to the wall. Note that in this case, the wall is not displayed.

Folded plate

This option is selected to use a folded plate as the section to connect the ladder to the wall. You can specify the thickness, height, fold radius, and leg length of the folded plate in the edit boxes available below this list. The distance of the ladder from the wall can be specified in the **Wall distance** edit box. Figure 15 shows a folded plate used to connect the ladder to the wall. Note that in this case also, the wall is not displayed.

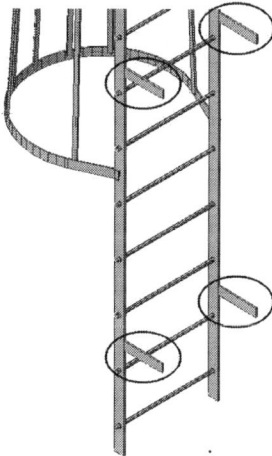

Figure 14 *Flat sections used to connect the ladder to the wall*

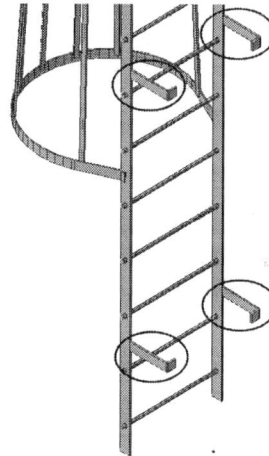

Figure 15 *Folded plates used to connect the ladder to the wall*

Plate

This option is selected to use a flat plate as the section to connect the ladder to the wall. You can specify the thickness and height of the plate in the edit boxes available below this list.

Positioning

This list is used to specify the position of the connection sections on the stringer.

Start distance

This edit box is used to specify the start distance of the wall connections. The list on the right of this edit box allows you to specify whether the start distance will be specified from the bottom of the ladder or the top.

Spacing

This edit box is used to specify the spacing between the connecting elements on the ladder.

Offset along ladder

This spinner is used to offset the specified connecting elements along the ladder. You can use the spinner to go to the connecting elements that needs to be offset and then enter the offset value in the edit box on the right of this spinner.

Ladder Category > Ladder exit Tab

The options available in the **Ladder exit** tab, shown in Figure 16, are used to specify the parameters at the exit of the ladder. These options are discussed next.

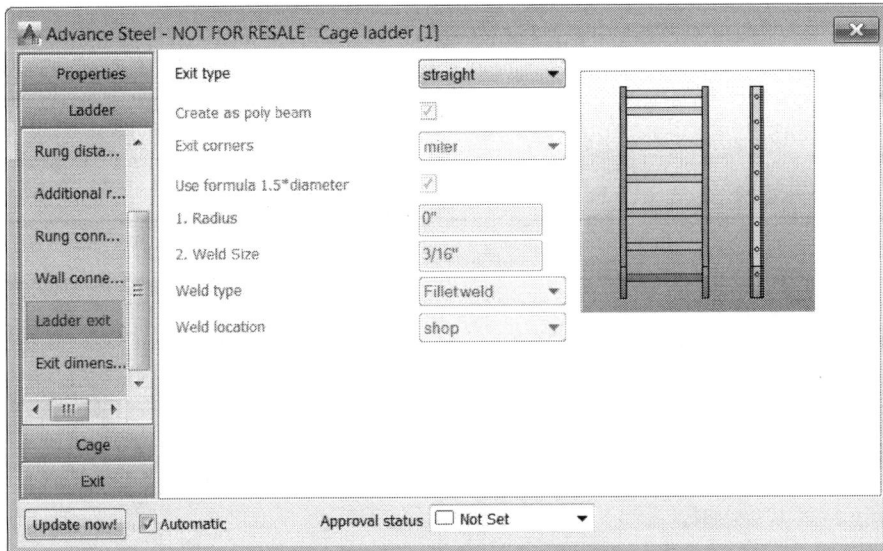

*Figure 16 The **Ladder exit** tab of the **Ladder** category*

Exit type

This list is used specify the exit type of the ladder. By default, **straight** is selected from this list. As a result, all the other options in this tab are disabled. There are five more types of exits you can select from this list. For all these exits, selecting the **Create as poly beam** tick box will ensure poly beams are used to create those exits. However, if you clear this tick box, you can specify the corner, radius, and weld sizes in the edit boxes available below this list. Figure 17 shows the **Type 2** exit and Figure 18 shows the **Type 5** exit.

Ladder Category > Exit dimensions Tab

The options available in the **Ladder exit** tab will vary, depending on the exit type selected from the **Exit type** tab. You can use these values to specify the sizes and dimensions of the exit type. Figure 19 shows the dimensions related to the **Type 5** exit.

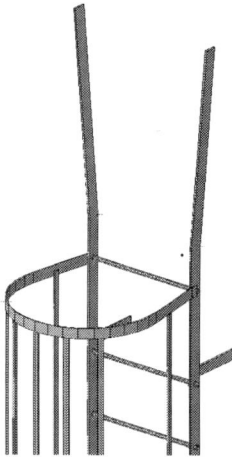

*Figure 17 The **Type 2** exit*

*Figure 18 The **Type 5** exit*

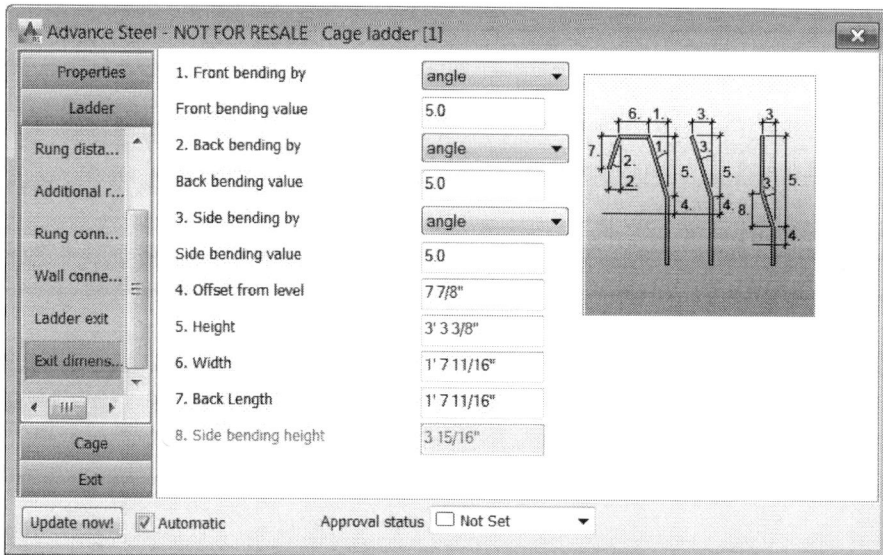

*Figure 19 The **Exit dimensions** tab of the **Ladder** category*

Cage Category > General Tab

The options available in the **General** tab, shown in Figure 20, are used to specify the parameters related to the cage type. These options are discussed next.

Cage

This list is used to specify the type of cage to be used. By default, **None** is selected from this list. As a result, all the options in this tab are disabled. There are four types of cages that can be selected from this list. Figure 21 shows the zoomed in view of **Type 1** cage and Figure 22 shows the zoomed in view of **Type 4** cage.

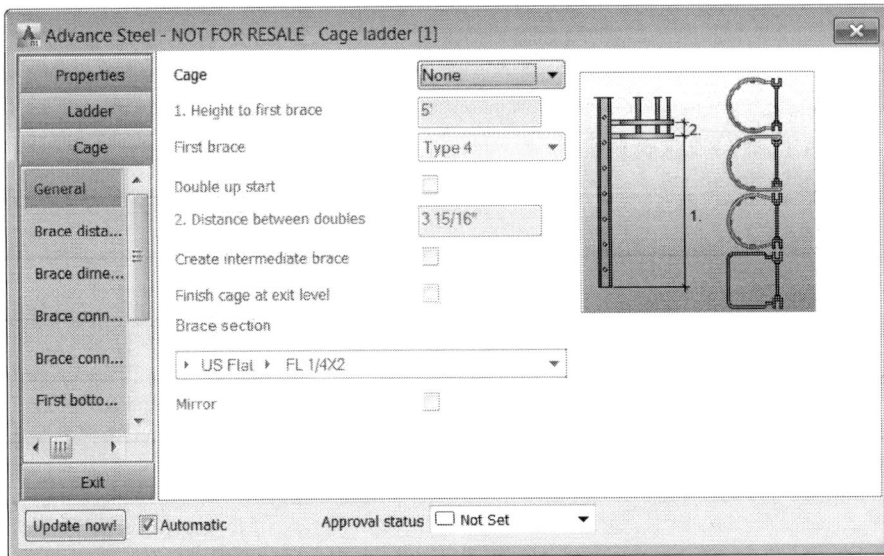

Figure 20 The **General** *tab of the* **Cage** *category*

Figure 21 The **Type 1** *cage*

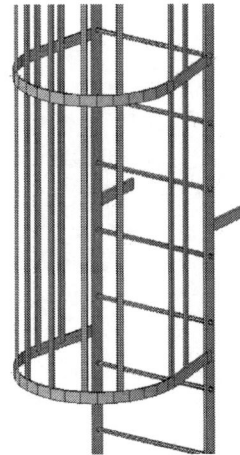

Figure 22 The **Type 4** *cage*

Height to first brace

This edit box is used to specify the height to the first brace of the bottom cage from the bottom of the ladder.

First brace

This list is used to specify a different cage for the first brace. Generally, the radius of the first brace is more than that of the rest of the braces. Therefore, it is better to not select **Same as other** from this list.

Double up start

If this tick box is selected, double braces will be inserted at the start cage. In this case, you can also define the distance between the two braces at the start using the **Distance between doubles** edit box available below this tick box. Figure 23 shows double braces at the start.

Figure 23 *Double braces at the start cage*

Create intermediate brace

This tick box is used to create an intermediate brace at the start.

Finish cage at exit level

In cases where you have extended the ladder using the **Height extension** edit box in the **Ladder** category > **Ladder** tab, you can select this tick box to finish the cage at the exit level. Figure 24 shows the zoomed in view of the top of the ladder with this tick box cleared and Figure 25 shows the zoomed in view of the top of the ladder with this tick box selected.

Brace section

This list is used to specify the section of the brace.

Cage Category > Brace distances Tab

The options available in the **Brace distances** tab, shown in Figure 26, are used to specify the brace distances. These options are discussed next.

Distance for first brace different

Selecting this tick box allows you to specify a different value for the first brace. In this case, you can specify the distance between the first braces in the **Distance between first braces** edit box available below this tick box.

Figure 24 *Cage not finishing at the exit level*

Figure 25 *Finishing cage at the exit level*

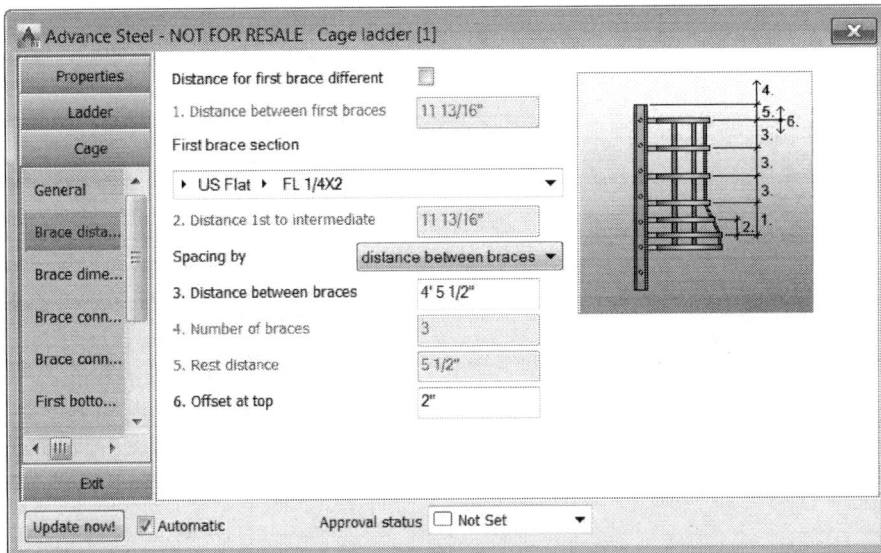

Figure 26 *The **Brace distances** tab of the **Cage** category*

Distance 1st to intermediate

This edit box is only available if you select the **Create intermediate brace** tick box from the **Cage** category > **General** tab. As the name suggests, this tick box is used to specify the distance between the first and the intermediate brace.

Spacing by

This list is used to specify whether the spacing between the braces will be defined using the distance between the braces of the total number of braces. You can specify their respective values in the edit boxes available below this list.

Offset at top

This edit box is used to specify the offset value for the top brace.

Cage Category > Brace dimensions Tab

The options available in the **Brace dimensions** tab, shown in Figure 27, are used to specify the brace size. Selecting the **Create as poly beam** tick box will ensure the brace is created as a single section. If you clear this tick box, the straight and curved sections of the brace will be created using separate sections that will be welded together.

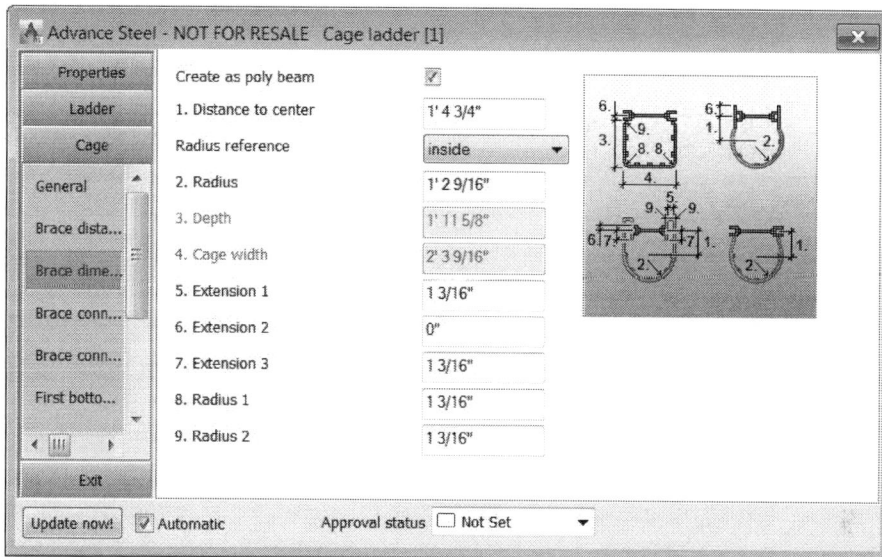

*Figure 27 The **Brace dimensions** tab of the **Cage** category*

Cage Category > Brace connection Tab

The options available in this tab are used to specify whether the braces will be welded or bolted to the stringers. These options are similar to those discussed in earlier chapters.

Cage Category > Brace connection details Tab

The options available in this tab are used to specify the bolt or weld parameters of the brace connections. These options are also similar to those discussed in earlier chapters.

Cage Category > First bottom brace dimension Tab

The options in this tab will only be available if you have not selected **Same as other** from the **First brace** list in the **Cage** category > **General** tab. In that case, you can select different dimension values for the bottom cage. As mentioned earlier, the radius of the bottom brace is generally bigger than the rest of the braces. So you can specify that radius value in this tab. Figure 28 shows the first brace the same as the rest and Figure 29 shows the first brace radius more than that of the rest.

Figure 28 *All brace radii the same*

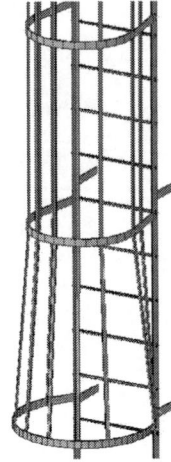

Figure 29 *The first brace with a bigger radius*

Note: The **First bottom brace connection** and **First brace connection details** tabs are similar to those discussed earlier.

Cage Category > Bands Tab

The options available in the **Bands** tab, shown in Figure 30, are used to specify the band parameters. These options are discussed next.

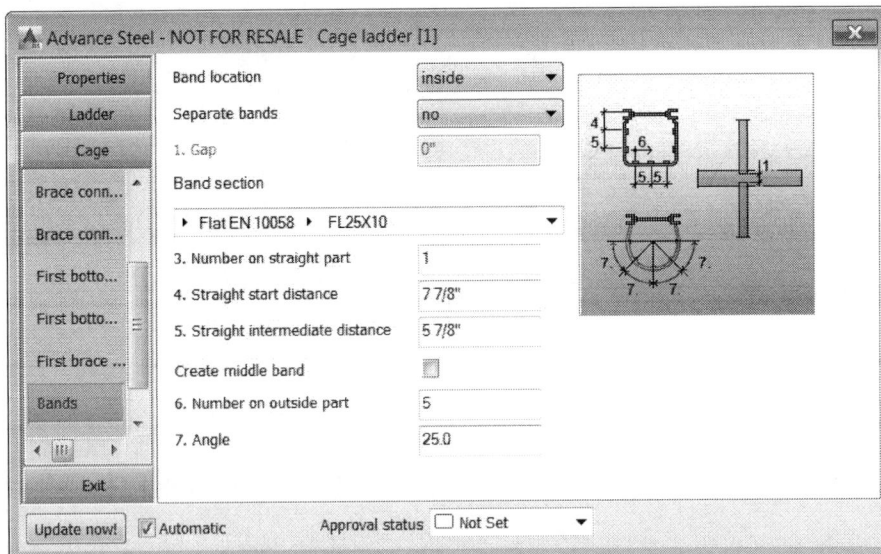

Figure 30 *The **Bands** tab of the **Cage** category*

Band location

This list is used to specify whether the bands will be located inside, outside, or in the middle of the braces.

Separate bands

This list is used to specify whether you want single pieces of bands running between all the cages, separate bands per brace, or separate bands per cage.

What I do

I generally prefer creating separate bands per cage, especially when my cage ladder has a cage at the exit as well. That way I can control the bands at the exit cage independent of the bands of the rest of the cages.

Gap

This edit box will only be available if you select **per brace** or **per cage** from the **Separate bands** list. In those cases, this edit box allows you to enter the gap at the start and end of the bands.

Band sections

This list is used to specify the type and size of the band sections.

Number on straight part

This edit box is used to specify the number of bands along the straight part of the brace section.

Start straight distance

This edit box is used to specify the start distance of the first band along the straight part of the brace section.

Straight intermediate distance

In case of creating more than one band along the straight part of the brace section, this edit box is used to specify the intermediate distance between the bands.

Create middle band

If you specify even number of bands in the **Number on outside part** edit box, selecting this tick box will create a middle band as well, making the total number or bands on the outside part to be an odd number.

Number on outside part

This edit box is used to specify total number of bands required on the outside part of the brace section. In case of **Type 2**, **Type 3**, and **Type 4** cages, these are the bands on the curved part of the brace section.

Angle

This edit box is used to specify the intermediate angle between the bands on the outside part.

Note: *The options available in the **Band connection** tab are similar to those discussed in earlier chapters.*

Exit Category

The options available in most tabs of this category are similar to those discussed earlier. However, some of the options in the **Top cage** and **Opening post** tabs are different. These tabs are discussed next.

Exit Category > Top cage Tab

The options available in the **Top cage** tab, shown in Figure 31, are only available if you select the **Finish cage at exit level** tick box from the **Cage** category > **General** tab.

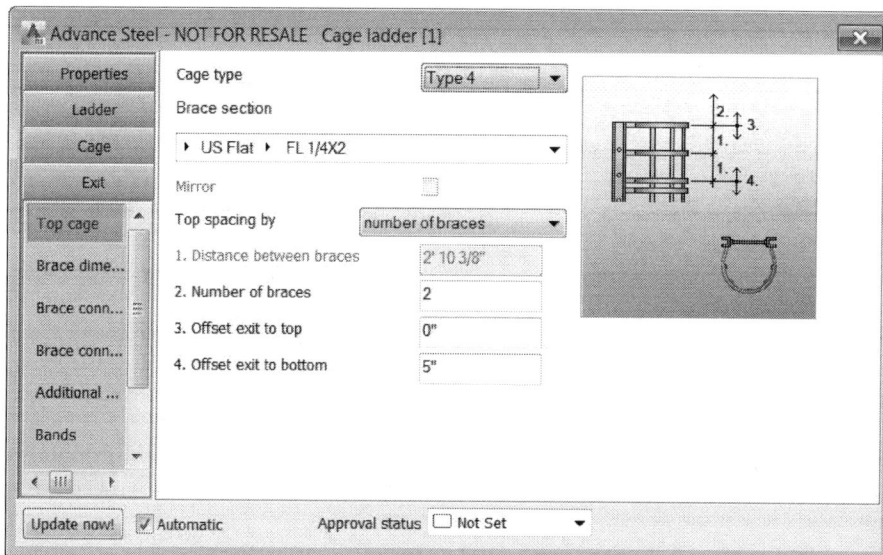

Figure 31 *The **Top cage** tab of the **Exit** category*

The options in this tab are used to define the top exit cage parameters. Most of the options in this tab are similar to those discussed earlier, except the **Cage** list. In addition to selecting the four cage types, this list also provides the options to select the cage with the right or left exit. Figure 32 shows the zoomed in view of the **Type 4** top cage with left exit and Figure 33 shows the same cage with the right exit.

Exit Category > Opening post Tab

The options available in the **Opening post** tab, shown in Figure 34, are only available if you select one of the cages with opening from the **Cage type** list in the **Top cage** tab. These options are discussed next.

Same as stringer

Select this tick box to use the same section type and size for the opening post as that of the

Figure 32 *Top cage with left exit*

Figure 33 *Top cage with right exit*

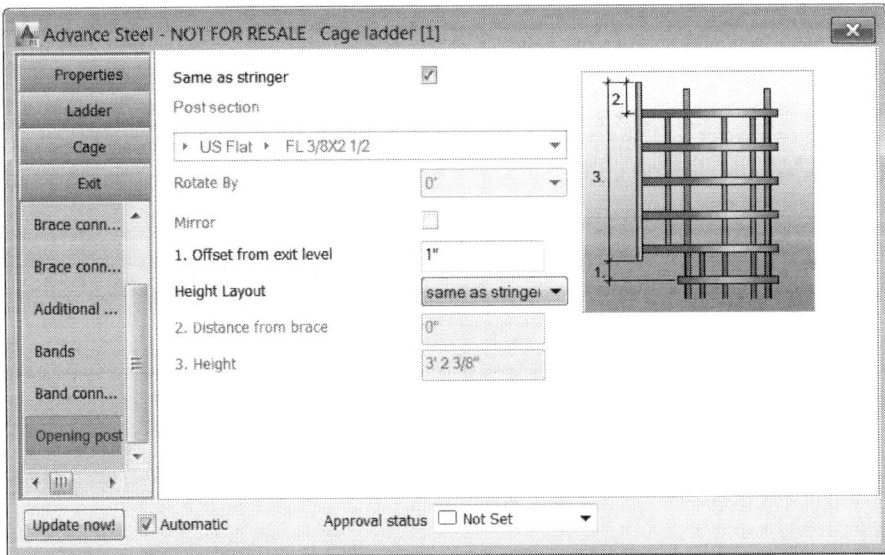

Figure 34 *The **Opening post** tab of the **Exit** category*

stringer. If you clear this tick box, you can select a different type and size for the opening post from the **Post section** list available below this tick box.

Offset from exit level

This edit box is used to define an offset value for the opening section from the exit level.

Height layout

This list is used to specify how the height of the opening post will be defined. These options are discussed next.

same as stringer
Select this option to ensure the opening post terminates at the same level as the stringer.

distance from brace
Select this option to define the height of the opening post in terms of its distance from the brace. The distance value can be specified in the **Distance from brace** edit box available below the **Height layout** list.

total
Select this option to specify the height of the opening post in terms of its total length value. This value can be defined in the **Height** edit box available below the **Height layout** list.

CREATING SPIRAL STAIRS

> **Extended Modeling Ribbon Tab > Structural Elements Ribbon Panel >
> Spiral staircase**

In Autodesk Advance Steel, you can use the **Spiral staircase** tool to create a spiral staircase by defining its center point, height, and the outside radius value. You can also specify the first or last step while defining the parameters of the spiral staircase. Figure 35 shows a spiral staircase with the handrails and Figure 36 shows the same staircase without the handrails.

Figure 35 *The spiral staircase with handrails*   **Figure 36** *The spiral staircase without handrails*

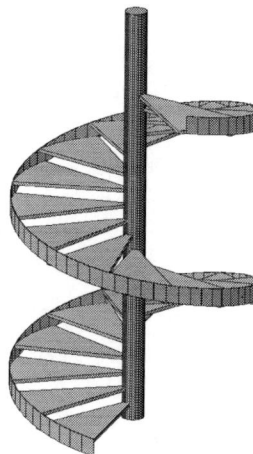

When you invoke this tool, you will be prompted to specify the center point of the spiral. Once you specify the start point, you will be prompted to select the point to define the height of the spiral. You can move the cursor and specify a point in the +Z direction to specify the height of the spiral. Once you specify the height, you will be prompted to specify whether you want to specify the first or last step. You can press ENTER at this prompt if you do not want to specify the steps. You will then be prompted to select a point that defines the direction and outside radius of the spiral. Once you specify all these parameters, the **Advance Steel - Spiral stairs [X]** dialog box will be displayed with the **Properties** category > **Properties** tab active. The options available in the tabs of this category and some of the other categories are similar to those discussed in earlier chapters. The rest of the categories are discussed next.

Column & Stringer Category > General Properties Tab

The options available in the **General Properties** tab, shown in Figure 37, are discussed next.

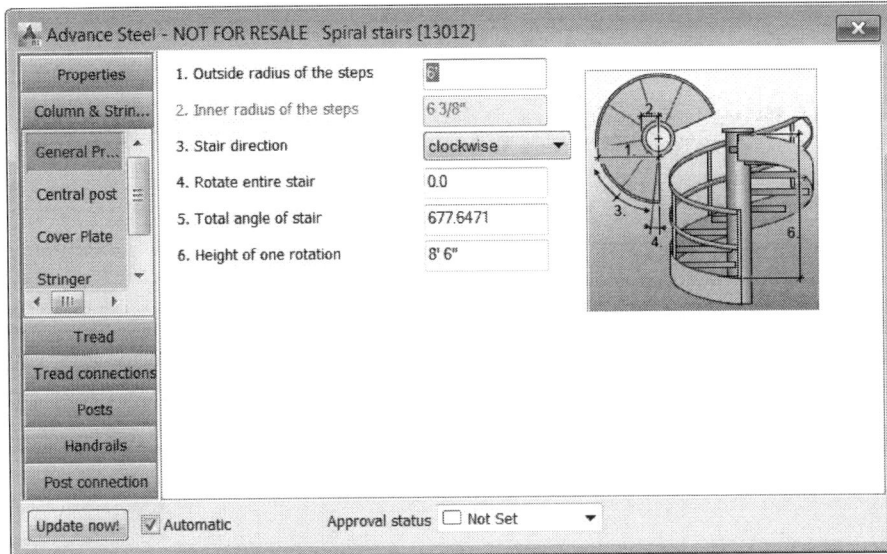

Figure 37 The General Properties tab of the Column & Stringer category

Outside radius of the steps

This edit box is used to override the outside radius value that was defined in the prompt sequence while creating the spiral stairs.

Inner radius of the steps

This edit box is only available if you clear the **Create central post** tick box from the **Column & Stringer** category > **Central post** tab. In that case, you can specify the inner radius of the steps in this edit box.

Stair direction

This list is used to specify whether the stair direction will be clockwise or counterclockwise.

Rotate enter stair

This edit box is used to rotate the entire staircase by the specified angle value.

Total angle of stair

This edit box is used to specify the total angle of the spiral stair. Note that the value in the **Height of one rotation** edit box is linked to that of this edit box. As a result, changing the total angle of the stair will also change the height of one rotation of the stair.

Height of one rotation

This edit box is used to specify the height of one rotation of the spiral stair. As mentioned earlier, this value is linked to the value of the total angle of the stair specified in the **Total angle of stair** edit box.

Column & Stringer Category > Central post Tab

The options available in the **Central post** tab, shown in Figure 38, are used to specify whether or not the central post will be created. If you decide to create the central post, you can also specify its section and offsets using the options available in this tab.

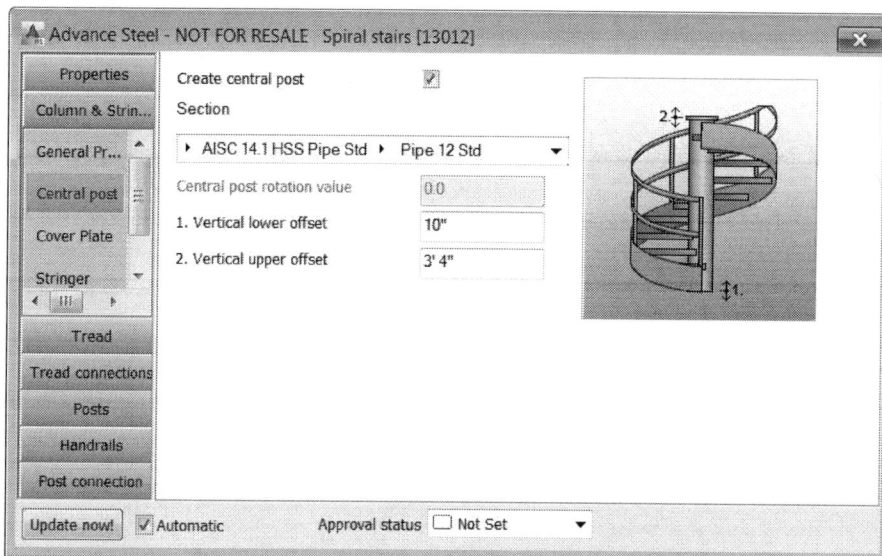

*Figure 38 The **Central post** tab of the **Column & Stringer** category*

Column & Stringer Category > Cover Plate Tab

The options available in the **Cover Plate** tab, shown in Figure 39, are used to specify whether or not the cover plate will be placed on top of the central post. If you decide to place the cover plate, you can specify its size layout and size using the options available in this tab. These options are similar to those discussed in earlier chapters.

Figure 40 shows a spiral staircase with the central post without the cover plate and Figure 41 shows the same staircase with the central post with the cover plate.

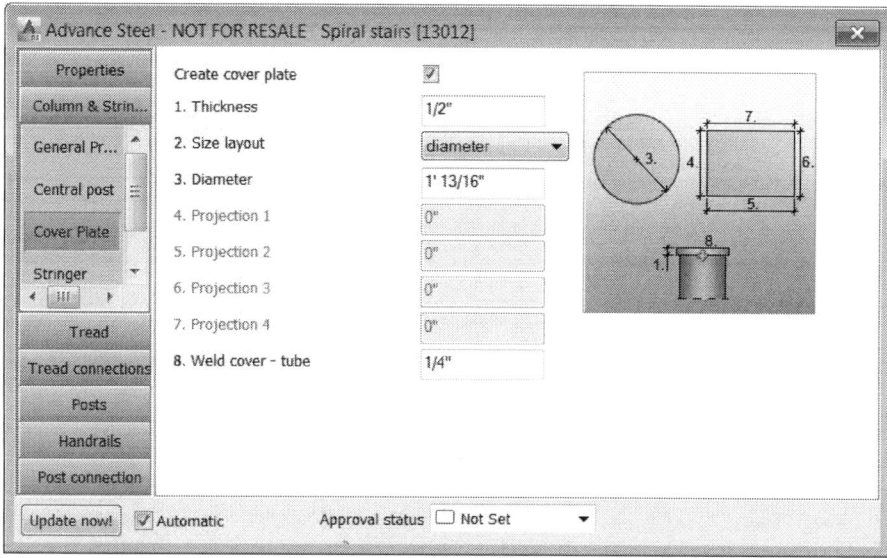

Figure 39 *The **Cover Plate** tab of the **Column & Stringer** category*

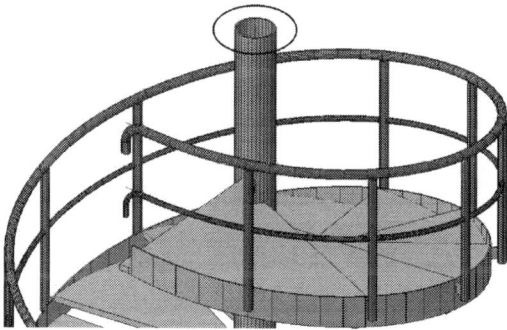

Figure 40 *The central post without the cover plate*

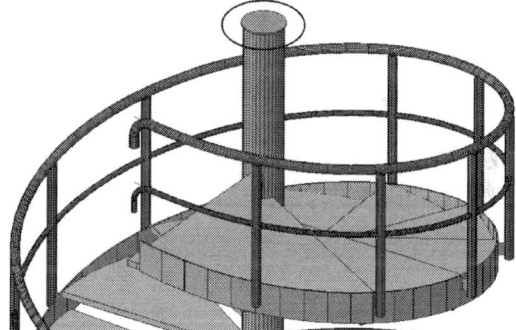

Figure 41 *The central post with the cover plate*

Column & Stringer Category > Stringer Tab

The options available in the **Stringer** tab, shown in Figure 42, are used to specify the stringer parameters. These options are discussed next.

Create stringer

This tick box is used to specify whether or not the stringer will be created.

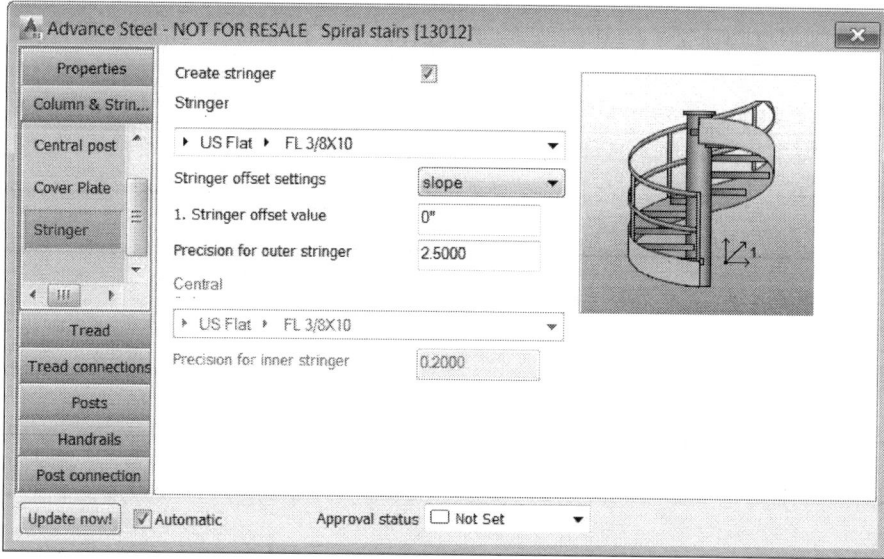

*Figure 42 The **Stringer** tab of the **Column & Stringer** category*

Stringer

This list is used to select the type and size of the stringer profile.

Stringer offset settings

This list is used to specify whether the stringer offset value should be calculated along the slope of the staircase, in the vertical direction, or the horizontal direction.

Stringer offset value

This edit box is used to offset the stringer. The direction of the offset will be based on the option selected from the **Stringer offset settings** list discussed above. Figure 43 shows stringer with no offset defined and Figure 44 shows the stringer with the offset defined along the slope.

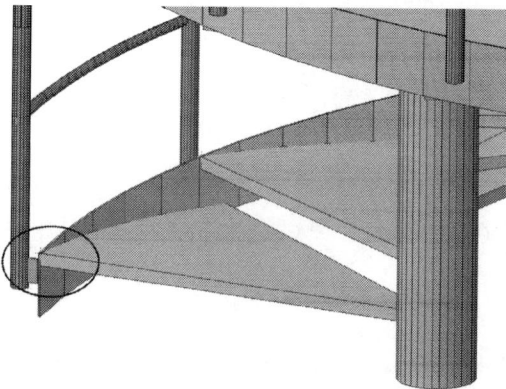

Figure 43 Stringer with no offset

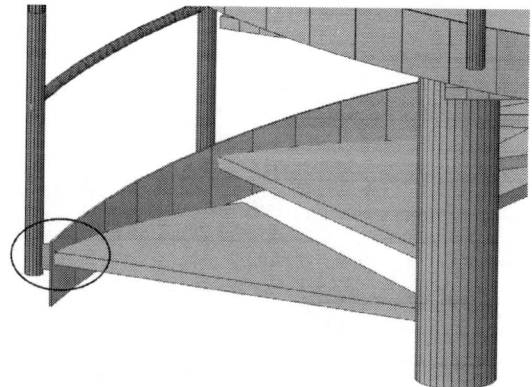

Figure 44 Stringer with offset along the slope

Prevision for outer stringer

This edit box allows you to enter a value for the precision of the outer stringer.

Central

This list will only be available when you are not creating the central post. In that case, the type and size of the central stringer can be specified using this list.

Precision for inner stringer

This edit box will also be available only when you are not creating the central post. In that case, you can specify the precision of the inner stringer using this edit box.

> **Note**: *The options available in the rest of the categories and tabs are similar to those discussed in earlier chapters.*

SELECTION TOOLS IN THE ADVANCE STEEL TOOL PALETTE

While creating a structure model, it is extremely important to make sure everything in the model is connected to each other using welds or bolts. Also, it is extremely common to check the objects that will be shop-connected to the object you select. The **Advance Steel Tool Palette Selection Tab** shown in Figure 45 provides a number of selection tools to perform these types of checks in the model. Some of these tools are discussed next.

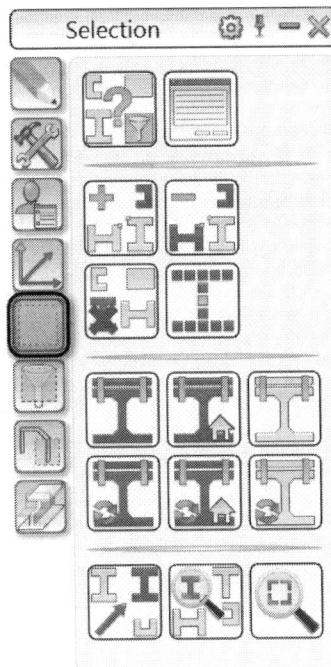

Figure 45 The Selection tools

Display connected objects

This tool is used to highlight and mark in Red, all the objects that are connected to the selected object. When you invoke this tool, you will be prompted to select objects. Once you select the object and press ENTER, all other objects that are directly or indirectly connected to the selected object by welds or bolts will be highlighted and marked in Red.

What I do

After completing my structural model, I use this button and select one of the objects in the model. On pressing ENTER, if everything in the model is highlighted, that means everything is connected either using welds or bolts. However, if anything in the model is not connected, it will not be highlighted. That allows me to find out why those objects are not connected so that I could manually connect them.

Clear marked objects

This tool is used to clear all the marked and highlighted objects. When you invoke this tool, all the Red highlights on the objects are removed and the objects are displayed with their original color.

Select all marked objects

This tool is used to select all the objects that are highlighted using any of the selection tools. The selected objects can then be isolated to view them easily.

*Tip: Once you highlight the connected objects, you can click on the **Select all marked objects** tool to select them. Then if you orbit the model by pressing and holding down the SHIFT key + Wheel Mouse Button, only the selected objects will be displayed in the orbit view. This is an easier way to isolate the objects temporarily to view them.*

Display objects connected in shop

As discussed earlier in various dialog boxes, while adding bolted or welded connections, you can specify whether those connections will be made in the shop or site. Selecting this button will highlight in Red all the objects that will be connected in the shop.

Display connection means

As button is used to highlight in Red the bolts or welds that are used to connect the selected object. When you invoke this tool, you will be prompted to identify object. Once you select the object and press ENTER, all the welds and bolts that are used to connect that object to the other objects will be highlighted in Red. As mentioned in the **Tip** earlier, you can then orbit the model by holding down the SHIFT key + Wheel Mouse Button to temporarily isolate those connection means and view them.

Remove marking + display connected objects

This tool clears all the previously highlighted objects and then prompts you to select an object. On selecting the object and pressing ENTER, all other objects connected to it will be highlighted and displayed in Red.

Remove marking + display objects connected in shop

This tool clears all the previously highlighted objects and then prompts you to select an object. On selecting the object and pressing ENTER, all other objects that will be shop connected to it will be highlighted and displayed in Red.

Remove marking + display connection means

This tool clears all the previously highlighted objects and then prompts you to select an object. On selecting the object and pressing ENTER, all welds and bolts that are used to connect the selected objects with other objects will be highlighted and displayed in Red.

What I do

*I regularly use the **Select all marked objects** tool discussed earlier with these other selection tools and orbit the model to temporarily isolate the highlighted objects. It provides a better understanding of the selected objects in the isolated display.*

CREATING BEAM CUT FEATURES

In Autodesk Advance Steel, you can use the **Features** tab of the **Advance Steel Tool Palette** to create beam cut features as well. Some of these features are similar to those discussed in plate cut features in earlier chapters. In this chapter, you will learn about the following two features highlighted in Figure 46.

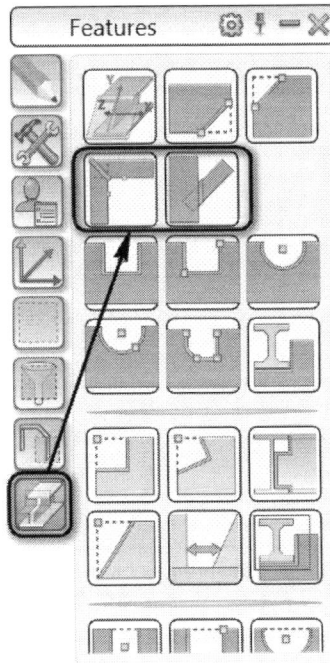

Figure 46 The Features tools

Creating Miter Cuts

The **Miter** tool is used to create miter cut between two sections. When you invoke this tool, you will be prompted to select the section to cut against. Once you select this section and press ENTER, you will be prompted to select the section to cut. On selecting this section, the miter cut will be created between the selected sections and the **Advance Steel - Miter [X]** dialog box will be displayed, as shown in Figure 47.

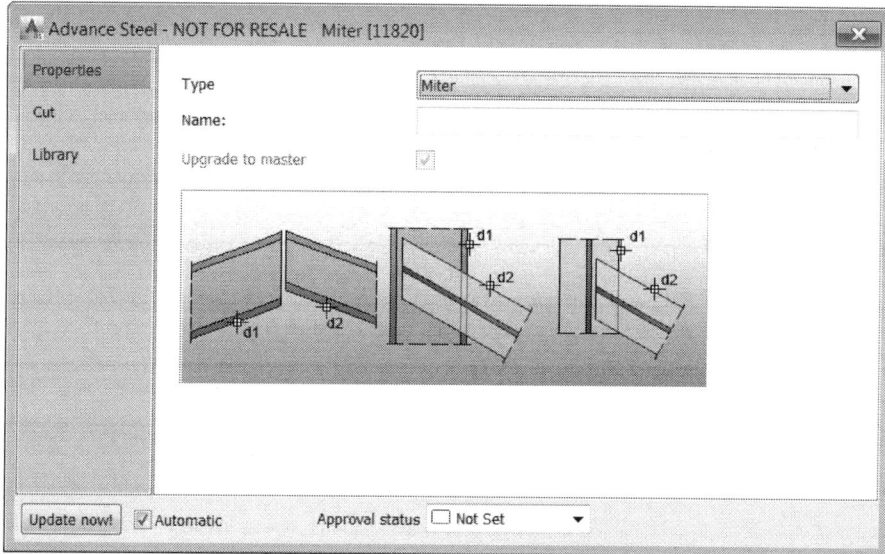

Figure 47   *The **Advance Steel - Miter [X]** dialog box*

Using the **Type** list in the **Properties** tab shown in Figure 47, you can also change this type of cut to saw cut to flange or saw cut to web. The options available on the **Cut** tab are used to specify if you want to define a gap between the two cut members and to specify whether you want the members to be welded together after the cut. Figure 48 shows two sections before the miter cut and Figure 49 shows the same sections after the miter cut.

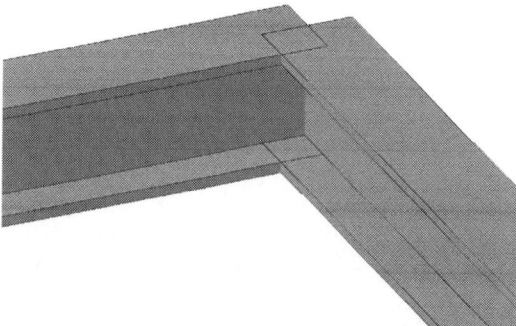

Figure 48   *Sections before miter cut*

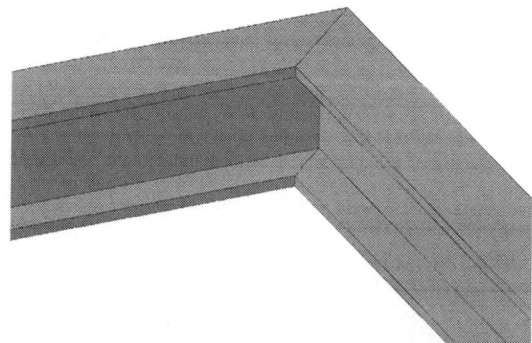

Figure 49   *Sections after miter cut*

Cutting at an Object

The **Cut at object** tool is used to cut the selected section against another section. It is important to note that you can only cut against one section at a time. If you want to cut a section against multiple other sections, then you will have to use this tool multiple times for each cut.

When you invoke this tool, you will be prompted to select the section to cut against. On selecting this section and pressing ENTER, you will be prompted to select the section to cut. Once you select the section to cut and press ENTER, the cut will be created and the **Advance Steel - Saw cut - Flange [X]** dialog box will be displayed, which is similar to the one shown in Figure 47 earlier. The only difference will be that from the **Type** list, **Saw cut - Flange** will be selected. Because of this, the options available on the **Cut** tab will be different, as shown in Figure 50.

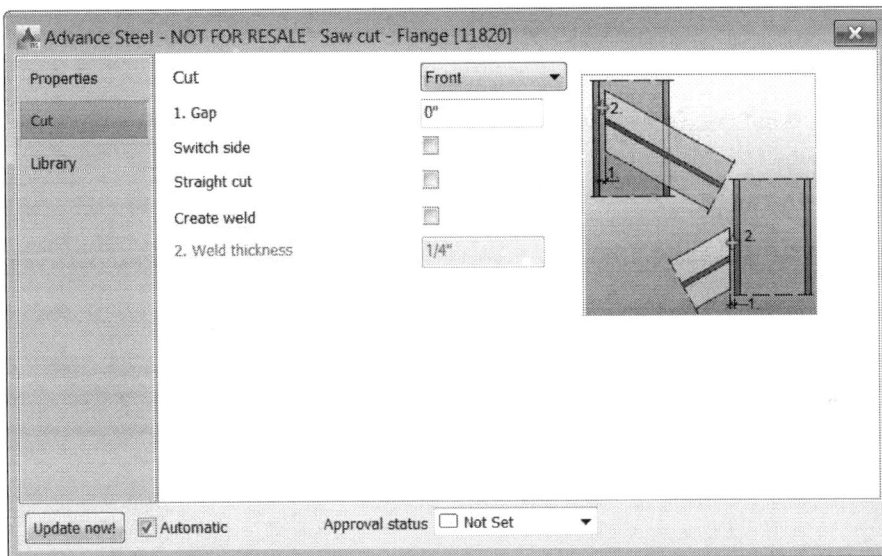

Figure 50 The Advance Steel - Saw cut - Flange [X] dialog box

These options are discussed next.

Cut

This list is used to specify whether the cut needs to be created to the front face of the object selected to cut against or the rear face. Figure 51 shows the object to cut against listed as **1** and the object to cut is listed as **2**. Figure 52 shows the cut created to the front face and Figure 53 shows the cut created to the rear face. In Figure 53, the camera is changed to show the rear face.

*Tip: You cannot edit the cut feature created between two sections and change it from front cut to rear cut. However, you can delete the cut feature between two sections and then recreate it. To do that, double-click on one of the sections and change the display type to **Features** to display the cut feature plane. Now, delete this cut feature plane to delete the feature.*

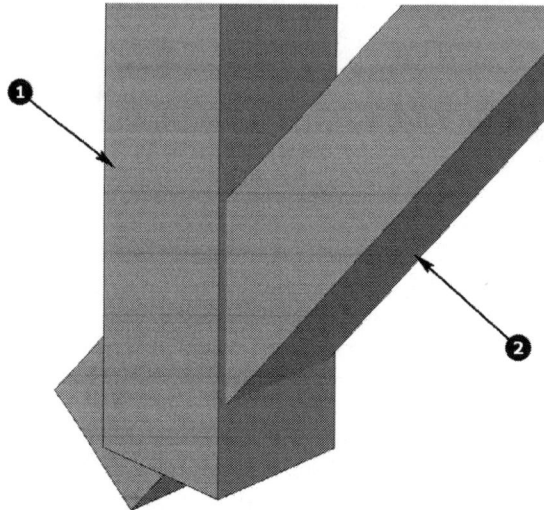

Figure 51 *The section to cut against and the section to cut*

Figure 52 *Section 2 cut to the front of section 1*

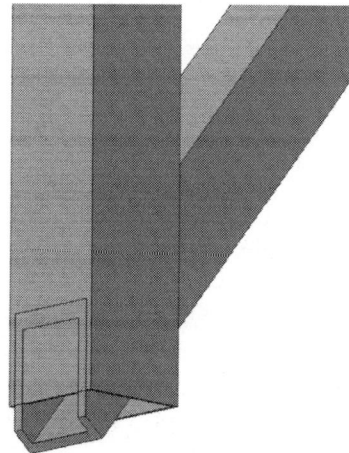

Figure 53 *Section 2 cut to the rear of section 1*

Gap

This edit box is used to specify a gap between the two sections after the cut. This gap is generally used for welding.

Switch sides

This tick box is used to switch sides of the cut feature.

Straight cut

Selecting this tick box will ensure the cut feature is created normal to the section selected to be cut.

Create weld

This tick box is selected to connect the two cut sections using weld. The weld value can be defined in the **Weld thickness** edit box available below this tick box. Note that the **Weld thickness** edit box will be activated only after you select this tick box.

MANUALLY CONNECTING PARTS

Autodesk Advance Steel allows you to manually connect parts using bolts or welds. This is generally used when you do not need to apply a joint between the parts to be connected. For example, Figure 54 shows a channel and a column. This channel needs to be connected to the column using bolts. However, there is no joint type that you can use for this. As a result, you can manually add a bolt pattern to connect these two, as shown in Figure 55.

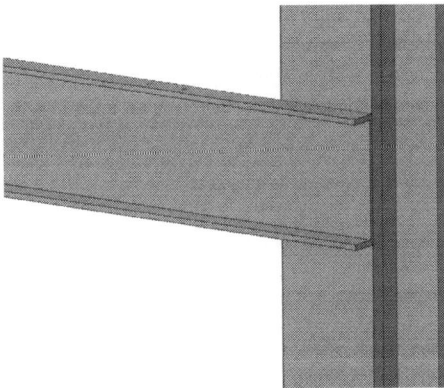

Figure 54 *The channel and the column to be connected*

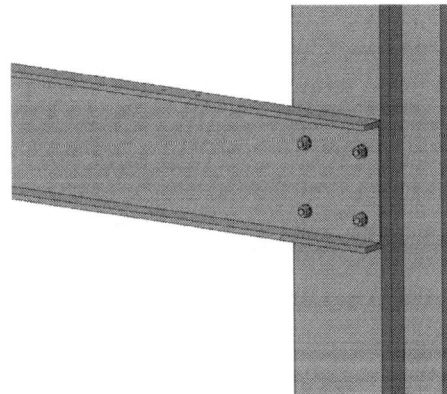

Figure 55 *Using the bolt pattern to connect the channel to the column*

Before you start inserting the bolt patterns, it is important to make sure the **bolts/anchors/ holes/shear studs** button in the **Objects** ribbon tab > **Switch** ribbon panel is set to the bolts preview, as shown in Figure 56. If this preview is set to the holes preview, only the holes will be created and not the bolts.

Figure 56 *The preview showing the bolts*

The tools available to manually connect objects are available on the **Object** ribbon tab > **Connection objects** ribbon panel, shown in Figure 57. It is important to note that the UCS needs to be aligned to the face of the object on which you want to insert the bolt pattern. The tools available on the **Connection objects** ribbon panel to connect parts are discussed next.

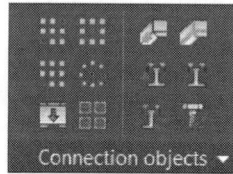

Figure 57 The **Connection objects** *ribbon panel*

Rectangular, 2 points

Objects Ribbon Tab > Connection objects Panel > Rectangular, 2 points

This tool is used to insert a rectangular bolt pattern by defining two corner points. It is important to remember that the UCS first needs to be aligned to the face of the part where you want to insert the bolt pattern. After aligning the UCS, when you invoke this tool, you will be prompted to select the parts to be connected. Once you select all the parts to be connected and press ENTER, you will be prompted to specify the lower left corner. On specifying this corner, you will be prompted to specify the upper right corner. On specifying this point, the bolt pattern will be inserted and the **Advance Steel - Bolts** dialog box will be displayed. You can use the options in various tabs of this dialog box to define the type, size, distances, naming, behavior, tolerance, and so on of the bolt pattern. These options are similar to those discussed in earlier chapters.

Rectangular, center point

Objects Ribbon Tab > Connection objects Panel > Rectangular, center point

This tool is used to insert a rectangular bolt pattern by defining the center point of the bolt pattern. When you invoke this tool, you will be prompted to select the parts to be connected. Once you select all the parts to be connected and press ENTER, you will be prompted to specify the central point. On specifying this central point, the bolt pattern will be inserted and the **Advance Steel - Bolts** dialog box will be displayed. As discussed earlier, you can use this dialog box to modify various parameters of the bolt pattern.

Rectangular, corner point

Objects Ribbon Tab > Connection objects Panel > Rectangular, corner point

This tool is used to insert a rectangular bolt pattern by defining the lower left corner point of the bolt pattern. When you invoke this tool, you will be prompted to select the parts to be connected. Once you select all the parts to be connected and press ENTER, you will be prompted to specify the start point. This is the lower left corner point of the hole pattern. On specifying this point, the bolt pattern will be inserted and the **Advance Steel - Bolts** dialog box will be displayed. As discussed earlier, you can use this dialog box to modify various parameters of the bolt pattern.

Circular, center point

Objects Ribbon Tab > Connection objects Panel > Circular, center point

This tool is used to insert a circular bolt pattern by defining the center point of the bolt pattern. When you invoke this tool, you will be prompted to select the parts to be connected. Once you select all the parts to be connected and press ENTER, you will be prompted to specify the center of the circle, which is the center of the circular pattern. On specifying this point, you will be prompted to specify the radius of the circle. This is the radius of the circular pattern. On specifying this point, the circular bolt pattern will be inserted and the **Advance Steel - Bolts** dialog box will be displayed. Using the options in the **Size** tab of this dialog box, you can override the radius value and also define the number of bolts in the circular pattern. Figure 58 shows a curved beam and a plate to be connected and Figure 59 shows the circular pattern used to connect the two parts.

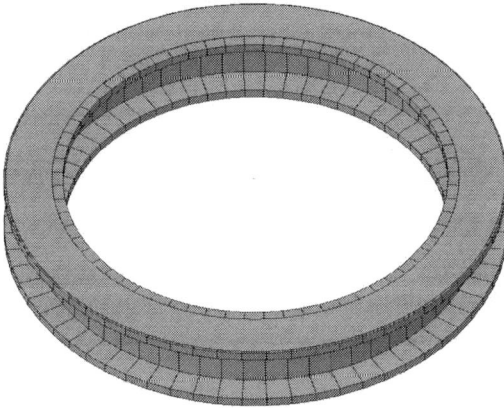

Figure 58 The two parts to be connected

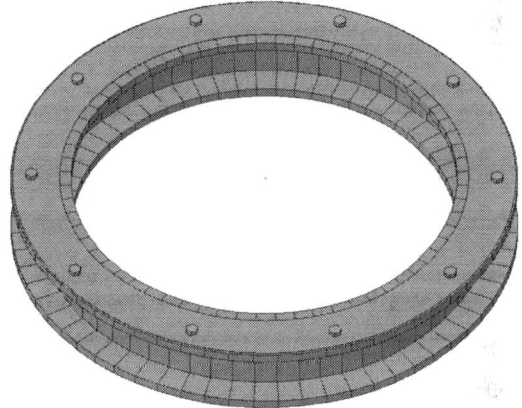

Figure 59 Circular bolt pattern inserted

Split bolt group

Objects Ribbon Tab > Connection objects Panel > Split bolt group

The bolts added using the rectangular or circular pattern tools discussed earlier are inserted as a group. As a result, you cannot individually modify those bolts. To do this, you will first need to explode the bolt group using the **Split bolt group** tool. When you invoke this tool, you will be prompted to select a pattern. When you select the bolt pattern and press ENTER, the group will be exploded and the bolts will become individual objects. You can then double-click on those bolts to individually modify them.

Weld point

Objects Ribbon Tab > Connection objects Panel > Weld point

This tool is used to weld the parts together and place a weld point to represent the welding. When you invoke this tool, you will be prompted to select the parts to be connected. Once you select the parts and press ENTER, you will be prompted to define insertion point for the weld. You can locate the weld point wherever you want. However, it is recommended to place it on the connecting faces to ensure the weld appears correctly in the assembly drawings.

Line of weld

Objects Ribbon Tab > Connection objects Panel > Line of weld

This tool is similar to the **Weld point** tool, with the difference that this tool allows you to insert a line to represent the welding instead of just inserting a point. When you invoke this tool and select the parts to be connected, you will be prompted to define the line of weld by specifying their start and end points. Once you have specified all the weld lines, you can press ENTER to terminate this tool and connect the parts.

Connection, add objects

Objects Ribbon Tab > Connection objects Panel > Connection, add objects

This tool is used to add additional parts to an existing bolted or welded connection. When you invoke this tool, you will be prompted to select a connecting bolt or weld element. Once you select the element and press ENTER, you will be prompted to select a beam or a plate to be inserted or removed. After selecting the part to be added, press ENTER; the selected part will be added to the existing bolt pattern or weld group. In case of bolt pattern, the grip length of the bolts will be automatically adjusted to include the new part.

*Tip: You can also add parts to an existing bolt pattern using the **Calculate grip length** tool available below the **Connection, add objects** tool.*

Connections, disconnect objects

Objects Ribbon Tab > Connection objects Panel > Connections, disconnect objects

This tool is opposite of the previous tool and is used to remove parts from the existing bolted or welded connection. When you invoke this tool, you will be prompted to select a connecting bolt or weld element. Once you select the element and press ENTER, you will be prompted to select a beam or a plate to be inserted or removed. After selecting the part to be removed, press ENTER; the selected part will be removed from the existing bolt pattern or weld group. In case of bolt pattern, the grip length of the bolts will be automatically adjusted to exclude the removed part.

Hands-on Tutorial (STRUC/BIM)	In this tutorial, you will complete the following tasks: 1. Insert the cage ladder, as shown in Figure 60. 2. Insert spiral staircase in the hospital model, as shown in Figure 61. 3. Create various beam features. 4. Insert rectangular bolt pattern to connect the channels to the columns, as shown in Figure 62.

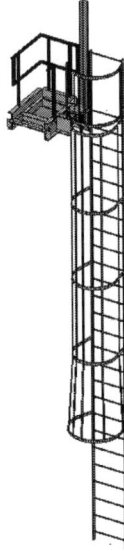

Figure 60 *The cage ladder to be inserted for the tutorial*

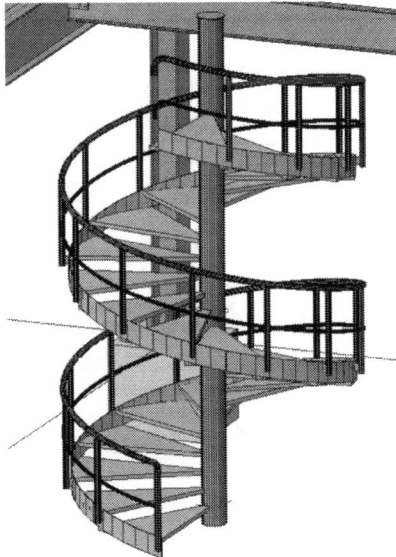

Figure 61 *The spiral staircase to be inserted for the tutorial*

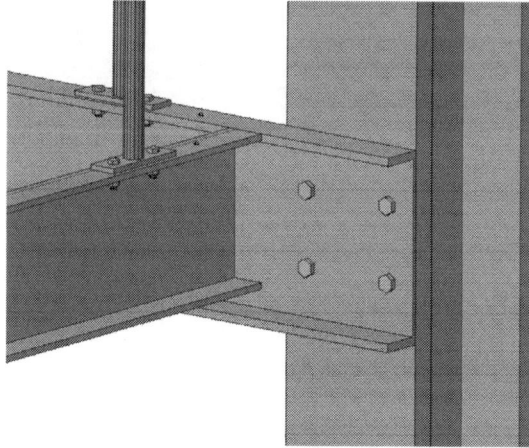

Figure 62 *The bolt pattern to connect the channel and column*

Section 1: Inserting Cage Ladder

In this section, you will open the **C09-Imperial-Cage-Ladder.dwg** or the **C09-Metric-Cage-Ladder.dwg** file and then insert the cage ladder.

1. From the **C09 > Struc-BIM** folder, open the **C09-Imperial-Cage-Ladder.dwg** or **C09-Metric-Cage-Ladder.dwg** file, based on the preferred units.

   To keep the size smaller, this file only has a part of the elevated platform, as shown in Figure 63, and a point created. The rest of the platform has been deleted.

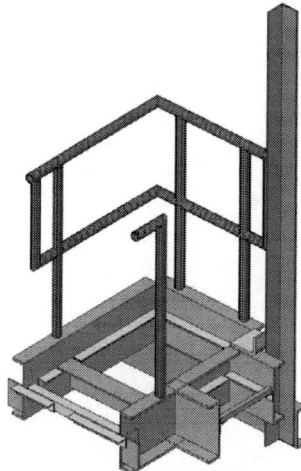

Figure 63 *The elevated platform to create the cage ladder (**Model courtesy Ilko Dimitrov, ACBS**)*

The orientation and cage of the ladder is dependent on the current UCS. Therefore, you will first rotate the UCS by -90-degrees around Z axis.

2. At the command line, type **UCS** and press ENTER; the following prompt sequence is displayed:

    Specify origin of UCS or [Face/NAmed/OBject/Previous/View/World/X/Y/Z/ZAxis]

3. Type **Z** and press ENTER; the following prompt sequence is displayed:

    Specify rotation angle about Z axis <90.0>:

4. Type **-90** and press ENTER; the UCS is rotated -90-degrees around the Z axis.

    You are now ready to insert the cage ladder.

5. From the **Home** ribbon tab > **Extended Modeling** ribbon panel, invoke the **Cage ladder** tool; you are prompted to specify the start point of the ladder.

6. Click on the point that was already created in this file; you are prompted to select the point to define the height of the ladder.

7. Move the cursor vertically up to track along the Z axis. Once the tracking line is displayed, type **25'9"** or **7850** as the value and press ENTER; a default ladder is inserted and the **Advance Steel - Cage ladder [X]** dialog box is displayed. If you are using the **Imperial** file, the default ladder will be as shown in Figure 64. However, if you are using the **Metric** file, the default ladder will be a lot shorter with a default cage.

Figure 64 *The default ladder created in the Imperial file*

You will now modify the parameters of this ladder and add the cage.

8. In the dialog box, activate the **Ladder** category > **Ladder** tab.

9. In the **Height** edit box, make sure it shows **25'9"** or **7850**.

10. In the **Height extension** edit box, enter **3' 3 3/8"** or **1000**.

11. Make sure **center** is selected from the **Width Layout** list.

12. In the **Width** edit box, enter **2'** or **610**.

13. Activate the **Ladder** category > **Sections** tab.

14. From the **Stringer section type** list, select **US Flat > FL 3/8X2 1/2** from Imperial or **Australian Flat - CF > 65x10 FL** for Metric.

15. From the **Rung section** list, select **US Round bar > RB 3/4** from Imperial or **Australian Round > RMS 20** for Metric. The zoomed in view at the top of the ladder in the Imperial file, after making these changes, is shown in Figure 65. The ladder in the Metric file will show the default cage also.

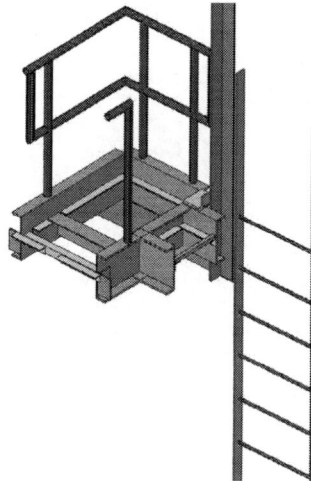

Figure 65 *Zoomed in view of the Imperial ladder after changing the sections*

16. Activate the **Ladder** category > **Rung distances** tab.

17. From the **Distance layout** list, make sure **from bottom** is selected.

18. From the **Position** list, select **middle**.

19. In the **Start distance** edit box, enter **6"** or **150**.

20. From the **Spacing by** list, make sure **distance between rungs** is selected.

21. In the **Distance between rungs** edit box, enter **1'** or **300**.

22. In the **Offset from top** edit box, enter **-3"** or **-175**.

23. Activate the **Ladder** category > **Additional rungs** tab.

24. Select the **Additional rungs** tick box.

25. In the **Start distance** edit box, enter **9"** or **150**.

26. From the **Spacing by** list, select **distance between rungs**.

27. In the **Distance between rungs** edit box, enter **1'** or **300**.

28. Activate the **Ladder** category > **Rung connection** tab.

29. From the **Connection at side 1** list, select **extended**.

    The **Same for other side** tick box is automatically selected, which is what you want.

30. In the **Extension at side 1** edit box, enter **1/2"** or **15**.

31. In the **Hole gap** edit box, enter **1/16"** or **1**.

    This ladder does not require any wall connection, so you will skip the **Wall connection** tab.

32. Activate the **Ladder** category > **Ladder exit** tab.

33. From the **Exit type** list, make sure **straight** is selected.

    Next, you will specify the cage parameters.

34. Activate the **Cage** category > **General** tab.

35. From the **Cage** list, select **Type 4**; type 4 cage is added to the ladder, as shown in the zoomed in view in Figure 66.

36. In the **Height to first brace** edit box, enter **7'6"** or **2200**.

37. From the **First brace** list, select **Type 4**. This will allow you to define the dimensions of the first brace independent of the remaining braces.

38. Select the **Finish cage at exit level** tick box.

39. From the **Brace section** list, select **US Flat > FL 1/4X2** for Imperial or **Australian Flat - CF > 50x6 FL** for Metric.

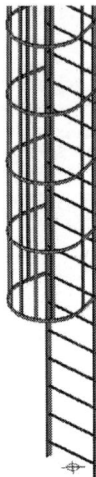

Figure 66 *Zoomed in view of the ladder after adding the* **Type 4** *cage*

40. Activate the **Cage** category > **Brace distances** tab.

41. Make sure the **Distance for first brace different** tick box is cleared.

42. From the **First brace section** list, select **US Flat > FL 1/4X2** for Imperial or **Australian Flat - CF > 50x6 FL** for Metric.

43. From the **Spacing by** list, select **distance between braces**.

44. In the **Distance between braces** edit box, enter **4'5"** or **1350**.

45. In the **Offset at top** edit box, enter **0**. You may not see any changes at this stage, but this will help other dimensions later.

46. Activate the **Cage** category > **Brace dimensions** tab.

47. Make sure the **Create as poly beam** tick box is selected.

48. In the **Distance to center** edit box, enter **1'5"** or **430**.

49. From the **Radius reference** list, make sure **inside** is selected.

50. In the **Radius** edit box, enter **1'2"** or **355**.

51. In the **Extension 3** edit box, enter **1 1/4"** or **35**.

The default settings for the brace connection will be accepted. So you can skip those tabs.

52. Activate the **Cage** category > **First bottom brace dimensions** tab.

53. Make sure the **Create as poly beam** tick box is selected.

54. In the **Distance to center** edit box, enter **1'6"** or **460**.

55. From the **Radius reference** list, make sure **inside** is selected.

56. In the **Radius** edit box, enter **1'6"** or **460**.

57. In the **Extension 3** edit box, enter **1 1/4"** or **35**. The bottom cage in the Imperial file now appears as shown in Figure 67.

Figure 67 *The Imperial bottom cage after modifying the parameters*

Rest of the parameters will be accepted as the default values. Next, you will specify the band parameters.

58. Activate the **Cage** category > **Bands** tab.

59. Make sure from the **Band location** list, **inside** is selected.

60. From the **Separate bands** list, select **no**.

61. From **Band section** list, select **US Flat > FL 3/8X1** for Imperial or **Australian Flat - CF > 25x10 FL** for Metric.

62. In the **Number on straight part** edit box, enter **1**.

63. In the **Straight start distance** edit box, enter **8"** or **200**.

64. Select the **Create middle band** tick box.

65. In the **Number on outside part** edit box, enter **6**.

66. In the **Angle** edit box, enter **25**.

The default settings of the band connection will be accepted. You will now go to the **Exit** category and specify the exit cage parameters.

67. Activate the **Exit** category > **Top cage** tab.

68. From the **Cage type** list, select **Type 4 left exit**.

69. From the **Brace section** list, select **US Flat > FL 1/4X2** for Imperial or **Australian Flat - CF > 50x6 FL** for Metric.

70. From the **Top spacing by** list, select **number of braces**.

71. In the **Number of braces** edit box, enter **2**.

72. In the **Offset exit to top** edit box, enter **-1"** or **-25**.

73. In the **Offset exit to bottom** edit box, enter **4"** or **100**.

74. Activate the **Exit** category > **Brace dimensions** tab.

75. Make sure the **Create as poly beam** tick box is selected.

76. In the **Distance to center** edit box, enter **1'5"** or **430**.

77. From the **Radius reference** list, make sure **inside** is selected.

78. In the **Radius** edit box, enter **1'2"** or **355**.

79. In the **Extension 3** edit box, enter **1 1/4"** or **35**.

80. In the **Dimension 1** edit box, enter **1' 10 1/2"** or **572**.

81. In the **Dimension 2** edit box, enter **2"** or **50**.

The default settings for the brace connection will be accepted. So you can skip those tabs.

82. Activate the **Exit** category > **Bands** tab.

83. Clear the **Same as other** tick box, if selected.

84. Make sure from the **Band location** list, **inside** is selected.

85. From **Band section** list, select **US Flat > FL 3/8X1** for Imperial or **Australian Flat - CF > 25x10 FL** for Metric.

86. In the **Number on straight part** edit box, enter **2**.

87. In the **Straight start distance** edit box, enter **6"** or **150**.

88. In the **Straight intermediate distance** edit box, enter **7"** or **175**.

89. Clear the **Create middle band** tick box.

90. In the **Number on outside part** edit box, enter **6**.

91. In the **Angle** edit box, enter **25**.

    The default values of the band connections will be accepted. You will now specify the parameters for the opening of the cage to exit the ladder.

92. Activate the **Exit** category > **Opening post** tab.

93. In the **Offset from exit level** edit box, enter **1"** or **24**.

94. From the **Height Layout** list, select **same as stringer**.

    With these settings, notice the spacing between the top band of the last cage and the bottom band of the exit cage, as shown in Figure 68.

Figure 68 The gap between the bands of the last cage and the exit cage

    You will now modify the offset of the top cage band that was set to **0** in step 45. You could not specify the correct value in step 45 as that causes dimensional issues in the ladder at that stage.

95. Activate the **Cage** category > **Brace distances** tab.

96. In the **Offset at top** edit box, enter **8"** or **275**. This fixes the gap, as shown in Figure 69.

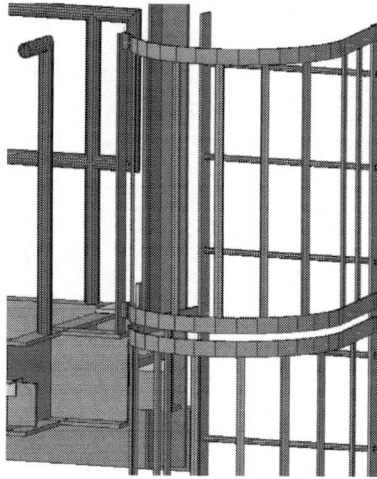

Figure 69 *The ladder after modifying the spacing between the braces of the top and exit cages*

This completes all the parameters related to this cage ladder.

97. Save the cage in the library and close the dialog box.

Section 2: Inserting Spiral Staircase

In this section, you will open the **C09-Imperial-Hospital.dwg** or the **C09-Metric-Hospital.dwg** file and then insert the spiral staircase.

1. From the **C09 > Struc-BIM** folder, open the **C09-Imperial-Hospital.dwg** or **C09-Metric-Hospital.dwg** file, based on the preferred units. Figure 70 shows the model in this file.

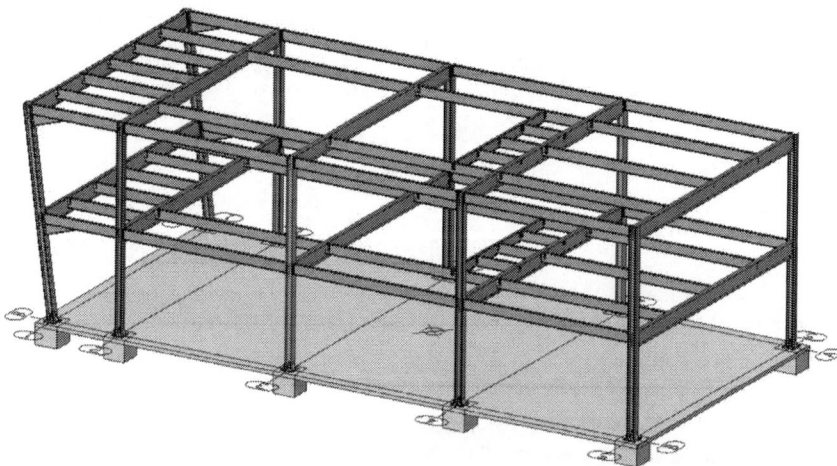

Figure 70 *The hospital model for inserting the spiral staircase*

As shown in Figure 71, this file has a point created that will be used to insert the spiral staircase.

> *Tip: Before you start creating the spiral staircase, it is better to hide the concrete slab and the structural sections that are obscuring the display of the staircase. You can use the **Advance Steel Tool Palette > Quick views** tab > **Selected objects off** tool for this.*

2. From the **Extended Modeling** ribbon tab > **Structure Elements** ribbon panel, invoke the **Spiral staircase** tool; you are prompted to select the center point of the spiral.

3. Click on the point that was already created in this file; you are prompted to select a point to define the spiral height.

4. Move the cursor vertically up to track along the Z axis. Once the tracking line is displayed, type **17'3"** or **5260** as the value and press ENTER; you are prompted to confirm if you want to specify the first or last step.

5. Press ENTER at this prompt; you are prompted to select a point which gives the direction and the outer radius value.

6. Move the cursor along the -Y direction to track along the 270-degrees angle. Once the tracking line is displayed, enter **6'** or **1825**; the default spiral staircase is created, as shown in Figure 71 and the **Advance Steel - Spiral stairs [X]** dialog box is displayed.

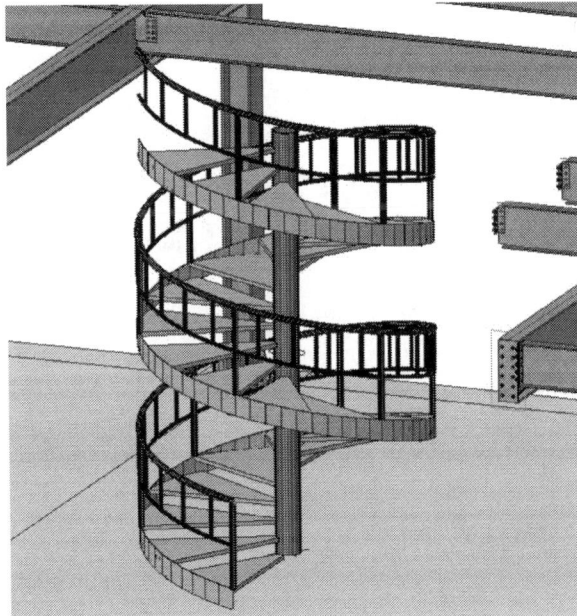

Figure 71 The default spiral staircase

You will now modify the parameters of this spiral staircase.

7. Activate the **Column & Stringer** category > **General Properties** tab (for Imperial) or **General** category > **General properties** tab.

8. In the **Total angle of stair** to **720**.

9. Activate the **Central post** tab.

10. From the **Section** list, select **AISC 14.1 HSS Pipe Std > Pipe 12 Std** for Imperial or **Australian Circular Hollow Section - CF C350L0 > 323.9x6.4 CHS** for Metric.

11. In the **Vertical lower offset** edit box, enter **10"** or **250**; the central post is extended at the bottom to touch the concrete floor.

12. In the **Vertical upper offset** edit box, enter **3'4"** or **1000**; the central post is extended at the top.

The staircase, after making these changes, is shown in Figure 72.

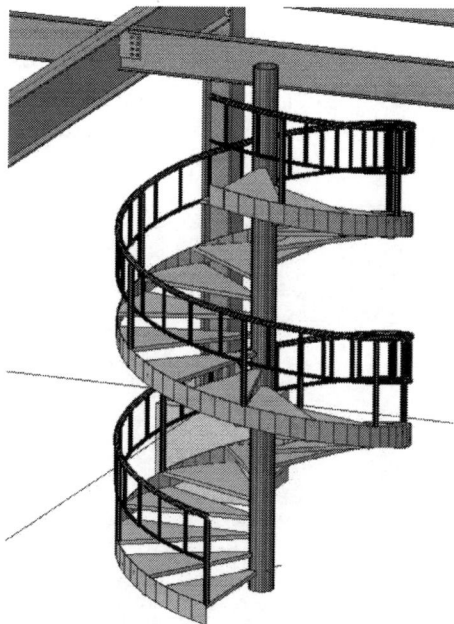

Figure 72 The staircase after changing the central post offset

13. Activate the **Cover Plate** tab.

14. Select the **Create cover plate** tick box; the cover plate is inserted at the top end of the central post.

15. In the **Thickness** edit box, enter **1/2"** or **12**.

16. From the **Size layout** list, select **diameter**.

17. In the **Diameter** edit box, enter **1'2"** or **350**.

18. Activate the **Stringer** tab.

19. From the **Stringer** list, select **US Flat > FL 3/8X10** for Imperial or **Australian Flat - CF > 250x10 FL** for Metric.

20. From the **Stringer offset settings** list, select **slope**.

21. In the **Stringer offset value** edit box, enter **1"** or **25**.

    Next, you will specify the tread parameters.

22. Activate the **Tread** category > **Step size** tab (for Imperial) of **Steps** category > **Tread size** tab (for Metric).

23. In the **Run line** edit box, enter **2'** or **600**.

24. Make sure the **Use formula (g+2h)** tick box is selected.

25. Activate the **Step size 2** tab (for Imperial) or **Tread size 2** tab (for Metric).

26. In the **Step/Tread overlap** edit box, enter **2"** or **50**.

27. Make sure the **Create first tread** tick box is selected.

28. Activate the **Tread Type** tab.

29. From the **Tread Type** list, select **Plate**.

30. From the **Tread Material** list, select **A36**.

31. Make sure the values in the two **Gap** edit boxes is set to **0**.

32. From the **Outer edge of the tread** list, select **curved**.

33. Activate the **Tread dimensions** tab.

34. In the **Tread Thickness** edit box, enter **2"** or **50**. The zoomed in view of the bottom staircase, after making these changes, is shown in Figure 73.

    For the tread connections, the default values will be accepted. You will now define the post parameters.

35. Activate the **Posts** category > **Post** tab.

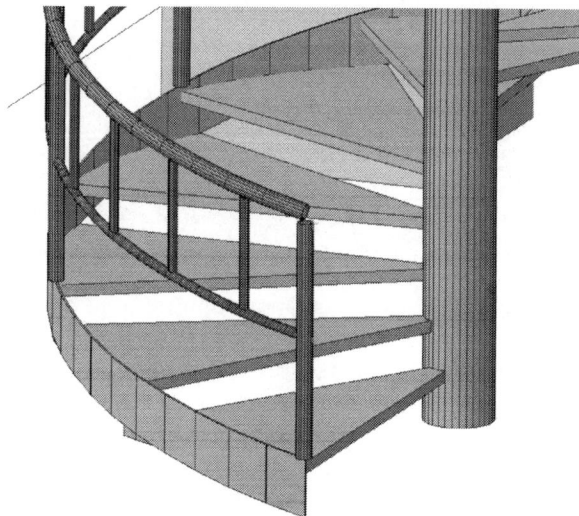

Figure 73 *The staircase, after defining the tread parameters*

36. From the **Section** list, select **AISC 14.1 HSS Pipe Std > Pipe 2 1/2 Std** for Imperial or **Australian Circular Hollow Section - CF C250L0 > 60.3x3.6 CHS** for Metric.

37. From the **Create posts at** list, select **each tread**.

38. From the **Alignment of posts** list, select **outside**.

39. Activate the **Post-Top Handrail** tab.

40. From the **Connection type** list, select **Aligned** (for Imperial) or **Flush** (for Metric).

41. Select the **Sloped connection** tick box.

42. Activate the **Post-Middle Handrail** tab.

43. From the **Connection type** list, select **Cut** (for Imperial) or **Fit** (for Metric).

44. Activate the **KickRail** tab (for Imperial) or **Kick plate** (for Metric).

45. Clear the **Create kickrail** tick box (for Imperial) or **Create kick plate** (for Metric).

    Next, you will define the handrail parameters.

46. Activate the **Handrails** category > **Top Handrail** tab.

47. Select the **Create handrail** tick box.

48. From the **Section** list, select **AISC 14.1 HSS Pipe Std > Pipe 2 1/2 Std** for Imperial or **Australian Circular Hollow Section - CF C250L0 > 60.3x3.6 CHS** for Metric.

49. In the **Height layout** list, select **from the top**.

50. In the **Height** edit box, enter **3'2"** or **965**.

51. Activate the **Middle Handrail** tab.

52. Select the **Create middle handrail** tick box.

53. From the **Section** list, select **AISC 14.1 HSS Pipe Std > Pipe 1 1/4 Std** for Imperial or **Australian Circular Hollow Section - CF C250L0 > 33.7x3.2 CHS** for Metric.

54. In the **Height layout** list, select **from the top**.

55. In the **Distance from top handrail** edit box, enter **1' 7 11/16"** or **500**.

56. In the **Number of middle handrails** edit box, make sure **1** is entered.

    You will not create any pickets/balusters, so you will turn off the option to create them.

57. Activate the **Pickets** tab (for Imperial) or **Balusters** (for Metric).

58. Clear the **Create pickets** tick box (for Imperial) or **Create balusters** tick box (for Metric).

59. Activate the **End of handrail** tab. This tab allows you to specify the handrail end parameters at the start side of the handrail.

60. From the **Connection type** list, select **Bent end** (for Imperial) or **Down stand** (for Metric).

61. Clear the **Same joint for both ends** tick box, if selected.

62. Select the **Use poly beams** tick box.

63. In the **Radius of top handrail** edit box, enter **2"** or **50**.

64. In the **Length** edit box, enter **2 1/2"** or **60**.

65. In the **Radius** edit box, enter **0**.

66. In the **Length** edit box, enter **0**.

67. Activate the **Handrails** category > **End of handrail (end)** tab. This tab allows you to specify the handrail end parameters at the end side of the handrail.

68. From the **Connection type** list, select **Bent end** (for Imperial) or **Down stand** (for Metric).

69. In the **Radius of top handrail** edit box, enter **5"** or **125**.

70. In the **Length** edit box, enter **9"** or **275**.

71. In the **Radius** edit box, enter **5"** or **125**.

72. In the **Length** edit box, enter **9"** or **275**.

This completes the handrail parameters. The zoomed in view at the start of the spiral staircase is shown in Figure 74.

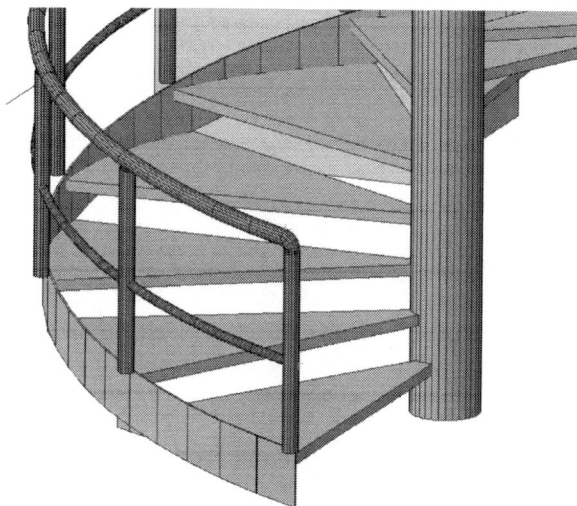

Figure 74 The start of the staircase after modifying the handrail parameters

You will now specify the post connection to the stringer.

73. Activate the **Post connection** category > **Post connection** tab (for Imperial) or **Fixing of post** tab (for Metric).

74. From the **Connection type** list, select **Perp. plate with weld**.

75. In the **Dist. from top** edit box, enter **6"** or **150**.

76. In the **Plate thickness** edit box, enter **3/8"** or **10**.

77. In the **Plate length** edit box, enter **2"** or **50**.

78. In the **Plate width** edit box, enter **4"** or **100**.

79. Activate the **Post connection** category > **Connection parameter** tab (for Imperial) or **Fixing parameters** tab (for Metric).

80. In the **Dist. edge of plate** edit box, enter **1"** or **25**.

81. Save the staircase in the library and close the dialog box. The finished staircase is shown in Figure 75.

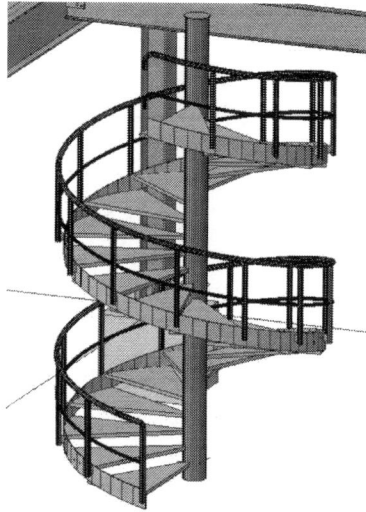

Figure 75 *The completed spiral staircase*

Section 3: Creating the Beam Cut Features

In this section, you will open the **C09-Imperial-Beam-Features.dwg** or the **C09-Metric-Beam-Features.dwg** file and then create various beam features. You will also weld the beams together during that process to connect all the beams.

1. From the **C09 > Struc-BIM** folder, open the **C09-Imperial-Beam-Features.dwg** or **C09-Metric-Beam-Features.dwg** file, based on the preferred units.

2. Zoom near the bottom of the model, as shown in Figure 76. This figure also shows various beams to be selected to create cut features.

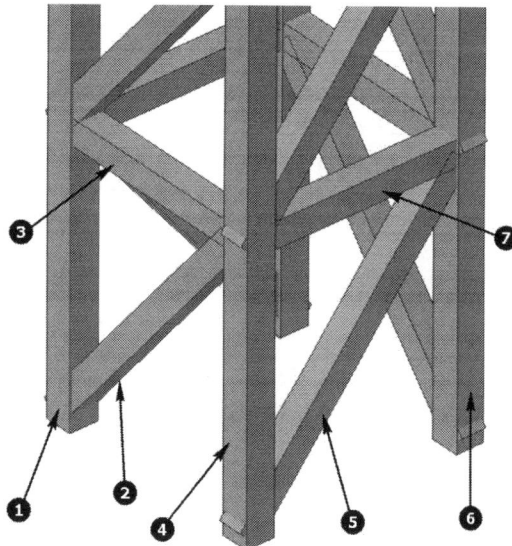

Figure 76 *Beams to be selected to create cut features*

3. From the **Advance Steel Tool Palette** > **Features** tab, click **Cut at object**, as shown in Figure 77; you are prompted to select the section to cut against.

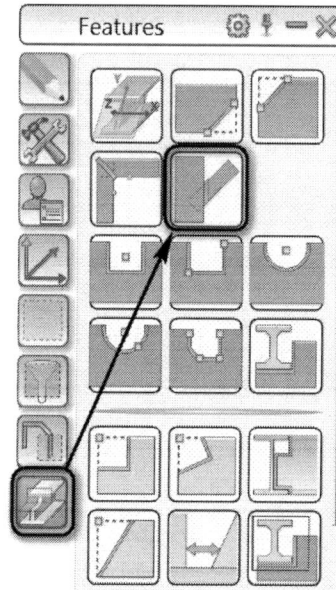

Figure 77 Invoking the Cut at object tool

It is important to note that this tool only allows you to cut against one section at a time. Therefore, you will have to create cuts by invoking this tool multiple times to get the desired result.

4. Select the section labeled as **1** in Figure 76 and press ENTER; you are prompted to select the section to cut.

5. Select the sectioned labeled as **2** in Figure 76 and press ENTER; the cut feature is created and the **Advance Steel - Saw cut - Flange [X]** dialog box is displayed.

   By default when you create the cut features, the welds are not created. However, in this case, you will weld the two sections together.

6. Activate the **Cut** tab in the dialog box.

7. Select the **Create weld** tick box.

8. In the **Weld thickness** edit box, enter **1/4"** or **6**.

   This completes this cut feature.

9. Close the dialog box. The model, after creating this cut feature, is shown in Figure 78.

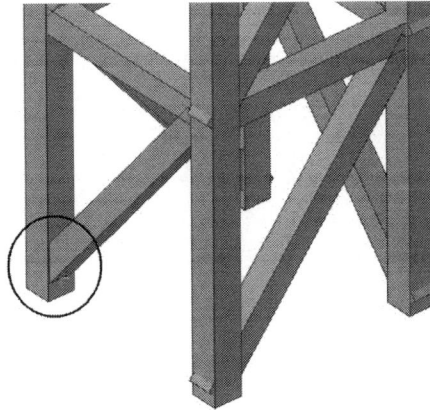

Figure 78 *The model after creating the first cut feature*

You will now repeat the same process to cut the section labeled as **2** against the section labeled as **3** in Figure 76.

10. Invoke the **Cut at object** tool again; you are prompted to select the section to cut against.

11. Select the section labeled as **3** in Figure 76 and press ENTER; you are prompted to select the section to cut.

12. Select the sectioned labeled as **2** in Figure 76 and press ENTER; the cut feature is created and the **Advance Steel - Saw cut - Flange [X]** dialog box is displayed.

13. Activate the **Cut** tab in the dialog box.

14. Select the **Create weld** tick box.

15. In the **Weld thickness** edit box, enter **1/4"** or **6**.

16. Close the dialog box.

17. Similarly, using the **Cut at object** tool, cut the section labeled as **2** against the section labeled as **4** in Figure 76. Figure 79 shows a zoomed in view of the last two cut features created in the model.

18. Similarly, create the cut features on the remaining sections. The model, after creating all the cut features, is shown in Figure 80.

*Tip: Instead of creating the cut features individually on all the sections in the model, you can create the cut features on the sections on the base level and then use the **Advance Copy** tool to copy them. Note that if you use the normal **Copy** tool, the sections will not be connected with welds. Therefore, it is important to use the **Advance Copy** tool.*

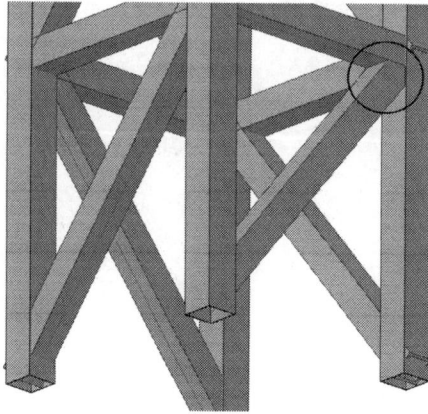

Figure 79 *The model after creating the two cut features*

Figure 80 *The model after creating all the cut features*

Next, you will check whether all the members are connected or not. This is done using the **Display connected objects** tool available on the **Advance Steel Tool Palette > Selection** tab.

19. From the **Advance Steel Tool Palette > Selection** tab, invoke the **Display connected objects** tool, as shown in Figure 81; you are prompted to select objects.

20. Select any section in the model and then press ENTER; all the connected objects are highlighted in Red.

*Note: If there are objects that are not highlighted in Red, that means they are not connected. You can manually weld them using the **Objects** ribbon tab > **Connection objects** ribbon panel > **Weld point** tool.*

You will now clear the highlight.

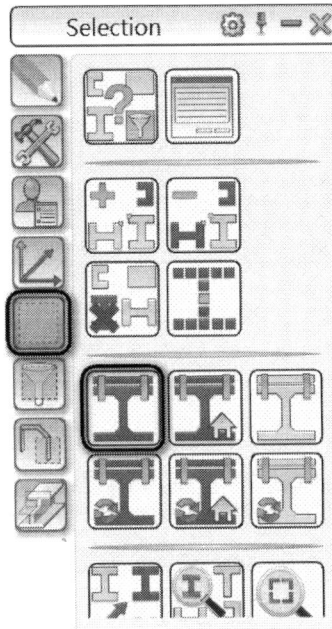

*Figure 81 Invoking the **Display connected objects** tool*

21. From the **Advance Steel Tool Palette > Selection** tab, invoke the **Clear marked objects** tool, which is available above the **Display connected objects** tool; the Red highlight on the objects is removed and their original color is restored.

Section 4: Connecting Sections using Bolt Patterns

In this section, you will open the **C09-Imperial-Elev-Tower.dwg** or the **C09-Metric-Elev-Tower.dwg** file and then connect the sections using the bolt patters.

1. From the **C09 > Struc-BIM** folder, open the **C09-Imperial-Elev-Tower.dwg** or **C09-Metric-Elev-Tower.dwg** file, based on the preferred units.

2. Zoom to the area shown in Figure 82. In this image, the display of circular bracing sections is turned off for better visibility.

   Before you start inserting the bolt pattern, you will check the objects that are not connected in the model.

3. From the **Advance Steel Tool Palette > Selection** tab, invoke the **Display connected objects** tool, as shown in Figure 81; you are prompted to select objects.

4. Select the column labeled as **1** in Figure 82 and then press ENTER; all the connected objects are highlighted in Red.

   Notice that the channels between the columns are not highlighted. Also, the stairs are not highlighted either. The reason is that these objects are not connected to the rest of the

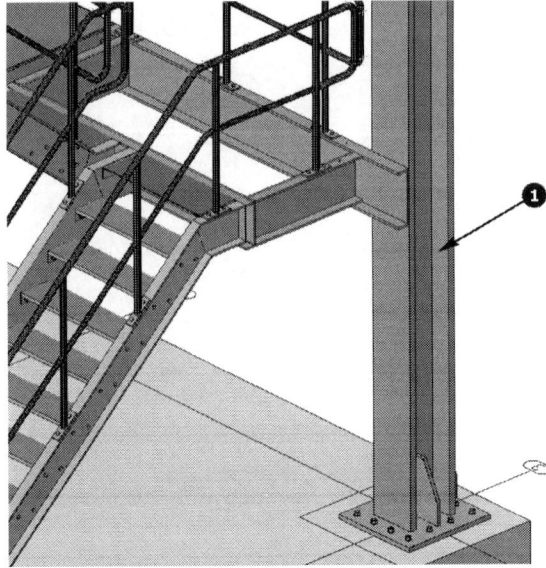

Figure 82 *Zoomed in view of the model*

model. So you need to connect the channel to the column using the bolt pattern. But first, you need to clear the marked objects.

5. From the **Advance Steel Tool Palette > Selection** tab, invoke the **Clear marked objects** tool, which is available above the **Display connected objects** tool; the Red highlight on the objects is removed and their original color is restored.

6. Using the **UCS 3 point** tool, align the UCS to the front face of the channel. Use the endpoint labeled as **1** in Figure 83 as the origin point, use the polar tracking labeled as **2** to define the point on positive X direction, and the endpoint labeled as **3** as the point along Y direction.

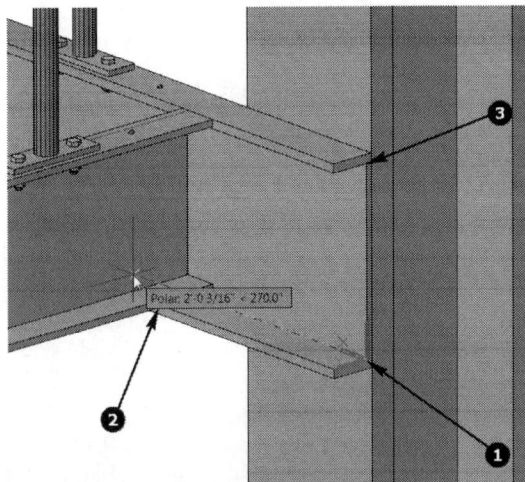

Figure 83 *Aligning the UCS using three points*

You are now ready to connect the channel to the column. You will use the **Rectangular, corner point** tool to insert the bolt pattern.

7. From the **Object** ribbon tab > **Connection objects** ribbon panel, invoke the **Rectangular, corner point** tool; you are prompted to select parts to be connected.

8. Select the column labeled as **1** and the channel labeled as **2** in Figure 84 and then press ENTER; you are prompted to specify the start point.

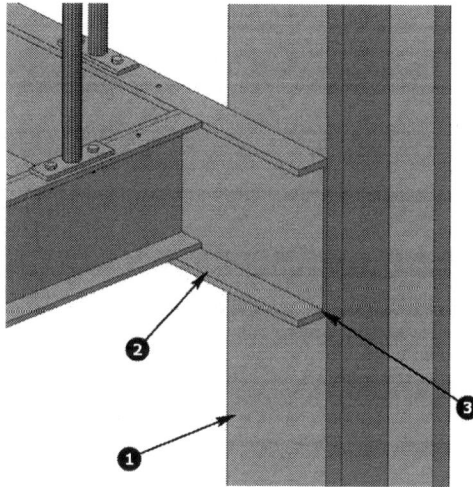

Figure 84 *Selecting the objects to connect*

9. Select the endpoint labeled as **3** in Figure 84 as the start point of the bolt pattern; the bolt pattern is inserted and the **Advance Steel - Bolts** dialog box is displayed.

10. In the **Definition** tab, select the **Inverted** tick box.

11. Activate the **Distance** tab

    Notice that the number of bolts in the **Number X** and **Number Y** edit boxes are set to 2.

12. In the **Intermediate distance X** edit box, enter **8"** or **200**.

13. In the **Intermediate distance Y** edit box, enter **6"** or **150**.

14. In the **Edge distance X** edit box, enter **2 5/16"** or **55.5**.

15. In the **Edge distance Y** edit box, enter **3 13/16"** or **97.5**.

    This completes the creation of the bolt pattern.

16. Close the dialog box. The model, after inserting the bolt pattern, is shown in Figure 85.

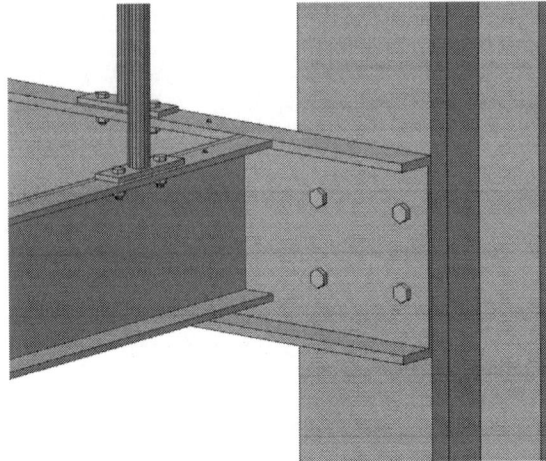

Figure 85 *The bolt pattern inserted*

Next, you will connect the other end of this channel to the column at **A1** grid intersection point. But before doing that, you need to rotate the UCS 90-degrees around the Y-axis two times. That will ensure the X-axis is pointing in the right direction.

17. From the **Advance Steel Tool Palette > UCS** tab, click on the **Rotate UCS around Y axis** button; the UCS rotates 90-degrees around the Y-axis.

18. Click on the same button again; the UCS rotates around the Y-axis by another 90-degrees.

19. From the **Object** ribbon tab > **Connection objects** ribbon panel, invoke the **Rectangular, corner point** tool; you are prompted to select parts to be connected.

20. Select the column labeled as **1** and the channel labeled as **2** in Figure 86 and then press ENTER; you are prompted to specify the start point.

21. Select the endpoint labeled as **3** in Figure 86 as the start point of the bolt pattern; the bolt pattern is inserted and the **Advance Steel - Bolts** dialog box is displayed.

    You will now modify the parameters of the bolt pattern.

22. In the **Definition** tab, clear the **Inverted** tick box, if selected.

23. Activate the **Distance** tab.

    Notice that the number of bolts in the **Number X** and **Number Y** edit boxes are set to 2.

24. In the **Intermediate distance X** edit box, enter **8"** or **200**.

25. In the **Intermediate distance Y** edit box, enter **6"** or **150**.

26. In the **Edge distance X** edit box, enter **2 5/16"** or **55.5**.

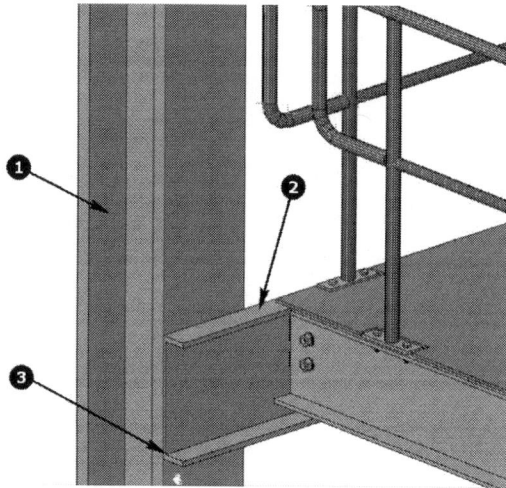

Figure 86 *Selecting the objects to connect*

27. In the **Edge distance Y** edit box, enter **3 13/16"** or **97.5**.

    This completes the creation of the bolt pattern. The model, after inserting the bolt pattern, is shown in Figure 87.

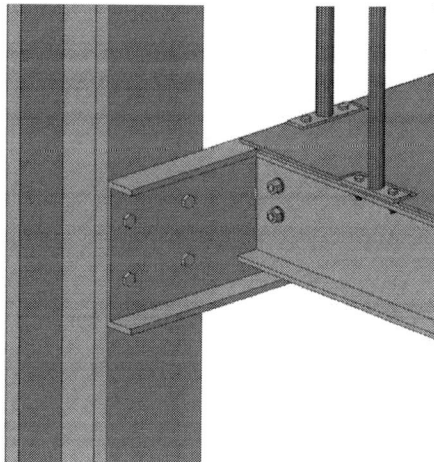

Figure 87 *The bolt pattern inserted*

28. Similarly, align the UCS on the other channel between the columns at **C1** and **C3** grid intersection points and connect it to the columns.

29. From the **Advance Steel Tool Palette > Selection** tab, invoke the **Display connected objects** tool, as shown in Figure 81; you are prompted to select objects.

30. Select one of the columns and then press ENTER; all the connected objects are highlighted in Red.

Notice that the channels are now displayed as connected.

31. From the **Advance Steel Tool Palette > Selection** tab, invoke the **Clear marked objects** tool; the Red highlight on the objects is removed and their original color is restored.

Section 5: Connecting Stairs and Remaining Channels (Optional)

1. Using various connection tools learned in earlier chapters, connect the remaining objects such as stairs and channels. After connecting all the objects, check the model using the **Display connected objects** tool to make sure everything, except the plates and gratings, is highlighted and marked in Red.

Skill Evaluation

Evaluate your skills to see how many questions you can answer correctly. The answers to these questions are given at the end of the book.

1. In Advance Steel, the **Cage ladder** tool can also be used to create a ladder without the cage. (True/False)

2. In Advance Steel, you can also cannot miter cut the sections. (True/False)

3. You cannot create spiral stairs without the handrails. (True/False)

4. In Advance Steel, you can easily check if a section is connected to the rest of the model or not. (True/False)

5. While creating the section cut features, you can also cut a section to the face of another section. (True/False)

6. Which tool is used to create a spiral staircase?

    (A) **Spiral**   (B) **Round**
    (C) **Spiral staircase**   (D) **Round staircase**

7. Which tool is used to check if the selected section is connected to the rest of the model?

    (A) **Display connected objects**   (B) **Display joints**
    (C) **Connect objects**   (D) **None**

8. Which tool is used to break the bolt pattern group into individual bolts?

    (A) **Split bolt groups**   (B) **Break group**
    (C) **Split bolts**   (D) **Break bolts**

9. Which tool is used to cut a section at the face of the other section?

    (A) **Center cut**   (B) **Cut at object**
    (C) **Part cut**   (D) **Face cut**

10. Which tool is used to create the circular pattern of bolts?

    (A) **Circular pattern**   (B) **Circular, center point**
    (C) **Polar pattern**   (D) **Circular array**

Class Test Questions

Answer the following questions:

1. Explain briefly the procedure of inserting a cage ladder.

2. What is the process of checking all the objects that will be shop connected?

3. Explain briefly the process of connecting objects using a weld point.

4. How do you clear all the marked objects that are connected together?

5. Explain briefly how to insert a spiral stair.

Chapter 10 – Adding Custom Connections

The objectives of this chapter are to:

√ *Explain why custom joints are required*
√ *Explain various tools required to model the custom connections*
√ *Explain the process of creating custom connections and saving them to the library*
√ *Explain the process of inserting custom connections*

CUSTOM CONNECTIONS

In the earlier chapters, you have seen how extensive the library of joints in the **Connection vault** is. However, most of the projects you will work with will require custom joints that cannot be found in the **Connection vault**. For example, Figures 1 and 2 show views from two directions of the channels connected using a custom joint. In Figure 2, the display type of the right channel is set to **Symbol** for better understanding of the joint.

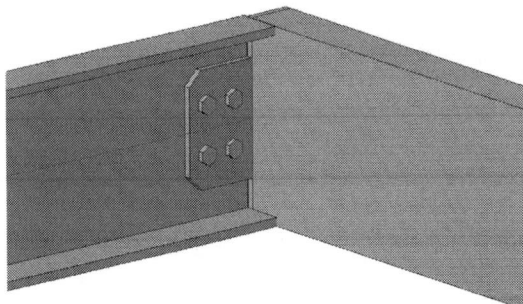

Figure 1 *Two channels to be connected*   *Figure 2* *Custom joint to connect the channels*

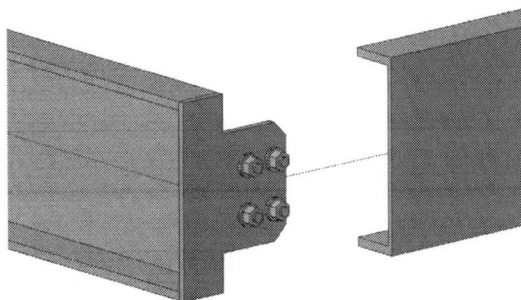

As you can see, the joint that is used to connect the channels is not one of the standard joints that you can use from the **Connection vault**. For situations such as this, Advance Steel allows you to create custom joints and add them to the library. You can then use these custom joints at different places in multiple projects.

To create plates to be used in the custom connections, you can create them using the tools discussed in Chapter 8 of this book. However, to simplify the process of creating various components of custom joints, Advance Steel also provides a number of tools that can be used to insert plates or bolts on beams. These tools are available on the **Custom connections** tab, as shown in Figure 3. This tab is divided into six areas, as marked in Figure 3. The following is a brief description of these areas.

> **Area 1**: This area provides the tools to create and insert the custom connections.
> **Area 2**: This area provides six tools to insert plates on a beam.
> **Area 3**: This area provides six tools to insert bolts, studs, or holes on the sections or plates.
> **Area 4**: This area provides three tools to insert plate on plates.
> **Area 5**: This area provides three tools to insert spacer plate, shim plate, or outside stiffeners.
> **Area 6**: This area provides three tools to insert beams, plates, or holes on a reference and also a fourth tool to insert column stiffener.

These tools are discussed next.

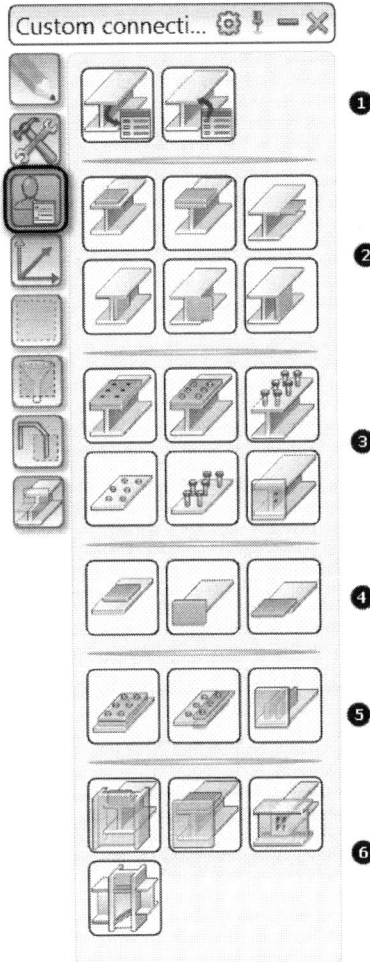

*Figure 3 The **Custom connections** tab*

Tip: *The buttons in areas 2 to 5 have animations associated with them. If you move the cursor on these buttons, an animation plays showing various options of inserting items using that tool.*

Inserting Plates Along the Beam Flange

The **Plate along beam flange** tool is used to insert a plate along the flange of a selected beam. When you invoke this tool, you will be prompted to select a beam. Once you select the beam, you will be prompted to specify whether or not you want to specify a reference point. If you specify a reference point, a sphere will be placed at that point. That sphere can be used to change the location of the inserted plate. Once you specify the reference point or press ENTER at the previous prompt, the plate will be inserted on the flange of the selected beam and the **Advance Steel - Plate along flange [X]** dialog box will be displayed, as shown in Figure 4. Most of the options in this dialog box are similar to those discussed in earlier chapters. The rest of them are discussed next.

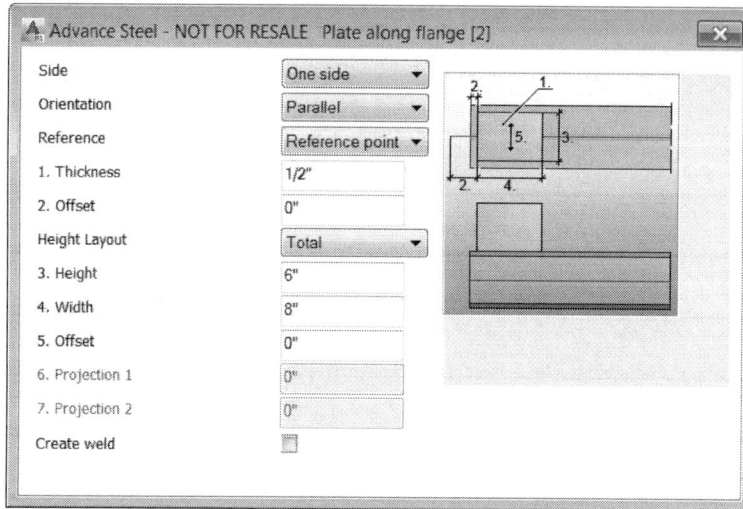

Figure 4 The Advance Steel - Plane along flange [X] dialog box

Side

This list is used to specify the side of the beam along which the plate needs to be inserted. You can also select the option to insert the plates along both sides.

Orientation

This list is used to specify whether the plates will be inserted parallel or perpendicular along the flange. Figure 5 shows plates inserted parallel along the beam flange on both sides of the beam and Figure 6 shows the same plates inserted perpendicular along the beam flange.

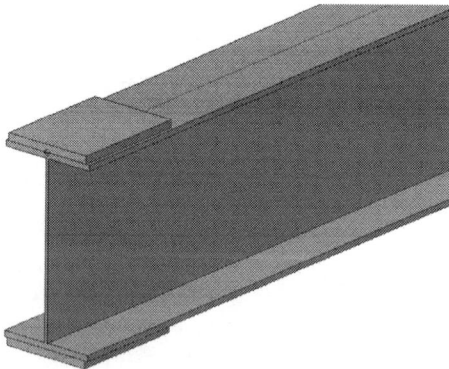

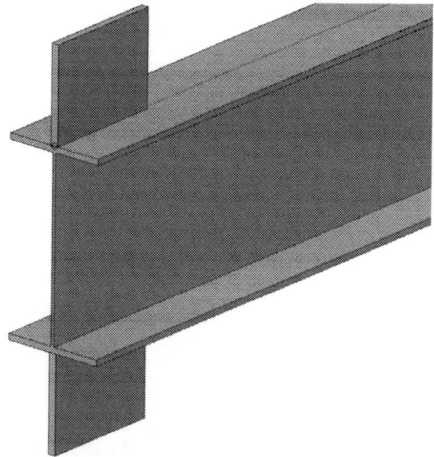

Figure 5 Parallel plates inserted along both sides of beam flange

Figure 6 Perpendicular plates inserted along both sides of beam flange

Reference

This list is used to specify whether the plates will be referenced to the point you defined in the prompt sequence, to the system axis of the main beam, or to the physical axis of the main beam.

Note: The rest of the options in this dialog box are similar to those discussed in earlier chapters.

Inserting Plates on the Beam Flange

The **Plate on beam flange** tool is similar to the **Plate along beam flange** tool, with the difference that if you decide to insert the plates in the perpendicular orientation, they will be inserted as shown in Figure 7 and not like the ones shown in Figure 6. The options available in this dialog box are similar to those discussed earlier.

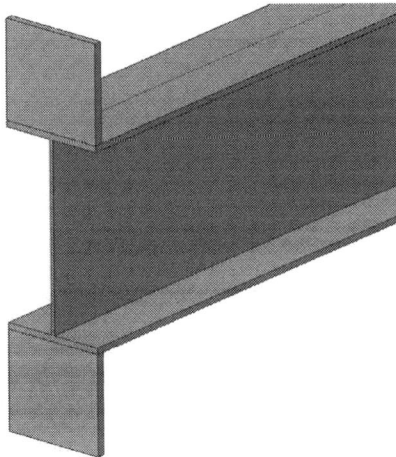

Figure 7 Perpendicular plates inserted on both sides of beam flange

Inserting Plates Parallel to the Beam Flange

The **Plate parallel beam flange** tool is used to insert a plate parallel to the flange of the beam. Similar to the previous tools, this tool also prompts you to select the beam and specify the reference point. On specifying all the required parameters in those prompts, a plate will be inserted parallel to the beam flange and the **Advance Steel - Plate parallel to flange [X]** dialog box will be displayed, as shown in Figure 8. Most of the options in this dialog box are similar to those discussed earlier. The rest of the options are discussed next.

Side

This list is used to specify the side of the beam on which the parallel plate needs to be inserted. You can also select the option to insert the parallel plates on both sides, as shown in Figure 9.

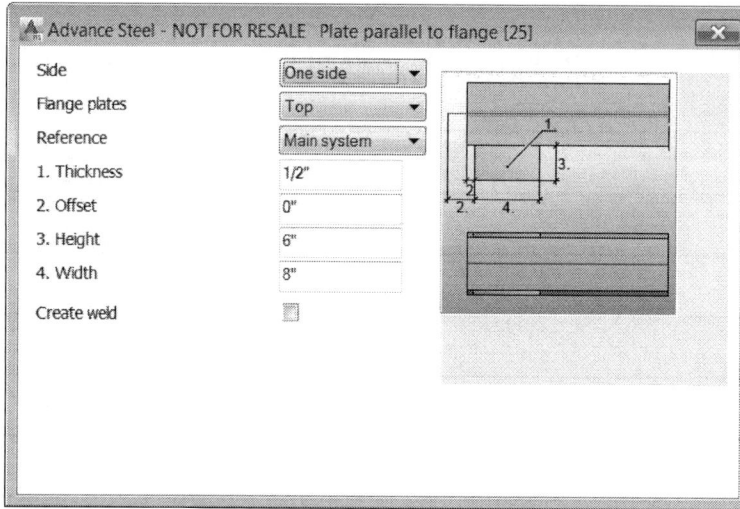

*Figure 8 The **Advance Steel - Plane parallel to flange [X]** dialog box*

Flange plates

This list is used to specify whether the plates will be inserted on the top flange, bottom flange, or both flanges. Figure 10 shows the beam with the plates inserted parallel to both sides of the top and bottom flanges.

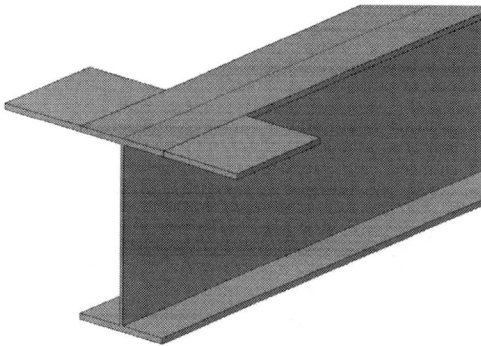

Figure 9 Plates inserted parallel to both sides of the top flange

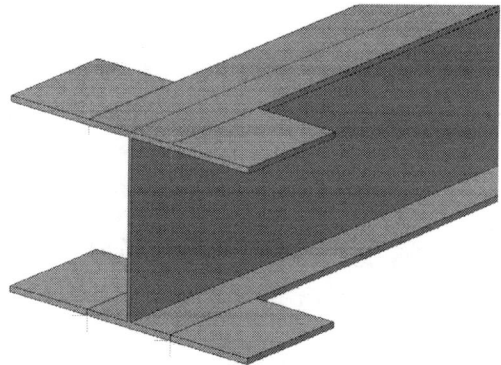

Figure 10 Plates inserted parallel to both sides of the top and the bottom flange

Tip: *To edit the inserted plates, right-click on one of the plates and then select **Advance Joint Properties** from the shortcut menu.*

Inserting Plates Along the Beam Web

The **Plate along beam web** tool is used to insert a plate along the web of a selected beam. This tool is similar to the **Plate along beam flange** tool, with the only difference being that the plates in this case will be inserted along the beam web. Figure 11 shows a plate inserted parallel along the beam web and Figure 12 shows a plate inserted perpendicular along the beam web.

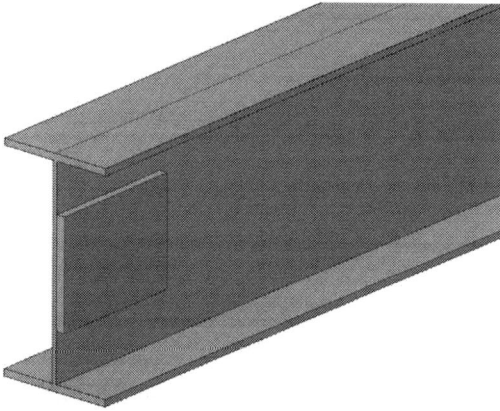

Figure 11 *A plate inserted parallel along the beam web*

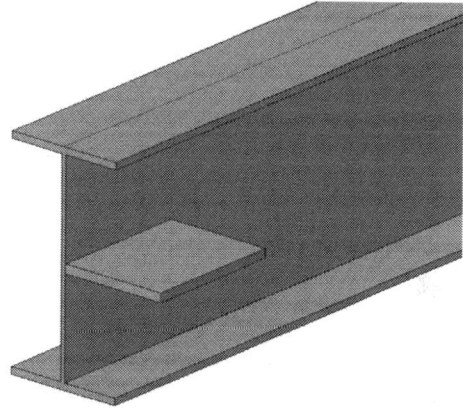

Figure 12 *A plate inserted perpendicular along the beam web*

Inserting Plates on the Beam Web

The **Plate on beam web** tool is similar to the **Plate on beam flange** tool, with the difference that the plates will be inserted on the beam web. Figure 13 shows the plates inserted on both the beam webs.

Inserting Plates Parallel to the Beam Web

The **Plate parallel beam web** tool is similar to the **Plate parallel beam flange** tool, with the difference that the plates will be inserted parallel to the beam web. Figure 14 shows plates inserted parallel to the web on both sides. In this case, the plate orientation is set to **Parallel** in the dialog box.

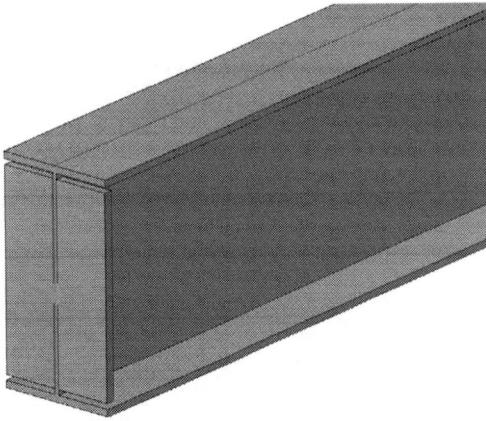

Figure 13 *Plates inserted on both sides of the beam web*

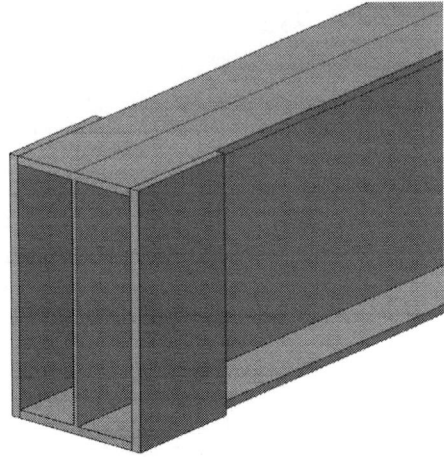

Figure 14 *Plates inserted parallel to both sides of beam web*

Inserting Bolts on Beams

The **Bolts on beam** tool is available in the third area of the **Custom connections** tab and is used to insert bolts on the selected beams to connect it to a plate. When you invoke this tool, you will be prompted to select a beam. On selecting the beam, you will be prompted to specify if you want to select a connecting object as well. In this case, you can select the plate to be bolted to the beam and then press ENTER; the bolt pattern will be inserted and the **Advance Steel - Bolts on beam [X]** dialog box will be displayed, as shown in Figure 15.

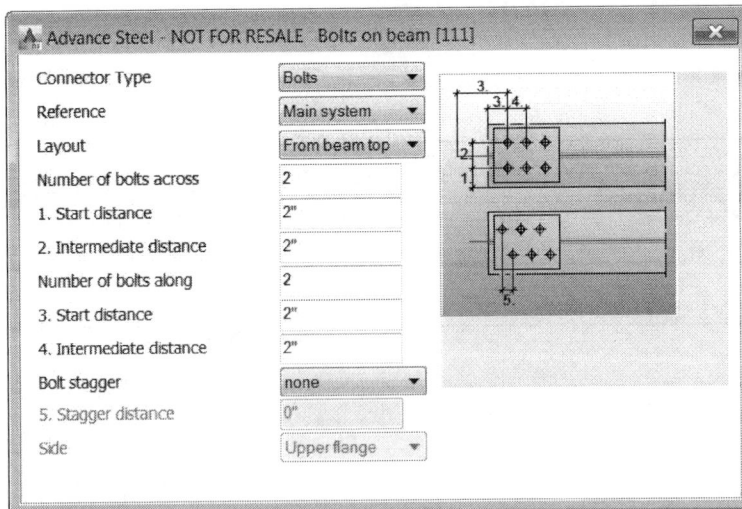

Figure 15 *The **Advance Steel - Bolts on beam [X]** dialog box*

Most of the options in this dialog box are similar to those discussed earlier. The rest of the options are discussed next.

Connector Type

This list is used to specify whether the beam will be connected to the plate using bolts or anchors, or only the holes will be created. Figure 16 shows the beam and plate connected using bolts and Figure 17 shows only the holes created.

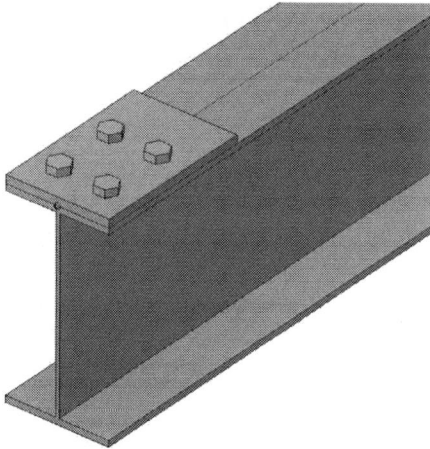

Figure 16 The beam connected to the plate using bolts

Figure 17 Only holes inserted between the beam and the plate

Tip: *To edit the bolt type and diameter, double-click on one of the bolts; the **Advance Steel - Bolts** dialog box will be displayed. You can change the type, grade, size, hole tolerance, and so on of the bolts using this dialog box. However, you cannot edit the number and spacing of the bolts in this dialog box. These parameters need to be edited by right-clicking on the bolts and selecting **Advance Joint Properties** from the shortcut menu.*

Layout

This list is used to specify whether the bolts will be placed centered on the flange or defined in terms of a distance from one of the beam side edges.

Bolt stagger

This list is used to specify whether or not the bolts will be staggered. By default, **none** is selected from this list. As a result, the bolts are not staggered. However, you can select to stagger the bolts on side 1 or side 2 by selecting their respective option from this list. If you decide to stagger the bolt, the stagger distance can be specified in the **Stagger distance** edit box. Figure 18 shows the bolts with no stagger and Figure 19 shows the bolts stagger along side 1 with a stagger value of 1" or 25mm.

Note: *The rest of the options in this dialog box are similar to those discussed earlier.*

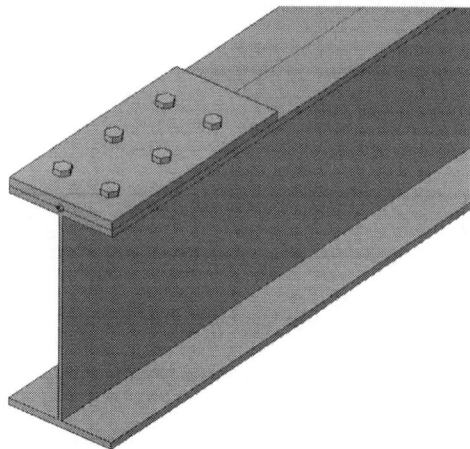

Figure 18   Bolts not staggered

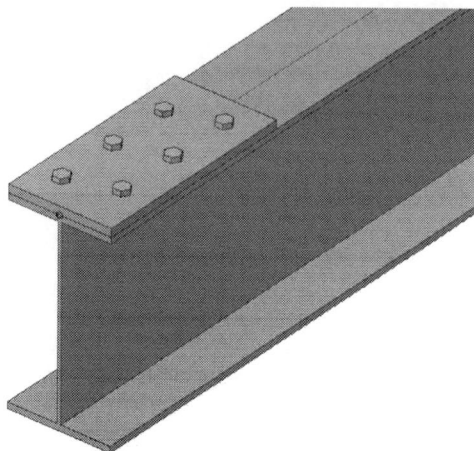

Figure 19   Bolts staggered on side 1

Inserting Bolts on Beam Gauge Line

The **Bolts on beam gauge line** tool is similar to the **Bolts on beam** tool and is used to insert the bolts on the gauge line of the selected beam. You can insert the bolts on one side of the gauge line, other side of the gauge line, or both sides of the gauge line.

Inserting Studs on the Beams

The **Studs on beam** tool is used to insert studs on the selected beam. In this case, you will not be prompted to select any other object to connect. You can insert the studs on one side or both sides of the beam. When you invoke this tool, you will be prompted to select a beam. On selecting the beam, the studs will be inserted with the default parameters and the **Advance Steel - Studs on beam [X]** dialog box will be displayed. The options in this dialog box are similar to those discussed earlier. Figure 20 shows a beam with studs inserted on one side and Figure 21 shows the same beam with the studs inserted on both sides.

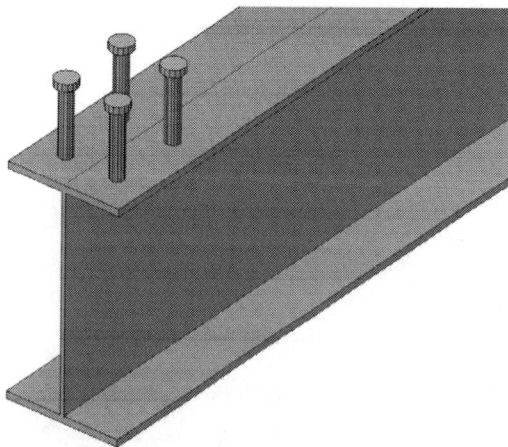

Figure 20   Studs inserted on one side

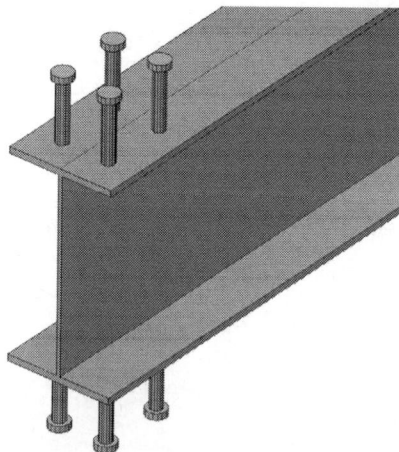

Figure 21   Studs inserted on both sides

Note: The **Bolts/Anchors/Holes on plate** tool work similar to the **Bolts on beam** tool, with the difference that in the **Bolts/Anchors/Holes on plate** tool, you will have to select the plate first. Same is true for the **Studs on beam** and **Studs on plate** tools.

Inserting Galvanizing Holes

The **Galvanizing holes** tool is used to insert galvanizing holes on an end plate of the beam or on the web of the beam. When you invoke this tool, you will be prompted to select beams and then select a plate. On making these two selections and pressing ENTER, the galvanizing holes with the default parameters will be inserted and the **Advance Steel - Galvanizing holes** dialog box will be displayed, as shown in Figure 22.

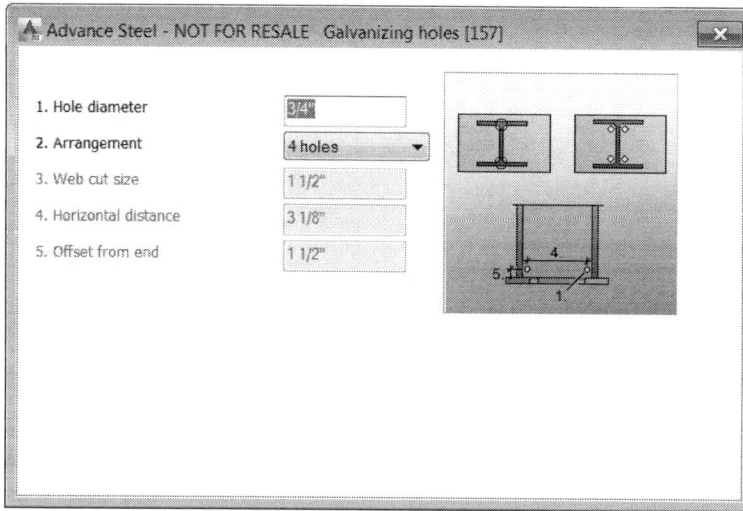

Figure 22 The Advance Steel - Galvanizing holes [X] dialog box

Most of the options in this dialog box are similar to those discussed earlier, except the **Arrangement** list. This is discussed next.

Arrangement

This list is used to specify the arrangement of the galvanizing holes on the end plate or the beam. These options are discussed next.

2 holes centered

This option inserts two galvanizing holes centered on the end plate, as shown in Figure 23. You can change the diameter of these holes in the **Hole diameter** edit box. However, the rest of the parameters are automatically controlled based on the plate and the beam selected.

4 holes

This option inserts four galvanizing holes on the end plate, as shown in Figure 24. You can change the diameter of these holes in the **Hole diameter** edit box. However, the rest of the parameters are automatically controlled based on the plate and the beam selected.

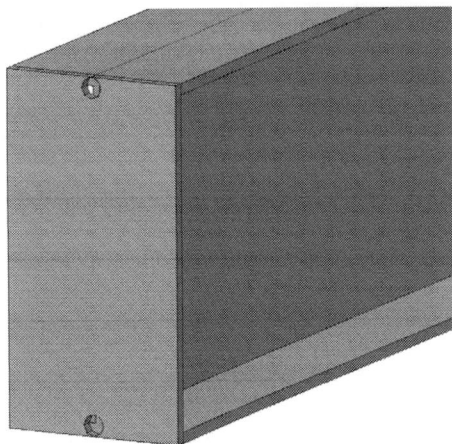

Figure 23 *Two galvanizing holes inserted centered on the end plate*

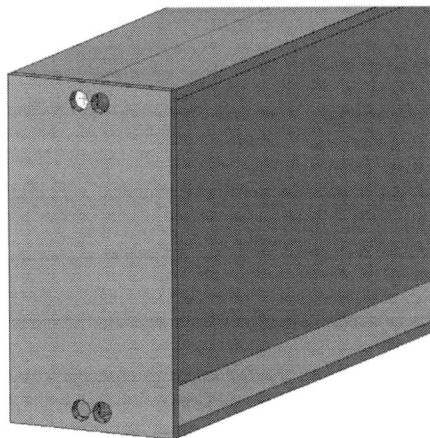

Figure 24 *Four galvanizing holes inserted on the end plate*

Web cut
This option inserts two cuts in the web of the beam, as shown in Figure 25. The size of the cut can be defined in the **Web cut size** edit box.

> *Tip: To edit the web cut feature on the beam, you will have to first change the display type of the beam to **Features**. Next, right-click on the web cut feature and select **Advance Joint Properties** to edit the galvanizing hole properties.*

Web holes
This option inserts two holes in the web of the beam, as shown in Figure 26. The size of the holes can be defined in various edit boxes in this dialog box.

Figure 25 *Two cuts created in the web of the beam*

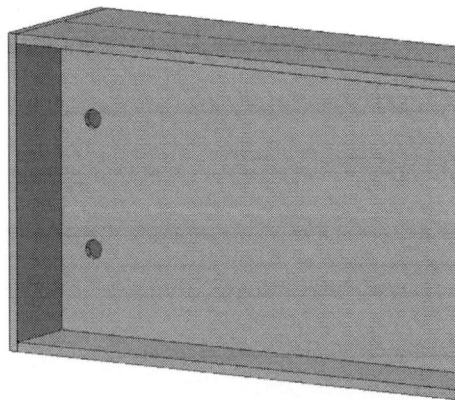

Figure 26 *Two holes created in the web of the beam*

2 holes diagonal
This option inserts two diagonal holes on the end plate on either side of the beam web.

Inserting a Plate on an Existing Plate

The **Plate on plate** tool is used to insert a plate on an existing plate. When you invoke this tool, you will be prompted to select a plate. On selecting the plate and pressing ENTER, a new plate with the default parameters will be inserted and the **Advance Steel - Plate on plate [X]** dialog box will be displayed. You can edit the size and location of the new plate using the options in this dialog box.

Inserting a Plate Perpendicular to an Existing Plate

The **Plate perpendicular to plate** tool is used to insert a plate perpendicular to an edge of an existing plate. Note that the new perpendicular plate will be inserted at the edge closest to the selection point on the existing plate. When you invoke this tool, you will be prompted to select a plate. On selecting the plate, a new plate will be inserted perpendicular to the edge closest to the selection point and the **Advance Steel - Plate perpendicular to plate [X]** dialog box will be displayed. You can edit the size and location of the new plate using the options in this dialog box.

Inserting a Plate at an Edge of the Existing Plate

The **Plate at plate edge** tool is used to insert a plate at an edge of an existing plate. Note that the new plate will be inserted at the edge closest to the selection point on the existing plate. When you invoke this tool, you will be prompted to select a plate. On selecting the plate, a new plate will be inserted at the edge closest to the selection point and the **Advance Steel - Plate at plate edge [X]** dialog box will be displayed. You can edit the size and location of the new plate using the options in this dialog box.

Note: The tools in areas 5 and 6 of the Custom connections tab are similar to those discussed earlier. As a result, they are not discussed in detail.

CREATING CUSTOM CONNECTIONS

You will now learn how to create custom connections. You can create one or multiple custom connections in the same file, if required. It is important to note that after you have created the custom connections, the file needs to be saved in the location listed below where 20XX represents the release of Advance Steel you are using:

C:\ProgramData\Autodesk\Advance Steel 20XX\Shared\ConnectionTemplates

Note: The Program Data folder is a hidden folder and so you will have to first turn on the option to show hidden files and folders to locate this folder.

To create the custom connection, invoke the **Create connection template** tool, which is the first tool in the **Custom connections** tab. When you invoke this tool, the **Choose the definition method** dialog box is displayed, as shown in Figure 27.

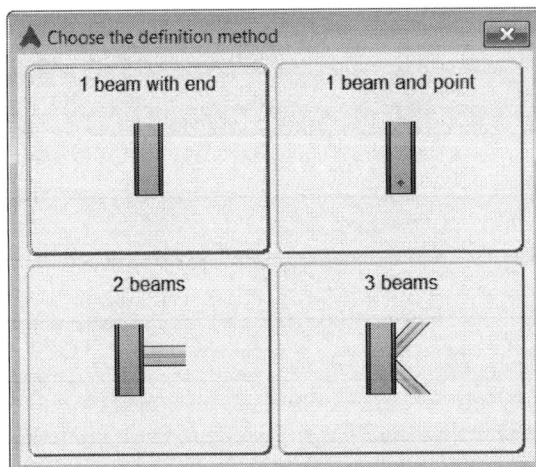

Figure 27 The **Choose the definition method** *dialog box*

This dialog box allows you to create custom connection using four definition methods. These four methods are:

1 beam with end: This definition method is used when you want to create end plate or base plate connections.

1 beam and point: This definition method is used when you want to create the connections by defining a reference point.

2 beams: This definition method is used to create a custom connection between two selected beams.

3 beams: This definition method is used to create a custom connection between three selected beams.

The procedures for creating custom joints using all these definition methods are discussed next.

Procedure for Creating Custom Connection using the 1 Beam with End Method

The following is the procedure for creating the custom connection using the **1 beam with end** definition method.

1. Create all the components of a custom joint. Figure 28 shows the hexagonal base plate, weld, and the circular anchor bolt pattern created for the custom connection at the bottom end of a column.

Tip: Make sure in the connections similar to the one shown in Figure 28, you add a weld between the plate and the column and include that weld as part of the custom connection. This will ensure when you insert this type of custom connection, the plate is automatically welded to the column.

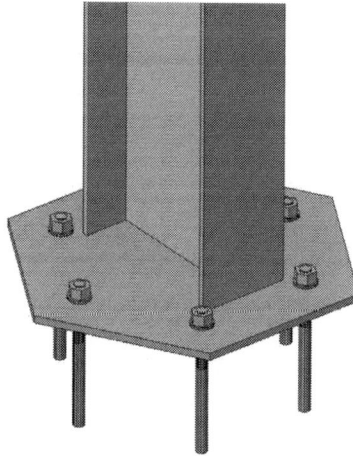

Figure 28 Hexagonal base plate, weld, and circular anchor bolt pattern for custom connection

2. Save the file for the custom connection in the following folder where 20XX represents the release of Advance Steel you are using:

    C:\ProgramData\Autodesk\Advance Steel 20XX\Shared\ConnectionTemplates

3. From the **Advance Steel Tool Palette** > **Custom Connections** tab, invoke the **Create connection template** tool; the **Choose the definition method** dialog box will be displayed.

4. Click on the **1 beam with end** button in the dialog box; the dialog box will be temporarily closed and you will be prompted to select the beam.

5. Select the beam; the **Advance Steel - User template** dialog box will be displayed, as shown in Figure 29.

6. In the **Name** edit box, enter the name of the new custom connection.

7. Click the **Reselect driven/output objects** button; the dialog box will be temporarily closed and you will be prompted to select the objects for connection template.

8. Using a crossing or a window, select all the objects that need to be a part of the custom connection. Make sure you carefully select all the welds that need to be a part of the connection. Next, press the ENTER key; a connection box will be placed around the custom connection and the dialog box will be redisplayed.

9. In the **Driver selection prompts** edit box, enter the prompt sequence that you want to see while inserting this connection.

10. Close the dialog box.

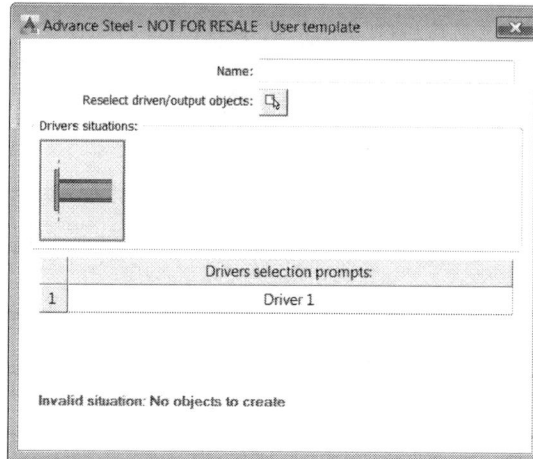

Figure 29   *The Advance Steel - User template* dialog box

11. Save the file. This completes the process of creating a custom joint using the **1 beam with end** method.

Tip: *Once the custom connection is created, you can double-click on the custom connection box created around the custom connection objects to edit the connection.*

Note: *The process of inserting custom connections will be discussed in detail later in this chapter.*

Procedure for Creating Custom connection using the 1 Beam and Point Method

The following is the procedure for creating the custom connection using the **1 beam and point** definition method.

1. Create all the components of a custom joint. Figure 30 shows the plates, welds, and bolt pattern created for the custom connection.

2. Save the file for the custom connection in the following folder where 20XX represents the release of Advance Steel you are using:

   C:\ProgramData\Autodesk\Advance Steel 20XX\Shared\ConnectionTemplates

3. From the **Advance Steel Tool Palette** > **Custom Connections** tab, invoke the **Create connection template** tool; the **Choose the definition method** dialog box will be displayed.

4. Click on the **1 beam and point** button in the dialog box; the dialog box will be temporarily closed and you will be prompted to select the beam.

5. Select the beam and press ENTER; you will be prompted to select the input point.

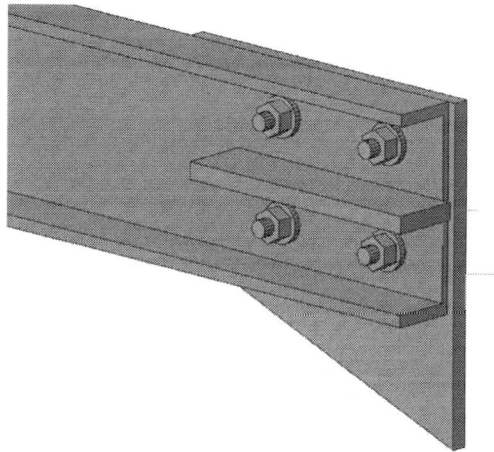

Figure 30 *Plates, welds, and bolt pattern created for the custom connection*

6. Select the reference point to be used for the custom connection; the **Advance Steel - User template** dialog box will be displayed, as shown in Figure 31.

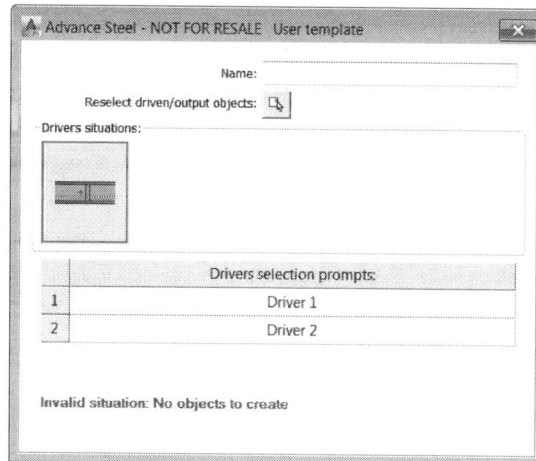

Figure 31 *The **Advance Steel - User template** dialog box*

7. In the **Name** edit box, enter the name of the new custom connection.

8. Click the **Reselect driven/output objects** button; the dialog box will be temporarily closed and you will be prompted to select the objects for the connection template.

9. Using a crossing or a window, select all the objects that need to be a part of the custom connection. Make sure you carefully select all the welds that need to be a part of the connection. Next, press the ENTER key; a connection box will be placed around the custom connection and the dialog box will be redisplayed.

Notice that there are two prompt rows displayed in the **Driver selection prompts** area. The first prompt is for the beam and the second prompt is for the reference point.

10. In the **Driver selection prompts** area > **Driver 1** edit box, enter the prompt sequence that you want to be displayed for selecting the beam.

11. In the **Driver selection prompts** area > **Driver 2** edit box, enter the prompt sequence that you want to be displayed for selecting the point.

12. Close the dialog box.

13. Save the file. This completes the process of creating a custom joint using the **1 beam and point** method.

Procedure for Creating Custom Connection using the 2 Beams Method

The following is the procedure for creating the custom connection using the **2 beams** definition method.

1. Create all the components of a custom joint. Figure 32 shows the plate, weld, and bolt pattern created for the custom connection. In this figure, the display type of the channel on the right is set to **Symbol** for better visibility of the connection.

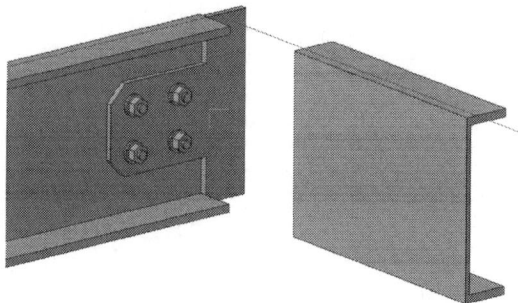

Figure 32 Plate, weld, and bolt pattern created for the custom connection

2. Save the file for the custom connection in the following folder where 20XX represents the release of Advance Steel you are using:

C:\ProgramData\Autodesk\Advance Steel 20XX\Shared\ConnectionTemplates

3. From the **Advance Steel Tool Palette** > **Custom Connections** tab, invoke the **Create connection template** tool; the **Choose the definition method** dialog box will be displayed.

4. Click on the **2 beams** button in the dialog box; the dialog box will be temporarily closed and you will be prompted to select the input beam.

5. Select the first beam and press ENTER; you will be prompted to select the input beam again.

6. Select the second beam and press ENTER; the **Advance Steel - User template** dialog box will be displayed, as shown in Figure 33.

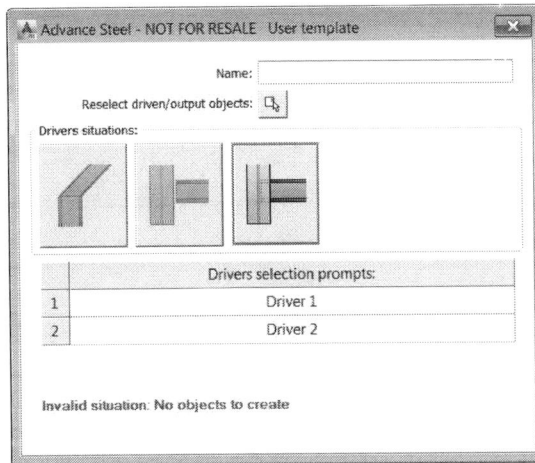

Figure 33 The Advance Steel - User template dialog box

Note: *Based on the beams selected, the button from the **Driver situation** is automatically selected. However, you can click any of the other two buttons to manually define a different driver situation.*

7. In the **Name** edit box, enter the name of the new custom connection.

8. Click the **Reselect driven/output objects** button; the dialog box will be temporarily closed and you will be prompted to select the objects for connection template.

9. Using a crossing or a window, select all the objects that need to be a part of the custom connection. Make sure you carefully select all the welds that need to be a part of the connection. Next, press the ENTER key; a connection box will be placed around the custom connection and the dialog box will be redisplayed.

   Notice that there are two prompt rows displayed in the **Driver selection prompts** area. The first prompt is for the first beam and the second prompt is for the second beam.

10. In the **Driver selection prompts** area > **Driver 1** edit box, enter the prompt sequence that you want to be displayed for selecting the first beam.

11. In the **Driver selection prompts** area > **Driver 2** edit box, enter the prompt sequence that you want to be displayed for selecting the second beam.

12. Close the dialog box.

13. Save the file. This completes the process of creating a custom joint using the **2 beams** method.

Procedure for Creating Custom Connection using the 3 Beams Method

The following is the procedure for creating the custom connection using the **3 beams** definition method.

1. Create all the components of a custom joint. Figure 34 shows the plate, weld, and bolt pattern created for the custom connection. In this figure, the display type of the channel on the right is set to **Symbol** for better visibility of the connection.

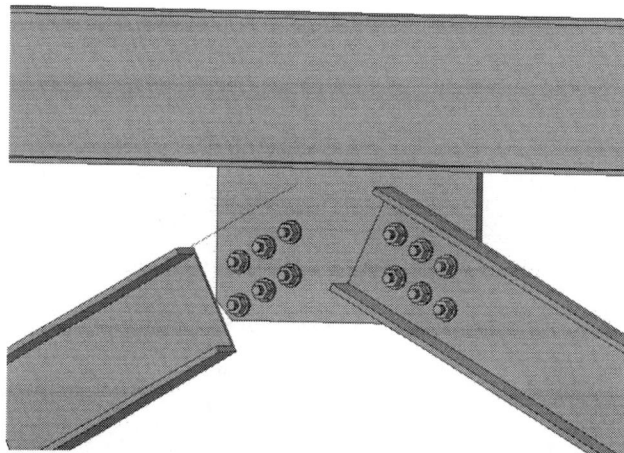

Figure 34   Plate, weld, and staggered bolts on beams created for the custom connection

2. Save the file for the custom connection in the following folder where 20XX represents the release of Advance Steel you are using:

C:\ProgramData\Autodesk\Advance Steel 20XX\Shared\ConnectionTemplates

3. From the **Advance Steel Tool Palette** > **Custom Connections** tab, invoke the **Create connection template** tool; the **Choose the definition method** dialog box will be displayed.

4. Click on the **3 beams** button in the dialog box; the dialog box will be temporarily closed and you will be prompted to select the input beam.

5. Select the first beam and press ENTER; you will be prompted to select input beam again.

6. Select the first beam and press ENTER; you will be prompted to select input beam the third time.

7. Select the third beam and press ENTER; the **Advance Steel - User template** dialog box will be displayed, as shown in Figure 35.

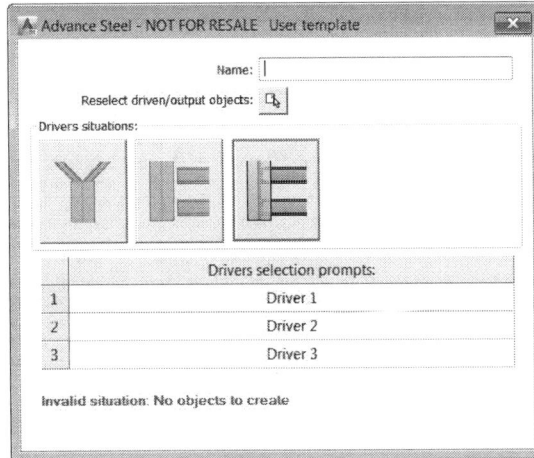

Figure 35 The Advance Steel - User template dialog box

8. In the **Name** edit box, enter the name of the new custom connection.

9. Click the **Reselect driven/output objects** button; the dialog box will be temporarily closed and you will be prompted to select the objects for the connection template.

10. Using a crossing or a window, select all the objects that need to be a part of the custom connection. Make sure you carefully select all the welds that need to be a part of the connection. Next, press the ENTER key; a connection box will be placed around the custom connection and the dialog box will be redisplayed.

    Notice that there are three prompt rows displayed in the **Driver selection prompts** area.

11. In the **Driver selection prompts** area > **Driver 1** edit box, enter the prompt sequence that you want to be displayed for selecting the first beam.

12. In the **Driver selection prompts** area > **Driver 2** edit box, enter the prompt sequence that you want to be displayed for selecting the second beam.

13. In the **Driver selection prompts** area > **Driver 3** edit box, enter the prompt sequence that you want to be displayed for selecting the third beam.

14. Close the dialog box.

15. Save the file. This completes the process of creating a custom joint using the **3 beams** method.

INSERTING CUSTOM CONNECTIONS

Once the custom connections are created and their files saved, you are ready to insert them. It is strongly recommended that you close the file in which you created the custom connection before you start inserting them. To insert the custom connection, you can use the **Insert connection template** tool available on the **Advance Steel Tool Palette** > **Custom connections** tab. When you invoke this tool, the **Connection template explorer** dialog box will be displayed, as shown in Figure 36.

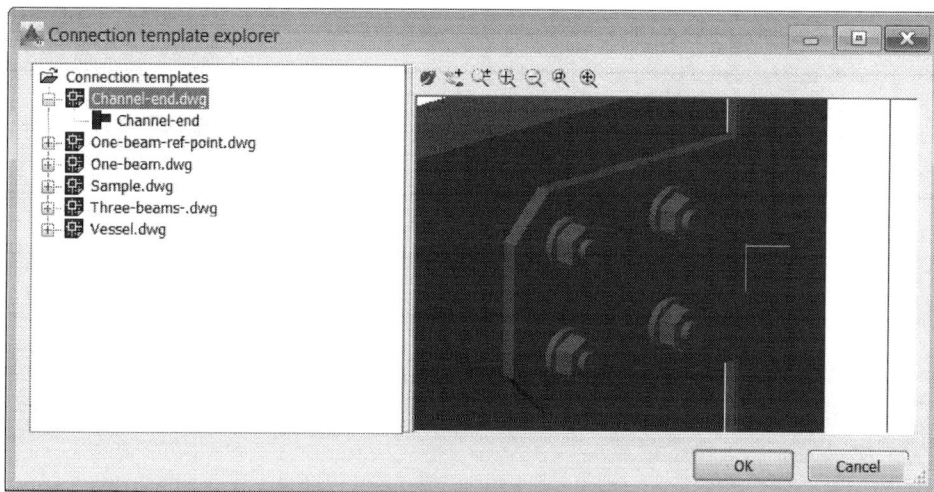

*Figure 36 The **Connection template explorer** dialog box*

All the custom connection files saved in the **C:\ProgramData\Autodesk\Advance Steel 20XX\Shared\ConnectionTemplates** folder will be listed in the left side of the dialog box. The right side of the dialog box displays the preview of the custom connection. You can use various navigation tools available above the preview window to navigate around the model. Once you select the connection you want to insert, click the **OK** button. On doing so, the prompt sequences to insert the selected custom joint will be displayed. Once all the selections related to the connection are made, the **Advance Steel - User template** dialog box will be displayed, as shown in Figure 37. In this dialog box, you can specify whether or not you want to allow connection object modification. If this tick box is selected, you can double-click on the connection objects such as plates or bolt patterns and edit them.

*Tip: If there is only one custom connection in the file listed on the left side of the **Connection template explorer** dialog box, then it is recommended to select the file name instead of the connection name and click **OK** in the dialog box. The reason is that in some cases, selecting the connection name and clicking **OK** does not insert the right type of connection.*

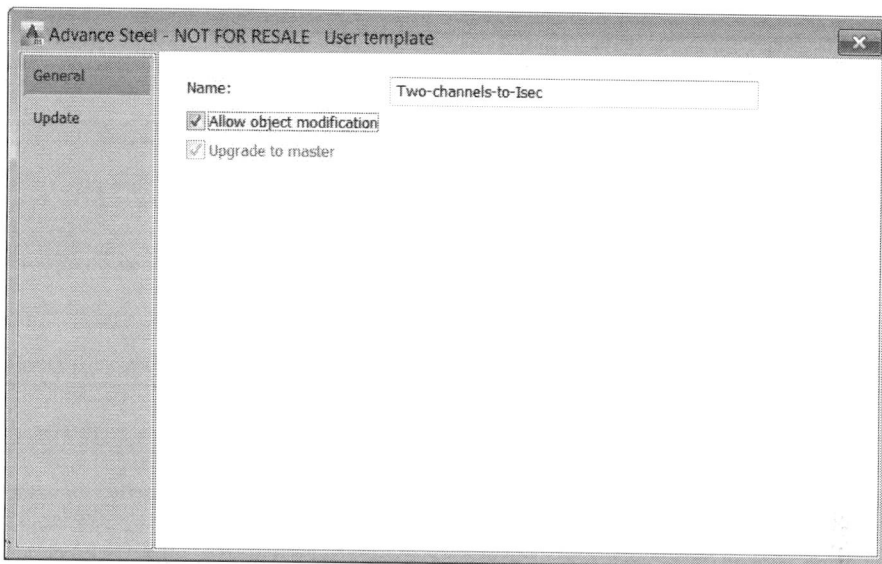

Figure 37 *The Advance Steel - User template* dialog box

Procedure for Inserting Custom Connections

The following is the procedure for inserting the custom connection.

1. Open the file in which you want to insert the custom connection.

2. From the **Advance Steel Tool Palette** > **Custom Connections** tab, invoke the **Insert connection template** tool; the **Choose the definition method** dialog box will be displayed.

3. Follow the prompt sequences and make all the selections required for the connection. Once all the selections are made, the connection will be inserted and the **Advance Steel - User template** dialog box will be displayed.

4. If required, select the **Allow object modification** tick box from the dialog box. This will ensure that you can double-click on the connection objects such as plates, bolt patterns, and so on, and modify them in their respective dialog boxes.

Hands-on Tutorial (STRUC/ BIM)	*In this tutorial, you will complete the following tasks:*
	1. Create the **1 beam with end** custom connection, as shown in Figure 38.
	2. Create the **1 beam and point** custom connection, as shown in Figure 39.
	3. Create the **2 beams** custom connection, as shown in Figure 40.
	4. One by one, insert all the custom connections.

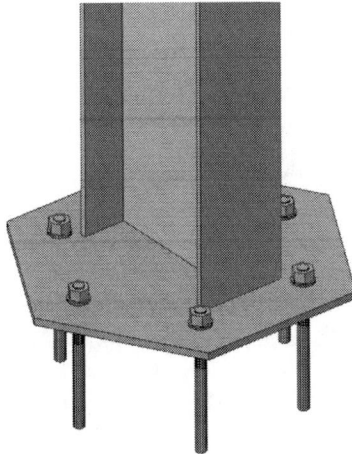

Figure 38 *The **1 beam with end** custom joint to be created*

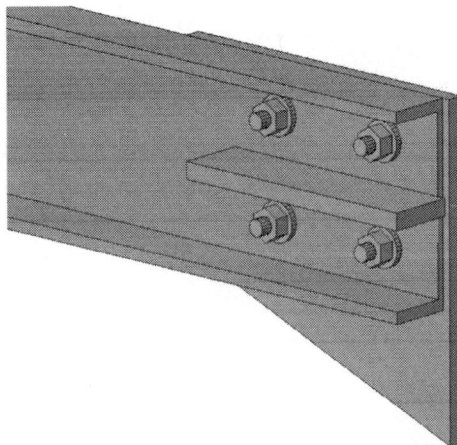

Figure 39 *The **1 beam and point** custom joint to be created*

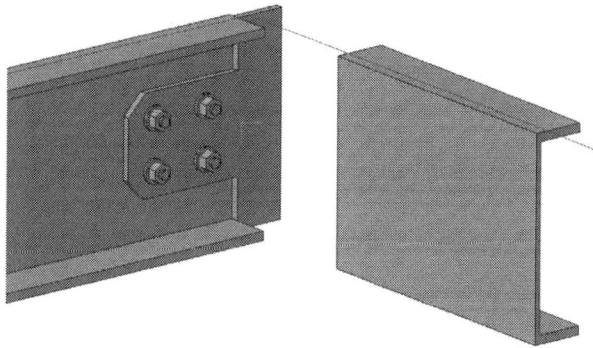

Figure 40 The 2 **beams** custom connection to be created

Section 1: Creating the Custom Connection using the 1 beam with end Method

In this section, you will open the **C10-Imperial-End-Plate.dwg** or the **C10-Metric-End-Plate.dwg** file and then insert the circular bolt pattern. Next, you will weld the end plate to the column and then create the custom connection using the **1 beam with end** method.

1. From the **C10 > Struc-BIM** folder, open the **C10-Imperial-End-Plate.dwg** or **C10-Metric-End-Plate.dwg** file, based on the preferred units. The model in this file appears similar to the one shown in Figure 41.

Figure 41 The model to create the custom connection

2. From the **Object** ribbon tab > **Switch** ribbon panel, click on the **bolts/anchors/holes/shear studs** button once so that it changes to display the preview of the anchor bolt.

3. Using the **Object** ribbon tab > **Connection objects** ribbon panel > **Circular, center point** tool, insert the circular pattern of anchor bolts on the end plate. The following are the parameters for this pattern:

Center of the Pattern:	**Use the node point at the bottom of the column**
Radius of circle:	**9" or 230 mm**
Anchor bolt type:	**US Normal Anchors (for Imperial) or HILTI HAS-E (for Metric)**
Diameter:	**3/4 inch or 20.00 mm**
Length:	**10" or 255.00**
Number of bolts:	**6**

The model, after inserting the anchor bolt pattern, is shown in Figure 42.

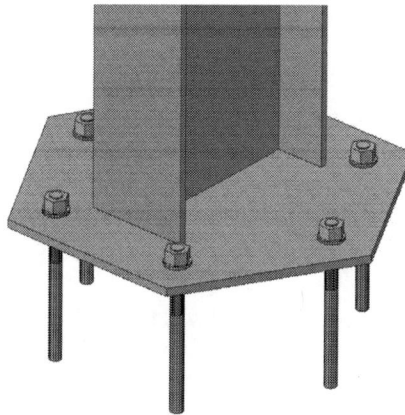

Figure 42 *The model, after inserting the anchor bolt pattern*

Next, you will connect the plate to the column using the **Weld point** tool.

4. Using the **Object** ribbon tab > **Connection objects** ribbon panel > **Weld point** tool, connect the end plate to the column. Place the weld symbol at the node point of the column. Use the default weld parameters.

> **Tip**: *To ensure all the custom joint objects are connected, it is a good idea to use the **Display connected objects** tool available on the **Advance Steel Tool Palette** > **Selection** tab to check it.*

Before you create the custom connection, you need to save a copy of the current file into the **C:\ProgramData\Autodesk\Advance Steel 20XX \Shared\ConnectionTemplates** folder, where **20XX** represents the release of Advance Steel you are using.

5. Using the **Save As** tool, save a copy of this file with the name **Custom-Base-Plate** in the **C:\ProgramData\Autodesk\Advance Steel 20XX\Shared\ConnectionTemplates** folder.

With this, you are now ready to create the custom connection.

6. From the **Advance Steel Tool Palette > Custom connections** tab, invoke the **Create connection template** tool; the **Choose the definition method** dialog box is displayed, as shown in Figure 43.

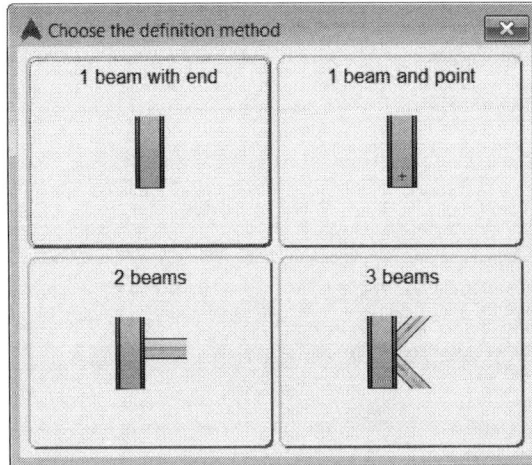

Figure 43 The **Choose the definition method** *dialog box*

7. Click the **1 beam with end** button; you are prompted to select the beam.

8. Select the column; the **Advance Steel - User template** dialog box is displayed, as shown in Figure 44.

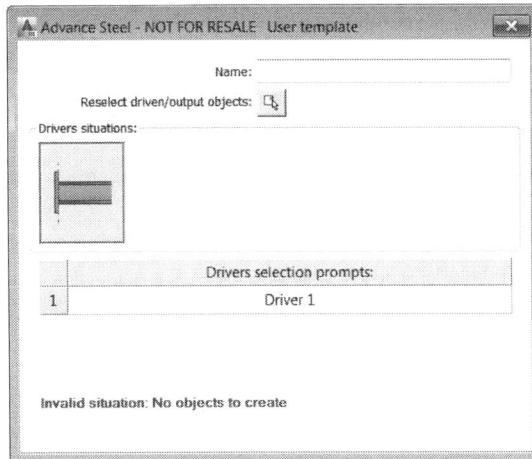

Figure 44 The **Advance Steel - User template** *dialog box*

9. Enter **Hex base plate** in the **Name** edit box.

10. Click the **Reselect driven/output objects** button; the dialog box is temporarily closed and you are prompted to select objects for the connection template.

11. Using a crossing, select everything in the model, including the column. It is not important to turn on the layer of the connection/joint box and select it.

12. Press ENTER; the dialog box is redisplayed and a custom connection box is created around the connection objects.

> **Note:** You can see the custom connection box only if the layer of the connection box is turned on.

13. In the **Driver selection prompts > Driver 1** edit box, type the following:

    the column

    The reason you types only **the column** is because Advance Steel automatically adds **Select** at the start of the prompt sequence while inserting the joint.

14. Close the dialog box.

15. Save the file and close it.

Section 2: Creating the Custom Connection using the 1 beam and point Method

In this section, you will open the **C10-Imperial-One-Beam-Ref-Point.dwg** or the **C10-Metric-One-Beam-Ref-Point.dwg** file and then insert a plate along the web of the channel. Next, you will insert a bolt pattern. Finally, you will use the **1 beam and point** method to create the custom connection.

1. From the **C10 > Struc-BIM** folder, open the **C10-Imperial-One-Beam-Ref-Point.dwg** or **C10-Metric-One-Beam-Ref-Point.dwg** file, based on the preferred units. The model in this file appears similar to the one shown in Figure 45.

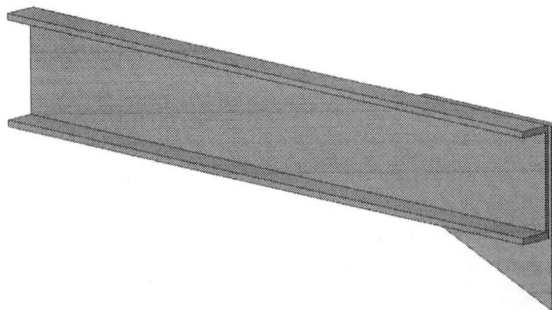

Figure 45 The model to create the custom connection

2. From the **Advance Steel Tool Palette > Custom connections** tab, invoke the **Plate along beam web** tool; you are prompted to select the beam.

3. Select the channel in the model; you are prompted to specify if you want to select a reference point.

4. Click **No** in the previous prompt; a plate is inserted on the web of the channel and the **Advance Steel - Plate along web [X]** dialog box is displayed, as shown in Figure 46.

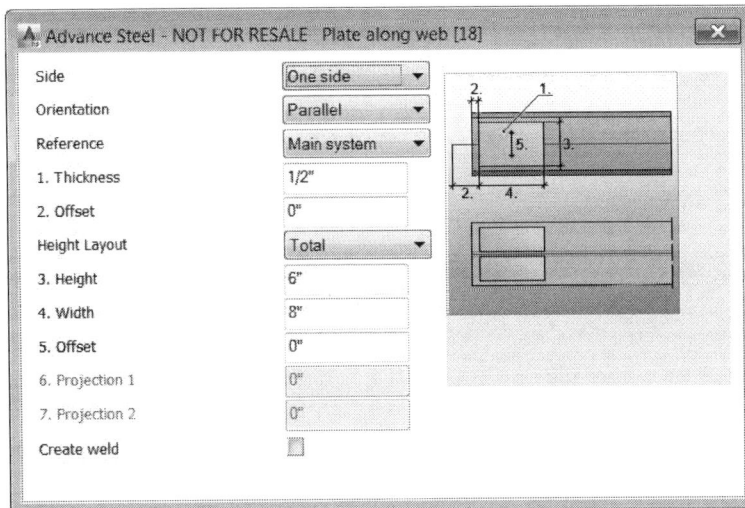

Figure 46 The Advance Steel - Plate along web [X] dialog box

5. From the **Orientation** list, select **Perpendicular**; the plate is oriented perpendicular to the web of the flange.

6. In the **Thickness** edit box, enter **3/8"** or **10**.

7. In the **Offset** edit box, enter **-3/16"** or **-5**.

8. In the **Height** edit box, enter **2"** or **70**.

9. In the **Width** edit box, enter **5"** or **150**.

10. Select the **Create weld** tick box; the plate is welded to the channel.

    This completes all the parameters for this plate.

11. Close the dialog box. The zoomed in view of the model, after inserting the plate, is shown in Figure 47.

    Next, you will insert the bolts to connect the channel to the plate that was already created in this model.

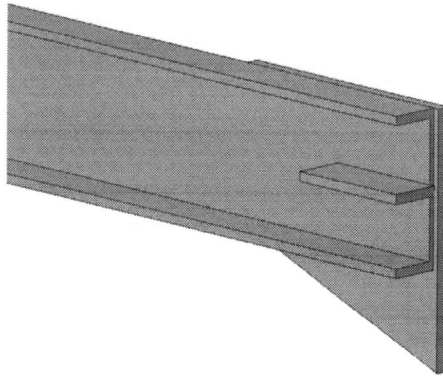

Figure 47 *The model after inserting the perpendicular plate*

12. From the **Advance Steel Tool Palette > Custom connections** tab, invoke the **Bolts on beam** tool; you are prompted to select the beam.

13. Select the channel in the model; you are prompted to specify if you want to select the connecting object.

14. Type **Y** in the prompt and press ENTER; you are prompted to select the objects to be connected.

15. Select the plate that was already created in this model and press ENTER; a rectangular bolt pattern is inserted connecting the channel and the plate, and the **Advance Steel - Bolts on beam [X]** dialog box is displayed, as shown in Figure 48.

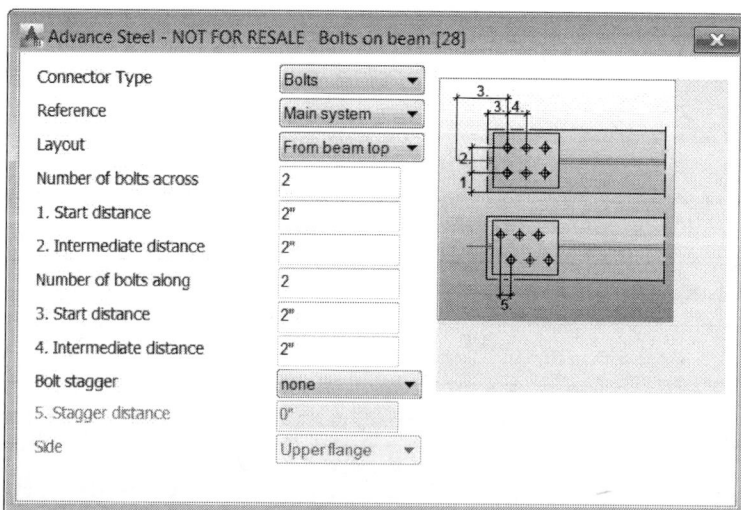

Figure 48 *The Advance Steel - Bolts on beam [X] dialog box*

*Tip: Notice that there is no parameter to change the bolt diameter in the **Advance Steel - Bolts along beam [X]** dialog box. This is because the bolt diameter can be changed later on by double-clicking on any of the bolts in the pattern.*

16. From the **Layout** list, select **Centered**; the bolt pattern is centered on the web of the channel.

17. In the **2. Intermediate distance** edit box, enter **2 1/2"** or **75**.

18. In the **Number of bolts along** edit box, enter **3**; three columns of bolts are now inserted in the pattern.

19. In the **Start distance** edit box, enter **1 1/2"** or **35**.

20. In the **4. Intermediate distance** edit box, enter **3 1/2"** or **90**.

21. Make sure **none** is selected from the **Bolt stagger** list.

    This completes all the parameters of this bolt pattern. However, you still need to change the bolt diameter.

22. Close the dialog box.

    You will now edit the bolt diameter.

23. Double-click on one of the bolts in the pattern; the **Advance Steel - Bolts** dialog box is displayed.

24. From the **Diameter** list, select **1/2 inch** (for Imperial) or **16.00 mm** (for Metric).

25. Close the dialog box. The zoomed in view of the model, after inserting the bolt pattern, is shown in Figure 49.

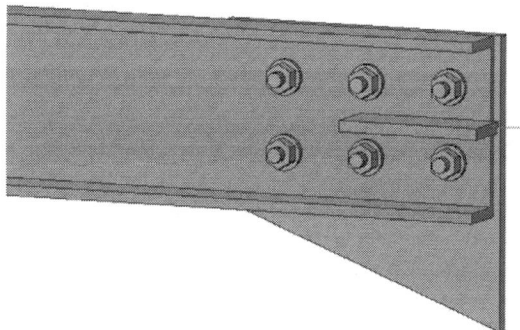

***Figure 49** The model after inserting the bolt pattern*

Before you create the custom connection, you need to save a copy of the current file into the **C:\ProgramData\Autodesk\Advance Steel 20XX \Shared\ConnectionTemplates** folder.

26. Using the **Save As** tool, save a copy of this file with the name **Custom-Vessel-Channel** in the **C:\ProgramData\Autodesk\Advance Steel 20XX\Shared\ConnectionTemplates** folder.

You are now ready to create the custom connection.

27. From the **Advance Steel Tool Palette > Custom connections** tab, invoke the **Create connection template** tool; the **Choose the definition method** dialog box is displayed.

28. Click the **1 beam and point** button; you are prompted to select input beam.

29. Select the channel and press ENTER; you are prompted select the input point.

30. Select the endpoint labeled as **1** in Figure 50 as the reference point; the **Advance Steel - User template** dialog box is displayed.

Figure 50 *The point to be selected*

31. Enter **Vessel platform base** in the **Name** edit box.

32. Click the **Reselect driven/output objects** button; the dialog box is temporarily closed and you are prompted to select objects for the connection template.

33. Using a crossing, select everything in the model, including the channel. There should be a total of eight objects selected excluding the connection boxes.

34. Press ENTER; the dialog box is redisplayed and a custom connection box is created around the connection objects.

35. In the **Driver selection prompts > Driver 1** edit box, type the following:

    the beam

36. In the **Driver selection prompts > Driver 2** edit box, type the following:

    the point

    This completes the process of creating the custom connection.

37. Close the dialog box.

38. Save the file and close it.

Section 3: Creating the Custom Connection using the 2 beams Method

In this section, you will open the **C10-Imperial-Channel-End.dwg** or the **C10-Metric-Channel-End.dwg** file and then insert the bolt patterns. Next, you will insert a point weld between the plates and the other connecting channels. Finally, you will use the **2 beams** method to create the right side and left side custom connections.

1. From the **C10 > Struc-BIM** folder, open the **C10-Imperial-Channel-End.dwg** or **C10-Metric-Channel-End.dwg** file, based on the preferred units. The model in this file appears similar to the one shown in Figure 51.

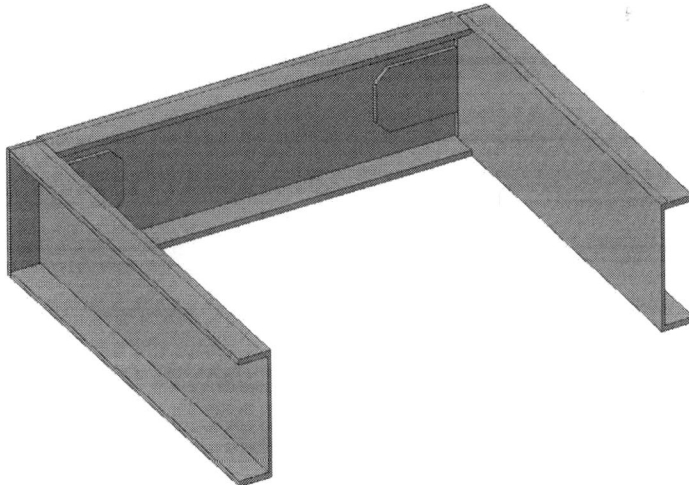

Figure 51 *The model to create the custom connection*

You will first align the UCS on the front face of the right plate and then insert the bolt pattern using the **Rectangular, corner point** tool.

2. Align the UCS to the front face of the plate using the three point shown in Figure 52. Note that the point labeled **2** is the midpoint of the edge.

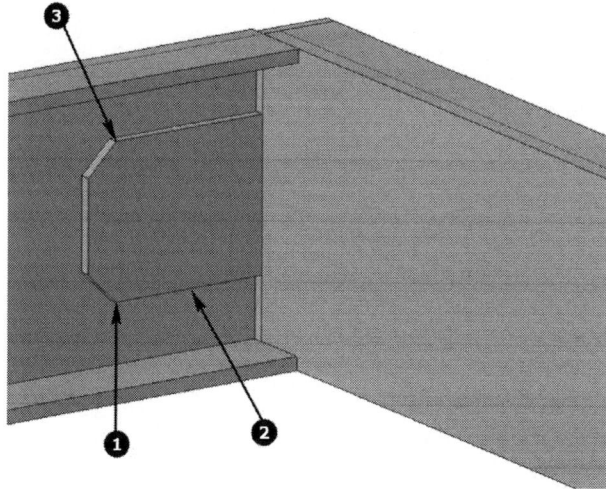

Figure 52 *The default bolt pattern inserted on the channel*

3. From the **Object** ribbon tab > **Switch** ribbon panel, click on the **bolts/anchors/holes/shear studs** button so that it changes to display the preview of the bolt with threads.

4. Using the **Object** ribbon tab > **Connection objects** ribbon panel > **Rectangular, corner point** tool, insert a bolt pattern to connect the plate and the channel behind it, as shown in Figure 53. The following are the parameters for the bolt pattern:

Start point:	**Point labeled as 1 in Figure 52**
Bolt Diameter:	**1/2 inch or 12.00 mm**
Intermediate X Distance:	**2 1/2" or 55 mm**
Intermediate Y Distance:	**2" or 40 mm**
Edge Distance X:	**1 1/4" or 20 mm**
Edge Distance Y:	**1 1/5" or 20 mm**

5. Similarly, align the UCS on the other plate and insert a bolt pattern. The model, after inserting the two bolt patterns, is shown in Figure 53.

    You will now insert a point weld between the right plate and the right channel.

6. From the **Objects** ribbon tab > **Connection objects** ribbon panel, invoke the **Weld point** tool; you are prompted to select the parts to be connected.

7. Select the channel labeled as **1** and the plate labeled as **2** in Figure 54 as the objects to be connected and press ENTER; you are prompted to define the insertion point for the weld.

8. Select the midpoint labeled as **3** in Figure 54 as the location of the weld symbol; the **Advance Steel - Weld** dialog box is displayed.

9. Configure the parameters to fillet weld of **1/4"** or **6.00 mm** thickness.

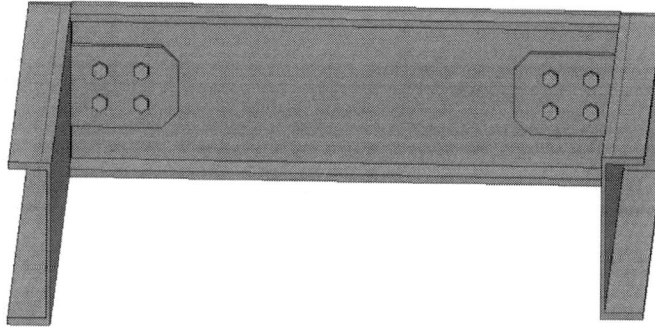

Figure 53 The model, after inserting the bolt patterns

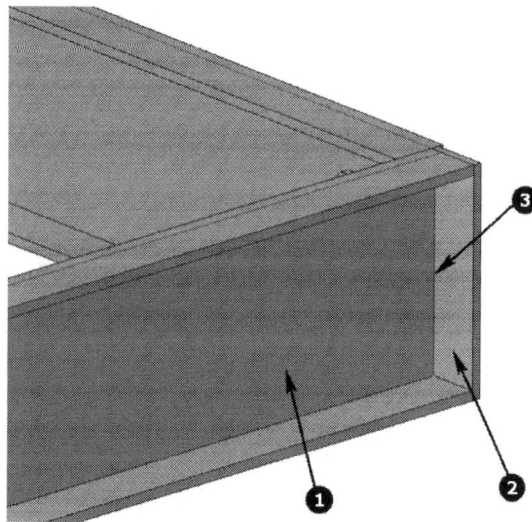

Figure 54 The objects to be welded

10. Close the dialog box.

11. Similarly, weld the left plate with the left channel.

    Before you create the custom connections, you need to save a copy of the current file into the custom connection folder.

12. Using the **Save As** tool, save a copy of this file with the name **Channel-End-Connection** in the **C:\ProgramData\Autodesk\Advance Steel 20XX\Shared\ConnectionTemplates** folder.

    You are now ready to create the custom connection.

13. From the **Advance Steel Tool Palette > Custom connections** tab, invoke the **Create connection template** tool; the **Choose the definition method** dialog box is displayed.

14. Click the **2 beams** button; you are prompted to select input beam.

15. Select the channel labeled as **1** in Figure 55 and press ENTER; you are prompted select the input beam again.

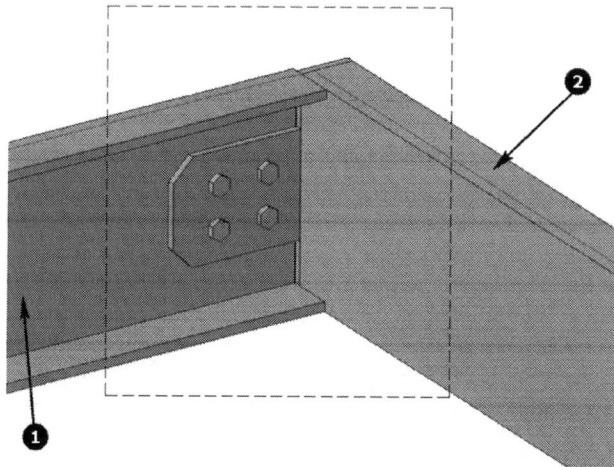

Figure 55 *Selecting the objects to create right side custom joint*

16. Select the channel labeled as **2** in Figure 55 and press ENTER; the **Advance Steel - User template** dialog box is displayed.

17. Enter **Channel end plate - right** in the **Name** edit box.

18. Click the **Reselect driven/output objects** button; the dialog box is temporarily closed and you are prompted to select objects for the connection template.

19. Using a crossing box similar to the one shown in Figure 55, select the objects that need to be a part of the custom connection. There should be a total of seven objects selected, excluding the connection boxes.

20. Press ENTER; the dialog box is redisplayed and a custom connection box is created around the connection objects.

21. In the **Driver selection prompts > Driver 1** edit box, type the following:

    the beam to bolt the plate

22. In the **Driver selection prompts > Driver 2** edit box, type the following:

    the beam to weld the plate

This completes the process of creating this custom connection.

23. Close the dialog box.

24. Similarly, create the left custom joint and name it **Channel end plate - left**.

25. Save the file and close it.

Section 4: Inserting Custom Connections

In this section, you will open various files and insert the custom connections you created in earlier sections. You will start with inserting the hexagonal base plate connection you created in the first section of the tutorial.

1. From the **C10 > Struc-BIM** folder, open the **C10-Imperial-Col-Insert.dwg** or **C10-Metric-Col-Insert.dwg** file, based on the preferred units. The model in this file appears similar to the one shown in Figure 56.

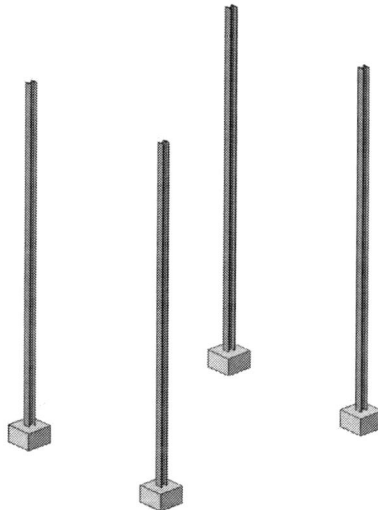

Figure 56 The file to insert the custom base plate connection

2. Zoom close to the bottom of one of the columns.

3. From the **Advance Steel Tool Palette > Custom connections tab**, invoke the **Insert custom template** tool, which is the second tool in this tab; the **Connection template explorer** dialog box is displayed, as shown in Figure 57.

    As mentioned earlier, if the file has only one custom joint, it is better to select the DWG file to insert the joint.

4. From the left side of the dialog box, click on the **Custom-Base-Plate.dwg** file.

5. Click **OK** in the dialog box; you are prompted to select the column.

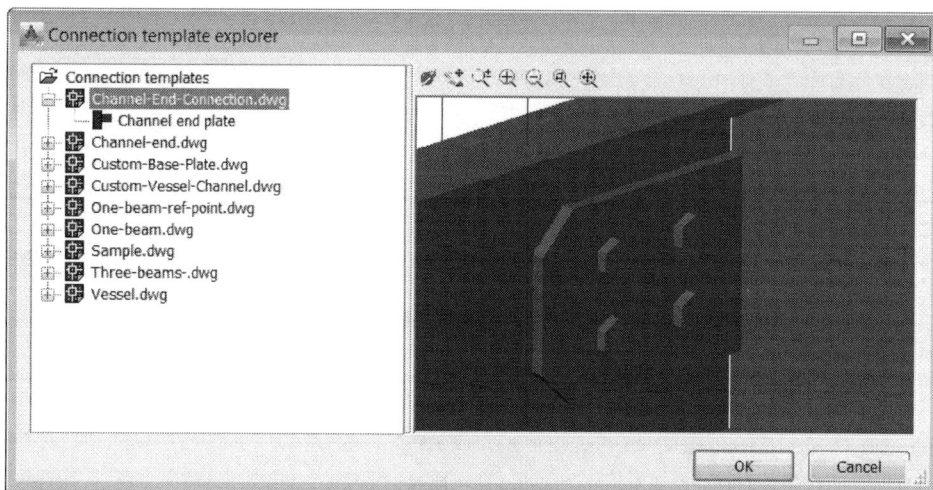

*Figure 57 The **Connection template explorer** dialog box*

Notice that Advance Steel automatically adds **Select** at the start of the prompt sequence. This is the reason you only types in partial prompts while creating the custom connections.

6.   Select the column; the custom base plate connection is inserted and the **Advance Steel - User template** dialog box is displayed.

7.   Select the **Allow object modification** tick box in the dialog box.

8.   Close the dialog box. The zoomed in view of the model, after inserting the joint, is shown in Figure 58. In this figure, the display of the isolated footing is turned off.

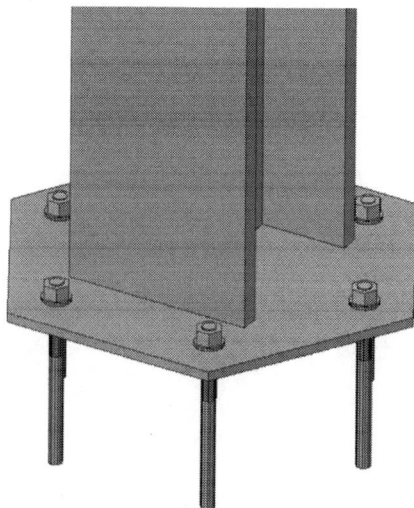

Figure 58 The model after inserting one of the joints

9. Using the **Advance Steel Tool Palette > Tools** tab > **Create joint in a group, multiple** tool, copy the inserted joint to the remaining three columns.

10. Save the file and close it.

Next, you will insert the **Vessel platform base** custom connection that you created in the **Custom-Vessel-Channel.dwg** file.

11. From the **C10 > Struc-BIM** folder, open the **C10-Imperial-Vessel.dwg** or **C10-Metric-Vessel.dwg** file, based on the preferred units. The model in this file appears similar to the one shown in Figure 59.

Figure 59 The model to insert the custom connection

The custom connections need to be inserted on the channels supporting the platform where the cage ladder ends. For easier selection of the channels, you will turn off the visibility of the grating and the angles around the grating and also turn off the **Vessel** layer.

12. Using the **Advance Steel Tool Palette > Quick views** tab > **Select objects off** tool, turn off the visibility of the grating and the angle sections supporting the grating.

13. Turn off the **Vessel** layer. The zoomed in view of the model, showing the channels on which the custom joint will be inserted, is shown in Figure 60.

You are now ready to insert the custom connection.

14. From the **Advance Steel Tool Palette > Custom connections tab**, invoke the **Insert custom template** tool; the **Connection template explorer** dialog box is displayed.

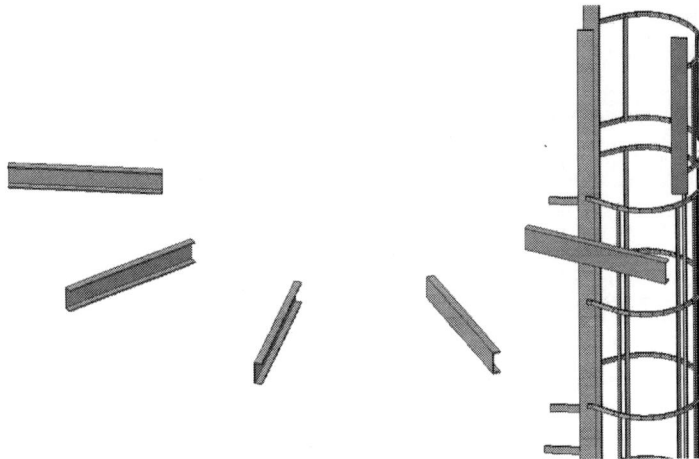

Figure 60 *The model after turning off the grating and the vessel*

15. From the left side of the dialog box, click on the **Custom-Vessel-Channel.dwg** file.

16. Click **OK** in the dialog box; you are prompted to select the beam.

17. Select the channel closest to the cage ladder labeled as **1** in Figure 61; you are prompted to select the point.

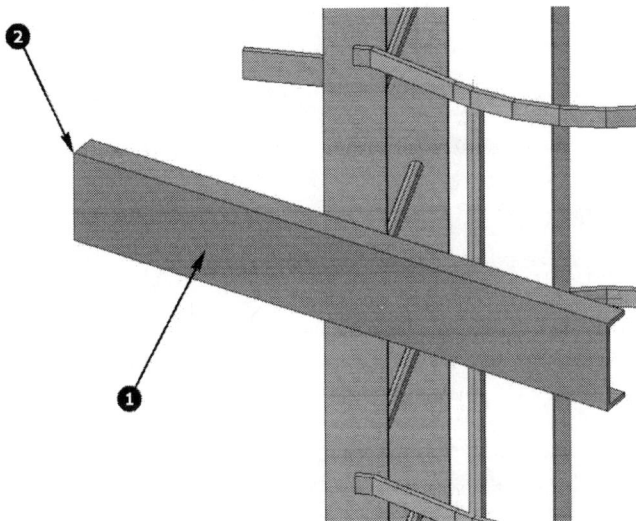

Figure 61 *The channel and the point to be selected*

18. Select the endpoint of the channel labeled as **2** in Figure 61; the custom connection is inserted and the **Advance Steel - User template** dialog box is displayed.

19. Select the **Allow object modification** tick box in the dialog box.

20. Close the dialog box. The zoomed in view of the model, after inserting the joint, is shown in Figure 62.

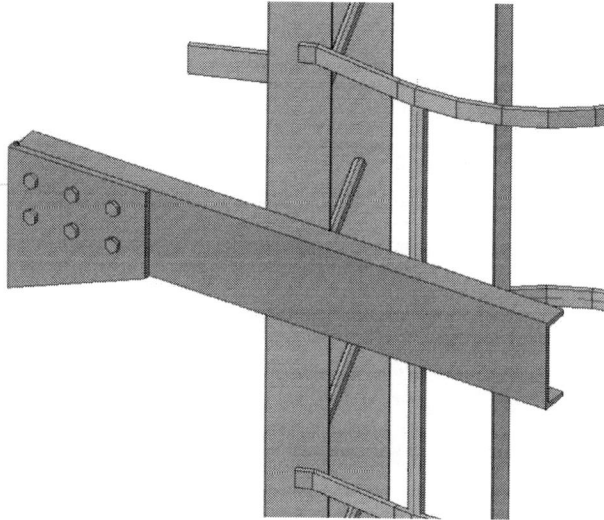

Figure 62 *The model after inserting the joint*

This joint cannot be copied at the remaining channels. Therefore, you will have to manually insert it at the remaining four channels.

21. Using the **Insert connection template** tool, insert the same connection at the remaining four channels.

Tip: To edit the joint, turn on the connection box/joint box layer and then double-click on the custom connection box created around the custom joint.

22. Turn on the visibility of all the objects and the **Vessel** layer.

23. Save and close the file.

Lastly, you will insert the **Channel end plate** connection that was created in the **Channel-End-Connection.dwg** file.

24. From the **C10 > Struc-BIM** folder, open the **C10-Imperial-Channel-Insert.dwg** or **C10-Metric-Channel-Insert.dwg** file, based on the preferred units. The model in this file appears similar to the one shown in Figure 63. This figure also shows the member labels to be selected to insert the connection.

25. From the **Advance Steel Tool Palette > Custom connections tab**, invoke the **Insert custom template** tool; the **Connection template explorer** dialog box is displayed.

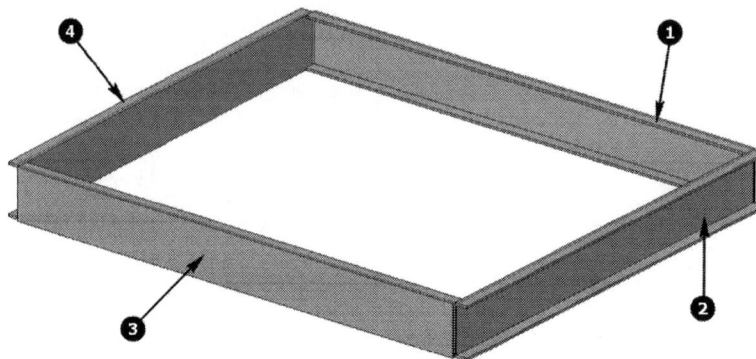

Figure 63 *The model to insert the custom connection*

26. From the left side of the dialog box, expand the **Channel-End-Connection.dwg** file and select **Channel end plate - right**.

27. Click **OK** in the dialog box; you are prompted to select the beam to bolt the plate.

28. Select the channel labeled as **1** in Figure 63; you are prompted to select the beam to weld the plate.

29. Select the channel labeled as **2** in Figure 63; the custom connection is inserted and the **Advance Steel - User template** dialog box is displayed.

30. Select the **Allow object modification** tick box in the dialog box.

31. Close the dialog box. The zoomed in view of the model, after inserting the joint, is shown in Figure 64.

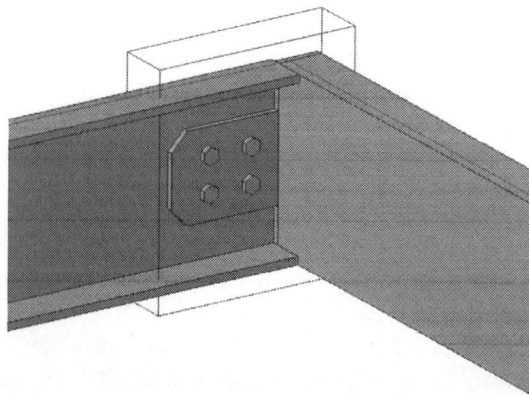

Figure 64 *The zoomed in view of the model after inserting the connection*

Next, you will insert the left side joint between the channels labeled as **1** and **4** in Figure 63.

32. Invoke the **Insert custom template** tool; the **Connection template explorer** dialog box is displayed.

33. From the left side of the dialog box, expand the **Channel-End-Connection.dwg** file and select **Channel end plate - left**.

34. Click **OK** in the dialog box; you are prompted to select the beam to bolt the plate.

35. Select the channel labeled as **1** in Figure 63; you are prompted to select the beam to weld the plate.

36. Select the channel labeled as **4** in Figure 63; the custom connection is inserted and the **Advance Steel - User template** dialog box is displayed.

37. Select the **Allow object modification** tick box in the dialog box.

38. Close the dialog box. The model, after inserting the left side connection, is shown in Figure 65.

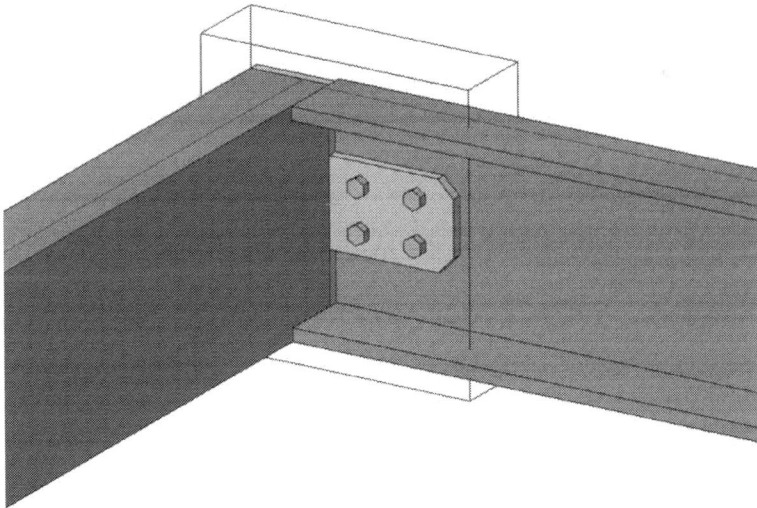

Figure 65 *The model after inserting the left side connection*

39. Similarly, insert **Channel end plate - left** connection between channels labeled as **3** and **2** in and the **Channel end plate - right** between channels labeled as **3** and **4**.

The model, after inserting all the connections, is shown in Figure 66.

40. Save and close the file.

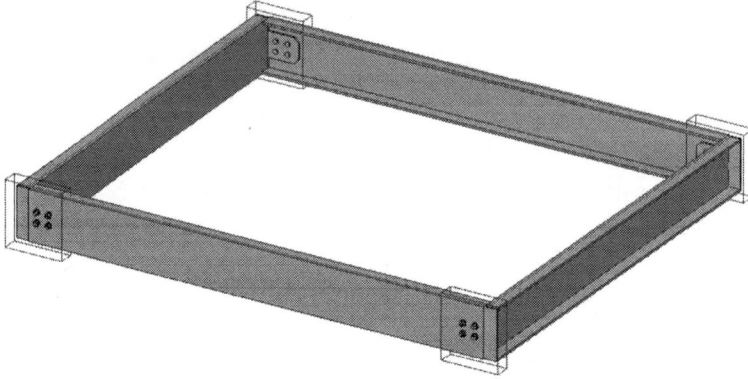

Figure 66 *The model after inserting all the connections*

Tip: You can use the **Display connected objects tool** to ensure all the sections are connected.

Skill Evaluation

Evaluate your skills to see how many questions you can answer correctly. The answers to these questions are given at the end of the book.

1. Advance Steel does not allow you to create custom connections. (True/False)

2. Advance Steel provides you with a special directory to save all your custom connections. (True/False)

3. Advance Steel provides you with special tools that act as the basic building blocks for your custom connections. (True/False)

4. In Advance Steel, the bolts on a beam can only be inserted using the tools available on the **Object** ribbon tab > **Connection objects** ribbon panel. (True/False)

5. You can connect the objects using bolts or welds and include them as part of the custom connection. (True/False)

6. Which tab of the **Advance Steel Tool Palette** provides the tools to create custom joints?

    (A) **Connection**   (B) **Custom Connections**
    (C) **New Connections**   (D) **Simple Connections**

7. Which tool is available in the third area of the **Custom connections** tab and is used to insert bolts on the selected beams to connect it to a plate?

    (A) **Insert Bolts**   (B) **Rectangular bolts**
    (C) **Bolts on beam**   (D) **Bolts on Channels**

8. Which tool is used to insert galvanizing holes on an end plate of the beam or on the web of the beam?

    (A) **Cut Holes**   (B) **Holes**
    (C) **Holes by Model**   (D) **Galvanizing holes**

9. Which tool is used to insert custom connection?

    (A) **Insert connection template**   (B) **Insert custom connection**
    (C) **Create connection template**   (D) **Insert joint**

10. Which tool is used to create custom connection?

    (A) **Create connection**   (B) **Custom connections create**
    (C) **Create joint**   (D) **Create connection template**

Class Test Questions
Answer the following questions:

1. Explain briefly the procedure of inserting a bolt pattern to connect a plate to the beam using the **Advance Steel Tool Palette**.

2. Which folder do you need to save the custom connections in?

3. Explain briefly the process of creating custom connections.

4. Explain briefly the process of inserting custom connections.

5. Explain briefly how to insert a plate normal along the web of a beam.

Index

Answers to Skill Evaluation

Chapter 1
1. T
2. T
3. F
4. T
5. F
6. (A) **Building Grid**
7. (A) Cylinder, (C) Block
8. (C) **Concrete curved beam**
9. (D) **Continuous footing**
10. (A) **Standard**

Chapter 2
1. F
2. T
3. F
4. T
5. T
6. (A) **Curved beam**
7. (D) **Match Properties**
8. (A) **Properties**
9. (D) **Column**
10. (A) **Positioning**

Chapter 3
1. T
2. F
3. F
4. T
5. T
6. (D) **Mono-pitch frame**
7. (D) **Distances**
8. (B) **Truss**
9. (A) **Portal/Gable Frame**
10. (A) **Warren**, (B) **Half Pratt**, (C) **Prat**

Chapter 4
1. F
2. F
3. F
4. T
5. T
6. (C) **Advance Steel**
7. (D) **Tube base plate**
8. (B) **Gable wall end plate**
9. (C) **Favorites**
10. (C) **Column-Beam**

Chapter 5
1. F
2. F
3. T
4. F
5. T
6. (D) **Purlin & Cold rolled**
7. (C) **Shear plate**
8. (A) **Clip angle**
9. (A) **Single purlin plate**, (D) **Double purlin splice plate**
10. (C) **Column-Beam**

Chapter 6
1. F
2. F
3. F
4. F
5. T
6. (B) **Hand-railing**
7. (A) **Start and end point**
8. (B) **Top Chords**, (D) **Support plates**
9. (C) **Glass**
10. (A) **Platform**

Chapter 7
1. T
2. F
3. F
4. F
5. F
6. (A) **Gusset plate to column and base plate**
7. (D) **Stair Anchor Angle Joint**
8. (C) **Railing joint handrail Joint**
9. (D) **Tube connection with sandwich plates - 2 diagonals**
10. (A) **Platform**

Chapter 8

1. F
2. F
3. T
4. T
5. T
6. (A) **Plate at poly line**
7. (C) **Merge plates**
8. (B) **Quick views**
9. (D) **Rectangular contour, center**
10. (A) **Casing cross section**, (D) **Exact cross section**

Chapter 9

1. T
2. F
3. F
4. T
5. T
6. (C) **Spiral staircase**
7. (A) **Display connected objects**
8. (A) **Split bolt groups**
9. (B) **Cut at object**
10. (B) **Circular, center point**

Chapter 10

1. F
2. T
3. T
4. F
5. T
6. (B) **Custom Connections**
7. (A) **Display connected objects**
8. (D) **Galvanizing holes**
9. (A) **Insert connection template**
10. (D) **Create connection template**

Made in the USA
San Bernardino, CA
13 June 2017